SOLUTIONS TO RED EXERCISES

ROXY WILSON

University of Illinois, Urbana–Champaign

TWELFTH EDITION

CHEMISTRY
THE CENTRAL SCIENCE

BROWN LeMAY BURSTEN MURPHY WOODWARD

Prentice Hall

Boston Columbus Indianapolis New York San Francisco Upper Saddle River
Amsterdam Cape Town Dubai London Madrid Milan Munich Paris Montréal Toronto
Delhi Mexico City São Paulo Sydney Hong Kong Seoul Singapore Taipei Tokyo

Editor in Chief, Chemistry: Adam Jaworski
Acquisitions Editor: Terry Haugen
Marketing Manager: Erin Gardner
Senior Project Editor: Jennifer Hart
Managing Editor, Chemistry and Geosciences: Gina M. Cheselka
Project Manager, Science: Shari Toron
Operations Specialist: Maura Zaldivar
Supplement Cover Designer: Paul Gourhan
Cover Credit: Graphene by Dr. Jannik C. Meyer of the University of Ulm, Germany

ISBN-13:   978-0-321-70548-8

ISBN-10:   0-321-70548-3

Prentice Hall
is an imprint of

www.pearsonhighered.com

Contents

Introduction

Chemistry: The Central Science, 12th edition, contains more than 2600 end-of-chapter exercises. Considerable attention has been given to these exercises because one of the best ways for students to master chemistry is by solving problems. Grouping the exercises according to subject matter is intended to aid the student in selecting and recognizing particular types of problems. Within each subject matter group, similar problems are arranged in pairs. This provides the student with an opportunity to reinforce a particular kind of problem. There are also a substantial number of general exercises in each chapter to supplement those grouped by topic. Visualizing Concepts, general exercises which require students to analyze visual data in order to formulate conclusions about chemical concepts, and Integrative Exercises, which require students to integrate concepts from several chapters, are continuing features of the 12th edition. Answers to the odd numbered topical exercises plus selected general exercises, about 1100 in all, are provided in the text. These appendix answers help to make the text a useful self-contained vehicle for learning.

This manual, **Solutions to Red Exercises** in **Chemistry: The Central Science, 12th edition**, was written to enhance the end-of-chapter exercises by providing documented solutions for those problems answered in the appendix of the text. The manual assists the instructor by saving time spent generating solutions for assigned problem sets and aids the student by offering a convenient independent source to check their understanding of the material. Most solutions have been worked in the same detail as the in-chapter sample exercises to help guide students in their studies.

To reinforce the '*Analyze, Plan, Solve, Check*' problem-solving method used extensively in the text, this strategy has also been incorporated into the Solution Manual. Solutions to most red topical exercises and selected Visualizing, Additional and Integrative exercises feature this four-step approach. We strongly encourage students to master this powerful and totally general method.

When using this manual, keep in mind that the numerical result of any calculation is influenced by the precision of the numbers used in the calculation. In this manual, for example, atomic masses and physical constants are typically expressed to four significant figures, or at least as precisely as the data given in the problem. If students use slightly different values to solve problems, their answers will differ slightly from those listed in the appendix of the text or this manual. This is a normal and a common occurrence when comparing results from different calculations or experiments.

Rounding methods are another source of differences between calculated values. In this manual, when a solution is given in steps, intermediate results will be rounded to the correct number of significant figures; however, unrounded numbers will be used in subsequent calculations. By following this scheme, calculators need not be cleared to re-enter rounded intermediate results in the middle of a calculation sequence. The final answer will appear with the correct number of significant figures. This may result in a small discrepancy in the last significant digit between student-calculated answers and those given in this manual. Variations due to rounding can occur in any analysis of numerical data.

The first step in checking your solution and resolving differences between your answer and the listed value is to look for similarities and differences in problem-solving methods. Ultimately, resolving the small numerical differences described above is less important than understanding the general method for solving a problem. The goal of this manual is to provide a reference for sound and consistent problem-solving methods in addition to accurate answers to text exercises.

Extraordinary efforts have been made to keep this manual as error-free as possible. All exercises were worked and proof-read by at least three chemists to ensure clarity in methods and accuracy in mathematics. The work and advice of Ms. Renee Rice, Ms. Kate Vigor, Dr. Christopher Musto, and Dr. Timothy Kucharski have been invaluable to this project. However, in a written work as technically challenging as this manual, typos and errors inevitably creep in. Please help us find and eliminate them. We hope that both instructors and students will find this manual accurate, helpful and instructive.

Roxy B. Wilson
1829 Maynard Dr.
Champaign, IL 61822
rbwilson@illinois.edu

1 Introduction: Matter and Measurement

Visualizing Concepts

1.1 *Pure elements* contain only one kind of atom. Atoms can be present singly or as tightly bound groups called molecules. *Compounds* contain two or more kinds of atoms bound tightly into molecules. *Mixtures* contain more than one kind of atom and/or molecule, not bound into discrete particles.

   (a) pure element: i

   (b) mixture of elements: v, vi

   (c) pure compound: iv

   (d) mixture of an element and a compound: ii, iii

1.3 To brew a cup of coffee, begin with ground coffee beans, a heterogeneous mixture, and water, a pure substance. Hot water contacts the coffee grounds and dissolves components of the coffee bean that are water-soluble. This creates a new heterogeneous mixture of undissolved coffee bean solids and liquid coffee solution; this mixture is separated by filtration. Undissolved grounds are left on the filter paper and liquid coffee, itself a homogeneous mixture, drips into the container below.

   Overall, two separations occur. Chemical differences among the components of the coffee bean allow certain compounds to dissolve in water, while other components remain insoluble. This kind of separation based on solubility differences is called *extraction*. The insoluble grounds are then separated from the coffee solution by *filtration*.

1.5 Density is the ratio of mass to volume. For a sphere, size is like volume; both are determined by the radius of the sphere.

   (a) For spheres of the same size or volume, the denominator of the density relationship is the same. The denser the sphere, the heavier it is. A list from lightest to heaviest is in order of increasing density and mass. The aluminum sphere (density = 2.70 g/cm³) is lightest, then nickel (density = 8.90 g/cm³), then silver (density = 10.409 g/cm³).

   (b) For spheres of equal mass, the numerator of the density relationship is the same. The denser the sphere, the smaller its volume or size. A list from smallest to largest is in order of decreasing density. The platinum sphere (density = 21.45 g/cm³) is smallest, then gold (density = 19.30 g/cm³), then lead (density = 11.35 g/cm³).

1

1.7   (a)   7.5 cm. There are two significant figures in this measurement; the number of cm can be read precisely, but there is some estimating (uncertainty) required to read tenths of a centimeter. Listing two significant figures is consistent with the convention that measured quantities are reported so that there is uncertainty in only the last digit.

      (b)   The speed is 72 mi/hr (inner scale, two significant figures) or 115 km/hr (outer scale, three significant figures). Both scales are read with certainty in the "hundreds" and "tens" place, and some uncertainty in the "ones" place. The km/hr speed has one more significant figure because its magnitude is in the hundreds.

1.9   When converting units, arrange the conversion factor so that the given unit cancels and the desired unit is in the correct position. For example, suppose a quantity is expressed in terms of centimeters, but the desired result is expressed in inches. If the given unit has 'cm' in the numerator, then the conversion factor must have 'cm' in its denominator. However, if the original unit has 'cm' in the denominator, the conversion factor must have 'cm' in the numerator. Ideally, this will lead to the desired units in the appropriate location, numerator or denominator. However, the inverse of the answer can be taken when necessary.

Classification and Properties of Matter (sections 1.2 and 1.3)

1.11 (a)   heterogeneous mixture

      (b)   homogeneous mixture (If there are undissolved particles, such as sand or decaying plants, the mixture is heterogeneous.)

      (c)   pure substance

      (d)   pure substance

1.13 (a) S   (b) Au   (c) K   (d) Cl   (e) Cu   (f) uranium

      (g)   nickel   (h) sodium   (i) aluminum   (j) silicon

1.15 $A(s) \rightarrow B(s) + C(g)$

      When solid carbon is burned in excess oxygen gas, the two elements combine to form a gaseous compound, carbon dioxide. Clearly substance C is this compound. Since C is produced when A is heated in the absence of oxygen (from air), both the carbon and oxygen in C must have been present in A originally. A is, therefore, a compound composed of two or more elements chemically combined. Without more information on the chemical or physical properties of B, we cannot determine absolutely whether it is an element or a compound. However, few if any elements exist as white solids, so B is probably also a compound.

1.17 *Physical properties*: silvery white (color); lustrous; melting point = 649°C; boiling point = 1105°C; density at 20°C = 1.738 g/cm^3; pounded into sheets (malleable); drawn into wires (ductile); good conductor. *Chemical properties*: burns in air to give intense white light; reacts with Cl_2 to produce brittle white solid.

1.19   (a)   chemical (b) physical (c) physical (d) chemical (e) chemical

1.21   (a)   Take advantage of the different water solubilities of the two solids. Add water to dissolve the sugar; filter this mixture, collecting the sand on the filter paper and the sugar water in the flask. Evaporate water from the flask to recover solid sugar.

        (b)   Take advantage of the different solubilities and densities of the two liquids. Allow the mixture to settle so that there are two distinct layers. Vinegar (a water solution) is denser and on the bottom; oil (the organic layer) is less dense and on top. Carefully pour off most of the top layer. After the layers reform, use a dropper to remove any remaining oil. Vinegar is in the original vessel and oil is in a second container.

Units and Measurement (section 1.4)

1.23   (a) 1×10^{-1} (b) 1×10^{-2} (c) 1×10^{-15} (d) 1×10^{-6} (e) 1×10^{6}

        (f) 1×10^{3} (g) 1×10^{-9} (h) 1×10^{-3} (i) 1×10^{-12}

1.25   (a) $°C = 5/9 (°F – 32°); 5/9 (72 – 32) = 22°C$

        (b) $°F = 9/5 (°C) + 32°; 9/5 (216.7) + 32 = 422.1°F$

        (c) $K = °C + 273.15; 233°C + 273.15 = 506 K$

        (d) $°C = 315 K – 273.15 = 41.85 = 42°C; °F = 9/5 (41.85°C) + 32 = 107°F$

        (e) $°C = 5/9 (°F – 32°); 5/9 (2500 – 32) = 1371°C; K = 1371°C + 273.15 = 1644 K$
               (assuming 2500 °F has 4 sig figs)

        (f) $°C = 0 K – 273.15 = –273.15°C; °F = 9/5 (–273.15°C) + 32 = –459.67°F$
               (assuming 0 K has infinite sig figs)

1.27   (a) $\text{density} = \dfrac{\text{mass}}{\text{volume}} = \dfrac{40.55\,g}{25.0\,mL} = 1.62\,g/mL \text{ or } 1.62\,g/cm^3$

               (The units cm^3 and mL will be used interchangeably in this manual.)

               Tetrachloroethylene, 1.62 g/mL, is more dense than water, 1.00 g/mL; tetrachloroethylene will sink rather than float on water.

        (b) $25.0\,cm^3 \times 0.469\,\dfrac{g}{cm^3} = 11.7\,g$

1.29   (a) $\text{density} = \dfrac{38.5\,g}{45\,mL} = 0.86\,g/mL$

               The substance is probably toluene, density = 0.866 g/mL.

        (b) $45.0\,g \times \dfrac{1\,mL}{1.114\,g} = 40.4\,mL \text{ ethylene glycol}$

        (c) $(5.00)^3\,cm^3 \times \dfrac{8.90\,g}{1\,cm^3} = 1.11 \times 10^3\,g\ (1.11\,kg) \text{ nickel}$

1.31 $31 \text{ billion tons} \times \dfrac{1 \times 10^9\,\text{tons}}{1\,\text{billion tons}} \times \dfrac{2000\,lb}{1\,ton} \times \dfrac{453.59\,g}{1\,lb} = 2.8 \times 10^{16}\,g$

The metric prefix for "1×10^{15}" is peta, abbreviated P.

$$2.8 \times 10^{16}\,g \; \times \; \frac{1\,Pg}{1 \times 10^{15}\,g} = 28\,Pg$$

Uncertainty in Measurement (section 1.5)

1.33   Exact: (c), (d), and (f) (All others depend on measurements and standards that have margins of error, e.g., the length of a week as defined by the earth's rotation.)

1.35   (a)   3   (b) 2   (c) 5   (d) 3   (e) 5   (f)   1 [See Sample Exercise 1.6 (c)]

1.37   (a) 1.025×10^2 (b) 6.570×10^5 (c) 8.543×10^{-3}

          (d) 2.579×10^{-4} (e) -3.572×10^{-2}

1.39   (a) $14.3505 + 2.65 = 17.0005 = 17.00$ (For addition and subtraction, the minimum number of decimal places, here two, determines decimal places in the result.)

          (b) $952.7 - 140.7389 = 812.0$

          (c) $(3.29 \times 10^4)(0.2501) = 8.23 \times 10^3$ (For multiplication and division, the minimum number of significant figures, here three, determines sig figs in the result.)

          (d) $0.0588 / 0.677 = 8.69 \times 10^{-2}$

1.41   The mass 21.427 g has 5 significant figures.

Dimensional Analysis (section 1.6)

1.43   In each conversion factor, the old unit appears in the denominator, so it cancels, and the new unit appears in the numerator.

          (a)   mm $\rightarrow$ nm : $\dfrac{1 \times 10^{-3}\,m}{1\,mm} \times \dfrac{1\,nm}{1 \times 10^{-9}\,m} = 1 \times 10^6\,nm/mm$

          (b)   mg $\rightarrow$ kg : $\dfrac{1 \times 10^{-3}\,g}{1\,mg} \times \dfrac{1\,kg}{1000\,g} = 1 \times 10^{-6}\,kg/mg$

          (c)   km $\rightarrow$ ft : $\dfrac{1000\,m}{1\,km} \times \dfrac{1\,cm}{1 \times 10^{-2}\,m} \times \dfrac{1\,in}{2.54\,cm} \times \dfrac{1\,ft}{12\,in} = 3.28 \times 10^3\,km/ft$

          (d)   in^3 $\rightarrow$ cm^3 : $\dfrac{(2.54)^3\,cm^3}{1^3\,in^3} = 16.4\ cm^3/in^3$

1.45   (a) $\dfrac{15.2\,m}{s} \times \dfrac{1\,km}{1000\,m} \times \dfrac{60\,s}{1\,min} \times \dfrac{60\,min}{1\,hr} = 54.7\,km/hr$

          (b) $5.0 \times 10^3\,L \times \dfrac{1\,gal}{3.7854\,L} = 1.3 \times 10^3\,gal$

          (c) $151\,ft \times \dfrac{1\,yd}{3\,ft} \times \dfrac{1\,m}{1.0936\,yd} = 46.025 = 46.0\,m$

          (d) $\dfrac{60.0\,cm}{d} \times \dfrac{1\,in}{2.54\,cm} \times \dfrac{1\,d}{24\,hr} = 0.984\,in/hr$

1.47   (a) $5.00 \, \text{days} \times \dfrac{24 \, \text{hr}}{1 \, \text{day}} \times \dfrac{60 \, \text{min}}{1 \, \text{hr}} \times \dfrac{60 \, \text{s}}{1 \, \text{min}} = 4.32 \times 10^5 \, \text{s}$

       (b) $0.0550 \, \text{mi} \times \dfrac{1.6093 \, \text{km}}{\text{mi}} \times \dfrac{1000 \, \text{m}}{1 \, \text{km}} = 88.5 \, \text{m}$

       (c) $\dfrac{\$1.89}{\text{gal}} \times \dfrac{1 \, \text{gal}}{3.7854 \, \text{L}} = \dfrac{\$0.499}{\text{L}}$

       (d) $\dfrac{0.510 \, \text{in}}{\text{ms}} \times \dfrac{2.54 \, \text{cm}}{1 \, \text{in}} \times \dfrac{1 \times 10^{-2} \, \text{m}}{1 \, \text{cm}} \times \dfrac{1 \, \text{km}}{1000 \, \text{m}} \times \dfrac{1 \, \text{ms}}{1 \times 10^{-3} \, \text{s}} \times \dfrac{60 \, \text{s}}{1 \, \text{min}} \times \dfrac{60 \, \text{min}}{1 \, \text{hr}} = 46.6 \, \dfrac{\text{km}}{\text{hr}}$

           Estimate: $0.5 \times 2.5 = 1.25$; $1.25 \times 0.01 \approx 0.01$; $0.01 \times 60 \times 60 \approx 36 \, \text{km/hr}$

       (e) $\dfrac{22.50 \, \text{gal}}{\text{min}} \times \dfrac{3.7854 \, \text{L}}{\text{gal}} \times \dfrac{1 \, \text{min}}{60 \, \text{s}} = 1.41953 = 1.420 \, \text{L/s}$

           Estimate: $20 \times 4 = 80$; $80/60 \approx 1.3 \, \text{L/s}$

       (f) $0.02500 \, \text{ft}^3 \times \dfrac{12^3 \, \text{in}^3}{1 \, \text{ft}^3} \times \dfrac{2.54^3 \, \text{cm}^3}{1 \, \text{in}^3} = 707.9 \, \text{cm}^3$

           Estimate: $10^3 = 1000$; $3^3 = 27$; $1000 \times 27 = 27{,}000$; $27{,}000/0.04 \approx 700 \, \text{cm}^3$

1.49   (a) $31 \, \text{gal} \times \dfrac{4 \, \text{qt}}{1 \, \text{gal}} \times \dfrac{1 \, \text{L}}{1.057 \, \text{qt}} = 1.2 \times 10^2 \, \text{L}$

           Estimate: $(30 \times 4)/1 \approx 120 \, \text{L}$

       (b) $\dfrac{6 \, \text{mg}}{\text{kg (body)}} \times \dfrac{1 \, \text{kg}}{2.205 \, \text{lb}} \times 185 \, \text{lb} = 5 \times 10^2 \, \text{mg}$

           Estimate: $6/2 = 3$; $3 \times 180 = 540 \, \text{mg}$

       (c) $\dfrac{400 \, \text{km}}{47.3 \, \text{L}} \times \dfrac{1 \, \text{mi}}{1.6093 \, \text{km}} \times \dfrac{1 \, \text{L}}{1.057 \, \text{qt}} \times \dfrac{4 \, \text{qt}}{1 \, \text{gal}} = \dfrac{19.9 \, \text{mi}}{\text{gal}}$

           $(2 \times 10^1 \, \text{mi/gal for 1 sig fig})$

           Estimate: $400/50 = 8$; $8/1.6 = 5$; $5/1 = 5$; $5 \times 4 \approx 20 \, \text{mi/gal}$

       (d) $\dfrac{50 \, \text{cups}}{1 \, \text{lb}} \times \dfrac{1 \, \text{qt}}{4 \, \text{cups}} \times \dfrac{1 \, \text{L}}{1.057 \, \text{qt}} \times \dfrac{1000 \, \text{mL}}{1 \, \text{L}} \times \dfrac{1 \, \text{lb}}{453.6 \, \text{g}} = \dfrac{26 \, \text{mL}}{\text{g}}$

           $(3 \times 10^1 \, \text{mL/L for 1 sig fig})$

           Estimate: $50/4 = 12$; $1000/500 = 2$; $(12 \times 2)/1 \approx 24 \, \text{mL/g}$

1.51 $14.5 \, \text{ft} \times 16.5 \, \text{ft} \times 8.0 \, \text{ft} = 1914 = 1.9 \times 10^3 \, \text{ft}^3$ (2 sig figs)

       $1914 \, \text{ft}^3 \times \dfrac{(1 \, \text{yd})^3}{(3 \, \text{ft})^3} \times \dfrac{(1 \, \text{m})^3}{(1.0936)^3 \, \text{yd}^3} \times \dfrac{10^3 \, \text{dm}^3}{1 \, \text{m}^3} \times \dfrac{1 \, \text{L}}{1 \, \text{dm}^3} \times \dfrac{1.19 \, \text{g}}{\text{L}} \times \dfrac{1 \, \text{kg}}{1000 \, \text{g}} = 64.4985 = 64 \, \text{kg air}$

       Estimate: $1900/27 \approx 60$; $(60 \times 1)/1 \approx 60 \, \text{kg}$

1.53   Select a common unit for comparison, in this case the cm.

       $1 \, \text{in} \approx 2.5 \, \text{cm}$, $1 \, \text{m} = 100 \, \text{cm}$; $57 \, \text{cm} = 57 \, \text{cm}$; $14 \, \text{in} \approx 35 \, \text{cm}$; $1.1 \, \text{m} = 110 \, \text{cm}$

       The order of length from shortest to longest is 14-in shoe < 57-cm string < 1.1-m pipe.

1.55　　Strategy: 1) Calculate volume of gold (Au) in cm^3 in the sheet

　　　　　　　　2) Mass = density × volume

　　　　　　　　3) Change g → troy oz and $

$$100\,ft \times 82\,ft \times \frac{(12)^2\,in^2}{1\,ft^2} \times 5\times10^{-6}\,in \times \frac{(2.54)^3\,cm^3}{1\,in^3} = 96.75 = 1\times10^2\,cm^3\,Au$$

$$96.75\,cm^3\,Au \times \frac{19.32\,g}{1\,cm^3} \times \frac{1\,troy\,oz}{31.1034768\,g} \times \frac{\$953}{troy\,oz} = \$57,272 = \$6 \times 10^4$$

(Strictly speaking, the datum 100 ft has 1 sig fig, so the result has 1 sig fig.)

Additional Exercises

1.59　　Any sample of vitamin C has the same relative amount of carbon and oxygen; the ratio of oxygen to carbon in the isolated sample is the same as the ratio in synthesized vitamin C.

$$\frac{2.00\,g\,O}{1.50\,g\,C} = \frac{x\,g\,O}{6.35\,g\,C};\ \ x = \frac{(2.00\,g\,O)(6.35\,g\,C)}{1.50\,g\,C} = 8.47\,g\,O$$

This calculation assumes the law of constant composition.

1.62　　(a)　volume　　(b)　area　　　(c)　volume　　　(d)　density

　　　　(e)　time　　　(f)　length　　(g)　temperature

1.65　　(a)　$575\,ft \times \dfrac{12\,in}{1\,ft} \times \dfrac{2.54\,cm}{1\,in} \times \dfrac{10\,mm}{1\,cm} \times \dfrac{1\,quarter}{1.55\,mm} = 1.1307\times10^5 = 1.13 \times 10^5$ quarters

　　　　(b)　1.1307×10^5 quarters $\times \dfrac{5.67\,g}{1\,quarter} = 6.41 \times 10^5$ g (641 kg)

　　　　(c)　1.1307×10^5 quarters $\times \dfrac{1\,dollar}{4\,quarters} = \$28,268 = \$2.83 \times 10^4$

　　　　(d)　$\$11,687,233,914,811.11 \times \dfrac{1\,stack}{\$28,268} = 4.13 \times 10^8$ stacks

1.68　　The most dense liquid, Hg, will sink; the least dense, cyclohexane, will float; H_2O will be in the middle.

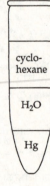

1.71   mass of toluene = 58.58 g – 32.65 g = 25.93 g

volume of toluene = $25.93 \text{ g} \times \dfrac{1 \text{ mL}}{0.864 \text{ g}} = 30.0116 = 30.0 \text{ mL}$

volume of solid = 50.00 mL – 30.0116 mL = 19.9884 = 20.0 mL

density of solid = $\dfrac{32.65 \text{ g}}{19.9884 \text{ mL}} = 1.63 \text{ g/mL}$

1.74   (a) $\dfrac{40 \text{ lb peat}}{14 \times 20 \times 30 \text{ in}^3} \times \dfrac{1 \text{ in}^3}{(2.54)^3 \text{ cm}^3} \times \dfrac{453.6 \text{ g}}{1 \text{ lb}} = 0.13 \text{ g/cm}^3 \text{ peat}$

$\dfrac{40 \text{ lb soil}}{1.9 \text{ gal}} \times \dfrac{1 \text{ gal}}{4 \text{ qt}} \times \dfrac{1.057 \text{ qt}}{1 \text{ L}} \times \dfrac{1 \times 10^{-3} \text{ L}}{1 \text{ mL}} \times \dfrac{1 \text{ mL}}{1 \text{ cm}^3} \times \dfrac{453.6 \text{ g}}{1 \text{ lb}} = 2.5 \text{ g/cm}^3 \text{ soil}$

No. Volume must be specified in order to compare mass. The densities tell us that a certain volume of peat moss is "lighter" (weighs less) than the same volume of top soil.

(b)   1 bag peat = $14 \times 20 \times 30 = 8.4 \times 10^3 \text{ in}^3$

$15.0 \text{ ft} \times 20.0 \text{ ft} \times 3.0 \text{ in} \times \dfrac{12^2 \text{ in}^2}{\text{ft}^2} = 129{,}600 = 1.3 \times 10^5 \text{ in}^3 \text{ peat needed}$

$129{,}600 \text{ in}^3 \times \dfrac{1 \text{ bag}}{8.4 \times 10^3 \text{ in}^3} = 15.4 = 15 \text{ bags (Buy 16 bags of peat.)}$

1.77 $11.86 \text{ g ethanol} \times \dfrac{1 \text{ cm}^3}{0.789 \text{ g ethanol}} = 15.0317 = 15.03 \text{ cm}^3, \text{ volume of cylinder}$

$V = \pi r^2 h; \ r = (V/\pi h)^{1/2} = \left[\dfrac{15.0317 \text{ cm}^3}{\pi \times 15.0 \text{ cm}} \right]^{1/2} = 0.5648 = 0.565 \text{ cm}$

d = 2r = 1.13 cm

1.79   The separation with distinctly separated red and blue spots is more successful. The procedure that produced the purple blur did not separate the two dyes. To quantify the characteristics of the separation, calculate a reference value for each spot that is

$$\frac{\text{distance travelled by spot}}{\text{distance travelled by solvent}}$$

If the values for the two spots are fairly different, the separation is successful. (One could measure the distance between the spots, but this would depend on the length of paper used and be different for each experiment. The values suggested above are independent of the length of paper.)

2 Atoms, Molecules, and Ions

Visualizing Concepts

2.1 (a) The path of the charged particle bends because the particle is repelled by the negatively charged plate and attracted to the positively charged plate.

(b) Like charges repel and opposite charges attract, so the sign of the electrical charge on the particle is negative.

(c) The greater the magnitude of the charges, the greater the electrostatic repulsion or attraction. As the charge on the plates is increased, the bending will increase.

(d) As the mass of the particle increases and speed stays the same, linear momentum (mv) of the particle increases and bending decreases. (See **A Closer Look**: The Mass Spectrometer.)

2.4 Since the number of electrons (negatively charged particles) does not equal the number of protons (positively charged particles), the particle is an ion. The charge on the ion is 2–.

Atomic number = number of protons = 16. The element is S, sulfur.

Mass number = protons + neutrons = 32

$^{32}_{16}S^{2-}$

2.6 Formula: IF_5 Name: iodine pentafluoride
Since the compound is composed of elements that are all nonmetals, it is molecular.

Atomic Theory and the Discovery of Atomic Structure (sections 2.1-2.2)

2.9 Postulate 4 of the atomic theory is the *law of constant composition*. It states that the relative number and kinds of atoms in a compound are constant, regardless of the source. Therefore, 1.0 g of pure water should always contain the same relative amounts of hydrogen and oxygen, no matter where or how the sample is obtained.

2.11 (a) $\dfrac{17.60 \text{ g oxygen}}{30.82 \text{ g nitrogen}} = \dfrac{0.5711 \text{ g O}}{1 \text{ g N}}$; $0.5711/0.5711 = 1.0$

$\dfrac{35.20 \text{ g oxygen}}{30.82 \text{ g nitrogen}} = \dfrac{1.142 \text{ g O}}{1 \text{ g N}}$; $1.142/0.5711 = 2.0$

$\dfrac{70.40 \text{ g oxygen}}{30.82 \text{ g nitrogen}} = \dfrac{2.284 \text{ g O}}{1 \text{ g N}}$; $2.284/0.5711 = 4.0$

$$\frac{88.00 \text{ g oxygen}}{30.82 \text{ g nitrogen}} = \frac{2.855 \text{ g O}}{1 \text{ g N}}; \quad 2.855/0.5711 = 5.0$$

(b)   These masses of oxygen per one gram nitrogen are in the ratio of 1:2:4:5 and thus obey the *law of multiple proportions*. Multiple proportions arise because atoms are the indivisible entities combining, as stated in Dalton's theory. Since atoms are indivisible, they must combine in ratios of small whole numbers.

2.13   Evidence that cathode rays were negatively charged particles was (1) that electric and magnetic fields deflected the rays in the same way they would deflect negatively charged particles and (2) that a metal plate exposed to cathode rays acquired a negative charge.

2.15   (a)   Most of the volume of an atom is empty space in which electrons move. Most alpha particles passed through this space. The path of the massive alpha particle would not be significantly altered by interaction with a "puny" electron.

       (b)   Most of the mass of an atom is contained in a very small, dense area called the nucleus. The few alpha particles that hit the massive, positively charged gold nuclei were strongly repelled and essentially deflected back in the direction they came from.

       (c)   The Be nuclei have a much smaller volume and positive charge than the Au nuclei; the charge repulsion between the alpha particles and the Be nuclei will be less, and there will be fewer direct hits because the Be nuclei have an even smaller volume than the Au nuclei. Fewer alpha particles will be scattered in general and fewer will be strongly back scattered.

Modern View of Atomic Structure; Atomic Weights (sections 2.3-2.4)

2.17   (a) $1.35 \text{ Å} \times \dfrac{1 \times 10^{-10} \text{ m}}{1 \text{ Å}} \times \dfrac{1 \text{ nm}}{1 \times 10^{-9} \text{ m}} = 0.135 \text{ nm}$

              $1.35 \text{ Å} \times \dfrac{1 \times 10^{-10} \text{ m}}{1 \text{ Å}} \times \dfrac{1 \text{ pm}}{1 \times 10^{-12} \text{ m}} = 1.35 \times 10^2 \text{ or } 135 \text{ pm } (1 \text{ Å} = 100 \text{ pm})$

       (b)   Aligned Au atoms have **diameters** touching. d = 2r = 2(1.35 Å) = 2.70 Å

              $1.0 \text{ mm} \times \dfrac{1 \text{ m}}{1000 \text{ mm}} \times \dfrac{1 \text{ Å}}{1 \times 10^{-10} \text{ m}} \times \dfrac{1 \text{ Au atom}}{2.70 \text{ Å}} = 3.70 \times 10^6 \text{ Au atoms}$

       (c) $V = 4/3 \pi r^3. \quad r = 1.35 \text{ Å} \times \dfrac{1 \times 10^{-10} \text{ m}}{1 \text{ Å}} \times \dfrac{100 \text{ cm}}{\text{m}} = 1.35 \times 10^{-8} \text{ cm}$

              $V = (4/3)(\pi)(1.35 \times 10^{-8})^3 \text{ cm}^3 = 1.03 \times 10^{-23} \text{ cm}^3$

2.19   (a)   proton, neutron, electron

       (b)   proton = +1, neutron = 0, electron = –1

       (c)   The neutron is most massive. (The neutron and proton have very similar masses).

       (d)   The electron least massive.

2.21   (a)   *Atomic number* is the number of protons in the nucleus of an atom. *Mass number* is the total number of nuclear particles, protons plus neutrons, in an atom.

        (b)   The mass number can vary without changing the identity of the atom, but the atomic number of every atom of a given element is the same.

2.23   p = protons, n = neutrons, e = electrons

        (a) ^{40}Ar has 18 p, 22 n, 18 e   (b) ^{65}Zn has 30 p, 35 n, 30 e

        (c) ^{70}Ga has 31 p, 39 n, 31 e   (d) ^{80}Br has 35 p, 45 n, 35 e

        (e) ^{184}W has 74 p, 110 n, 74 e   (f) ^{243}Am has 95 p, 148 n, 95 e

2.25

Symbol	^{52}Cr	^{55}Mn	^{112}Cd	^{222}Rn	^{207}Pb
Protons	24	25	48	86	82
Neutrons	28	30	64	136	125
Electrons	24	25	48	86	82
Mass no.	52	55	112	222	207

2.27   (a) $^{196}_{78}$Pt   (b) $^{84}_{36}$Kr   (c) $^{75}_{33}$As   (c) $^{24}_{12}$Mg

2.29   (a) $^{12}_{6}$C

        (b)   Atomic weights are really average atomic masses, the sum of the mass of each naturally occurring isotope of an element times its fractional abundance. Each B atom will have the mass of one of the naturally occurring isotopes, while the "atomic weight" is an average value. The naturally occurring isotopes of B, their atomic masses, and relative abundances are:

        ^{10}B, 10.012937, 19.9%; ^{11}B, 11.009305, 80.1%.

2.31   Atomic weight (average atomic mass) = Σ fractional abundance × mass of isotope

        Atomic weight = 0.6917(62.9296) + 0.3083(64.9278) = 63.5456 = 63.55 amu

2.33   (a)   Compare Figures 2.4 and 2.12, referring to Solution 2.14. In Thomson's cathode ray experiments and in mass spectrometry a stream of charged particles is passed through a magnetic field. The charged particles are deflected by the magnetic field according to their mass and charge. For a constant magnetic field strength and speed of the particles, the lighter particles experience a greater deflection.

        (b)   The x-axis label (independent variable) is atomic weight and the y-axis label (dependent variable) is signal intensity.

        (c)   Uncharged particles are not deflected in a magnetic field. The effect of the magnetic field on moving, *charged* particles is the basis of their separation by mass.

2.35   (a)   Average atomic mass = 0.7899(23.98504) + 0.1000(24.98584) + 0.1101(25.98259)

            = 24.31 amu

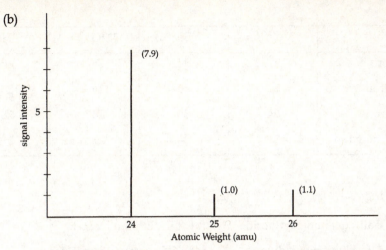

(b)

The relative intensities of the peaks in the mass spectrum are the same as the relative abundances of the isotopes. The abundances and peak heights are in the ratio ^{24}Mg: ^{25}Mg: ^{26}Mg as $7.8 : 1.0 : 1.1$.

The Periodic Table; Molecules and Ions (sections 2.5-2.7)

2.37 (a) Cr, 24 (metal) (b) He, 2 (nonmetal) (c) P, 15 (nonmetal)

(d) Zn, 30 (metal) (e) Mg, 12 (metal) (f) Br, 35 (nonmetal)

(g) As, 33 (metalloid)

2.39 (a) K, alkali metals (metal) (b) I, halogens (nonmetal)

(c) Mg, alkaline earth metals (metal) (d) Ar, noble gases (nonmetal)

(e) S, chalcogens (nonmetal)

2.41 An *empirical formula* shows the simplest ratio of the different atoms in a molecule. A *molecular formula* shows the exact number and kinds of atoms in a molecule. A *structural formula* shows how these atoms are arranged.

2.43 (a) $AlBr_3$ (b) C_4H_5 (c) C_2H_4O (d) P_2O_5

2.47 (e) C_3H_2Cl (f) BNH_2

2.45 (a) 6 (b) 6 (c) 12

2.47 (a) C_2H_6O

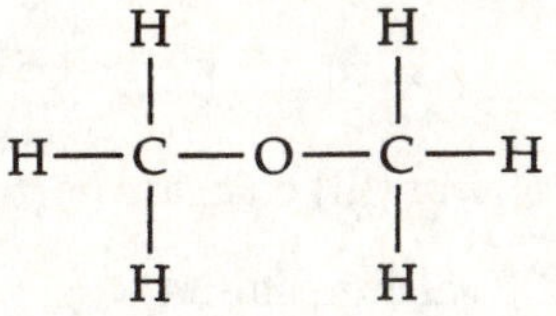

$$\begin{array}{ccccc} & H & & H & \\ & | & & | & \\ H- & C & -O- & C & -H \\ & | & & | & \\ & H & & H & \end{array}$$

(b) C_2H_6O

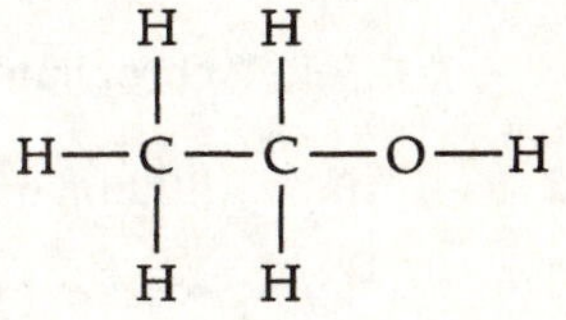

$$\begin{array}{ccccc} & H & H & & \\ & | & | & & \\ H- & C & - C & -O-H \\ & | & | & & \\ & H & H & & \end{array}$$

(c) CH_4O

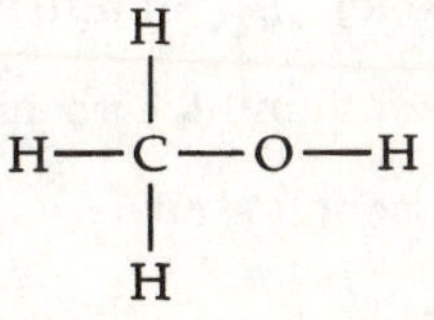

$$\begin{array}{ccc} & H & \\ & | & \\ H- & C & -O-H \\ & | & \\ & H & \end{array}$$

(d) PF_3

$$\begin{array}{ccc} F- & P & -F \\ & | & \\ & F & \end{array}$$

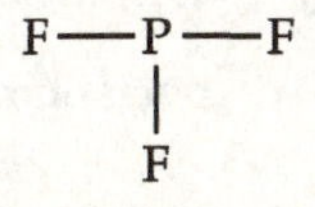

2.49

Symbol	$^{59}Co^{3+}$	$^{80}Se^{2-}$	$^{192}Os^{2+}$	$^{200}Hg^{2+}$
Protons	27	34	76	80
Neutrons	32	46	116	120
Electrons	24	36	74	78
Net Charge	3+	2–	2+	2+

2.51 (a) Mg^{2+} (b) Al^{3+} (c) K^+ (d) S^{2-} (e) F^-

2.53 (a) GaF_3, gallium(III) fluoride   (b) LiH, lithium hydride

(c) AlI_3, aluminum iodide   (d) K_2S, potassium sulfide

2.55 (a) $CaBr_2$ (b) K_2CO_3 (c) $Al(CH_3COO)_3$ (d) $(NH_4)_2SO_4$ (e) $Mg_3(PO_4)_2$

2.57

Ion	K^+	NH_4^+	Mg^{2+}	Fe^{3+}
Cl^-	KCl	NH_4Cl	$MgCl_2$	$FeCl_3$
OH^-	KOH	NH_4OH^*	$Mg(OH)_2$	$Fe(OH)_3$
CO_3^{2-}	K_2CO_3	$(NH_4)_2CO_3$	$MgCO_3$	$Fe_2(CO_3)_3$
PO_4^{3-}	K_3PO_4	$(NH_4)_3PO_4$	$Mg_3(PO_4)_2$	$FePO_4$

*Equivalent to $NH_3(aq)$.

2.59 Molecular (all elements are nonmetals):

(a) B_2H_6 (b) CH_3OH (f) NOCl   (g) NF_3

Ionic (formed by a cation and an anion, usually contains a metal cation):

(c) $LiNO_3$ (d) Sc_2O_3 (e) CsBr   (h) Ag_2SO_4

Naming Inorganic Compounds; Organic Molecules (sections 2.8-2.9)

2.61 (a) ClO_2^- (b) Cl^- (c) ClO_3^- (d) ClO_4^- (e) ClO^-

2.63 (a) calcium, 2+; oxide, 2–   (b) sodium, 1+; sulfate, 2–

(c) potassium, 1+; perchlorate, 1–   (d) iron, 2+; nitrate, 1–

(e) chromium, 3+; hydroxide, 1–

2.65 (a) lithium oxide   (b) iron(III) chloride (ferric chloride)

(c) sodium hypochlorite   (d) calcium sulfite

(e) copper(II) hydroxide (cupric hydroxide)   (f) iron(II) nitrate (ferrous nitrate)

(g) calcium acetate   (h) chromium(III) carbonate (chromic carbonate)

(i) potassium chromate   (j) ammonium sulfate

2.67   (a) $Al(OH)_3$ (b) K_2SO_4 (c) Cu_2O (d) $Zn(NO_3)_2$

          (e) $HgBr_2$ (f) $Fe_2(CO_3)_3$ (g) $NaBrO$

2.69   (a)   bromic acid   (b)   hydrobromic acid   (c)   phosphoric acid

          (d) $HClO$ (e) HIO_3 (f) H_2SO_3

2.71   (a)   sulfur hexafluoride   (b)   iodine pentafluoride   (c)   xenon trioxide

          (d) N_2O_4 (e) HCN (f) P_4S_6

2.73   (a) $ZnCO_3$, ZnO, CO_2 (b) HF, SiO_2, SiF_4, H_2O (c) SO_2, H_2O, H_2SO_3

          (d) PH_3 (e) $HClO_4$, Cd, $Cd(ClO_4)_2$ (f) VBr_3

2.75   (a)   A hydrocarbon is a compound composed of the elements hydrogen and carbon only.

          (b)

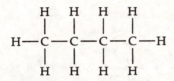

                molecular: C_4H_{10}

                empirical: C_2H_5

2.77   (a)   *Functional groups* are groups of specific atoms that are constant from one molecule to the next. For example, the alcohol functional group is an –OH. Whenever a molecule is called an alcohol, it contains the –OH group.

          (b)   —OH   (c)

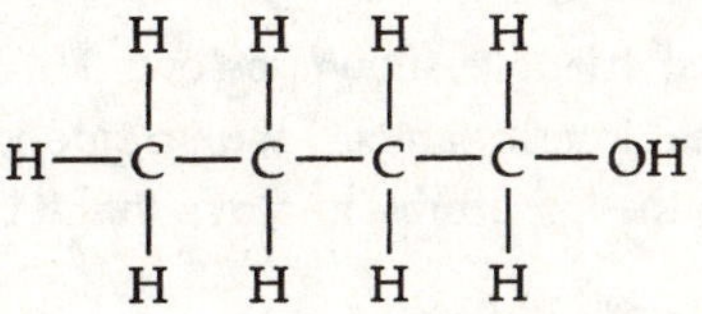

                                                        H   H   H   H
                                               H—C—C—C—C—OH
                                                  H   H   H   H

2.79   (a)

                 H   Cl   H   H   H   H
           H—C—C—C—H   H—C—C—C—Cl
                 H   H   H   H   H   H

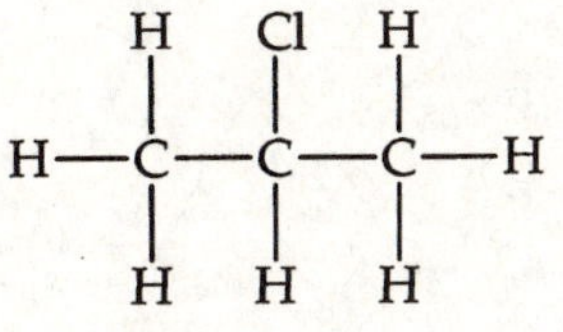

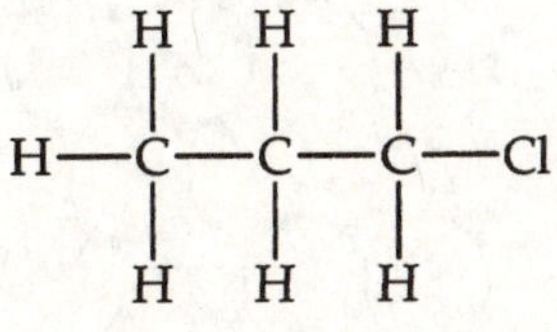

          (b)   1-chloropropane   2-chloropropane

Additional Exercises

2.82   (a) ^{3}He has 2 protons, 1 neutron, and 2 electrons.

         (b) ^{3}H has 1 proton, 2 neutrons, and 1 electron.

^{3}He: $2(1.672621673 \times 10^{-24} \text{ g}) + 1.674927211 \times 10^{-24} \text{ g} + 2(9.10938215 \times 10^{-28} \text{ g})$

$$= 5.021992 \times 10^{-24} \text{ g}$$

^{3}H: $1.672621673 \times 10^{-24} \text{ g} + 2(1.674927211 \times 10^{-24} \text{ g}) + 9.10938215 \times 10^{-28} \text{ g}$

$$= 5.023387 \times 10^{-24} \text{ g}$$

Tritium, ^{3}H, is more massive.

         (c)   The masses of the two particles differ by 0.0014×10^{-24} g. Each particle loses 1 electron to form the +1 ion, so the difference in the masses of the ions is still 1.4×10^{-27}. A mass spectrometer would need precision to 1×10^{-27} g to differentiate ^{3}He$^+$ and ^{3}H.

2.84   (a)   In arrangement A the number of atoms in 1 cm^2 is just the square of the number that fit linearly in 1 cm.

$$1.0 \text{ cm} \times \frac{1 \text{ atom}}{4.95 \text{ Å}} \times \frac{1 \times 10^{10} \text{ Å}}{1 \text{ m}} \times \frac{1 \text{ m}}{100 \text{ cm}} = 2.02 \times 10^7 = 2.0 \times 10^7 \text{ atoms/cm}$$

$$1.0 \text{ cm}^2 = (2.02 \times 10^7)^2 = 4.081 \times 10^{14} = 4.1 \times 10^{14} \text{ atoms/cm}^2$$

         (b)   In arrangement B, the atoms in the horizontal rows are touching along their diameters, as in arrangement A. The number of Rb atoms in a 1.0 cm row is then 2.0×10^7 Rb atoms. Relative to arrangement A, the vertical rows are offset by 1/2 of an atom. Atoms in a "column" are no longer touching along their vertical diameter. We must calculate the vertical distance occupied by a row of atoms, which is now less than the diameter of one Rb atom.

Consider the triangle shown below. This is an isosceles triangle (equal side lengths, equal interior angles) with a side-length of 2d and an angle of 60°. Drop a bisector to the uppermost angle so that it bisects the opposite side.

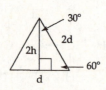

The result is a right triangle with two known side lengths. The length of the unknown side (the angle bisector) is 2h, two times the vertical distance occupied by a row of atoms. Solve for h, the "height" of one row of atoms.

$(2h)^2 + d^2 = (2d)^2; 4h^2 = 4d^2 - d^2 = 3d^2; h^2 = 3d^2/4$

$h = (3d^2/4)^{1/2} = (3(4.95\text{ Å})^2/4)^{1/2} = 4.2868 = 4.29\text{ Å}$

The number of rows of atoms in 1 cm is then

$$1.0\text{ cm} \times \frac{1\text{ row}}{4.2868\text{ Å}} \times \frac{1 \times 10^{10}\text{ Å}}{1\text{ m}} \times \frac{1\text{ m}}{100\text{ cm}} = 2.333 \times 10^7 = 2.3 \times 10^7$$

The number of atoms in a 1.0 cm² square area is then

$$\frac{2.020 \times 10^7\text{ atoms}}{1\text{ row}} \times 2.333 \times 10^7\text{ rows} = 4.713 \times 10^{14} = 4.7 \times 10^{14}\text{ atoms}$$

Note that we have ignored the loss of "1/2" atom at the end of each horizontal row. Out of 2.0×10^7 atoms per row, one atom is not significant.

(c) The ratio of atoms in arrangement B to arrangement A is then 4.713×10^{14} atoms$/4.081 \times 10^{14} = 1.555 = 1.2:1$. Clearly, arrangement B results in less empty space per unit area or volume. If extended to three dimensions, arrangement B would lead to a greater density for Rb metal.

2.87 (a) $^{16}_{8}O$, $^{17}_{8}O$, $^{18}_{8}O$

(b) All isotopes are atoms of the same element, oxygen, with the same atomic number (Z = 8), 8 protons in the nucleus and 8 electrons. Elements with similar electron arrangements have similar chemical properties (Section 2.5). Since the 3 isotopes all have 8 electrons, we expect their electron arrangements to be the same and their chemical properties to be very similar, perhaps identical. Each has a different number of neutrons (8, 9, or 10), a different mass number (A = 16, 17, or 18) and thus a different atomic mass.

2.90 (a) The 68.926 amu isotope has a mass number of 69, with 31 protons, 38 neutrons and the symbol $^{69}_{31}Ga$. The 70.925 amu isotope has a mass number of 71, 31 protons, 40 neutrons, and symbol $^{71}_{31}Ga$. (All Ga atoms have 31 protons.)

(b) The average mass of a Ga atom is 69.72 amu. Let x = abundance of the lighter isotope, 1–x = abundance of the heavier isotope. Then x(68.926) + (1–x)(70.925) = 69.72; x = 0.6028 = 0.603, ^{69}Ga = 60.3%, ^{71}Ga = 39.7%.

2.93 (a) Five significant figures. $^1H^+$ is a bare proton with mass 1.0073 amu. ^1H is a hydrogen atom, with 1 proton and 1 electron. The mass of the electron is 5.486×10^{-4} or 0.0005486 amu. Thus the mass of the electron is significant in the fourth decimal place or fifth significant figure in the mass of ^1H.

(b) Mass of ^1H = 1.0073 amu   (proton)

          <u>0.0005486 amu</u>   (electron)

          1.0078 amu   (We have not rounded up to 1.0079 since 49 < 50 in the final sum.)

$$\text{Mass \% of electron} = \frac{\text{mass of } e^-}{\text{mass of } {}^1\text{H}} \times 100 = \frac{5.486 \times 10^{-4} \text{ amu}}{1.0078 \text{ amu}} \times 100 = 0.05444\%$$

2.96   Strontium is an alkaline earth metal, similar in chemical properties to calcium and magnesium. Calcium is ubiquitous in biological organisms, humans included. It is a vital nutrient required for formation and maintenance of healthy bones and teeth. As such, there are efficient pathways for calcium uptake and distribution in the body, pathways that are also available to chemically similar strontium. Harmful strontium imitates calcium and then behaves badly when the body tries to use it as it uses calcium.

2.98   (a)   nickel(II) oxide, 2+   (b)   manganese(IV) oxide, 4+

       (c)   chromium(III) oxide, 3+   (d)   molybdenium(VI) oxide, 6+

2.101 (a)   perbromate ion   (b)   selenite ion

       (c) $AsO_4{}^{3-}$ (d) $HTeO_4{}^{-}$

2.104 (a)   potassium nitrate   (b) sodium carbonate   (c) calcium oxide

       (d)   hydrochloric acid   (e) magnesium sulfate   (f) magnesium hydroxide

3 Stoichiometry: Calculations with Chemical Formulas and Equations

Visualizing Concepts

3.1   Reactant A = blue, reactant B = red

Overall, 4 blue A_2 molecules + 4 red B atoms $\rightarrow$ 4 A_2B molecules

Since 4 is a common factor, this equation reduces to equation (a).

3.3   (a)   There are twice as many O atoms as N atoms, so the empirical formula of the original compound is NO_2.

(b)   No, because we have no way of knowing whether the empirical and molecular formulas are the same. NO_2 represents the simplest ratio of atoms in a molecule but not the only possible molecular formula.

3.5   (a)   *Analyze.* Given the molecular model, write the molecular formula.

Plan. Use the colors of the atoms (spheres) in the model to determine the number of atoms of each element.

Solve. Observe 2 gray C atoms, 5 white H atoms, 1 blue N atom, 2 red O atoms. $C_2H_5NO_2$

(b)   *Plan.* Follow the method in Sample Exercise 3.9. Calculate formula weight in amu and molar mass in grams.

2 C atoms = 2(12.0 amu) = 24.0 amu

5 H atoms = 5(1.0 amu) = 5.0 amu

1 N atoms = 1(14.0 amu) = 14.0 amu

2 O atoms = 2(16.0 amu) = <u>32.0 amu</u>

75.0 amu

Formula weight = 75.0 amu, molar mass = 75.0 g/mol

(c)   *Plan.* The molar mass of a substance provides the factor for converting moles to grams (or grams to moles).

Solve. $3 \text{ mol glycine} \times \dfrac{75.0 \text{ g glycine}}{\text{mol}} = 225 \text{ g glycine}$

(d)   *Plan.* Use the definition of mass % and the results from parts (a) and (b) above to find mass % N in glycine.

17

Solve. $\text{mass}\% \, N = \dfrac{g \, N}{g \, C_2H_5NO_2} \times 100$

Assume 1 mol $C_2H_5NO_2$. From the molecular formula of glycine [part (a)], there is 1 mol N/mol glycine.

$$\text{mass}\% \, N = \frac{1 \times (\text{molar mass N})}{\text{molar mass glycine}} \times 100 = \frac{14.0 \, g}{75.0 \, g} \times 100 = 18.7\%$$

3.7   *Analyze.* Given a box diagram and formulas of reactants, draw a box diagram of products.

Plan. Write and balance the chemical equation. Determine combining ratios of elements and decide on limiting reactant. Draw a box diagram of products, containing the correct number of product molecules and only excess reactant.

Solve. $N_2 + 3H_2 \longrightarrow 3NH_3.$ $N_2 = $ ⬤⬤ , $NH_3 = $ ◯⬤◯

Each N atom (1/2 of an N_2 molecule) reacts with 3 H atoms (1.5 H_2 molecules) to form an NH_3 molecule. Eight N atoms (4 N_2 molecules) require 24 H atoms (12 H_2 molecules) for complete reaction. Only 9 H_2 molecules are available, so H_2 is the limiting reactant. Nine H_2 molecules (18 H atoms) determine that 6 NH_3 molecules are produced. One N_2 molecule is in excess.

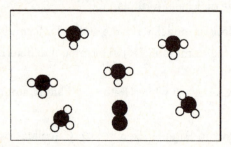

Check. Verify that mass is conserved in your solution, that the number and kinds of atoms are the same in reactant and product diagrams. In this example, there are 8 N atoms and 18 H atoms in both diagrams, so mass is conserved.

Balancing Chemical Equations (section 3.1)

3.9   (a)   In balancing chemical equations, the law of conservation of mass, that atoms are neither created nor destroyed during the course of a reaction, is observed. This means that the **number** and **kinds** of atoms on both sides of the chemical equation must be the same.

(b)   Subscripts in chemical formulas should not be changed when balancing equations, because changing the subscript changes the identity of the compound (law of constant composition).

(c)   liquid water = $H_2O(l)$; water vapor = $H_2O(g)$; aqueous sodium chloride = $NaCl(aq)$; solid sodium chloride = $NaCl(s)$

3.11 (a) $2CO(g) + O_2(g) \rightarrow 2CO_2(g)$

(b) $N_2O_5(g) + H_2O(l) \rightarrow 2HNO_3(aq)$

(c) $CH_4(g) + 4Cl_2(g) \rightarrow CCl_4(l) + 4HCl(g)$

(d) $Al_4C_3(s) + 12H_2O(l) \rightarrow 4Al(OH)_3(s) + 3CH_4(g)$

(e) $2C_5H_{10}O_2(l) + 13O_2(g) \rightarrow 10CO_2(g) + 10H_2O(g)$

(f) $2Fe(OH)_3(s) + 3H_2SO_4(aq) \rightarrow Fe_2(SO_4)_3(aq) + 6H_2O(l)$

(g) $Mg_3N_2(s) + 4H_2SO_4(aq) \rightarrow 3MgSO_4(aq) + (NH_4)_2SO_4(aq)$

3.13 (a) $CaC_2(s) + 2H_2O(l) \rightarrow Ca(OH)_2(aq) + C_2H_2(g)$

(b) $2KClO_3(s) \overset{\Delta}{\rightarrow} 2KCl(s) + 3O_2(g)$

(c) $Zn(s) + H_2SO_4(aq) \rightarrow H_2(g) + ZnSO_4(aq)$

(d) $PCl_3(l) + 3H_2O(l) \rightarrow H_3PO_3(aq) + 3HCl(aq)$

(e) $3H_2S(g) + 2Fe(OH)_3(s) \rightarrow Fe_2S_3(s) + 6H_2O(g)$

Patterns of Chemical Reactivity (section 3.2)

3.15 (a) When a metal reacts with a nonmetal, an ionic compound forms. The combining ratio of the atoms is such that the total positive charge on the metal cation(s) is equal to the total negative charge on the nonmetal anion(s). Determine the formula by balancing the positive and negative charges in the ionic product. All ionic compounds are solids. $2Na(s) + Br_2(l) \rightarrow 2NaBr(s)$

(b) The second reactant is oxygen gas from the air, $O_2(g)$. The products are $CO_2(g)$ and $H_2O(l)$. $2C_6H_6(l) + 15O_2(g) \rightarrow 12CO_2(g) + 6H_2O(l)$.

3.17 (a) $Mg(s) + Cl_2(g) \rightarrow MgCl_2(s)$

(b) $BaCO_3(s) \overset{\Delta}{\rightarrow} BaO(s) + CO_2(g)$

(c) $C_8H_8(l) + 10O_2(g) \rightarrow 8CO_2(g) + 4H_2O(l)$

(d) CH_3OCH_3 is C_2H_6O. $C_2H_6O(g) + 3O_2(g) \rightarrow 2CO_2(g) + 3H_2O(l)$

3.19 (a) $2C_3H_6(g) + 9O_2(g) \rightarrow 6CO_2(g) + 6H_2O(g)$ combustion

(b) $NH_4NO_3(s) \rightarrow N_2O(g) + 2H_2O(g)$ decomposition

(c) $C_5H_6O(l) + 6O_2(g) \rightarrow 5CO_2(g) + 3H_2O(g)$ combustion

(d) $N_2(g) + 3H_2(g) \rightarrow 2NH_3(g)$ combination

(e) $K_2O(s) + H_2O(l) \rightarrow 2KOH(aq)$ combination

Formula Weights (section 3.3)

3.21 *Analyze.* Given molecular formula or name, calculate formula weight.

Plan. If a name is given, write the correct molecular formula. Then, follow the method in Sample Exercise 3.5. *Solve.*

(a) HNO_3: $1(1.0) + 1(14.0) + 3(16.0) = 63.0$ amu

(b) $KMnO_4$: $1(39.1) + 1(54.9) + 4(16.0) = 158.0$ amu

(c) $Ca_3(PO_4)_2$: $3(40.1) + 2(31.0) + 8(16.0) = 310.3$ amu

(d) SiO_2: $1(28.1) + 2(16.0) = 60.1$ amu

(e) Ga_2S_3: $2(69.7) + 3(32.1) = 235.7$ amu

(f) $Cr_2(SO_4)_3$: $2(52.0) + 3(32.1) + 12(16.0) = 392.3$ amu

(g) PCl_3: $1(31.0) + 3(35.5) = 137.5$ amu

3.23   *Plan.* Calculate the formula weight (FW), then the mass % oxygen in the compound. *Solve.*

(a) $C_{17}H_{19}NO_3$: FW = $17(12.0) + 19(1.0) + 1(14.0) + 3(16.0) = 285.0$ amu

$$\% O = \frac{3(16.0)\,\text{amu}}{285.0\,\text{amu}} \times 100 = 16.842 = 16.8\%$$

(b) $C_{18}H_{21}NO_3$: FW = $18(12.0) + 21(1.0) + 1(14.0) + 3(16.0) = 299.0$ amu

$$\% O = \frac{3(16.0)\,\text{amu}}{299.0\,\text{amu}} \times 100 = 16.054 = 16.1\%$$

(c) $C_{17}H_{21}NO_4$: FW = $17(12.0) + 21(1.0) + 1(14.0) + 4(16.0) = 303.0$ amu

$$\% O = \frac{4(16.0)\,\text{amu}}{303.0\,\text{amu}} \times 100 = 21.122 = 21.1\%$$

(d) $C_{22}H_{24}N_2O_8$: FW = $22(12.0) + 24(1.0) + 2(14.0) + 8(16.0) = 444.0$ amu

$$\% O = \frac{8(16.0)\,\text{amu}}{444.0\,\text{amu}} \times 100 = 28.829 = 28.8\%$$

(e) $C_{41}H_{64}O_{13}$: FW = $4(12.0) + 64(1.0) + 13(16.0) = 764.0$ amu

$$\% O = \frac{13(16.0)\,\text{amu}}{764\,\text{amu}} \times 100 = 27.225 = 27.2\%$$

(f) $C_{66}H_{75}Cl_2N_9O_{24}$: FW = $66(12.0)+75(1.0)+2(35.5)+9(14.0)+24(16.0) = 1448.0$ amu

$$\% O = \frac{24(16.0)\,\text{amu}}{1448.0\,\text{amu}} \times 100 = 26.519 = 26.5\%$$

3.25   *Plan.* Follow the logic for calculating mass % C given in Sample Exercise 3.6.   *Solve.*

(a) C_7H_6O: FW = $7(12.0) + 6(1.0) + 1(16.0) = 106.0$ amu

$$\% C = \frac{7(12.0)\,\text{amu}}{106.0\,\text{amu}} \times 100 = 79.2\%$$

(b) $C_8H_8O_3$: FW = $8(12.0) + 8(1.0) + 3(16.0) = 152.0$ amu

$$\% C = \frac{8(12.0)\,\text{amu}}{152.0\,\text{amu}} \times 100 = 63.2\%$$

(c) $C_7H_{14}O_2$: FW = $7(12.0) + 14(1.0) + 2(16.0) = 130.0$ amu

$$\% C = \frac{7(12.0)\,\text{amu}}{130.0\,\text{amu}} \times 100 = 64.6\%$$

Avogadro's Number and the Mole (section 3.4)

3.27   (a) 6.022×10^{23}. This is the number of objects in a mole of anything.

(b)   The formula weight of a substance in amu has the same numerical value as the molar mass expressed in grams.

3.29   *Plan.* Since the mole is a counting unit, use it as a basis of comparison; determine the total moles of atoms in each given quantity.   *Solve.*

23 g Na contains 1 mol of atoms

0.5 mol H_2O contains (3 atoms $\times$ 0.5 mol) = 1.5 mol atoms

6.0×10^{23} N_2 molecules contains (2 atoms $\times$ 1 mol) = 2 mol atoms

3.31   *Analyze.* Given: 160 lb/person; Avogadro's number of people, 6.022×10^{23} people. Find: mass in kg of Avogadro's number of people; compare with mass of Earth.

Plan. people $\rightarrow$ mass in lb $\rightarrow$ mass in kg; mass of people /mass of Earth

Solve. 6.022×10^{23} people $\times \dfrac{160\,\text{lb}}{\text{person}} \times \dfrac{1\,\text{kg}}{2.2046\,\text{lb}} = 4.370 \times 10^{25} = 4.37 \times 10^{25}$ or 4.4×10^{25} kg

$\dfrac{4.370 \times 10^{25}\ \text{kg of people}}{5.98 \times 10^{24}\ \text{kg Earth}} = 7.31$ or 7.3

One mole of people weighs 7.31 times as much as Earth.

Check. This mass of people is reasonable since Avogadro's number is large.

Estimate: 160 lb $\approx$ 70 kg; $6 \times 10^{23} \times 70 = 420 \times 10^{23} = 4.2 \times 10^{25}$ kg

3.33   (a)   *Analyze.* Given: 0.105 mol sucrose, $C_{12}H_{22}O_{11}$. Find: mass in g.

Plan. Use molar mass (g/mol) of $C_{12}H_{22}O_{11}$ to find g $C_{12}H_{22}O_{11}$.

Solve. molar mass = 12(12.0107) + 22(1.00794) + 11(15.9994) = 342.296 = 342.30

0.105 mol sucrose $\times \dfrac{342.30\,\text{g}}{1\,\text{mol}} = 35.942 = 35.9$ g $C_{12}H_{22}O_{11}$

Check. 0.1(342) = 34.2 g. The calculated result is reasonable.

(b)   *Analyze.* Given: mass. Find: moles. *Plan.* Use molar mass of $Zn(NO_3)_2$.

Solve. molar mass = 1(65.39) + 2(14.0067) + 6(15.9994) = 189.3998 = 189.40

143.50 g $Zn(NO_3)_2 \times \dfrac{1\,\text{mol}}{189.40\ \text{g } Zn(NO_3)_2} = 0.75766$ mol $Zn(NO_3)_2$

Check. 140/180 $\approx$ 7/9 = 0.78 mol

(c)   *Analyze.* Given: moles. Find: molecules. *Plan.* Use Avogadro's number.

Solve. 1.0×10^{-6} mol $CH_3CH_2OH \times \dfrac{6.022 \times 10^{23}\ \text{molecules}}{1\,\text{mol}} = 6.022 \times 10^{17}$

$= 6.0 \times 10^{17}$ CH_3CH_2OH molecules

Check. $(1.0 \times 10^{-6})(6 \times 10^{23}) = 6 \times 10^{17}$

(d)   *Analyze.* Given: mol NH_3. Find: N atoms.

Plan. mol $NH_3 \rightarrow$ mol N atoms $\rightarrow$ N atoms

Solve. 0.410 mol $NH_3 \times \dfrac{1\,\text{mol N atoms}}{1\,\text{mol } NH_3} \times \dfrac{6.022 \times 10^{23}\ \text{atoms}}{1\,\text{mol}}$

$= 2.47 \times 10^{23}$ N atoms

Check. $(0.4)(6 \times 10^{23}) = 2.4 \times 10^{23}$.

3.35   *Analyze/Plan.* See Solution 3.33 for stepwise problem-solving approach.   *Solve.*

(a) $(NH_4)_3PO_4$ molar mass = $3(14.007) + 12(1.008) + 1(30.974) + 4(16.00) = 149.091$

$$= 149.1 \text{ g/mol}$$

$$2.50 \times 10^{-3} \text{ mol } (NH_4)_3PO_4 \times \frac{149.1 \text{ g } (NH_4)_3PO_4}{1 \text{ mol}} = 0.373 \text{ g } (NH_4)_3PO_4$$

(b) $AlCl_3$ molar mass = $26.982 + 3(35.453) = 133.341 = 133.34 \text{ g/mol}$

$$0.2550 \text{ g } AlCl_3 \times \frac{1 \text{ mol}}{133.34 \text{ g } AlCl_3} \times \frac{3 \text{ mol } Cl^-}{1 \text{ mol } AlCl_3} = 5.737 \times 10^{-3} \text{ mol } Cl^-$$

(c) $C_8H_{10}N_4O_2$ molar mass = $8(12.01) + 10(1.008) + 4(14.01) + 2(16.00) = 194.20$

$$= 194.2 \text{ g/mol}$$

$$7.70 \times 10^{20} \text{ molecules} \times \frac{1 \text{ mol}}{6.022 \times 10^{23} \text{ molecules}} \times \frac{194.2 \text{ g } C_8H_{10}N_4O_2}{1 \text{ mol caffeine}}$$

$$= 0.248 \text{ g } C_8H_{10}N_4O_2$$

(d) $\dfrac{0.406 \text{ g cholesterol}}{0.00105 \text{ mol}} = 387 \text{ g cholesterol/mol}$

3.37   (a) $C_6H_{10}OS_2$ molar mass = $6(12.01) + 10(1.008) + 1(16.00) + 2(32.07) = 162.28$

$$= 162.3 \text{ g/mol}$$

(b)   *Plan.* mg $\rightarrow$ g $\rightarrow$ mol   *Solve.*

$$5.00 \text{ mg allicin} \times \frac{1 \times 10^{-3} \text{ g}}{1 \text{ mg}} \times \frac{1 \text{ mol}}{162.3 \text{ g}} = 3.081 \times 10^{-5} = 3.08 \times 10^{-5} \text{ mol allicin}$$

Check. 5.00 mg is a small mass, so the small answer is reasonable.

$(5 \times 10^{-3})/200 = 2.5 \times 10^{-5}$

(c)   *Plan.* Use mol from part (b) and Avogadro's number to calculate molecules.

Solve. $3.081 \times 10^{-5} \text{ mol allicin} \times \dfrac{6.022 \times 10^{23} \text{ molecules}}{\text{mol}} = 1.855 \times 10^{19}$

$$= 1.86 \times 10^{19} \text{ allicin molecules}$$

Check. $(3 \times 10^{-5})(6 \times 10^{23}) = 18 \times 10^{18} = 1.8 \times 10^{19}$

(d)   *Plan.* Use molecules from part (c) and molecular formula to calculate S atoms.

Solve. $1.855 \times 10^{19} \text{ allicin molecules} \times \dfrac{2 \text{ S atoms}}{1 \text{ allicin molecule}} = 3.71 \times 10^{19} \text{ S atoms}$

Check. Obvious.

3.39   (a)   *Analyze.* Given: $C_6H_{12}O_6$, 1.250×10^{21} C atoms. Find: H atoms.

Plan. Use molecular formula to determine number of H atoms that are present with 1.250×10^{21} C atoms.   *Solve.*

$$\frac{12 \text{ H atoms}}{6 \text{ C atoms}} = \frac{2 \text{ H}}{1 \text{ C}} \times 1.250 \times 10^{21} \text{ C atoms} = 2.500 \times 10^{21} \text{ H atoms}$$

Check. $(2 \times 1 \times 10^{21}) = 2 \times 10^{21}$

(b) *Plan.* Use molecular formula to find the number of glucose molecules that contain 1.250×10^{21} C atoms. *Solve.*

$$\frac{1 \, C_6H_{12}O_6 \text{ molecule}}{6 \, C \text{ atoms}} \times 1.250 \times 10^{21} \, C \text{ atoms} = 2.0833 \times 10^{20}$$

$$= 2.083 \times 10^{20} \, C_6H_{12}O_6 \text{ molecules}$$

Check. $(12 \times 10^{20}/6) = 2 \times 10^{20}$

(c) *Plan.* Use Avogadro's number to change molecules $\rightarrow$ mol. *Solve.*

$$2.0833 \times 10^{20} \, C_6H_{12}O_6 \text{ molecules} \times \frac{1 \, mol}{6.022 \times 10^{23} \text{ molecules}}$$

$$= 3.4595 \times 10^{-4} = 3.460 \times 10^{-4} \, mol \, C_6H_{12}O_6$$

Check. $(2 \times 10^{20})/(6 \times 10^{23}) = 0.33 \times 10^{-3} = 3.3 \times 10^{-4}$

(d) *Plan.* Use molar mass to change mol $\rightarrow$ g. *Solve.*

1 mole of $C_6H_{12}O_6$ weighs 180.0 g (Sample Exercise 3.9)

$$3.4595 \times 10^{-4} \, mol \, C_6H_{12}O_6 \times \frac{180.0 \, g \, C_6H_{12}O_6}{1 \, mol} = 0.06227 \, g \, C_6H_{12}O_6$$

Check. $3.5 \times 180 = 630; 630 \times 10^{-4} = 0.063$

3.41 *Analyze.* Given: $g \, C_2H_3Cl/L$. Find: mol/L, molecules/L.

Plan. The /L is constant throughout the problem, so we can ignore it. Use molar mass for g $\rightarrow$ mol, Avogadro's number for mol $\rightarrow$ molecules. *Solve.*

$$\frac{2.0 \times 10^{-6} \, g \, C_2H_3Cl}{1 \, L} \times \frac{1 \, mol \, C_2H_3Cl}{62.50 \, g \, C_2H_3Cl} = 3.20 \times 10^{-8} = 3.2 \times 10^{-8} \, mol \, C_2H_3Cl/L$$

$$\frac{3.20 \times 10^{-8} \, mol \, C_2H_3Cl}{1 \, L} \times \frac{6.022 \times 10^{23} \text{ molecules}}{1 \, mol} = 1.9 \times 10^{16} \text{ molecules/L}$$

Check. $(200 \times 10^{-8})/60 = 2.5 \times 10^{-8} \, mol$

$(2.5 \times 10^{-8}) \times (6 \times 10^{23}) = 15 \times 10^{15} = 1.5 \times 10^{16}$

Empirical Formulas (section 3.5)

3.43 (a) *Analyze.* Given: moles. Find: empirical formula.

Plan. Find the **simplest ratio of moles** by dividing by the smallest number of moles present.

Solve. 0.0130 mol C / 0.0065 = 2

0.0390 mol H / 0.0065 = 6

0.0065 mol O / 0.0065 = 1

The empirical formula is C_2H_6O.

Check. The subscripts are simple integers.

(b)   *Analyze.* Given: grams. Find: empirical formula.

Plan. Calculate the moles of each element present, then the simplest ratio of moles.

Solve. $11.66 \text{ g Fe} \times \dfrac{1 \text{ mol Fe}}{55.85 \text{ g Fe}} = 0.2088 \text{ mol Fe}; \; 0.2088 / 0.2088 = 1$

$5.01 \text{ g O} \times \dfrac{1 \text{ mol O}}{16.00 \text{ g O}} = 0.3131 \text{ mol O}; \; 0.3131 / 0.2088 \approx 1.5$

Multiplying by two, the integer ratio is 2 Fe : 3 O; the empirical formula is Fe_2O_3.

Check. The subscripts are simple integers.

(c)   *Analyze.* Given: mass %. Find: empirical formulas.

Plan. Assume 100 g sample, calculate moles of each element, find the simplest ratio of moles.

Solve. $40.0 \text{ g C} \times \dfrac{1 \text{ mol C}}{12.01 \text{ g C}} = 3.33 \text{ mol C}; \; 3.33 / 3.33 = 1$

$6.7 \text{ g H} \times \dfrac{1 \text{ mol H}}{1.008 \text{ mol H}} = 6.65 \text{ mol H}; \; 6.65 / 3.33 \approx 2$

$53.3 \text{ g O} \times \dfrac{1 \text{ mol O}}{16.00 \text{ mol O}} = 3.33 \text{ mol O}; \; 3.33 / 3.33 = 1$

The empirical formula is CH_2O.

Check. The subscripts are simple integers.

3.45   *Analyze/Plan.* The procedure in all these cases is to assume 100 g of sample, calculate the number of moles of each element present in that 100 g, then obtain the ratio of moles as smallest whole numbers.   *Solve.*

(a) $10.4 \text{ g C} \times \dfrac{1 \text{ mol C}}{12.01 \text{ g C}} = 0.866 \text{ mol C}; \; 0.866 / 0.866 = 1$

$27.8 \text{ g S} \times \dfrac{1 \text{ mol S}}{32.07 \text{ g S}} = 0.867 \text{ mol S}; \; 0.867 / 0.866 \approx 1$

$61.7 \text{ g Cl} \times \dfrac{1 \text{ mol Cl}}{35.45 \text{ g Cl}} = 1.74 \text{ mol Cl}; \; 1.74 / 0.866 \approx 2$

The empirical formula is $CSCl_2$.

(b) $21.7 \text{ g C} \times \dfrac{1 \text{ mol C}}{12.01 \text{ g C}} = 1.81 \text{ mol C}; \; 1.81 / 0.600 \approx 3$

$9.6 \text{ g O} \times \dfrac{1 \text{ mol O}}{16.00 \text{ g O}} = 0.600 \text{ mol O}; \; 0.600 / 0.600 = 1$

$68.7 \text{ g F} \times \dfrac{1 \text{ mol F}}{19.00 \text{ g F}} = 3.62 \text{ mol F}; \; 3.62 / 0.600 \approx 6$

The empirical formula is C_3OF_6.

(c)   The mass of F is [100 g total – (32.79 g Na + 13.02 Al)] = 54.19 g F

$$32.79 \text{ g Na} \times \frac{1 \text{ mol Na}}{22.99 \text{ g Na}} = 1.426 \text{ mol Na}; \quad 1.426 / 0.4826 \approx 3$$

$$13.02 \text{ g Al} \times \frac{1 \text{ mol Al}}{26.98 \text{ g Al}} = 0.4826 \text{ mol Al}; \quad 0.4826 / 0.4826 = 1$$

$$54.19 \text{ g F} \times \frac{1 \text{ mol F}}{19.00 \text{ g F}} = 2.852 \text{ mol F}; \quad 2.852 / 0.4826 \approx 6$$

The empirical formula is $Na_3 AlF_6$.

3.47   *Analyze.* Given: mass% F; empirical formula XF_3 implies 3:1 ratio of mol F to mol X. Find: atomic mass (AM) of X.

Plan. Calculate mol F. This is 3 times mol X. mol X = 35 g X/AM X.
mol F/3 = 35 g X/AM X. Solve for AM X.

Solve. Mol F = 65/19.0 = 3.421 = 3.4; mol X = 3.421/3 = 1.14035 = 1.1

1.14035 mol X = 35 g X/AM X; AM X = 35 g X/1.14035 mol X = 30.69 = 31 g/mol

3.49   *Analyze.* Given: empirical formula, molar mass. Find: molecular formula.

Plan. Calculate the empirical formula weight (FW); divide FW by molar mass (MM) to calculate the integer that relates the empirical and molecular formulas. Check. If FW/MM is an integer, the result is reasonable. *Solve.*

(a)   FW $CH_2 = 12 + 2(1) = 14$. $\dfrac{MM}{FW} = \dfrac{84}{14} = 6$

The subscripts in the empirical formula are multiplied by 6. The molecular formula is $C_6 H_{12}$.

(b)   FW $NH_2 Cl = 14.01 + 2(1.008) + 35.45 = 51.48$. $\dfrac{MM}{FW} = \dfrac{51.5}{51.5} = 1$

The empirical and molecular formulas are $NH_2 Cl$.

3.51   *Analyze.* Given: mass %, molar mass. Find: molecular formula.

Plan. Use the plan detailed in Solution 3.45 to find an empirical formula from mass % data. Then use the plan detailed in 3.47 to find the molecular formula. Note that some indication of molar mass must be given, or the molecular formula cannot be determined. *Check.* If there is an integer ratio of moles and MM/ FW is an integer, the result is reasonable.   *Solve.*

(a) $92.3 \text{ g C} \times \dfrac{1 \text{ mol C}}{12.01 \text{ g C}} = 7.685 \text{ mol C}; \quad 7.685/7.639 = 1.006 \approx 1$

$$7.7 \text{ g H} \times \frac{1 \text{ mol H}}{1.008 \text{ g H}} = 7.639 \text{ mol H}; \quad 7.639/7.639 = 1$$

The empirical formula is CH, FW = 13.

$$\frac{MM}{FW} = \frac{104}{13} = 8; \text{ the molecular formula is } C_8 H_8.$$

(b) $49.5 \text{ g C} \times \dfrac{1 \text{ mol C}}{12.01 \text{ g C}} = 4.12 \text{ mol C}; \; 4.12/1.03 \approx 4$

$5.15 \text{ g H} \times \dfrac{1 \text{ mol H}}{1.008 \text{ g H}} = 5.11 \text{ mol H}; \; 5.11/1.03 \approx 5$

$28.9 \text{ g N} \times \dfrac{1 \text{ mol N}}{14.01 \text{ g N}} = 2.06 \text{ mol N}; \; 2.06/1.03 \approx 2$

$16.5 \text{ g O} \times \dfrac{1 \text{ mol O}}{16.00 \text{ g O}} = 1.03 \text{ mol O}; \; 1.03/1.03 = 1$

Thus, $C_4H_5N_2O$, FW = 97. If the molar mass is about 195, a factor of 2 gives the molecular formula $C_8H_{10}N_4O_2$.

(c) $35.51 \text{ g C} \times \dfrac{1 \text{ mol C}}{12.01 \text{ g C}} = 2.96 \text{ mol C}; \; 2.96/0.592 = 5$

$4.77 \text{ g H} \times \dfrac{1 \text{ mol H}}{1.008 \text{ g H}} = 4.73 \text{ mol H}; \; 4.73/0.592 = 7.99 \approx 8$

$37.85 \text{ g O} \times \dfrac{1 \text{ mol O}}{16.00 \text{ g O}} = 2.37 \text{ mol O}; \; 2.37/0.592 = 4$

$8.29 \text{ g N} \times \dfrac{1 \text{ mol N}}{14.01 \text{ g N}} = 0.592 \text{ mol N}; \; 0.592/0.592 = 1$

$13.60 \text{ g Na} \times \dfrac{1 \text{ mol Na}}{22.99 \text{ g Na}} = 0.592 \text{ mol Na}; \; 0.592/0.592 = 1$

The empirical formula is $C_5H_8O_4NNa$, FW = 169 g. Since the empirical formula weight and molar mass are approximately equal, the empirical and molecular formulas are both $NaC_5H_8O_4N$.

3.53   (a)   *Analyze.* Given: mg CO_2, mg H_2O Find: empirical formula of hydrocarbon, C_xH_y

Plan. Upon combustion, all $C \rightarrow CO_2$, all $H \rightarrow H_2O$.

mg $CO_2 \rightarrow$ g $CO_2 \rightarrow$ mol C; mg $H_2O \rightarrow$ g H_2O, mol H

Find simplest ratio of moles and empirical formula.   *Solve.*

$5.86 \times 10^{-3} \text{ g CO}_2 \times \dfrac{1 \text{ mol CO}_2}{44.01 \text{ g CO}_2} \times \dfrac{1 \text{ mol C}}{1 \text{ mol CO}_2} = 1.33 \times 10^{-4} \text{ mol C}$

$1.37 \times 10^{-3} \text{ g H}_2\text{O} \times \dfrac{1 \text{ mol H}_2\text{O}}{18.02 \text{ g H}_2\text{O}} \times \dfrac{2 \text{ mol H}}{1 \text{ mol H}_2\text{O}} = 1.52 \times 10^{-4} \text{ mol H}$

Dividing both values by 1.33×10^{-4} gives C:H of 1:1.14. This is not "close enough" to be considered 1:1. No obvious multipliers (2, 3, 4) produce an integer ratio. Testing other multipliers (trial and error!), the correct factor seems to be 7. The empirical formula is C_7H_8.

Check. See discussion of C:H ratio above.

(b)   *Analyze.* Given: g of menthol, g CO_2, g H_2O, molar mass. Find: molecular formula.

Plan/Solve. Calculate mol C and mol H in the sample.

$$0.2829 \text{ g } CO_2 \times \frac{1 \text{ mol } CO_2}{44.01 \text{ g } CO_2} \times \frac{1 \text{ mol } C}{1 \text{ mol } CO_2} = 0.0064281 = 0.006428 \text{ mol C}$$

$$0.1159 \text{ g } H_2O \times \frac{1 \text{ mol } H_2O}{18.02 \text{ g } H_2O} \times \frac{2 \text{ mol } H}{1 \text{ mol } H_2O} = 0.012863 = 0.01286 \text{ mol H}$$

Calculate g C, g H and get g O by subtraction.

$$0.0064281 \text{ mol C} \times \frac{12.01 \text{ g } C}{1 \text{ mol } C} = 0.07720 \text{ g C}$$

$$0.012863 \text{ mol H} \times \frac{1.008 \text{ g } H}{1 \text{ mol } H} = 0.01297 \text{ g H}$$

mass O = 0.1005 g sample – (0.07720 g C + 0.01297 g H) = 0.01033 g O

Calculate mol O and find integer ratio of mol C: mol H: mol O.

$$0.01033 \text{ g O} \times \frac{1 \text{ mol } O}{16.00 \text{ g } O} = 6.456 \times 10^{-4} \text{ mol O}$$

Divide moles by 6.456×10^{-4}.

$$C: \frac{0.006428}{6.456 \times 10^{-4}} \approx 10; \quad H: \frac{0.01286}{6.456 \times 10^{-4}} \approx 20; \quad O: \frac{6.456 \times 10^{-4}}{6.456 \times 10^{-4}} = 1$$

The empirical formula is $C_{10}H_{20}O$.

$$FW = 10(12) + 20(1) + 16 = 156; \quad \frac{M}{FW} = \frac{156}{156} = 1$$

The molecular formula is the same as the empirical formula, $C_{10}H_{20}O$.

Check. The mass of O wasn't negative or greater than the sample mass; empirical and molecular formulas are reasonable.

3.55   *Analyze.* Given: mass gain in H_2O absorber = mass of H_2O; mass gain in CO_2 absorber = mass CO_2; molecular model of valproic acid. Find: empirical formula of valproic acid from combustion data and compare to molecular formula from model.

Plan. Calculate mol C and mol H, then g C and g H; get g O by subtraction.

Solve.

$$0.403 \text{ g } CO_2 \times \frac{1 \text{ mol } CO_2}{44.01 \text{ g } CO_2} \times \frac{1 \text{ mol } C}{1 \text{ mol } CO_2} = 9.157 \times 10^{-3} = 9.16 \times 10^{-3} \text{ mol C}$$

$$0.166 \text{ g } H_2O \times \frac{1 \text{ mol } H_2O}{18.02 \text{ g } H_2O} \times \frac{2 \text{ mol } H}{1 \text{ mol } H_2O} = 0.018424 = 0.0184 \text{ mol H}$$

$$9.157 \times 10^{-3} \text{ mol C} \times \frac{12.01 \text{ g } C}{1 \text{ mol } C} = 0.10998 \text{ g C} = 0.110 \text{ g C}$$

$$0.018424 \text{ mol H} \times \frac{1.008 \text{ g } H}{1 \text{ mol } H} = 0.01857 \text{ g H} = 0.0186 \text{ g H}$$

mass of O = 0.165 mg sample – (0.10998 g C + 0.01857 mg H) = 0.03645 = 0.036 g O

$$0.03645\,g\,O \times \frac{1\,mol\,O}{16.00\,g\,O} = 2.278 \times 10^{-3} = 2.3 \times 10^{-3}\,mol\,O.\ \text{Divide moles by } 2.278 \times 10^{-3}.$$

$$C:\ \frac{9.157 \times 10^{-3}}{2.278 \times 10^{-3}} \approx 4;\quad H:\ \frac{0.018424}{2.278 \times 10^{-3}} \approx 8;\quad O:\ \frac{2.278 \times 10^{-3}}{2.278 \times 10^{-3}} = 1$$

The empirical formula is C_4H_8O.

On the molecular model, there are 8 grey C atoms, 16 white H atoms, and 2 red O atoms. The molecular formula is $C_8H_{16}O_2$. Dividing by the common factor 2, the empirical formula is C_4H_8O, which matches the empirical formula derived from combustion data.

3.57   *Analyze.* Given 2.558 g $Na_2CO_3 \cdot xH_2O$, 0.948 g Na_2CO_3. Find: x.

Plan. The reaction involved is $Na_2CO_3 \cdot xH_2O(s) \rightarrow Na_2CO_3(s) + xH_2O(g)$.

Calculate the mass of H_2O lost and then the mole ratio of Na_2CO_3 and H_2O.

Solve. g H_2O lost = 2.558 g sample – 0.948 g Na_2CO_3 = 1.610 g H_2O

$$0.948\,g\,Na_2CO_3 \times \frac{1\,mol\,Na_2CO_3}{106.0\,g\,Na_2CO_3} = 0.00894\,mol\,Na_2CO_3$$

$$1.610\,g\,H_2O \times \frac{1\,mol\,H_2O}{18.02\,g\,H_2O} = 0.08935\,mol\,H_2O$$

The formula is $Na_2CO_3 \cdot \underline{\mathbf{10}}\,H_2O$.

Check. x is an integer.

Calculations Based on Chemical Equations (section 3.6)

3.59   The mole ratios implicit in the coefficients of a balanced chemical equation express the fundamental relationship between amounts of reactants and products. If the equation is not balanced, the mole ratios will be incorrect and lead to erroneous calculated amounts of products.

3.61 $Na_2SiO_3(s) + 8HF(aq) \rightarrow H_2SiF_6(aq) + 2NaF(aq) + 3H_2O(l)$

(a)   *Analyze.* Given: mol Na_2SiO_3. Find: mol HF. *Plan.* Use the mole ratio 8HF:1Na_2SiO_3 from the balanced equation to relate moles of the two reactants.
Solve.

$$0.300\,mol\,Na_2SiO_3 \times \frac{8\,mol\,HF}{1\,mol\,Na_2SiO_3} = 2.40\,mol\,HF$$

Check. Mol HF should be greater than mol Na_2SiO_3.

(b)   *Analyze.* Given: mol HF. Find: g NaF. *Plan.* Use the mole ratio 2NaF:8HF to change mol HF to mol NaF, then molar mass to get NaF.   *Solve.*

$$0.500\,mol\,HF \times \frac{2\,mol\,NaF}{8\,mol\,HF} \times \frac{41.99\,g\,NaF}{1\,mol\,NaF} = 5.25\,g\,NaF$$

Check. (0.5/4) = 0.125; 0.13 × 42 > 4 g NaF

(c)   *Analyze.* Given: g HF Find: g Na_2SiO_3.

Plan. $g\,HF \;\rightarrow\; mol\,HF \left(\dfrac{mol}{ratio}\right) \;\rightarrow\; mol\,Na_2SiO_3 \;\rightarrow\; g\,Na_2SiO_3$

The mole ratio is at the heart of every stoichiometry problem. Molar mass is used to change to and from grams.   *Solve.*

$$0.800\,g\,HF \times \frac{1\,mol\,HF}{20.01\,g\,HF} \times \frac{1\,mol\,Na_2SiO_3}{8\,mol\,HF} \times \frac{122.1\,g\,Na_2SiO_3}{1\,mol\,Na_2SiO_3} = 0.610\,g\,Na_2SiO_3$$

Check. 0.8 (120/160) < 0.75 mol

3.63   (a) $Al(OH)_3(s) + 3HCl(aq) \rightarrow AlCl_3(aq) + 3H_2O(l)$

(b)   *Analyze.* Given mass of one reactant, find stoichiometric mass of other reactant and products.

Plan. Follow the logic in Sample Exercise 3.16. Calculate mol $Al(OH)_3$ in 0.500 g $Al(OH_3)_3$ separately, since it will be used several times.

Solve. $0.500\,g\,Al(OH)_3 \times \dfrac{1\,mol\,Al(OH)_3}{78.00\,g\,Al(OH)_3} = 6.410 \times 10^{-3} = 6.41 \times 10^{-3}\,mol\,Al(OH)_3$

$6.410 \times 10^{-3}\,mol\,Al(OH)_3 \times \dfrac{3\,mol\,HCl}{1\,mol\,Al(OH)_3} \times \dfrac{36.46\,g\,HCl}{1\,mol\,HCl} = 0.7012 = 0.701\,g\,HCl$

(c) $6.410 \times 10^{-3}\,mol\,Al(OH)_3 \times \dfrac{1\,mol\,HCl}{1\,mol\,Al(OH)_3} \times \dfrac{133.34\,g\,AlCl_3}{1\,mol\,AlCl_3} = 0.8547$

$$= 0.855\,g\,AlCl_3$$

$6.410 \times 10^{-3}\,mol\,Al(OH)_3 \times \dfrac{3\,mol\,H_2O}{1\,mol\,Al(OH)_3} \times \dfrac{18.02\,g\,H_2O}{1\,mol\,H_2O} = 0.3465 = 0.347\,g\,H_2O$

(d)   Conservation of mass: mass of products = mass of reactants

reactants: $Al(OH)_3$ + HCl, 0.500 g + 0.701 g = 1.201 g

products: $AlCl_3 + H_2O$, 0.855 g + 0.347 g = 1.202 g

The 0.001 g difference is due to rounding (0.8547 + 0.3465 = 1.2012). This is an excellent *check* of results.

3.65   (a) $Al_2S_3(s) + 6H_2O(l) \rightarrow 2Al(OH)_3(s) + 3H_2S(g)$

(b)   *Plan.* g A → mol A → mol B → g B. See Solution 3.61 (c).   *Solve.*

$$14.2\,g\,Al_2S_3 \times \frac{1\,mol\,Al_2S_3}{150.2\,g\,Al_2S_3} \times \frac{2\,mol\,Al(OH)_3}{1\,mol\,Al_2S_3} \times \frac{78.00\,g\,Al(OH)_3}{1\,mol\,Al(OH)_3} = 14.7\,g\,Al(OH)_3$$

Check. $14\left(\dfrac{2 \times 78}{150}\right) \approx 14(1) \approx 14\,g\,Al(OH)_3$

3.67   (a)   *Analyze.* Given: mol NaN_3. Find: mol N_2.

Plan. Use mole ratio from balanced equation.   *Solve.*

$$1.50 \text{ mol NaN}_3 \times \frac{3 \text{ mol N}_2}{2 \text{ mol NaN}_3} = 2.25 \text{ mol N}_2$$

Check. The resulting mol N_2 should be greater than mol NaN_3, (the N_2:NaN_3 ratio is > 1), and it is.

     (b)   *Analyze.* Given: $g \ N_2$ Find: $g \ NaN_3$.

Plan. Use molar masses to get from and to grams, mol ratio to relate moles of the two substances.   *Solve.*

$$10.0 \text{ g N}_2 \times \frac{1 \text{ mol N}_2}{28.01 \text{ g N}_2} \times \frac{2 \text{ mol NaN}_3}{3 \text{ mol N}_2} \times \frac{65.01 \text{ g NaN}_3}{1 \text{ mol NaN}_3} = 15.5 \text{ g NaN}_3$$

Check. Mass relations are less intuitive than mole relations. Estimating the ratio of molar masses is sometimes useful. In this case, $65 \text{ g NaN}_3 / 28 \text{ g N}_2 \approx 2.25$ Then, $(10 \times 2/3 \times 2.25) \approx 15 \text{ g NaN}_3$. The calculated result looks reasonable.

     (c)   *Analyze.* Given: vol N_2 in ft^3, density N_2 in g/L. Find: $g \ NaN_3$.

Plan. First determine how many $g \ N_2$ are in $10.0 \ ft^3$, using the density of N_2. Then proceed as in part (b).

Solve.

$$\frac{1.25 \text{ g}}{1 \text{ L}} \times \frac{1 \text{ L}}{1000 \text{ cm}^3} \times \frac{(2.54)^3 \text{ cm}^3}{1 \text{ in}^3} \times \frac{(12)^3 \text{ in}^3}{1 \text{ ft}^3} \times 10.0 \text{ ft}^3 = 354.0 = 354 \text{ g N}_2$$

$$354.0 \text{ g N}_2 \times \frac{1 \text{ mol N}_2}{28.01 \text{ g N}_2} \times \frac{2 \text{ mol NaN}_3}{3 \text{ mol N}_2} \times \frac{65.01 \text{ g NaN}_3}{1 \text{ mol NaN}_3} = 548 \text{ g NaN}_3$$

Check. $1 \ ft^3 \sim 28$ L; $10 \ ft^3 \sim 280$ L; 280 L $\times 1.25 \sim 350 \text{ g N}_2$

Using the ratio of molar masses from part (b), $(350 \times 2/3 \times 2.25) \approx 525 \text{ g NaN}_3$

3.69   (a)   *Analyze.* Given: dimensions of Al foil. Find: mol Al.

Plan. Dimensions $\rightarrow$ vol $\xrightarrow{\text{density}}$ mass $\xrightarrow{\frac{\text{molar}}{\text{mass}}}$ mol Al

Solve. $1.00 \text{ cm} \times 1.00 \text{ cm} \times 0.550 \text{ mm} \times \frac{1 \text{ cm}}{10 \text{ mm}} = 0.0550 \text{ cm}^3 \text{ Al}$

$$0.0550 \text{ cm}^3 \text{ Al} \times \frac{2.699 \text{ g Al}}{1 \text{ cm}^3} \times \frac{1 \text{ mol Al}}{26.98 \text{ g Al}} = 5.502 \times 10^{-3} = 5.50 \times 10^{-3} \text{ mol Al}$$

Check. $2.699/26.98 \approx 0.1$; $(0.055 \text{ cm}^3 \times 0.1) = 5.5 \times 10^{-3} \text{ mol Al}$

     (b)   *Plan.* Write the balanced equation to get a mole ratio; change mol Al $\rightarrow$ mol $AlBr_3 \rightarrow g \ AlBr_3$.

Solve. $2Al(s) + 3Br_2(l) \rightarrow 2AlBr_3(s)$

$$5.502 \times 10^{-3} \text{ mol Al} \times \frac{2 \text{ mol AlBr}_3}{2 \text{ mol Al}} \times \frac{266.69 \text{ g AlBr}_3}{1 \text{ mol AlBr}_3} = 1.467 = 1.47 \text{ g AlBr}_3$$

Check. $(0.006 \times 1 \times 270) \approx 1.6 \text{ g AlBr}_3$

Limiting Reactants; Theoretical Yields (section 3.7)

3.71   (a)   The *limiting reactant* determines the maximum number of product moles resulting from a chemical reaction; any other reactant is an *excess reactant*.

(b)   The limiting reactant regulates the amount of products, because it is completely used up during the reaction; no more product can be made when one of the reactants is unavailable.

(c)   Combining ratios are molecule and mole ratios. Since different molecules have different masses, equal masses of different reactants will not have equal numbers of molecules. By comparing initial moles, we compare numbers of available reactant molecules, the fundamental combining units in a chemical reaction.

3.73   (a)   Each bicycle needs 2 wheels, 1 frame, and 1 set of handlebars. A total of 4815 wheels corresponds to 2407.5 pairs of wheels. This is more than the number of frames or handlebars. The 2255 handlebars determine that 2255 bicycles can be produced.

(b)   2305 frames – 2255 bicycles = 50 frames left over

2407.5 pairs of wheels – 2255 bicycles = 152.5 pairs of wheels left over 2(152.5) = 305 wheels left over

(c)   The handlebars are the "limiting reactant" in that they determine the number of bicycles that can be produced.

3.75   *Analyze.* Given: 1.85 mol NaOH, 1.00 mol CO_2. Find: mol Na_2CO_3.

Plan. Amounts of more than one reactant are given, so we must determine which reactant regulates (limits) product. Then apply the appropriate mole ratio from the balanced equation.

Solve. The mole ratio is $2NaOH:1CO_2$, so 1.00 mol CO_2 requires 2.00 mol NaOH for complete reaction. Less than 2.00 mol NaOH are present, so NaOH is the limiting reactant.

$$1.85 \text{ mol NaOH} \times \frac{1 \text{ mol Na}_2\text{CO}_3}{2 \text{ mol NaOH}} = 0.925 \text{ mol Na}_2\text{CO}_3 \text{ can be produced}$$

The $Na_2CO_3:CO_2$ ratio is 1:1, so 0.925 mol Na_2CO_3 produced requires 0.925 mol CO_2 consumed. (Alternately, 1.85 mol NaOH × 1 mol CO_2/2 mol NaOH = 0.925 mol CO_2 reacted). 1.00 mol CO_2 initial – 0.925 mol CO_2 reacted = 0.075 mol CO_2 remain.

Check.	$2NaOH(s)$	+	$CO_2(g)$	$\rightarrow$	$Na_2CO_3(s)$	+	$H_2O(l)$
initial	1.85 mol		1.00 mol		0 mol		
change (reaction)	–1.85 mol		–0.925 mol		+0.925 mol		
final	0 mol		0.075 mol		0.925 mol		

Note that the "change" line (but not necessarily the "final" line) reflects the mole ratios from the balanced equation.

3.77 $3NaHCO_3(aq) + H_3C_6H_5O_7(aq) \rightarrow 3CO_2(g) + 3H_2O(l) + Na_3C_6H_5O_7(aq)$

(a)   *Analyze/Plan.* Abbreviate citric acid as H_3Cit. Follow the approach in Sample Exercise 3.19. *Solve.*

$$1.00 \text{ g NaHCO}_3 \times \frac{1 \text{ mol NaHCO}_3}{84.01 \text{ g NaHCO}_3} = 1.190 \times 10^{-2} = 1.19 \times 10^{-2} \text{ mol NaHCO}_3$$

$$1.00 \text{ g H}_3C_6H_5O_7 \times \frac{1 \text{ mol H}_3Cit}{192.1 \text{ g H}_3Cit} = 5.206 \times 10^{-3} = 5.21 \times 10^{-3} \text{ mol H}_3Cit$$

But $NaHCO_3$ and H_3Cit react in a 3:1 ratio, so 5.21×10^{-3} mol H_3Cit require $3(5.21 \times 10^{-3}) = 1.56 \times 10^{-2}$ mol $NaHCO_3$. We have only 1.19×10^{-2} mol $NaHCO_3$, so $NaHCO_3$ is the limiting reactant.

(b) 1.190×10^{-2} mol $NaHCO_3 \times \dfrac{3 \text{ mol CO}_2}{3 \text{ mol NaHCO}_3} \times \dfrac{44.01 \text{ g CO}_2}{1 \text{ mol CO}_2} = 0.524 \text{ g CO}_2$

(c) 1.190×10^{-2} mol $NaHCO_3 \times \dfrac{1 \text{ mol H}_3Cit}{3 \text{ mol NaHCO}_3} = 3.968 \times 10^{-3}$

$$= 3.97 \times 10^{-3} \text{ mol H}_3Cit \text{ react}$$

5.206×10^{-3} mol $H_3Cit - 3.968 \times 10^{-3}$ mol react $= 1.238 \times 10^{-3}$

$$= 1.24 \times 10^{-3} \text{ mol H}_3Cit \text{ remain}$$

$$1.238 \times 10^{-3} \text{ mol H}_3Cit \times \frac{192.1 \text{ g H}_3Cit}{\text{mol H}_3Cit} = 0.238 \text{ g H}_3Cit \text{ remain}$$

3.79   *Analyze.* Given: initial g Na_2CO_3, g $AgNO_3$. Find: final g Na_2CO_3, $AgNO_3$, Ag_2CO_3, $NaNO_3$

Plan. Write balanced equation; determine limiting reactant; calculate amounts of excess reactant remaining and products, based on limiting reactant.

Solve. $2AgNO_3(aq) + Na_2CO_3(aq) \rightarrow Ag_2CO_3(s) + 2NaNO_3(aq)$

$$3.50 \text{ g Na}_2CO_3 \times \frac{1 \text{ mol Na}_2CO_3}{106.0 \text{ g Na}_2CO_3} = 0.03302 = 0.0330 \text{ mol Na}_2CO_3$$

$$5.00 \text{ g AgNO}_3 \times \frac{1 \text{ mol AgNO}_3}{169.9 \text{ g AgNO}_3} = 0.02943 = 0.0294 \text{ mol AgNO}_3$$

$$0.02943 \text{ mol AgNO}_3 \times \frac{1 \text{ mol Na}_2CO_3}{2 \text{ mol AgNO}_3} = 0.01471 = 0.0147 \text{ mol Na}_2CO_3 \text{ required}$$

$AgNO_3$ is the limiting reactant and Na_2CO_3 is present in excess.

	$2AgNO_3(aq)$	+	$Na_2CO_3(aq)$	$\rightarrow$	$Ag_2CO_3(s)$	+	$2NaNO_3(aq)$
initial	0.0294 mol		0.0330 mol		0 mol		0 mol
reaction	−0.0294 mol		−0.0147 mol		+0.0147 mol		+0.0294 mol
final	0 mol		0.0183 mol		0.0147 mol		0.0294 mol

0.01830 mol $Na_2CO_3 \times 106.0$ g/mol $= 1.940 = 1.94$ g Na_2CO_3

0.01471 mol $Ag_2CO_3 \times 275.8$ g/mol $= 4.057 = 4.06$ g Ag_2CO_3

0.02943 mol $NaNO_3$ × 85.00 g/mol = 2.502 = 2.50 g $NaNO_3$

Check. The initial mass of reactants was 8.50 g, and the final mass of excess reactant and products is 8.50 g; mass is conserved.

3.81   *Analyze.* Given: amounts of two reactants. Find: theoretical yield.

Plan. Determine the limiting reactant and the maximum amount of product it could produce. Then calculate % yield.   *Solve.*

(a) $30.0 \text{ g } C_6H_6 \times \dfrac{1 \text{ mol } C_6H_6}{78.11 \text{ g } C_6H_6} = 0.3841 = 0.384 \text{ mol } C_6H_6$

$65.0 \text{ g } Br_2 \times \dfrac{1 \text{ mol } Br_2}{159.8 \text{ g } Br_2} = 0.4068 = 0.407 \text{ mol } Br_2$

Since C_6H_6 and Br_2 react in a 1:1 mole ratio, C_6H_6 is the limiting reactant and determines the theoretical yield.

$0.3841 \text{ mol } C_6H_6 \times \dfrac{1 \text{ mol } C_6H_5Br}{1 \text{ mol } C_6H_6} \times \dfrac{157.0 \text{ g } C_6H_5Br}{1 \text{ mol } C_6H_5Br} = 60.30 = 60.3 \text{ g } C_6H_5Br$

Check. 30/78 ~ 3/8 mol C_6H_6. 65/160 ~ 3/8 mol Br_2. Since moles of the two reactants are similar, a precise calculation is needed to determine the limiting reactant. 3/8 × 160 ≈ 60 g product

(b) $\% \text{ yield} = \dfrac{42.3 \text{ g } C_6H_5Br \text{ actual}}{60.3 \text{ g } C_6H_5Br \text{ theoretical}} \times 100 = 70.149 = 70.10\%$

3.83   *Analyze.* Given: g of two reactants, % yield. Find: g S_8.

Plan. Determine limiting reactant and theoretical yield. Use definition of % yield to calculate actual yield.   *Solve.*

$30.0 \text{ g } H_2S \times \dfrac{1 \text{ mol } H_2S}{34.08 \text{ g } H_2S} = 0.8803 = 0.880 \text{ mol } H_2S$

$50.0 \text{ g } O_2 \times \dfrac{1 \text{ mol } O_2}{32.00 \text{ g } H_2S} = 1.5625 = 1.56 \text{ mol } O_2$

$0.8803 \text{ mol } H_2S \times \dfrac{4 \text{ mol } O_2}{8 \text{ mol } H_2S} = 0.4401 = 0.440 \text{ mol } O_2 \text{ required}$

Since there is more than enough O_2 to react exactly with 0.880 mol H_2S, O_2 is present in excess and H_2S is the limiting reactant.

$0.8803 \text{ mol } H_2S \times \dfrac{1 \text{ mol } S_8}{8 \text{ mol } H_2S} \times \dfrac{256.56 \text{ g } S_8}{1 \text{ mol } S_8} = 28.231 = 28.2 \text{ g } S_8 \text{ theoretical yield}$

Check. 30/34 ≈ 1 mol H_2S; 50/32 ≈ 1.5 mol O_2. Twice as many mol H_2S as mol O_2 are required, so H_2S limits. 1 × (260/8) ≈ 30 g S_8 theoretical.

$\% \text{ yield} = \dfrac{\text{actual}}{\text{theoretical}} \times 100; \quad \dfrac{\% \text{ yield} \times \text{theoretical}}{100} = \text{actual yield}$

$\dfrac{98\%}{100} \times 28.231 \text{ g } S_8 = 27.666 = 28 \text{ g } S_8 \text{ actual}$

Additional Exercises

3.85   (a) $CH_3COOH = C_2H_4O_2$. At room temperature and pressure, pure acetic acid is a liquid. $C_2H_4O_2(l) + 2\,O_2(g) \rightarrow 2\,CO_2(g) + 2\,H_2O(l)$

       (b) $Ca(OH)_2(s) \rightarrow CaO(s) + H_2O(g)$

       (c) $Ni(s) + Cl_2(g) \rightarrow NiCl_2(s)$

3.89   (a)   *Analyze.* Given: diameter of Si sphere (dot), density of Si. Find: mass of dot.

          Plan. Calculate volume of sphere in cm^3, use density to calculate mass of the sphere (dot).

          Solve. $V = 4/3\pi\, r^3$; $r = d/2$

$$\text{radius of dot} = \frac{4\,\text{nm}}{2} \times \frac{1 \times 10^{-9}\,\text{m}}{1\,\text{nm}} \times \frac{1\,\text{cm}}{1 \times 10^{-2}\,\text{m}} = 2 \times 10^{-7}\,\text{cm}$$

$$\text{volume of dot} = (4/3) \times \pi \times (2 \times 10^{-7})^3 = 3.35 \times 10^{-20} = 3 \times 10^{-20}\,\text{cm}^3$$

$$3.35 \times 10^{-20}\,\text{cm}^3 \times \frac{2.3\,\text{g Si}}{\text{cm}^3} = 7.707 \times 10^{-20} = 8 \times 10^{-20}\,\text{g Si in dot}$$

       (b)   *Plan.* Change g Si to mol Si using molar mass, then mol Si to atoms Si using Avogadro's number. *Solve.*

$$7.707 \times 10^{-20}\,\text{g Si} \times \frac{1\,\text{mol Si}}{28.0855\,\text{g Si}} \times \frac{6.022 \times 10^{23}\,\text{Si atoms}}{\text{mol Si}} = 1.653 \times 10^3$$

$$= 2 \times 10^3\,\text{Si atoms}$$

       (c)   *Plan.* A 4 nm quantum dot of Ge also has a volume of $3 \times 10^{-20}\,cm^3$. Use density of Ge and Avogadro's number to calculate the number of Ge atoms in a 4 nm spherical quantum dot.

$$3.35 \times 10^{-20}\,\text{cm}^3 \times \frac{5.325\,\text{g Ge}}{\text{cm}^3} \times \frac{1\,\text{mol Ge}}{72.64\,\text{g Ge}} \times \frac{6.022 \times 10^{23}\,\text{Ge atoms}}{\text{mol Ge}}$$

$$= 1.479 \times 10^3 = 1 \times 10^3\,\text{Ge atoms}$$

          Strictly speaking, the result has 1 sig fig (from 4 nm). A more meaningful comparison might be 1700 Si atoms vs. 1500 Ge atoms. Although Ge has greater molar mass, it is also more than twice as dense as Si, so the numbers of atoms in the Si and Ge dots are similar.

3.93   Since all the C in the vanillin must be present in the CO_2 produced, get g C from g CO_2.

$$2.43\,\text{g CO}_2 \times \frac{1\,\text{mol CO}_2}{44.01\,\text{g CO}_2} \times \frac{12.01\,\text{g C}}{1\,\text{mol C}} = 0.6631 = 0.663\,\text{g C}$$

Since all the H in vanillin must be present in the H_2O produced, get g H from g H_2O.

$$0.50\,\text{g H}_2\text{O} \times \frac{1\,\text{mol H}_2\text{O}}{18.02\,\text{g H}_2\text{O}} \times \frac{2\,\text{mol H}}{1\,\text{mol H}_2\text{O}} \times \frac{1.008\,\text{g H}}{1\,\text{mol H}} = 0.0559 = 0.056\,\text{g H}$$

Get g O by subtraction. (Since the analysis was performed by combustion, an unspecified amount of O_2 was a reactant, and thus not all the O in the CO_2 and H_2O produced came from vanillin.) 1.05 g vanillin – 0.663 g C – 0.056 g H = 0.331 g O

$$0.6631 \text{ g C} \times \frac{1 \text{ mol C}}{12.01 \text{ g C}} = 0.0552 \text{ mol C}; \ 0.0552 / 0.0207 = 2.67$$

$$0.0559 \text{ g H} \times \frac{1 \text{ mol H}}{1.008 \text{ g H}} = 0.0555 \text{ mol C}; \ 0.0555 / 0.0207 = 2.68$$

$$0.331 \text{ g O} \times \frac{1 \text{ mol O}}{16.00 \text{ g O}} = 0.0207 \text{ mol O}; \ 0.0207 / 0.0207 = 1.00$$

Multiplying the numbers above by **3** to obtain an integer ratio of moles, the empirical formula of vanillin is $C_8H_8O_3$.

3.97　　$O_3(g) + 2NaI(aq) + H_2O(l) \rightarrow O_2(g) + I_2(s) + 2NaOH(aq)$

(a)　　$5.95 \times 10^{-6} \text{ mol O}_3 \times \dfrac{2 \text{ mol NaI}}{1 \text{ mol O}_3} = 1.19 \times 10^{-5} \text{ mol NaI}$

(b)　　$1.3 \text{ mg O}_3 \times \dfrac{1 \times 10^{-3} \text{ g}}{1 \text{ mg}} \times \dfrac{1 \text{ mol O}_3}{48.00 \text{ g O}_3} \times \dfrac{2 \text{ mol NaI}}{1 \text{ mol O}_3} \times \dfrac{149.9 \text{ g NaI}}{1 \text{ mol NaI}}$

$$= 8.120 \times 10^{-3} = 8.1 \times 10^{-3} \text{ g NaI} = 8.1 \text{ mg NaI}$$

3.101　　$N_2(g) + 3H_2(g) \rightarrow 2NH_3(g)$

Determine the moles of N_2 and H_2 required to form the 3.0 moles of NH_3 present after the reaction has stopped.

$$3.0 \text{ mol NH}_3 \times \frac{3 \text{ mol H}_2}{2 \text{ mol NH}_3} = 4.5 \text{ mol H}_2 \text{ reacted}$$

$$3.0 \text{ mol NH}_3 \times \frac{1 \text{ mol N}_2}{2 \text{ mol NH}_3} = 1.5 \text{ mol N}_2 \text{ reacted}$$

mol H_2 initial = 3.0 mol H_2 remain + 4.5 mol H_2 reacted = 7.5 mol H_2

mol N_2 initial = 3.0 mol N_2 remain + 1.5 mol N_2 reacted = 4.5 mol N_2

In tabular form:	$N_2(g)$	+	$3H_2(g)$	$\rightarrow$	$2NH_3(g)$
initial	4.5 mol		7.5 mol		0 mol
reaction	–1.5 mol		–4.5 mol		+3.0 mol
final	3.0 mol		3.0 mol		3.0 mol

(Tables like this will be extremely useful for solving chemical equilibrium problems in Chapter 15.)

Integrative Exercises

3.105　　*Plan.*　Volume cube $\xrightarrow{\text{density}}$ mass $CaCO_3 \rightarrow$ moles $CaCO_3 \rightarrow$ moles O $\rightarrow$ O atoms

Solve. $(2.005)^3 \text{ in}^3 \times \dfrac{(2.54)^3 \text{ cm}^3}{1 \text{ in}^3} \times \dfrac{2.71 \text{ g CaCO}_3}{1 \text{ cm}^3} \times \dfrac{1 \text{ mol CaCO}_3}{100.1 \text{ g CaCO}_3} \times \dfrac{3 \text{ mol O}}{1 \text{ mol CaCO}_3}$

$$\times \frac{6.022 \times 10^{23} \text{ O atoms}}{1 \text{ mol O}} = 6.46 \times 10^{24} \text{ O atoms}$$

3.107 (a) *Analyze.* Given: gasoline = C_8H_{18}, density = 0.69 g/mL, 20.5 mi/gal, 225 mi.

Find: kg CO_2.

Plan. Write and balance the equation for the combustion of octane. Change mi → gal octane → mL → g octane. Use stoichiometry to calculate g and kg CO_2 from g octane.

Solve. $2C_8H_{18}(l) + 25O_2(g) \rightarrow 16CO_2(g) + 18H_2O(l)$

$$225\,\text{mi} \times \frac{1\,\text{gal}}{20.5\,\text{mi}} \times \frac{3.7854\,\text{L}}{1\,\text{gal}} \times \frac{1\,\text{mL}}{1\times10^{-3}\,\text{L}} \times \frac{0.69\,\text{g octane}}{1\,\text{mL}} = 2.8667\times10^4\,\text{g}$$

$$= 29\,\text{kg octane}$$

$$2.8667\times10^4\,\text{g}\,C_8H_{18} \times \frac{1\,\text{mol}\,C_8H_{18}}{114.2\,\text{g}\,C_8H_{18}} \times \frac{16\,\text{mol}\,CO_2}{2\,\text{mol}\,C_8H_{18}} \times \frac{44.01\,\text{g}\,CO_2}{1\,\text{mol}\,CO_2} = 8.8382\times10^4\,\text{g}$$

$$= 88\,\text{kg}\,CO_2$$

Check. $\left(\dfrac{225\times4\times0.7}{20}\right) \times 10^3 = (45\times0.7)\times10^3 = 30\times10^3\,\text{g} = 30\,\text{kg octane}$

$\dfrac{44}{114} \approx \dfrac{1}{3}$; $\dfrac{30\,\text{kg}\times8}{3} \approx 80\,\text{kg}\,CO_2$

(b) *Plan.* Use the same strategy as part (a). *Solve.*

$$225\,\text{mi} \times \frac{1\,\text{gal}}{5\,\text{mi}} \times \frac{3.7854\,\text{L}}{1\,\text{gal}} \times \frac{1\,\text{mL}}{1\times10^{-3}\,\text{L}} \times \frac{0.69\,\text{g octane}}{1\,\text{mL}} = 1.1754\times10^5$$

$$= 1\times10^2\,\text{kg octane}$$

$$1.1754\times10^5\,\text{g}\,C_8H_{18} \times \frac{1\,\text{mol}\,C_8H_{18}}{114.2\,\text{g}\,C_8H_{18}} \times \frac{16\,\text{mol}\,CO_2}{2\,\text{mol}\,C_8H_{18}} \times \frac{44.01\,\text{g}\,CO_2}{1\,\text{mol}\,CO_2} =$$

$$3.624\times10^5\,\text{g} = 4\times10^2\,\text{kg}\,CO_2$$

Check. Mileage of 5 mi/gal requires ~4 times as much gasoline as mileage of 20.5 mi/gal, so it should produce ~4 times as much CO_2. 90 kg CO_2 [from (a)] × 4 = 360 = 4×10^2 kg CO_2 [from (b)].

3.109 (a) $S(s) + O_2(g) \rightarrow SO_2(g)$; $SO_2(g) + CaO(s) \rightarrow CaSO_3(s)$

(b) If the coal contains 2.5% S, then 1 g coal contains 0.025 g S.

$$\frac{2000\,\text{tons coal}}{\text{day}} \times \frac{2000\,\text{lb}}{1\,\text{ton}} \times \frac{1\,\text{kg}}{2.20\,\text{lb}} \times \frac{1000\,\text{g}}{1\,\text{kg}} \times \frac{0.025\,\text{g S}}{1\,\text{g coal}} \times \frac{1\,\text{mol S}}{32.07\,\text{g S}}$$

$$\times \frac{1\,\text{mol}\,SO_2}{1\,\text{mol S}} \times \frac{1\,\text{mol CaO}}{1\,\text{mol}\,SO_2} \times \frac{56.08\,\text{g CaO}}{1\,\text{mol CaO}} \times \frac{1\,\text{kg CaO}}{1000\,\text{g CaO}} =$$

$79,485 = 7.9\times10^4\,\text{kg CaO or }7.9\times10^7\,\text{g CaO}$

(c) 1 mol CaO = 1 mol $CaSO_3$

$$7.9485\times10^7\,\text{g CaO} \times \frac{1\,\text{mol CaO}}{56.08\,\text{g CaO}} \times \frac{1\,\text{mol}\,CaSO_3}{1\,\text{mol CaO}} \times \frac{120.14\,\text{g}\,CaSO_3}{1\,\text{mol}\,CaSO_3}$$

$$= 1.703\times10^8 = 1.7\times10^8\,\text{g}\,CaSO_3$$

This corresponds to about 190 tons of $CaSO_3$ per day as a waste product.

4 Reactions in Aqueous Solution

Visualizing Concepts

4.1   *Analyze*. Correlate the formula of the solute with the charged spheres in the diagrams.

Plan. Determine the electrolyte properties of the solute and the relative number of cations, anions, or neutral molecules produced when the solute dissolves.

Solve. Li_2SO_4 is a strong electrolyte, a soluble ionic solid that dissociates into separate Li^+ and SO_4^{2-} when it dissolves in water. There are twice as many Li^+ cations as SO_4^{2-} anions. Diagram (c) represents the aqueous solution of a 2:1 electrolyte.

4.3   *Analyze/Plan*. From the molecular representations, write molecular formulas for the compounds. Using Table 4.2 and molecular formulas (there are no ionic compounds in this exercise), classify the compounds as strong acid, strong base, weak acid, weak base (NH_3) or nonelectrolyte. Strong acids and bases are strong electrolytes, weak acids and bases are weak electrolytes. *Solve*.

(a)   HCOOH.   The molecule has a –COOH group; it is a weak acid and weak electrolyte (it is not one of the strong acids listed in Table 4.2).

(b) HNO_3. The molecule is a strong acid (Table 4.2) and a strong electrolyte.

(c) CH_3CH_2OH. The molecule is neither an acid nor a base; it is a nonelectrolyte.

4.5   *Analyze/Plan*. From the names and/or formulas of the three possible solids, determine which exhibits the described solubility properties. Use Table 4.1.

Solve. The three possible compounds are $BaCl_2$, $PbCl_2$, and $ZnCl_2$. $PbCl_2$ does not dissolve in water to give a clear solution, so it can be eliminated. Of the remaining possibilities, Ba^{2+} has a sulfate precipitate, but Zn^{2+} does not. The compound is indeed $BaCl_2$.

4.7   *Analyze*. Given the formulas of some ions, determine whether these ions ever form precipitates in aqueous solution. *Plan*. Use Table 4.1 to determine if the given ions can form precipitates. If not, they will always be spectator ions.   *Solve*.

(a) Cl^- can form precipitates with Ag^+, Hg_2^{2+}, Pb^{2+}.

(b) NO_3^- never forms precipitates, so it is always a spectator.

(c) NH_4^+ never forms precipitates, so it is always a spectator.

(d) S^{2-} usually forms precipitates.

(e) SO_4^{2-} can form precipitates with Sr^{2+}, Ba^{2+}, Hg_2^{2+}, Pb^{2+}.

Check. NH_4^+ is a soluble exception for sulfides, phosphates, and carbonates, which usually form precipitates, so all rules indicate that it is a perpetual spectator.

4.9　　In a redox reaction, one reactant loses electrons and a different reactant gains electrons; electrons are transferred. Acids ionize in aqueous solution to produce (donate) hydrogen ions (H^+, protons). Bases are substances that react with or accept protons (H^+). In an acid-base reaction, protons are transferred from an acid to a base. We characterize redox reactions by tracking electron transfer using oxidation numbers. We characterize acid-base reactions by tracking H^+ (proton) transfer via molecular formulas of reactants and products.

General Properties of Aqueous Solutions (section 4.1)

4.11　　No. Electrolyte solutions conduct electricity because the dissolved ions carry charge through the solution (from one electrode to the other).

4.13　　Although H_2O molecules are electrically neutral, there is an unequal distribution of electrons throughout the molecule. There are more electrons near O and fewer near H, giving the O end of the molecule a partial negative charge and the H end of the molecule a partial positive charge. Ionic compounds are composed of positively and negatively charged ions. The partially positive ends of H_2O molecules are attracted to the negative ions (anions) in the solid, while the partially negative ends are attracted to the positive ions (cations). Thus, both cations and anions in an ionic solid are surrounded and separated (dissolved) by H_2O molecules.

We do not expect ionic compounds to be soluble in molecular liquids such as $Br_2(l)$ or $Hg(l)$. There is a symmetrical charge distribution in Hg atoms and Br_2 molecules. There is no permanent full or partial charge on these solvent atoms or molecules, no attractive forces to stabilize the separated ions of an ionic solid.

4.15　　*Analyze/Plan.* Given the solute formula, determine the separate ions formed upon dissociation. *Solve.*

(a)　　$ZnCl_2(aq) \rightarrow Zn^{2+}(aq) + 2Cl^-(aq)$

(b)　　$HNO_3(aq) \rightarrow H^+(aq) + NO_3^-(aq)$

(c)　　$(NH_4)_2SO_4(aq) \rightarrow 2NH_4^+(aq) + SO_4^{2-}(aq)$

(d)　　$Ca(OH)_2(aq) \rightarrow Ca^{2+}(aq) + 2OH^-(aq)$

4.17　　*Analyze/Plan.* Apply the definition of a weak electrolyte to HCOOH.

Solve. When HCOOH dissolves in water, neutral HCOOH molecules, H^+ ions and $HCOO^-$ ions are all present in the solution. $HCOOH(aq) \rightleftharpoons H^+(aq) + HCOO^-(aq)$

Precipitation Reactions (section 4.2)

4.19　　*Analyze.* Given: formula of compound. Find: solubility.

Plan. Follow the guidelines in Table 4.1, in light of the anion present in the compound and notable exceptions to the "rules." *Solve.*

(a)　　Mg**Br**$_2$: soluble

(b)　　Pb**I**$_2$: insoluble, Pb^{2+} is an exception to soluble iodides

(c)　　$(NH_4)_2$**CO**$_3$: soluble, NH_4^+ is an exception to insoluble carbonates

(d) $Sr(OH)_2$: soluble, Sr^{2+} is an exception to insoluble hydroxides

(e) $ZnSO_4$: soluble

4.21   *Analyze.* Given: formulas of reactants. Find: balanced equation including precipitates.

Plan. Follow the logic in Sample Exercise 4.3.

Solve. In each reaction, the precipitate is in bold type.

(a) $Na_2CO_3(aq) + 2AgNO_3(aq) \rightarrow \mathbf{Ag_2CO_3(s)} + 2NaNO_3(aq)$

(b)   No precipitate (all nitrates and most sulfates are soluble).

(c) $FeSO_4(aq) + Pb(NO_3)_2(aq) \rightarrow \mathbf{PbSO_4(s)} + Fe(NO_3)_2(aq)$

4.23   *Analyze/Plan.* Follow the logic in Sample Exercise 4.4. From the complete ionic equation, identify the ions that don't change during the reaction; these are the spectator ions. *Solve.*

(a) $2Na^+(aq) + CO_3{}^{2-}(aq) + Mg^{2+}(aq) + SO_4{}^{2-}(aq) \rightarrow MgCO_3(s) + 2Na^+(aq) + SO_4{}^{2-}(aq)$
       Spectators: Na^+, $SO_4{}^{2-}$

(b) $Pb^{2+}(aq) + 2NO_3{}^-(aq) + 2Na^+(aq) + S^{2-}(aq) \rightarrow PbS(s) + 2Na^+(aq) + 2NO_3{}^-(aq)$
       Spectators: Na^+, $NO_3{}^-$

(c) $6NH_4{}^+(aq) + 2PO_4{}^{3-}(aq) + 3Ca^{2+}(aq) + 6Cl^-(aq) \rightarrow Ca_3(PO_4)_2(s) + 6NH_4{}^+(aq) + 6Cl^-(aq)$ Spectators: $NH_4{}^+$, Cl^-

4.25   *Analyze.* Given: reactions of unknown with HBr, H_2SO_4, $NaOH$. Find: The unknown contains a single salt. Is K^+ or Pb^{2+} or Ba^{2+} present?

Plan. Analyze solubility guidelines for Br^-, $SO_4{}^{2-}$ and OH^- and select the cation that produces a precipitate with each of the anions.

Solve. K^+ forms no precipitates with any of the anions. $BaSO_4$ is insoluble, but $BaCl_2$ and $Ba(OH)_2$ are soluble. Since the unknown forms precipitates with all three anions, it must contain Pb^{2+}.

Check. $PbBr_2$, $PbSO_4$, and $Pb(OH)_2$ are all insoluble according to Table 4.1, so our process of elimination is confirmed by the insolubility of the Pb^{2+} compounds.

4.27   *Analyze.* Given: three possible salts in an unknown solution react with $Ba(NO_3)_2$ and then $NaCl$. Find: Can the results identify the unknown salt? Do the three possible unknowns give distinctly different results with $Ba(NO_3)_2$ and $NaCl$?

Plan. Using Table 4.1, determine whether each of the possible unknowns will form a precipitate with $Ba(NO_3)_2$ and $NaCl$. *Solve.*

Compound	$Ba(NO_3)_2$ result	NaCl result
$AgNO_3(aq)$	no ppt	AgCl ppt
$CaCl_2(aq)$	no ppt	no ppt
$Al_2(SO_4)_3$	$BaSO_4$ ppt	no ppt

This sequence of tests would definitively identify the contents of the bottle, because the results for each compound are unique.

Acids, Bases, and Neutralization Reactions (section 4.3)

4.29   *Analyze.* Given: solute and concentration of three solutions. Find: solution with greatest concentration of solvated protons.

Plan: See Sample Exercise 4.6. Determine whether solutes are strong or weak acids or bases or nonelectrolytes. For solutions of equal concentration, strong acids will have greatest concentration of solvated protons. Take varying concentration into consideration when evaluating the same class of solutions.

Solve. LiOH is a strong base, HI is a strong acid, CH_3OH is a molecular compound and nonelectrolyte. The strong acid HI will have the greatest concentration of solvated protons.

Check. The solution concentrations weren't needed to answer the question.

4.31   (a)   A *monoprotic acid* has one ionizable (acidic) H and a *diprotic acid* has two.

(b)   A *strong acid* is completely ionized in aqueous solution, whereas only a fraction of *weak acid* molecules are ionized.

(c)   An *acid* is an H^+ donor, a substance that increases the concentration of H^+ in aqueous solution. A *base* is an H^+ acceptor and thus increases the concentration of OH^- in aqueous solution.

4.33   When each of the strong acids in Table 4.2 dissociates, the anions formed are the same ones that normally form soluble ionic compounds (Table 4.1). Although ClO_3^- and ClO_4^- don't appear in Table 4.1, their ionic compounds are typically water soluble, but less common than compounds containing anions that do appear in the table. The one anion that typically forms soluble compounds that is not the anion of a strong acid is acetate, CH_3COO^-; it is the anion of a weak acid.

4.35   *Analyze.* Given: chemical formulas. Find: classify as acid, base, salt; strong, weak, or nonelectrolyte.

Plan. See Table 4.3. Ionic or molecular? Ionic, soluble: OH^-, strong base and strong electrolyte; otherwise, salt, strong electrolyte. Molecular: NH_3, weak base and weak electrolyte; H-first, acid; strong acid (Table 4.2), strong electrolyte; otherwise weak acid and weak electrolyte. *Solve.*

(a)   HF: acid, mixture of ions and molecules (weak electrolyte)

(b) CH_3CN: none of the above, entirely molecules (nonelectrolyte)

(c) $NaClO_4$: salt, entirely ions (strong electrolyte)

(d) $Ba(OH)_2$: base, entirely ions (strong electrolyte)

4.37   *Analyze.* Given: chemical formulas. Find: electrolyte properties.

Plan. In order to classify as electrolytes, formulas must be identified as acids, bases, or salts as in Solution 4.35.   *Solve.*

(a) H_2SO_3: H first, so acid; not in Table 4.2, so weak acid; therefore, weak electrolyte

(b) C_2H_5OH: not acid, not ionic (no metal cation), contains OH group, but not as anion so not a base; therefore, nonelectrolyte

(c) NH_3: common weak base; therefore, weak electrolyte

(d) $KClO_3$: ionic compound, so strong electrolyte

(e) $Cu(NO_3)_2$: ionic compound, so strong electrolyte

4.39   *Plan.* Follow Sample Exercise 4.7.   *Solve.*

(a) $2HBr(aq) + Ca(OH)_2(aq) \rightarrow CaBr_2(aq) + 2H_2O(l)$

       $H^+(aq) + OH^-(aq) \rightarrow H_2O(l)$

(b) $Cu(OH)_2(s) + 2HClO_4(aq) \rightarrow Cu(ClO_4)_2(aq) + 2H_2O(l)$

       $Cu(OH)_2(s) + 2H^+(aq) \rightarrow 2H_2O(l) + Cu^{2+}(aq)$

(c) $Al(OH)_3(s) + 3HNO_3(aq) \rightarrow Al(NO_3)_3(aq) + 3H_2O(l)$

       $Al(OH)_3(s) + 3H^+(aq) \rightarrow 3H_2O(l) + Al^{3+}(aq)$

4.41   *Analyze.* Given: names of reactants. Find: gaseous products.

Plan. Write correct chemical formulas for the reactants, complete and balance the metathesis reaction, and identify either H_2S or CO_2 products as gases.   *Solve.*

(a) $CdS(s) + H_2SO_4(aq) \rightarrow CdSO_4(aq) + H_2S(g)$

       $CdS(s) + 2H^+(aq) \rightarrow H_2S(g) + Cd^{2+}(aq)$

(b) $MgCO_3(s) + 2HClO_4(aq) \rightarrow Mg(ClO_4)_2(aq) + H_2O(l) + CO_2(g)$

       $MgCO_3(s) + 2H^+(aq) \rightarrow H_2O(l) + CO_2(g) + Mg^{2+}(aq)$

4.43   *Analyze.* Given the formulas or names of reactants, write balanced molecular and net ionic equations for the reactions.

Plan. Write correct chemical formulas for all reactants. Predict products of the neutralization reactions by exchanging ion partners. Balance the complete molecular equation, identify spectator ions by recognizing strong electrolytes, write the corresponding net ionic equation (omitting spectators).   *Solve.*

(a) $MgCO_3(s) + 2HCl(aq) \rightarrow MgCl_2(aq) + H_2O(l) + CO_2(g)$

       $MgCO_3(s) + 2H^+(aq) \rightarrow Mg^{2+}(aq) + H_2O(l) + CO_2(g)$

       $MgO(s) + 2HCl(aq) \rightarrow MgCl_2(aq) + H_2O(l)$

       $MgO(s) + 2H^+(aq) \rightarrow Mg^{2+}(aq) + H_2O(l)$

       $Mg(OH)_2(s) + 2HCl(aq) \rightarrow MgCl_2(aq) + 2H_2O(l)$

       $Mg(OH)_2(s) + 2H^+(aq) \rightarrow Mg^{2+}(aq) + 2H_2O(l)$

(b)   Yes. The reaction involving magnesium carbonate, $MgCO_3(s)$, produces $CO_2(g)$ which appears as bubbles. The other two reactions are calm.

(c)   If excess $HCl(aq)$ is added in each case, the identity of the ions in the clear product solution is the same. Each different reaction produces $Mg^{2+}(aq)$, along with Cl^-(aq) spectator ions. There will be $H^+(aq)$ [and more $Cl^-(aq)$] from the excess acid.

Oxidation-Reduction Reactions (section 4.4)

4.45   (a)   In terms of electron transfer, *oxidation* is the loss of electrons by a substance, and *reduction* is the gain of electrons (LEO says GER).

(b)   Relative to oxidation numbers, when a substance is oxidized, its oxidation number increases. When a substance is reduced, its oxidation number decreases.

4.47   *Analyze.* Given the labeled periodic chart, determine which region is most readily oxidized and which is most readily reduced.

Plan. Review the definition of oxidation and apply it to the properties of elements in the indicated regions of the chart.

Solve. Oxidation is loss of electrons. Elements readily oxidized form positive ions; these are metals. Elements not readily oxidized tend to gain electrons and form negative ions; these are nonmetals. Elements in regions A, B, and C are metals, and their ease of oxidation is shown in Table 4.5. Metals in region A, Na, Mg, K, and Ca are most easily oxidized. Elements in region D are nonmetals and are least easily oxidized.

4.49   *Analyze.* Given the chemical formula of a substance, determine the oxidation number of a particular element in the substance.

Plan. Follow the logic in Sample Exercise 4.8. *Solve.*

(a) +4   (b) +4   (c) +7   (d) +1   (e) 0   (f) –1 (O_2^{2-} is peroxide ion)

4.51   *Analyze.* Given: chemical reaction. Find: element oxidized or reduced. *Plan.* Assign oxidation numbers to all species. The element whose oxidation number increases (becomes more positive) is oxidized; the one whose oxidation number decreases (becomes more negative) is reduced.   *Solve.*

(a) $N_2(g)$ [N, 0] → $2NH_3(g)$ [N, –3], N is reduced; $3H_2(g)$ [H, 0] → $2NH_3(g)$ [H, +1], H is oxidized.

(b) Fe^{2+} → Fe, Fe is reduced; Al → Al^{3+}, Al is oxidized

(c) Cl_2 → $2Cl^-$, Cl is reduced; $2I^-$ → I_2, I is oxidized

(d) S^{2-} → SO_4^{2-}(S, +6), S is oxidized; H_2O_2 (O, –1) → H_2O (O, –2); O is reduced

4.53   *Analyze.* Given: reactants. Find: balanced molecular and net ionic equations.

Plan. Metals oxidized by H^+ form cations. Predict products by exchanging cations and balance. The anions are the spectator ions and do not appear in the net ionic equations. *Solve.*

(a) $Mn(s) + H_2SO_4(aq) → MnSO_4(aq) + H_2(g)$; $Mn(s) + 2H^+(aq) → Mn^{2+}(aq) + H_2(g)$

Products with the metal in a higher oxidation state are possible, depending on reaction conditions and acid concentration.

(b) $2Cr(s) + 6HBr(aq) → 2CrBr_3(aq) + 3H_2(g)$; $2Cr(s) + 6H^+(aq) → 2Cr^{3+}(aq) + 3H_2(g)$

(c) $Sn(s) + 2HCl(aq) → SnCl_2(aq) + H_2(g)$; $Sn(s) + 2H^+(aq) → Sn^{2+}(aq) + H_2(g)$

(d) $2Al(s) + 6HCOOH(aq) → 2Al(HCOO)_3(aq) + 3H_2(g)$;

$2Al(s) + 6HCOOH(aq) → 2Al^{3+}(aq) + 6HCOO^-(aq) + 3H_2(g)$

4.55   *Analyze.* Given: a metal and an aqueous solution. Find: balanced equation.

Plan. Use Table 4.5. If the metal is above the aqueous solution, reaction will occur; if the aqueous solution is higher, NR. If reaction occurs, predict products by exchanging cations (a metal ion or H^+), then balance the equation.   *Solve.*

(a) $Fe(s) + Cu(NO_3)_2(aq) \rightarrow Fe(NO_3)_2(aq) + Cu(s)$

(b) $Zn(s) + MgSO_4(aq) \rightarrow NR$

(c) $Sn(s) + 2HBr(aq) \rightarrow SnBr_2(aq) + H_2(g)$

(d) $H_2(g) + NiCl_2(aq) \rightarrow NR$

(e) $2Al(s) + 3CoSO_4(aq) \rightarrow Al_2(SO_4)_3(aq) + 3Co(s)$

4.57   (a)   i. $Zn(s) + Cd^{2+}(aq) \rightarrow Cd(s) + Zn^{2+}(aq)$

ii. $Cd(s) + Ni^{2+}(aq) \rightarrow Ni(s) + Cd^{2+}(aq)$

(b)   According to Table 4.5, the most active metals are most easily oxidized, and Zn is more active than Ni. Observation (i) indicates that Cd is less active than Zn; observation (ii) indicates that Cd is more active than Ni. Cd is between Zn and Ni on the activity series.

(c)   Place an iron strip in $CdCl_2(aq)$. If Cd(s) is deposited, Cd is less active than Fe; if there is no reaction, Cd is more active than Fe. Do the same test with Co if Cd is less active than Fe or with Cr if Cd is more active than Fe.

Concentrations of Solutions (section 4.5)

4.59   (a)   *Concentration* is an *intensive* property; it is the **ratio** of the amount of solute present in a certain quantity of solvent or solution. This ratio remains constant regardless of how much solution is present.

(b)   The term *0.50 mol HCl* defines an amount (~18 g) of the pure substance HCl. The term 0.50 M HCl is a ratio; it indicates that there are 0.50 mol of HCl solute in 1.0 liter of solution. This same ratio of moles solute to solution volume is present regardless of the volume of solution under consideration.

4.61   *Analyze/Plan.* Follow the logic in Sample Exercises 4.11 and 4.13.   *Solve.*

(a) $M = \dfrac{mol\ solute}{L\ solution}; \dfrac{0.175\ mol\ ZnCl_2}{150\ mL} \times \dfrac{1000\ mL}{1\ L} = 1.17\ M\ ZnCl_2$

Check. $(0.175 / 0.150) > 1.0\ M$

(b) $mol = M \times L; \dfrac{4.50\ mol\ HNO_3}{1\ L} \times 0.0350\ L = 0.158\ mol\ HNO_3$

Check. $(4.5 \times .04) \approx 0.16\ mol$

(c) $L = \dfrac{mol}{M}; \dfrac{0.325\ mol\ NaOH}{6.00\ mol\ NaOH/L} = 0.0542\ L\ or\ 54.2\ mL\ of\ 6.00\ M\ NaOH$

Check. $(0.325/6.0) > 0.50\ L.$

4.63   *Analyze.* Given molarity, M, and volume, L, find mass of Na^+(aq) in the blood.

Plan. Calculate moles Na^+(aq) using the definition of molarity: $M = \dfrac{mol}{L}$; $mol = M \times L$.

Calculate mass Na^+(aq) using the definition moles: $mol = g/MM$; $g = mol \times MM$. (MM is the symbol for molar mass in this manual.)

Solve. $\dfrac{0.135\ mol}{L} \times 5.0\ L \times \dfrac{23.0\ g\ Na^+}{mol\ Na^+} = 15.525 = 16\ g\ Na^+(aq)$

Check. Since there are more than 0.1 mol/L and we have 5.0 L, there should be more than half a mol (11.5 g) of Na^+. The calculation agrees with this estimate.

4.65   *Analyze.* Given: g alcohol/100 mL blood; molecular formula of alcohol. Find: molarity (mol/L) of alcohol. *Plan.* Use the molar mass (MM) of alcohol to change (g/100) mL to (mol/100 mL) then mL to L.

Solve. MM of alcohol = 2(12.01) + 6(1.1008) + 1(16.00) = 46.07 g alcohol/mol

$BAC = \dfrac{0.08\ g\ alcohol}{100\ mL\ blood} \times \dfrac{1\ mol\ alcohol}{46.07\ g\ alcohol} \times \dfrac{1000\ mL}{1\ L} = 0.0174 = 0.02\ M\ alcohol$

4.67   *Plan.* Proceed as in Sample Exercises 4.11 and 4.13.

$M = \dfrac{mol}{L}$; $mol = \dfrac{g}{MM}$ (MM is the symbol for molar mass in this manual.)

Solve.

(a) $\dfrac{0.175\ M\ KBr}{1\ L} \times 0.250\ L \times \dfrac{119.0\ g\ KBr}{1\ mol\ KBr} = 5.21\ g\ KBr$

     Check. $(0.25 \times 120) \approx 30$; $30 \times 0.18 \approx 5.4$ g KBr

(b) $14.75\ g\ Ca(NO_3)_2 \times \dfrac{1\ mol\ Ca(NO_3)_2}{164.09\ g\ Ca(NO_3)_2} \times \dfrac{1}{1.375\ L} = 0.06537\ M\ Ca(NO_3)_2$

     Check. $(15/1.5) \approx 10$; $10/160 = 5/80 \approx 0.06\ M\ Ca(NO_3)_2$

(c) $2.50\ g\ Na_3PO_4 \times \dfrac{1\ mol\ Na_3PO_4}{163.9\ g\ Na_3PO_4} \times \dfrac{1\ L}{1.50\ mol\ Na_3PO_4} \times \dfrac{1000\ mL}{1\ L}$

                                               $= 10.2$ mL solution

     Check. $[25/(160 \times 1.5)] \approx 2.5/240 \approx 1/100 \approx 0.01\ L = 10$ mL

4.69   *Analyze.* Given: formula and concentration of each solute. Find: concentration of K^+ in each solution. *Plan.* Note mol K^+/mol solute and compare concentrations or total moles. *Solve.*

(a) $KCl \rightarrow K^+ + Cl^-$; 0.20 M KCl = 0.20 M K^+

     $K_2CrO_4 \rightarrow 2\ K^+ + CrO_4{}^{2-}$; 0.15 M K_2CrO_4 = 0.30 M K^+

     $K_3PO_4 \rightarrow 3\ K^+ + PO_4{}^{3-}$; 0.080 M K_3PO_4 = 0.24 M K^+

     0.15 M K_2CrO_4 has the highest K^+ concentration.

(b) K_2CrO_4: 0.30 M K^+ × 0.0300 L = 0.0090 mol K^+

K_3PO_4: 0.24 M K^+ × 0.0250 L = 0.0060 mol K^+

30.0 mL of 0.15 M K_2CrO_4 has more K^+ ions.

4.71   *Analyze.* Given: molecular formula and solution molarity. Find: concentration (M) of each ion.

Plan. Follow the logic in Sample Exercise 4.12.

Solve.

(a) $NaNO_3 \rightarrow Na^+, NO_3^-$; 0.25 M $NaNO_3$ = 0.25 M Na^+, 0.25 M NO_3^-

(b) $MgSO_4 \rightarrow Mg^{2+}, SO_4^{2-}$; 1.3 × 10^{-2} M $MgSO_4$ = 1.3 × 10^{-2} M Mg^{2+}, 1.3 × 10^{-2} M SO_4^{2-}

(c) $C_6H_{12}O_6 \rightarrow C_6H_{12}O_6$ (molecular solute); 0.0150 M $C_6H_{12}O_6$ = 0.0150 M $C_6H_{12}O_6$

(d)   *Plan.* There is no reaction between NaCl and $(NH_4)_2CO_3$, so this is just a dilution problem, $M_1V_1 = M_2V_2$. Then account for ion stoichiometry.

Solve. 45.0 mL + 65.0 mL = 110.0 mL total volume

$$\frac{0.272\ M\ NaCl \times 45.0\ mL}{110.0\ mL} = 0.111\ M\ NaCl;\ 0.111\ M\ Na^+, 0.111\ M\ Cl^-$$

$$\frac{0.0247\ M\ (NH_4)_2CO_3 \times 65.0\ mL}{110.0\ mL} = 0.0146\ M\ (NH_4)_2CO_3$$

$2 \times (0.0146\ M) = 0.0292\ M\ NH_4^+, 0.0146\ M\ CO_3^{2-}$

Check. By adding the two solutions (with no common ions or chemical reaction), we have approximately doubled the solution volume, and reduced the concentration of each ion by approximately a factor of two.

4.73   *Analyze/Plan.* Follow the logic of Sample Exercise 4.14.

Solve.

(a) $V_1 = M_2V_2/M_1$; $\dfrac{0.250\ M\ NH_3 \times 1000.0\ mL}{14.8\ M\ NH_3} = 16.89 = 16.9\ mL\ 14.8\ M\ NH_3$

Check. 250/15 ≈ 15 M

(b) $M_2 = M_1V_1/V_2$; $\dfrac{14.8\ M\ NH_3 \times 10.0\ mL}{500\ mL} = 0.296\ M\ NH_3$

Check. 150/500 ≈ 0.30 M

4.75   (a)   *Plan/Solve.* Follow the logic in Sample Exercise 4.13. The number of moles of sucrose needed is $\dfrac{0.250\ mol}{1\ L} \times 0.250\ L = 0.06250 = 0.0625\ mol$

Weigh out 0.0625 mol $C_{12}H_{22}O_{11} \times \dfrac{342.3\ g\ C_{12}H_{22}O_{11}}{1\ mol\ C_{12}H_{22}O_{11}} = 21.4\ g\ C_{12}H_{22}O_{11}$

Add this amount of solid to a 250 mL volumetric flask, dissolve in a small volume of water, and add water to the mark on the neck of the flask. Agitate thoroughly to ensure total mixing.

(b) *Plan/Solve.* Follow the logic in Sample Exercise 4.14. Calculate the moles of solute present in the final 350.0 mL of 0.100 M $C_{12}H_{22}O_{11}$ solution:

$$\text{moles } C_{12}H_{22}O_{11} = M \times L = \frac{0.100 \text{ mol } C_{12}H_{22}O_{11}}{1 \text{ L}} \times 0.3500 \text{ L} = 0.0350 \text{ mol } C_{12}H_{22}O_{11}$$

Calculate the volume of 1.50 M glucose solution that would contain 0.0350 mol $C_{12}H_{22}O_{11}$:

$$L = \text{moles}/M; 0.0350 \text{ mol } C_{12}H_{22}O_{11} \times \frac{1 \text{ L}}{1.50 \text{ mol } C_{12}H_{22}O_{11}} = 0.02333 = 0.0233 \text{ L}$$

$$0.02333 \text{ L} \times \frac{1000 \text{ mL}}{1 \text{ L}} = 23.3 \text{ mL}$$

Thoroughly rinse, clean, and fill a 50 mL buret with the 1.50 M $C_{12}H_{22}O_{11}$. Dispense 23.3 mL of this solution into a 350 mL volumetric container, add water to the mark, and mix thoroughly. (23.3 mL is a difficult volume to measure with a pipette.)

4.77　*Analyze.* Given: density of pure acetic acid, volume pure acetic acid, volume new solution. Find: molarity of new solution. *Plan.* Calculate the mass of acetic acid, CH_3COOH, present in 20.0 mL of the pure liquid.　　　*Solve.*

$$20.00 \text{ mL acetic acid} \times \frac{1.049 \text{ g acetic acid}}{1 \text{ mL acetic acid}} = 20.98 \text{ g acetic acid}$$

$$20.98 \text{ g } CH_3COOH \times \frac{1 \text{ mol } CH_3COOH}{60.05 \text{ g } CH_3COOH} = 0.349375 = 0.3494 \text{ mol } CH_3COOH$$

$$M = \text{mol}/L = \frac{0.349375 \text{ mol } CH_3COOH}{0.2500 \text{ L solution}} = 1.39750 = 1.398 \text{ M } CH_3COOH$$

Check. $(20 \times 1) \approx 20$ g acid; $(20/60) \approx 0.33$ mol acid; $(0.33/0.25 = 0.33 \times 4) \approx 1.33$ M

Solution Stoichiometry and Chemcial Analysis (section 4.6)

4.79　*Analyze.* Given: volume and molarity $AgNO_3$. Find: mass KCl.

Plan. $M \times L = \text{mol } AgNO_3 = \text{mol } Ag^+$; balanced equation gives ratio mol KCl/mol $AgNO_3$; mol KCl → g KCl.　　　*Solve.*

$$\frac{0.200 \text{ mol } AgNO_3}{1 \text{ L}} \times 0.0150 \text{ L} = 3.00 \times 10^{-3} \text{ mol } AgNO_3 \text{ (aq)}$$

$$AgNO_3 \text{(aq)} + KCl \text{(aq)} \rightarrow AgCl \text{(s)} + KNO_3 \text{(aq)}$$

$$\text{mol KCl} = \text{mol } AgNO_3 = 3.00 \times 10^{-3} \text{ mol KCl}$$

$$3.00 \times 10^{-3} \text{ mol KCl} \times \frac{74.55 \text{ g KCl}}{1 \text{ mol KCl}} = 0.224 \text{ g KCl}$$

Check. $(0.2 \times 0.015) = 0.003$ mol; $(0.003 \times 75) \approx 0.225$ g KCl

4.81　(a)　*Analyze.* Given: M and vol base, M acid. Find: vol acid

Plan/Solve. Write the balanced equation for the reaction in question:

$$HClO_4 \text{(aq)} + NaOH \text{(aq)} \rightarrow NaClO_4 \text{(aq)} + H_2O \text{(l)}$$

Calculate the moles of the known substance, in this case NaOH.

$$\text{moles NaOH} = M \times L = \frac{0.0875 \text{ mol NaOH}}{1\,L} \times 0.0500\,L = 0.004375 = 0.00438 \text{ mol NaOH}$$

Apply the mole ratio (mol unknown/mol known) from the chemical equation.

$$0.004375 \text{ mol NaOH} \times \frac{1 \text{ mol HClO}_4}{1 \text{ mol NaOH}} = 0.004375 \text{ mol HClO}_4$$

Calculate the desired quantity of unknown, in this case the volume of 0.115 M HClO$_4$ solution.

$$L = \text{mol}/M; \ \ L = 0.004375 \text{ mol HClO}_4 \times \frac{1\,L}{0.115 \text{ mol HClO}_4} = 0.0380\,L = 38.0\,mL$$

Check. $(0.09 \times 0.05) = 0.0045$ mol; $(0.0045/0.11) \approx 0.040\,L \approx 40$ mL

(b) Following the logic outlined in part (a):

$$2HCl(aq) + Mg(OH)_2(s) \rightarrow MgCl_2(aq) + 2H_2O(l)$$

$$2.87 \text{ g Mg(OH)}_2 \times \frac{1 \text{ mol Mg(OH)}_2}{58.32 \text{ g Mg(OH)}_2} = 0.049211 = 0.0492 \text{ mol Mg(OH)}_2$$

$$0.0492 \text{ mol Mg(OH)}_2 \times \frac{2 \text{ mol HCl}}{1 \text{ mol Mg(OH)}_2} = 0.0984 \text{ mol HCl}$$

$$L = \text{mol}/M = 0.09840 \text{ mol HCl} \times \frac{1\,L \text{ HCl}}{0.128 \text{ mol HCl}} = 0.769\,L = 769\,mL$$

(c) $AgNO_3(aq) + KCl(aq) \rightarrow AgCl(s) + KNO_3(aq)$

$$785 \text{ mg KCl} \times \frac{1 \times 10^{-3} \text{ g}}{1 \text{ mg}} \times \frac{1 \text{ mol KCl}}{74.55 \text{ g KCl}} \times \frac{1 \text{ mol AgNO}_3}{1 \text{ mol KCl}} = 0.01053 = 0.0105 \text{ mol AgNO}_3$$

$$M = \text{mol}/L = \frac{0.01053 \text{ mol AgNO}_3}{0.0258\,L} = 0.408\,M \text{ AgNO}_3$$

(d) $HCl(aq) + KOH(aq) \rightarrow KCl(aq) + H_2O(l)$

$$\frac{0.108 \text{ mol HCl}}{1\,L} \times 0.0453\,L \times \frac{1 \text{ mol KOH}}{1 \text{ mol HCl}} \times \frac{56.11 \text{ g KOH}}{1 \text{ mol KOH}} = 0.275 \text{ g KOH}$$

4.83 *Analyze/Plan.* See Exercise 4.81(a) for a more detailed approach. *Solve.*

$$\frac{6.0 \text{ mol H}_2SO_4}{1\,L} \times 0.027\,L \times \frac{2 \text{ mol NaHCO}_3}{1 \text{ mol H}_2SO_4} \times \frac{84.01 \text{ g NaHCO}_3}{1 \text{ mol NaHCO}_3} = 27 \text{ g NaHCO}_3$$

4.85 *Analyze.* Given: M and vol HCl. Find: MM of base, an alkali metal hydroxide.

Plan. Alkali metal ions have a 1+ charge, so the general mormula of an alkali metal hydroxide is MOH. One mol of MOH requires one mol of HCl for neutralization.

$$M \times L = \text{mol HCl} = \text{mol MOH.} \ \ \text{MM} = \frac{\text{g MOH}}{1 \text{ mol MOH}}. \ \ \textit{Solve.}$$

(a) $\dfrac{2.50 \text{ mol HCl}}{1 \text{ L}} \times 0.0170 \text{ L} = 0.0425 \text{ mol HCl} = 0.0425 \text{ mol MOH}$

MM of MOH $= \dfrac{4.36 \text{ g MOH}}{0.0425 \text{ mol MOH}} = 102.59 = 103 \text{ g/mol}$

(b)   MM of alkali metal = MM of MOH – (17.01 g)   *Solve.*

MM of alkali metal = (102.59 g/mol – 17.01 g/mol) = 85.58 = 86 g/mol

The experimental molar mass most closely fits that of Rb^+, 85.47 g/mol

Check. The experimental molar mass matches one of the alkali metals.

4.87   (a) $NiSO_4(aq) + 2KOH(aq) \rightarrow Ni(OH)_2(s) + K_2SO_4(aq)$

(b)   The precipitate is $Ni(OH)_2$.

(c)   *Plan.* Compare mol of each reactant; mol = $M \times L$

Solve. 0.200 M KOH $\times$ 0.1000 L KOH = 0.0200 mol KOH

0.150 M NiSO$_4$ $\times$ 0.2000 L KOH = 0.0300 mol NiSO$_4$

1 mol NiSO$_4$ requires 2 mol KOH, so 0.0300 mol NiSO$_4$ requires 0.0600 mol KOH. Since only 0.0200 mol KOH is available, KOH is the limiting reactant.

(d)   *Plan.* The amount of the limiting reactant (KOH) determines amount of product, in this case $Ni(OH)_2$.

Solve. 0.0200 mol KOH $\times \dfrac{1 \text{ mol Ni(OH)}_2}{2 \text{ mol KOH}} \times \dfrac{92.71 \text{ g Ni(OH)}_2}{1 \text{ mol Ni(OH)}_2} = 0.927 \text{ g Ni(OH)}_2$

(e)   *Plan/Solve.* Limiting reactant: OH^-: no excess OH^- remains in solution.

Excess reactant: Ni^{2+}: M Ni^{2+} remaining = mol Ni^{2+} remaining/L solution

0.0300 mol Ni^{2+} initial – 0.0100 mol Ni^{2+} reacted = 0.0200 mol Ni^{2+} remaining

0.0200 mol Ni^{2+}/0.3000 L = 0.0667 M Ni^{2+}(aq)

Spectators: SO_4^{2-}, K^+. These ions do not react, so the only change in their concentration is dilution. The final volume of the solution is 0.3000 L.

$M_2 = M_1 V_1 / V_2$: 0.200 M K^+ $\times$ 0.1000 L/0.3000 L = 0.0667 M K^+(aq)

0.150 M SO_4^{2-} $\times$ 0.2000 L/0.3000 L = 0.100 M SO_4^{2-}(aq)

4.89   *Analyze.* Given: mass impure $Mg(OH)_2$; M and vol **excess** HCl; M and vol NaOH.

Find: mass % $Mg(OH)_2$ in sample. *Plan/Solve.* Write balanced equations.

$Mg(OH)_2(s) + 2HCl(aq) \rightarrow MgCl_2(aq) + 2H_2O(l)$

$HCl(aq) + NaOH(aq) \rightarrow NaCl(aq) + H_2O(l)$

Calculate total moles HCl = M HCl $\times$ L HCl

$\dfrac{0.2050 \text{ mol HCl}}{1 \text{ L soln}} \times 0.1000 \text{ L} = 0.02050 \text{ mol HCl total}$

mol excess HCl = mol NaOH used = M NaOH $\times$ L NaOH

$$\frac{0.1020 \text{ mol NaOH}}{1 \text{ L soln}} \times 0.01985 \text{ L} = 0.0020247 = 0.002025 \text{ mol NaOH}$$

mol HCl reacted with $Mg(OH)_2$ = total mol HCl – excess mol HCl

0.02050 mol total – 0.0020247 mol excess = 0.0184753 = 0.01848 mol HCl reacted

(The result has 5 decimal places and 4 sig. figs.)

Use mol ratio to get mol $Mg(OH)_2$ in sample, then molar mass of $Mg(OH)_2$ to get g pure $Mg(OH)_2$.

$$0.0184753 \text{ mol HCl} \times \frac{1 \text{ mol Mg(OH)}_2}{2 \text{ mol HCl}} \times \frac{58.32 \text{ g Mg(OH)}_2}{1 \text{ mol Mg(OH)}_2} = 0.5387 \text{ Mg(OH)}_2$$

$$\text{mass \% Mg(OH)}_2 = \frac{\text{g Mg(OH)}_2}{\text{g sample}} \times 100 = \frac{0.5388 \text{ g Mg(OH)}_2}{0.5895 \text{ g sample}} \times 100 = 91.40\% \text{ Mg(OH)}_2$$

Additional Exercises

4.92   The precipitate is $CdS(s)$. $Na^+(aq)$ and $NO_3^-(aq)$ are spectator ions and remain in solution. Any excess reactant ions also remain in solution. The net ionic equation is:

$Cd^{2+}(aq) + S^{2-}(aq) \rightarrow CdS(s)$.

4.94   (a,b)   Expt. 1 No reaction

Expt. 2 $2Ag^+(aq) + CrO_4^{2-}(aq) \rightarrow Ag_2CrO_4(s)$ red precipitate

Expt. 3 $Ca^{2+}(aq) + CrO_4^{2-}(aq) \rightarrow CaCrO_4(s)$ yellow precipitate

Expt. 4 $2Ag^+(aq) + C_2O_4^{2-}(aq) \rightarrow Ag_2C_2O_4(s)$ white precipitate

Expt. 5 $Ca^{2+}(aq) + C_2O_4^{2-}(aq) \rightarrow CaC_2O_4(s)$ white precipitate

Expt. 6 $Ag^+(aq) + Cl^-(aq) \rightarrow AgCl(s)$ white precipitate

4.96 $4NH_3(g) + 5O_2(g) \rightarrow 4NO(g) + 6H_2O(g)$.
        N = –3   O = 0   N = +2   O = –2

(a)   redox reaction   (b)   N is oxidized, O is reduced

$2NO(g) + O_2(g) \rightarrow 2NO_2(g)$.
N = +2   O = 0   N = +4, O = –2

(a)   redox reaction   (b)   N is oxidized, O is reduced

$3NO_2(g) + H_2O(l) \rightarrow HNO_3(aq) + NO(g)$.
N = +4   N = +5   N = +2

(a)   redox reaction

(b)   N is oxidized ($NO_2 \rightarrow HNO_3$), N is reduced ($NO_2 \rightarrow NO$). A reaction where the same element is both oxidized and reduced is called disproportionation.

4.99   *Plan.* Calculate moles KBr from the two quantities of solution (mol = $M \times$ L), then new molarity (M = mol/L). KBr is nonvolatile, so no solute is lost when the solution is evaporated to reduce the total volume. *Solve.*

$$1.00\ M\ \text{KBr} \times 0.0350\ \text{L} = 0.0350\ \text{mol KBr}; \quad 0.600\ M\ \text{KBr} \times 0.060\ \text{L} = 0.0360\ \text{mol KBr}$$

$$0.0350\ \text{mol KBr} + 0.0360\ \text{mol KBr} = 0.0710\ \text{mol KBr total}$$

$$\frac{0.0710\ \text{mol KBr}}{0.0500\ \text{L soln}} = 1.42\ M\ \text{KBr}$$

4.100   (a) $\dfrac{50\ \text{pg}}{1\ \text{mL}} \times \dfrac{1 \times 10^{-12}\ \text{g}}{1\ \text{pg}} \times \dfrac{1 \times 10^{3}\ \text{mL}}{\text{L}} \times \dfrac{1\ \text{mol Na}}{23.0\ \text{g Na}} = 2.17 \times 10^{-9} = 2.2 \times 10^{-9}\ M\ \text{Na}^{+}$

(b) $\dfrac{2.17 \times 10^{-9}\ \text{mol Na}}{1\ \text{L soln}} \times \dfrac{1\ \text{L}}{1 \times 10^{3}\ \text{cm}^{3}} \times \dfrac{6.02 \times 10^{23}\ \text{Na atom}}{1\ \text{mol Na}}$

$$= 1.3 \times 10^{12}\ \text{atom or Na}^{+}\ \text{ions/cm}^{3}$$

4.103   (a) $10.45\ \text{g Sr(OH)}_2 \times \dfrac{1\ \text{mol Sr(OH)}_2}{121.64\ \text{g Sr(OH)}_2} \times \dfrac{1}{0.05000\ \text{L}} = 1.7182 = 1.718\ M\ \text{Sr(OH)}_2$

(b) $2\,\text{HNO}_3(\text{aq}) + \text{Sr(OH)}_2(\text{s}) \rightarrow \text{Sr(NO}_3)_2(\text{aq}) + 2\,\text{H}_2\text{O(l)}$

(c)   *Plan.* mol $\text{Sr(OH)}_2 = M \times \text{L} \rightarrow$ mol ratio $\rightarrow$ mol $\text{HNO}_3 \rightarrow M\ \text{HNO}_3$.   *Solve.*

$$\frac{1.7182\ \text{mol Sr(OH)}_2}{\text{L}} \times 0.0239\ \text{L Sr(OH)}_2 \times \frac{2\ \text{mol HNO}_3}{1\ \text{mol Sr(OH)}_2} \times \frac{1}{0.0315\ \text{L HNO}_3}$$

$$= 2.6073 = 2.61\ M\ \text{HNO}_3$$

Integrative Exercises

4.106   (a)   At the equivalence point of a titration, mol NaOH added = mol H^{+} present

$$M_{\text{NaOH}} \times \text{L}_{\text{NaOH}} = \frac{\text{g acid}}{\text{MM acid}}\ \text{(for an acid with 1 acidic hydrogen)}$$

$$\text{MM acid} = \frac{\text{g acid}}{M_{\text{NaOH}} \times \text{L}_{\text{NaOH}}} = \frac{0.2053\ \text{g}}{0.1008\ M \times 0.0150\ \text{L}} = 136\ \text{g/mol}$$

(b)   Assume 100 g of acid.

$$70.6\ \text{g C} \times \frac{1\ \text{mol C}}{12.01\ \text{g C}} = 5.88\ \text{mol C};\ 5.88 / 1.47 \approx 4$$

$$5.89\ \text{g H} \times \frac{1\ \text{mol H}}{1.008\ \text{g H}} = 5.84\ \text{mol H};\ 5.84 / 1.47 \approx 4$$

$$23.5\ \text{g O} \times \frac{1\ \text{mol O}}{16.00\ \text{g O}} = 1.47\ \text{mol O};\ 1.47 / 1.47 = 1$$

The empirical formula is $\text{C}_4\text{H}_4\text{O}$.

$$\frac{\text{MM}}{\text{FW}} = \frac{136}{68.1} = 2;\ \text{the molecular formula is } 2 \times \text{the empirical formula.}$$

The molecular formula is $\text{C}_8\text{H}_8\text{O}_2$.

4.109   (a) $Mg(OH)_2(s) + 2HNO_3(aq) \rightarrow Mg(NO_3)_2(aq) + 2H_2O(l)$

       (b) $5.53 \text{ g Mg(OH)}_2 \times \dfrac{1 \text{ mol Mg(OH)}_2}{58.32 \text{ g Mg(OH)}_2} = 0.09482 = 0.0948 \text{ mol Mg(OH)}_2$

          $0.200 \, M \text{ HNO}_3 \times 0.0250 \text{ L} = 0.00500 \text{ mol HNO}_3$

          The 0.00500 mol HNO_3 would neutralize 0.00250 mol $Mg(OH)_2$ and much more $Mg(OH)_2$ is present, so HNO_3 is the limiting reactant.

       (c)   Since HNO_3 limits, 0 mol HNO_3 is present after reaction.

          0.00250 mol $Mg(NO_3)_2$ is produced.

          0.09482 mol $Mg(OH)_2$ initial – 0.00250 mol $Mg(OH)_2$ react

                                       $= 0.0923 \text{ mol Mg(OH)}_2 \text{ remain}$

4.112 $Ag^+(aq) + Cl^-(aq) \rightarrow AgCl(s)$

       $\dfrac{0.2997 \text{ mol Ag}^+}{1 \text{ L}} \times 0.04258 \text{ L} \times \dfrac{1 \text{ mol Cl}^-}{1 \text{ mol Ag}^+} \times \dfrac{35.453 \text{ g Cl}^-}{1 \text{ mol Cl}^-} = 0.45242 = 0.4524 \text{ g Cl}^-$

       $25.00 \text{ mL seawater} \times \dfrac{1.025 \text{ g}}{\text{mL}} = 25.625 = 25.63 \text{ g seawater}$

       $\text{mass \% Cl}^- = \dfrac{0.45242 \text{ g Cl}^-}{25.625 \text{ g seawater}} \times 100 = 1.766\% \text{ Cl}^-$

4.114   *Analyze.* Given 10 ppb $AsO_4{}^{3-}$, find mass Na_3AsO_4 in 1.00 L of drinking water.

       Plan. Use the definition of ppb to calculate g $AsO_4{}^{3-}$ in 1.0 L of water. Convert g $AsO_4{}^{3-} \rightarrow$ g Na_3AsO_4 using molar masses. Assume the density of H_2O is 1.00 g/mL.

       Solve. 1 billion $= 1 \times 10^9$; $1 \text{ ppb} = \dfrac{1 \text{ g solute}}{1 \times 10^9 \text{ g solution}}$

       $\dfrac{1 \text{ g solute}}{1 \times 10^9 \text{ g solution}} \times \dfrac{1 \text{ g solution}}{1 \text{ mL solution}} \times \dfrac{1 \times 10^3 \text{ mL}}{1 \text{ L solution}} = \dfrac{\text{g AsO}_4{}^{3-}}{1 \times 10^6 \text{ L H}_2\text{O}}$

       $10 \text{ ppb AsO}_4{}^{3-} = \dfrac{10 \text{ g AsO}_4}{1 \times 10^6 \text{ L H}_2\text{O}} \times 1 \text{ L H}_2\text{O} = 1.0 \times 10^{-5} \text{ g As/L.}$

       $1.0 \times 10^{-5} \text{ g AsO}_4{}^{3-} \times \dfrac{1 \text{ mol AsO}_4{}^{3-}}{138.92 \text{ g AsO}_4{}^{3-}} \times \dfrac{1 \text{ mol Na}_3\text{AsO}_4}{1 \text{ mol As}} \times \dfrac{207.89 \text{ g Na}_3\text{AsO}_4{}^{3-}}{1 \text{ mol Na}_3\text{AsO}_4}$

                                       $= 1.5 \times 10^{-5} \text{ g Na}_3\text{AsO}_4 \text{ in 1.00 L H}_2\text{O}$

5 Thermochemistry

Visualizing Concepts

5.1 The book's potential energy is due to the opposition of gravity by an object of mass m at a distance d above the surface of the earth. Kinetic energy is due to the motion of the book. As the book falls, d decreases and potential energy changes into kinetic energy.

The first law states that the total energy of a system is conserved. At the instant before impact, all potential energy has been converted to kinetic energy, so the book's total kinetic energy is 71 J, assuming no transfer of energy as heat.

For an object under the influence of gravity, both potential ($E_p = mgd$) and kinetic energy ($E_k = \frac{1}{2} mv^2$) are directly proportional to mass of the object. A heavier book falling from the same shelf has greater kinetic energy when it hits the floor.

5.5 (a). No. This distance traveled to the top of a mountain depends on the path taken by the hiker. Distance is a path function, not a state function.

(b) Yes. Change in elevation depends only on the location of the base camp and the height of the mountain, not on the path to the top. Change in elevation is a state function, not a path function.

5.8 (a) $w = -P\Delta V$. Since ΔV for the process is (–), the sign of w is (+).

(b) $\Delta E = q + w$. At constant pressure, $\Delta H = q$. If the reaction is endothermic, the signs of ΔH and q are (+). From (a), the sign of w is (+), so the sign of ΔE is (+). The internal energy of the system increases during the change. (This situation is described by the diagram (ii) in Exercise 5.4.)

5.11 (a) $\Delta H_A = \Delta H_B + \Delta H_C$. The net enthalpy change associated with going from the initial state to the final state does not depend on path. The diagram shows that the change can be accomplished via reaction A, or via two successive reactions, B then C, with the same net enthalpy change. $\Delta H_A = \Delta H_B + \Delta H_C$ because ΔH is a state function, independent of path.

(b) $\Delta H_Z = \Delta H_X + \Delta H_Y$. The diagram indicates that Reaction Z can be written as the sum of reactions X and Y.

(c) Hess's law states that the enthalpy change for a net reaction is the sum of the enthalpy changes of the component steps, regardless of whether the reaction actually occurs via this path. The diagrams are a visual statement of Hess's law.

The Nature of Energy (section 5.1)

5.13   An object can possess energy by virtue of its motion or position. Kinetic energy, the energy of motion, depends on the mass of the object and its velocity. Potential energy, stored energy, depends on the position of the object relative to the body with which it interacts.

5.15   (a)   *Plan.* $E_k = 1/2 \, mv^2$; $m = 1,200 \, kg$; $v = 18 \, m/s$; $1 \, kg\text{-}m^2/s^2 = 1 \, J$

               Solve. $E_k = 1/2 \times 1,200 \, kg \times (18)^2 \, m^2/s^2 = 1.944 \times 10^5 = 1.9 \times 10^5 \, J$

      (b) $1 \, cal = 4.184 \, J$; $1.944 \times 10^5 \, J \times \dfrac{1 \, cal}{4.184 \, J} = 4.646 \times 10^4 \, cal = 4.6 \times 10^4 \, cal$

      (c)   As the automobile brakes to a stop, its speed (and hence its kinetic energy) drops to zero. Brakes stop a moving vehicle, so the kinetic energy of the automobile is primarily transferred to friction between brakes and wheels, and somewhat to deformation of the tire and friction between the tire and road.

5.17   *Analyze.* Given: heat capacity of water $= 1 \, Btu/lb\text{-}°F$ Find: J/Btu

      Plan. heat capacity of water $= 4.184 \, J/g\text{-}°C$; $\dfrac{J}{g\text{-}°C} \rightarrow \dfrac{J}{lb\text{-}°F} \rightarrow \dfrac{J}{Btu}$

      This strategy requires changing °F to °C. Since this involves the magnitude of a degree on each scale, rather than a specific temperature, the 32 in the temperature relationship is not needed. $100 \, °C = 180 \, °F$; $5 \, °C = 9 \, °F$

      Solve. $\dfrac{4.184 \, J}{g\text{-}°C} \times \dfrac{453.6 \, g}{lb} \times \dfrac{5 \, °C}{9 \, °F} \times \dfrac{1 \, lb\text{-}°F}{1 \, Btu} = 1054 \, J/Btu$

5.19   (a)   In thermodynamics, the *system* is the well-defined part of the universe whose energy changes are being studied.

      (b)   A *closed system* can exchange energy but not mass with its surroundings.

      (c)   Any part of the universe not part of the system is called the surroundings.

5.21   (a)   *Work* is a force applied over a distance.

      (b)   The amount of work done is the magnitude of the force times the distance over which it is applied. $w = F \times d$.

5.23   (a)   Gravity; work is done because the force of gravity is opposed and the pencil is lifted a distance above the desk.

      (b)   Mechanical force; work is done because the force of the coiled spring is opposed as the spring is compressed over a distance.

The First Law of Thermodynamics (section 5.2)

5.25   (a)   In any chemical or physical change, energy can be neither created nor destroyed, but it can be changed in form.

      (b)   The total *internal energy* (E) of a system is the sum of all the kinetic and potential energies of the system components.

(c)   The internal energy of a closed system (where no mass exchange with surroundings occurs) increases when work is done on the system by the surroundings and/or when heat is transferred to the system from the surroundings (the system is heated).

5.27   *Analyze.* Given: heat and work. Find: magnitude and sign of ΔE.

Plan. In each case, evaluate q and w in the expression $\Delta E = q + w$. For an exothermic process, q is negative; for an endothermic process, q is positive.   *Solve.*

(a)   q = 0.763 kJ, w = –840 J = 0.840 kJ. ΔE = 0.763 kJ – 0.840 kJ = –0.077 kJ. The process is endothermic.

(b)   q is negative because the system releases heat, and w is positive because work is done on the system. ΔE = –66.1 kJ + 44.0 kJ = –22.1 kJ. The process is exothermic.

(c)   q is positive because the system absorbs heat. No work is done, since $w = -P\Delta V$ and $\Delta V = 0$ (assuming that only P-V work can be done).
ΔE = 7.25 kJ + 0 kJ = 7.25 kJ. The process is endothermic.

5.29   *Analyze.* How do the different physical situations (cases) affect the changes to heat and work of the system upon addition of 100 J of energy?

Plan. Use the definitions of heat and work and the First Law to answer the questions.

Solve. If the piston is allowed to move, case (1), the heated gas will expand and push the piston up, doing work on the surroundings. If the piston is fixed, case (2), most of the electrical energy will be manifested as an increase in heat of the system.

(a)   Since little or no work is done by the system in case (2), the gas will absorb most of the energy as heat; the case (2) gas will have the higher temperature.

(b)   In case (2), w = 0 and q = 100 J. In case (1), a significant amount of energy will be used to do work on the surroundings (–w), but some will be absorbed as heat (+q). (The transfer of electrical energy into work is never completely efficient!)

(c) ΔE is greater for case (2), because the entire 100 J increases the internal energy of the system, rather than a part of the energy doing work on the surroundings.

5.31   (a)   A *state function* is a property of a system that depends only on the physical state (pressure, temperature, etc.) of the system, not on the route used by the system to get to the current state.

(b)   Internal energy and enthalpy **are** state functions; heat **is not** a state function.

(c)   Volume **is** a state function. The volume of a system depends only on conditions (pressure, temperature, amount of substance), not the route or method used to establish that volume.

Enthalpy (sections 5.3 and 5.4)

5.33   (a)   Change in enthalpy (ΔH) is usually easier to measure than change in internal energy (ΔE) because, at constant pressure, $\Delta H = q_p$. The heat flow associated with a process at constant pressure can easily be measured as a change in temperature. Measuring ΔE requires a means to measure both q and w.

(b) H describes the enthalpy of a system at a certain set of conditions; the value of H depends only on these conditions. q describes energy transferred as heat, an energy *change* which, in the general case, does depend on how the change occurs. We can equate change in enthalpy, ΔH, with heat, q_p, only for the specific conditions of constant pressure and exclusively P-V work.

(c) If ΔH is positive, the enthalpy of the system increases and the process is endothermic.

5.35 At constant pressure, $\Delta E = \Delta H - P\Delta V$. In order to calculate ΔE, more information about the conditions of the reaction must be known. For an ideal gas at constant pressure and temperature, $P\Delta V = RT\Delta n$. The values of either P and ΔV or T and Δn must be known to calculate ΔE from ΔH.

5.37 *Analyze/Plan.* q = 824 J = 0.824 kJ (heat is absorbed by the system), w = 0.65 kJ (work is done on the system). *Solve.*

$\Delta E = q + w = 0.824 \text{ kJ} + 0.65 \text{ kJ} = 1.47 \text{ kJ}.$ $\Delta H = q = 0.824 \text{ kJ}$ (at constant pressure).

Check. The reaction is endothermic.

5.39 (a) $C_2H_5OH(l) + 3O_2(g) \rightarrow 3H_2O(g) + 2CO_2(g)$ $\Delta H = -1235 \text{ kJ}$

(b) *Analyze.* How are reactants and products arranged on an enthalpy diagram?

Plan. The substances (reactants or products, collectively) with higher enthalpy are shown on the upper level, and those with lower enthalpy are shown on the lower level.

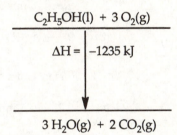

Solve. For this reaction, ΔH is negative, so the products have lower enthalpy and are shown on the lower level; reactants are on the upper level. The arrow points in the direction of reactants to products and is labeled with the value of ΔH.

5.41 *Analyze/Plan.* Consider ΔH for the exothermic reaction as written. *Solve.*

(a) $\Delta H = -284.6 \text{ kJ} / 2 \text{ mol } O_3(g) = -142.3 \text{ kJ/mol } O_3(g)$

(b) Since ΔH is negative, the reactants, $2O_3(g)$ has the higher enthalpy.

5.43 *Analyze/Plan.* Follow the strategy in Sample Exercise 5.4.   *Solve.*

(a) Exothermic (ΔH is negative)

(b) $3.55 \text{ g Mg} \times \dfrac{1 \text{ mol Mg}}{24.305 \text{ g Mg}} \times \dfrac{-1204 \text{ kJ}}{2 \text{ mol Mg}} = -87.9 \text{ kJ heat transferred}$

Check. The units of kJ are correct for heat. The negative sign indicates heat is evolved.

(c) $-234 \text{ kJ} \times \dfrac{2 \text{ mol MgO}}{-1204 \text{ kJ}} \times \dfrac{40.30 \text{ g MgO}}{1 \text{ mol Mg}} = 15.7 \text{ g MgO produced}$

Check. Units are correct for mass. $(200 \times 2 \times 40/1200) \approx (16000/1200) > 10 \text{ g}$

(d) $2\text{MgO(s)} \rightarrow 2\text{Mg(s)} + \text{O}_2\text{(g)}$ $\Delta H = +1204 \text{ kJ}$

This is the reverse of the reaction given above, so the sign of ΔH is reversed.

$40.3 \text{ g MgO} \times \dfrac{1 \text{ mol MgO}}{40.30 \text{ g MgO}} \times \dfrac{1204 \text{ kJ}}{2 \text{ mol MgO}} = +602 \text{ kJ heat absorbed}$

Check. 40.3 g MgO is just 1 mol MgO, so the calculated value is the heat absorbed per mol of MgO, 1204 kJ/2 mol MgO = 602 kJ.

5.45   *Analyze.* Given: balanced thermochemical equation, various quantities of substances and/or enthalpy. *Plan.* Enthalpy is an extensive property; it is "stoichiometric." Use the mole ratios implicit in the balanced thermochemical equation to solve for the desired quantity. Use molar masses to change mass to moles and vice versa where appropriate. *Solve.*

(a) $0.450 \text{ mol AgCl} \times \dfrac{-65.5 \text{ kJ}}{1 \text{ mol AgCl}} = -29.5 \text{ kJ}$

Check. Units are correct; sign indicates heat evolved.

(b) $9.00 \text{ g AgCl} \times \dfrac{1 \text{ mol AgCl}}{143.3 \text{ g AgCl}} \times \dfrac{-65.5 \text{ kJ}}{1 \text{ mol AgCl}} = -4.11 \text{ kJ}$

Check. Units correct; sign indicates heat evolved.

(c) $9.25 \times 10^{-4} \text{ mol AgCl} \times \dfrac{+65.5 \text{ kJ}}{1 \text{ mol AgCl}} = 0.0606 \text{ kJ} = 60.6 \text{ J}$

Check. Units correct; sign of ΔH reversed; sign indicates heat is absorbed during the reverse reaction.

5.47   *Analyze.* Given: balanced thermochemical equation. *Plan.* Follow the guidelines given in Section 5.4 for evaluating thermochemical equations.   *Solve.*

(a)   When a chemical equation is reversed, the sign of ΔH is reversed.

$\text{CO}_2\text{(g)} + 2\text{H}_2\text{O(l)} \rightarrow \text{CH}_3\text{OH(l)} + 3/2 \text{ O}_2\text{(g)}$ $\Delta H = +726.5 \text{ kJ}$

(b)   Enthalpy is extensive. If the coefficients in the chemical equation are multiplied by 2 to obtain all integer coefficients, the enthalpy change is also multiplied by 2.

$2\text{CH}_3\text{OH(l)} + 3\text{O}_2\text{(g)} \rightarrow 2\text{CO}_2\text{(g)} + 4\text{H}_2\text{O(l)}$ $\Delta H = 2(-726.5) \text{ kJ} = -1453 \text{ kJ}$

(c)   The exothermic forward reaction is more likely to be thermodynamically favored.

(d)   Decrease. Vaporization (liquid $\rightarrow$ gas) is endothermic. If the product were $\text{H}_2\text{O(g)}$, the reaction would be more endothermic and would have a smaller negative ΔH. (Depending on temperature, the enthalpy of vaporization for 2 mol H_2O is about +88 kJ, not large enough to cause the overall reaction to be endothermic.)

Calorimetry (section 5.5)

The specific heat of water to four significant figures, **4.184 J/g - K,** will be used in many of the following exercises; temperature units of K and °C will be used interchangeably.

5.49   (a)   J/mol-K or J/mol-°C. Heat capacity is the amount of heat in J required to raise the temperature of an object or a certain amount of substance 1 °C or 1 K. Molar heat capacity is the heat capacity of one mole of substance.

        (b) $\dfrac{J}{g \cdot {}^\circ C}$ or $\dfrac{J}{g \cdot K}$ Specific heat is a particular kind of heat capacity where the amount of substance is 1 g.

        (c)   To calculate heat capacity from specific heat, the **mass** of the particular piece of copper pipe must be known.

5.51   *Plan.* Manipulate the definition of specific heat to solve for the desired quantity, paying close attention to units. $C_s = q/(m \times \Delta t)$.   *Solve.*

        (a) $\dfrac{4.184\,J}{1g \cdot K}$ or $\dfrac{4.184\,J}{1g \cdot {}^\circ C}$

        (b) $\dfrac{4.184\,J}{1g \cdot {}^\circ C} \times \dfrac{18.02\,g\,H_2O}{1\,mol\,H_2O} = \dfrac{75.40\,J}{mol \cdot {}^\circ C}$

        (c) $\dfrac{185\,g\,H_2O \times 4.184\,J}{1g \cdot {}^\circ C} = 774\,J/{}^\circ C$

        (d) $10.00\,kg\,H_2O \times \dfrac{1000\,g}{1\,kg} \times \dfrac{4.184\,J}{1g \cdot {}^\circ C} \times \dfrac{1\,kJ}{1000\,J} \times (46.2{}^\circ C - 24.6{}^\circ C) = 904\,kJ$

        Check. $(10 \times 4 \times 20) \approx 800$ kJ; the units are correct. Note that the conversion factors for kg → g and J → kJ cancel. An equally correct form of specific heat would be kJ/kg - °C

5.53   *Analyze/Plan.* Follow the logic in Sample Exercise 5.5.   *Solve.*

        (a) $80.0\,g\,C_8H_{18} \times \dfrac{2.22\,J}{g \cdot K} \times (25.0{}^\circ C - 10.0{}^\circ C) = 2.66 \times 10^3\,J\ (or\ 2.66\,kJ)$

        (b)   *Plan.* Calculate the molar heat capacity of octane and compare it with the molar heat capacity of water, 75.40 J/mol-°C, as calculated in Exercise 5.51(b). *Solve.*

        $\dfrac{2.22\,J}{g \cdot K} \times \dfrac{114.2\,g\,C_8H_{18}}{1\,mol\,C_8H_{18}} = \dfrac{253.58\,J}{mol \cdot K} = \dfrac{254\,J}{mol \cdot K}$

        The molar heat capacity of $C_8H_{18}(l)$, 254 J/mol-K, is greater than that of $H_2O(l)$, so it will require more heat to increase the temperature of octane than to increase the temperature of water.

5.55   *Analyze.* Since the temperature of the water increases, the dissolving process is exothermic and the sign of ΔH is negative. The heat lost by the NaOH(s) dissolving equals the heat gained by the solution.

Plan/Solve. Calculate the heat gained by the solution. The temperature change is 37.8 – 21.6 = 16.2°C. The total mass of solution is (100.0 g H_2O + 6.50 g NaOH) = 106.5 g.

$$106.5 \text{ g solution} \times \frac{4.184 \text{ J}}{1 \text{ g} \cdot {}^{\circ}\text{C}} \times 16.2 {}^{\circ}\text{C} \times \frac{1 \text{ kJ}}{1000 \text{ J}} = 7.2187 = 7.22 \text{ kJ}$$

This is the amount of heat lost when 6.50 g of NaOH dissolves.

The heat loss per mole NaOH is

$$\frac{-7.2187 \text{ kJ}}{6.50 \text{ g NaOH}} \times \frac{40.00 \text{ g NaOH}}{1 \text{ mol NaOH}} = -44.4 \text{ kJ/mol} \quad \Delta H = q_p = -44.4 \text{ kJ/mol NaOH}$$

Check. $(-7/7 \times 40) \approx -40$ kJ; the units and sign are correct.

5.57   *Analyze/Plan.* Follow the logic in Sample Exercise 5.7.   *Solve.*

$q_{bomb} = -q_{rxn}$; $\Delta T = 30.57°C - 23.44°C = 7.13°C$

$$q_{bomb} = \frac{7.854 \text{ kJ}}{1 {}^{\circ}\text{C}} \times 7.13 {}^{\circ}\text{C} = 56.00 = 56.0 \text{ kJ}$$

At constant volume, $q_v = \Delta E$. ΔE and ΔH are very similar.

$$\Delta H_{rxn} \approx \Delta E_{rxn} = q_{rxn} = -q_{bomb} = \frac{-56.0 \text{ kJ}}{2.200 \text{ g } C_6H_4O_2} = -25.454 = -25.5 \text{ kJ/g } C_6H_4O_2$$

$$\Delta H_{rxn} = \frac{-25.454 \text{ kJ}}{1 \text{ g } C_6H_4O_2} \times \frac{108.1 \text{ g } C_6H_4O_2}{1 \text{ mol } C_6H_4O_2} = -2.75 \times 10^3 \text{ kJ/mol } C_6H_4O_2$$

5.59   *Analyze.* Given: specific heat and mass of glucose, ΔT for calorimeter. Find: heat capacity, C, of calorimeter. *Plan.* All heat from the combustion raises the temperature of the calorimeter. Calculate heat from combustion of glucose, divide by ΔT for calorimeter to get kJ/°C. $\Delta T = 24.72°C - 20.94°C = 3.78°C$ *Solve.*

(a) $C_{total} = 3.500 \text{ g glucose} \times \dfrac{15.57 \text{ kJ}}{1 \text{ g glucose}} \times \dfrac{1}{3.78 {}^{\circ}\text{C}} = 14.42 = 14.4 \text{ kJ/}{}^{\circ}\text{C}$

(b)   Qualitatively, assuming the same exact initial conditions in the calorimeter, twice as much glucose produces twice as much heat, which raises the calorimeter temperature by twice as many °C. Quantitatively,

$$7.000 \text{ g glucose} \times \frac{15.57 \text{ kJ}}{1 \text{ g glucose}} \times \frac{1 {}^{\circ}\text{C}}{14.42 \text{ kJ}} = 7.56 {}^{\circ}\text{C}$$

Check. Units are correct. ΔT is twice as large as in part (a). The result has 3 sig figs, because the heat capacity of the calorimeter is known to 3 sig figs.

Hess's Law (section 5.6)

5.61   Hess's Law is a consequence of the fact that enthalpy is a state function. Since ΔH is independent of path, we can describe a process by any series of steps that add up to the overall process and ΔH for the process is the sum of the ΔH values for the steps.

5.63 *Analyze/Plan.* Follow the logic in Sample Exercise 5.8. Manipulate the equations so that "unwanted" substances can be canceled from reactants and products. Adjust the corresponding sign and magnitude of ΔH. *Solve.*

$$
\begin{array}{ll}
P_4O_6(s) \rightarrow P_4(s) + 3O_2(g) & \Delta H = 1640.1\text{ kJ} \\
\underline{P_4(s) + 5O_2(g) \rightarrow P_4O_{10}(s)} & \underline{\Delta H = -2940.1\text{ kJ}} \\
P_4O_6(s) + 2O_2(g) \rightarrow P_4O_{10}(s) & \Delta H = -1300.0\text{ kJ}
\end{array}
$$

Check. We have obtained the desired reaction.

5.65 *Analyze/Plan.* Follow the logic in Sample Exercise 5.9. Manipulate the equations so that "unwanted" substances can be canceled from reactants and products. Adjust the corresponding sign and magnitude of ΔH. *Solve.*

$$
\begin{array}{ll}
C_2H_4(g) \rightarrow 2H_2(g) + 2C(s) & \Delta H = -52.3\text{ kJ} \\
2C(s) + 4F_2(g) \rightarrow 2CF_4(g) & \Delta H = 2(-680\text{ kJ}) \\
\underline{2H_2(g) + 2F_2(g) \rightarrow 4HF(g)} & \underline{\Delta H = 2(-537\text{ kJ})} \\
C_2H_4(g) + 6F_2(g) \rightarrow 2CF_4(g) + 4HF(g) & \Delta H = -2.49 \times 10^3\text{ kJ}
\end{array}
$$

Check. We have obtained the desired reaction.

Enthalpies of Formation (section 5.7)

5.67 (a) *Standard conditions* for enthalpy changes are usually P = 1 atm and T = 298 K. For the purpose of comparison, standard enthalpy changes, $\Delta H°$, are tabulated for reactions at these conditions.

(b) *Enthalpy of formation*, ΔH_f, is the enthalpy change that occurs when a compound is formed from its component elements.

(c) *Standard enthalpy of formation*, ΔH_f^o, is the enthalpy change that accompanies formation of one mole of a substance from elements in their standard states.

5.69 (a) $1/2\,N_2(g) + O_2(g) \rightarrow NO_2(g)$ $\Delta H_f^o = 33.84\text{ kJ}$

(b) $S(s) + 3/2\,O_2(g) \rightarrow SO_3(g)$ $\Delta H_f^o = -395.2\text{ kJ}$

(c) $Na(s) + 1/2\,Br_2(l) \rightarrow NaBr(s)$ $\Delta H_f^o = -361.4\text{ kJ}$

(d) $Pb(s) + N_2(g) + 3O_2(g) \rightarrow Pb(NO_3)_2(s)$ $\Delta H_f^o = -451.9\text{ kJ}$

5.71 *Plan.* $\Delta H_{rxn}^o = \Sigma n\Delta H_f^o \text{ (products)} - \Sigma n\Delta H_f^o \text{ (reactants)}$. Be careful with coefficients, states, and signs. *Solve.*

$\Delta H_{rxn}^o = \Delta H_f^o\,Al_2O_3(s) + 2\Delta H_f^o\,Fe(s) - \Delta H_f^o\,Fe_2O_3(s) - 2\Delta H_f^o\,Al(s)$

$\Delta H_{rxn}^o = (-1669.8\text{ kJ}) + 2(0) - (-822.16\text{ kJ}) - 2(0) = -847.6\text{ kJ}$

5.73 *Plan.* $\Delta H_{rxn}^o = \Sigma n\Delta H_f^o \text{ (products)} - \Sigma n\Delta H_f^o \text{ (reactants)}$. Be careful with coefficients, states and signs. *Solve.*

(a) $\Delta H_{rxn}^o = 2\Delta H_f^o\,SO_3(g) - 2\Delta H_f^o\,SO_2(g) - \Delta H_f^o\,O_2(g)$

$= 2(-395.2\text{ kJ}) - 2(-296.9\text{ kJ}) - 0 = -196.6\text{ kJ}$

(b) $\Delta H_{rxn}^{\circ} = \Delta H_f^{\circ}\ MgO(s) + \Delta H_f^{\circ}\ H_2O(l) - \Delta H_f^{\circ}\ Mg(OH)_2(s)$

$$= -601.8\ kJ + (-285.83\ kJ) - (-924.7\ kJ) = 37.1\ kJ$$

(c) $\Delta H_{rxn}^{\circ} = 4\Delta H_f^{\circ}\ H_2O(g) + \Delta H_f^{\circ}\ N_2(g) - \Delta H_f^{\circ}\ N_2O_4(g) - 4\Delta H_f^{\circ}\ H_2(g)$

$$= 4(-241.82\ kJ) + 0 - (9.66\ kJ) - 4(0) = -976.94\ kJ$$

(d) $\Delta H_{rxn}^{\circ} = \Delta H_f^{\circ}\ SiO_2(s) + 4\Delta H_f^{\circ}\ HCl(g) - \Delta H_f^{\circ}\ SiCl_4(l) - 2\Delta H_f^{\circ}\ H_2O(l)$

$$= -910.9\ kJ + 4(-92.30\ kJ) - (-640.1\ kJ) - 2(-285.83\ kJ) = -68.3\ kJ$$

5.75   *Analyze.* Given: combustion reaction, enthalpy of combustion, enthalpies of formation for most reactants and products. Find: enthalpy of formation for acetone.

Plan. Rearrange the expression for enthalpy of reaction to calculate the desired enthalpy of formation. *Solve.*

$$\Delta H_{rxn}^{\circ} = 3\Delta H_f^{\circ}\ CO_2(g) + 3\Delta H_f^{\circ}\ H_2O(l) - \Delta H_f^{\circ}\ C_3H_6O(l) - 4\Delta H_f^{\circ}\ O_2(g)$$

$$-1790\ kJ = 3(-393.5\ kJ) + 3(-285.83\ kJ) - \Delta H_f^{\circ}\ C_3H_6O(l) - 4(0)$$

$$\Delta H_f^{\circ}\ C_3H_6O(l) = 3(-393.5\ kJ) + 3(-285.83\ kJ) + 1790\ kJ = -248\ kJ$$

5.77   (a) $C_8H_{18}(l) + 25/2\ O_2(g) \rightarrow 8CO_2(g) + 9H_2O(g)$ $\Delta H^{\circ} = -5064.9\ kJ$

(b) $8C(s) + 9H_2(g) \rightarrow C_8H_{18}(l)$ $\Delta H_f^{\circ} = ?$

(c)   *Plan.* Follow the logic in Solution 5.75 and 5.76. *Solve.*

$$\Delta H_{rxn}^{\circ} = 8\Delta H_f^{\circ}\ CO_2(g) + 9\Delta H_f^{\circ}\ H_2O(g) - \Delta H_f^{\circ}\ C_8H_{18}(l) - 25/2\ \Delta H_f^{\circ}\ O_2(g)$$

$$-5064.9\ kJ = 8(-393.5\ kJ) + 9(-241.82\ kJ) - \Delta H_f^{\circ}\ C_8H_{18}(l) - 25/2(0)$$

$$\Delta H_f^{\circ}\ C_8H_{18}(l) = 8(-393.5\ kJ) + 9(-241.82\ kJ) + 5064.9\ kJ = -259.5\ kJ$$

5.79   (a) $C_2H_5OH(l) + 3O_2(g) \rightarrow 2CO_2(g) + 3H_2O(l)$

(b) $\Delta H_{rxn}^{\circ} = 2\Delta H_f^{\circ}\ CO_2(g) + 3\Delta H_f^{\circ}\ H_2O(g) - \Delta H_f^{\circ}\ C_2H_5OH(l) - 3\Delta H_f^{\circ}\ O_2(g)$

$$= 2(-393.5\ kJ) + 3(-241.82\ kJ) - (-277.7\ kJ) - 3(0) = -1234.76 = -1234.8\ kJ$$

(c)   *Plan.* The enthalpy of combustion of ethanol [from part (b)] is $-1234.8\ kJ/mol$. Change mol to mass using molar mass, then mass to volume using density. *Solve.*

$$\frac{-1234.76\ kJ}{mol\ C_2H_5OH} \times \frac{1\ mol\ C_2H_5OH}{46.06844\ g} \times \frac{0.789\ g}{mL} \times \frac{1000\ mL}{L} = -21{,}147 = -2.11 \times 10^4\ kJ/L$$

Check. $(1200/50) \approx 25;\ 25 \times 800 \approx 20{,}000$

(d)   *Plan.* The enthalpy of combustion corresponds to any of the molar amounts in the equation as written. Production of $-1234.76\ kJ$ also produces 2 mol CO_2. Use this relationship to calculate mass CO_2/kJ.

$$\frac{2\ mol\ CO_2}{-1234.76\ kJ} \times \frac{44.0095\ g\ CO_2}{mol} = 0.071284\ g\ CO_2/kJ\ emitted$$

Check. The negative sign associated with enthalpy indicates that energy is emitted.

Foods and Fuels (section 5.8)

5.81 (a) *Fuel value* is the amount of energy produced when 1 gram of a substance (fuel) is combusted.

(b) The fuel value of fats is 9 kcal/g and of carbohydrates is 4 kcal/g. Therefore, 5 g of fat produce 45 kcal, while 9 g of carbohydrates produce 36 kcal; 5 g of fat are a greater energy source.

(c) These products of metabolism are expelled as waste via the alimentary tract, $H_2O(l)$ primarily in urine and feces, and $CO_2(g)$ as gas.

5.83 (a) *Plan.* Calculate the Cal (kcal) due to each nutritional component of the soup, then sum. *Solve.*

$$2.5 \text{ g fat} \times \frac{38 \text{ kJ}}{1 \text{ g fat}} = 95.0 \text{ or } 0.95 \times 10^2 \text{ kJ}$$

$$14 \text{ g carbohydrates} \times \frac{17 \text{ kJ}}{1 \text{ g carbohydrate}} = 238 \text{ or } 2.4 \times 10^2 \text{ kJ}$$

$$7 \text{ g protein} \times \frac{17 \text{ kJ}}{1 \text{ g protein}} = 119 \text{ or } 1 \times 10^2 \text{ kJ}$$

total energy = 95.0 kJ + 238 kJ + 119 kJ = 452 or 5×10^2 kJ

$$452 \text{ kJ} \times \frac{1 \text{ kcal}}{4.184 \text{ kJ}} \times \frac{1 \text{ Cal}}{1 \text{ kcal}} = 108.03 \text{ or } 1 \times 10^2 \text{ Cal/serving}$$

Check. 100 Cal/serving is a reasonable result; units are correct. The data and the result have 1 sig fig.

(b) Sodium does not contribute to the calorie content of the food, because it is not metabolized by the body; it enters and leaves as Na^+.

5.85 *Plan.* g → mol → kJ → Cal *Solve.*

$$16.0 \text{ g } C_6H_{12}O_6 \times \frac{1 \text{ mol } C_6H_{12}O_6}{180.2 \text{ g } C_6H_{12}O_6} \times \frac{2812 \text{ kJ}}{\text{mol } C_6H_{12}O_6} \times \frac{1 \text{ Cal}}{4.184 \text{ kJ}} = 59.7 \text{ Cal}$$

Check. 60 Cal is a reasonable result for most of the food value in an apple.

5.87 *Plan.* Use enthalpies of formation to calculate molar heat (enthalpy) of combustion using Hess's Law. Use molar mass to calculate heat of combustion per kg of hydrocarbon. *Solve.*

Propyne: $C_3H_4(g) + 4O_2(g) \rightarrow 3CO_2(g) + 2H_2O(g)$

(a) $\Delta H_{rxn}^\circ = 3(-393.5 \text{ kJ}) + 2(-241.82 \text{ kJ}) - (185.4 \text{ kJ}) - 4(0) = -1849.5$
$$= -1850 \text{ kJ/mol } C_3H_4$$

(b) $$\frac{-1849.5 \text{ kJ}}{1 \text{ mol } C_3H_4} \times \frac{1 \text{ mol } C_3H_4}{40.065 \text{ g } C_3H_4} \times \frac{1000 \text{ g } C_3H_4}{1 \text{ kg } C_3H_4} = -4.616 \times 10^4 \text{ kJ/kg } C_3H_4$$

Propylene: $C_3H_6(g) + 9/2\, O_2(g) \rightarrow 3CO_2(g) + 3H_2O(g)$

(a) $\Delta H^{\circ}_{rxn} = 3(-393.5 \text{ kJ}) + 3(-241.82 \text{ kJ}) - (20.4 \text{ kJ}) - 9/2(0) = -1926.4$

$= -1926 \text{ kJ/mol } C_3H_6$

(b) $\dfrac{-1926.4 \text{ kJ}}{1 \text{ mol } C_3H_6} \times \dfrac{1 \text{ mol } C_3H_6}{42.080 \text{ g } C_3H_6} \times \dfrac{1000 \text{ g } C_3H_6}{1 \text{ kg } C_3H_6} = -4.578 \times 10^4 \text{ kJ/kg } C_3H_6$

Propane: $C_3H_8(g) + 5O_2(g) \rightarrow 3CO_2(g) + 4H_2O(g)$

(a) $\Delta H^{\circ}_{rxn} = 3(-393.5 \text{ kJ}) + 4(-241.82 \text{ kJ}) - (-103.8 \text{ kJ}) - 5(0) = -2044.0$

$= -2044 \text{ kJ/mol } C_3H_8$

(b) $\dfrac{-2044.0 \text{ kJ}}{1 \text{ mol } C_3H_8} \times \dfrac{1 \text{ mol } C_3H_8}{44.096 \text{ g } C_3H_8} \times \dfrac{1000 \text{ g } C_3H_8}{1 \text{ kg } C_3H_8} = -4.635 \times 10^4 \text{ kJ/kg } C_3H_8$

(c)   These three substances yield nearly identical quantities of heat per unit mass, but propane is marginally higher than the other two.

5.89   *Analyze/Plan.* Given population, Cal/person/day and kJ/mol glucose, calculate kg glucose/yr. Calculate kJ/yr, then kg/yr. 1 billion = 1×10^9. 365 day = 1 yr. 1 Cal = 1 kcal, 4.184 kJ = 1 kcal = 1 Cal. *Solve.*

$6.8 \times 10^9 \text{ persons} \times \dfrac{1500 \text{ Cal}}{\text{person} - \text{day}} \times \dfrac{365 \text{ day}}{1 \text{ yr}} \times \dfrac{4.184 \text{ kJ}}{1 \text{ Cal}} = 1.5577 \times 10^{16} = 1.6 \times 10^{16} \text{ kJ/yr}$

$\dfrac{1.5577 \times 10^{16} \text{ kJ}}{\text{yr}} \times \dfrac{1 \text{ mol } C_6H_{12}O_6}{2803 \text{ kJ}} \times \dfrac{180.2 \text{ g } C_6H_{12}O_6}{1 \text{ mol } C_6H_{12}O_6} \times \dfrac{1 \text{ kg}}{1000 \text{ g}} = 1.0 \times 10^{12} \text{ kg } C_6H_{12}O_6 / \text{yr}$

Check. 1×10^{12} kg is 1 *trillion* kg of glucose.

Additional Exercises

5.91   (a) $\text{mi/hr} \rightarrow \text{m/s}$

$1050 \dfrac{\text{mi}}{\text{hr}} \times \dfrac{1.6093 \text{ km}}{1 \text{ mi}} \times \dfrac{1000 \text{ m}}{1 \text{ km}} \times \dfrac{1 \text{ hr}}{3600 \text{ s}} = 469.38 = 469.4 \text{ m/s}$

(b)   Find the mass of one N_2 molecule in kg.

$\dfrac{28.0134 \text{ g } N_2}{1 \text{ mol}} \times \dfrac{1 \text{ mol}}{6.022 \times 10^{23} \text{ molecules}} \times \dfrac{1 \text{ kg}}{1000 \text{ g}} = 4.6518 \times 10^{-26}$

$= 4.652 \times 10^{-26} \text{ kg}$

$E_k = 1/2\, mv^2 = 1/2 \times 4.6518 \times 10^{-26} \text{ kg} \times (469.38 \text{ m/s})^2$

$= 5.1244 \times 10^{-21} \dfrac{\text{kg} \cdot \text{m}^2}{\text{s}^2} = 5.124 \times 10^{-21} \text{ J}$

(c) $\dfrac{5.1244 \times 10^{-21} \text{ J}}{\text{molecule}} \times \dfrac{6.022 \times 10^{23} \text{ molecules}}{1 \text{ mol}} = 3086 \text{ J/mol} = 3.086 \text{ kJ/mol}$

5.93   Like the combustion of $H_2(g)$ and $O_2(g)$ described in Section 5.4, the reaction that inflates airbags is spontaneous after initiation. Spontaneous reactions are usually exothermic, $-\Delta H$. The airbag reaction occurs at constant atmospheric pressure, $\Delta H = q_p$;

both are likely to be large and negative. When the bag inflates, work is done by the system on the surroundings, so the sign of w is negative.

5.97 $\Delta E = q + w = +38.95 \text{ kJ} - 2.47 \text{ kJ} = +36.48 \text{ kJ}$

$\Delta H = q_p = +38.95 \text{ kJ}$

5.102   (a) $Mg(s) + 2H_2O(l) \rightarrow Mg(OH)_2(s) + H_2(g)$

$\Delta H^\circ_{rxn} = \Delta H^\circ_f \, Mg(OH)_2(s) + \Delta H^\circ_f \, H_2(g) - 2\Delta H^\circ_f \, H_2O(l) - \Delta H^\circ_f \, Mg(s)$

$= -924.7 \text{ kJ} + 0 - 2(-285.83 \text{ kJ}) - 0 = -353.04 = -353.0 \text{ kJ}$

(b)   Use the specific heat of water, 4.184 J/g-°C, to calculate the energy required to heat the water. Use the density of water at 25°C to calculate the mass of H_2O to be heated. (The change in density of H_2O going from 21°C to 79°C does not substantially affect the strategy of the exercise.) Then use the 'heat stoichiometry' in (a) to calculate mass of Mg(s) needed.

$75 \text{ mL} \times \dfrac{0.997 \text{ g } H_2O}{mL} \times \dfrac{4.184 \text{ J}}{g\text{-}^\circ C} \times 58^\circ C \times \dfrac{1 \text{ kJ}}{1000 \text{ J}} = 18.146 \text{ kJ} = 18 \text{ kJ required}$

$18.146 \text{ kJ} \times \dfrac{1 \text{ mol Mg}}{353.04 \text{ kJ}} \times \dfrac{24.305 \text{ g Mg}}{1 \text{ mol Mg}} = 1.249 \text{ g} = 1.2 \text{ g Mg needed}$

5.106   (a) $3C_2H_2(g) \rightarrow C_6H_6(l)$

$\Delta H^\circ_{rxn} = \Delta H^\circ_f \, C_6H_6(l) - 3\Delta H^\circ_f \, C_2H_2(g) = 49.0 \text{ kJ} - 3(226.77 \text{ kJ}) = -631.31 = -631.3 \text{ kJ}$

(b)   Since the reaction is exothermic (ΔH is negative), the reactant, 3 moles of $C_2H_2(g)$, has more enthalpy than the product, 1 mole of $C_6H_6(l)$.

(c)   The fuel value of a substance is the amount of heat (kJ) produced when 1 gram of the substance is burned. Calculate the molar heat of combustion (kJ/mol) and use this to find kJ/g of fuel.

$C_2H_2(g) + 5/2 \, O_2(g) \rightarrow 2CO_2(g) + H_2O(l)$

$\Delta H^\circ_{rxn} = 2\Delta H^\circ_f \, CO_2(g) + \Delta H^\circ_f \, H_2O(l) - \Delta H^\circ_f \, C_2H_2(g) - 5/2 \, \Delta H^\circ_f \, O_2(g)$

$= 2(-393.5 \text{ kJ}) + (-285.83 \text{ kJ}) - 226.77 \text{ kJ} - 5/2 \, (0) = -1299.6 \text{ kJ/mol } C_2H_2$

$\dfrac{-1299.6 \text{ kJ}}{1 \text{ mol } C_2H_2} \times \dfrac{1 \text{ mol } C_2H_2}{26.036 \text{ g } C_2H_2} = 49.916 = 50 \text{ kJ/g } C_2H_2$

$C_6H_6(l) + 15/2 \, O_2(g) \rightarrow 6CO_2(g) + 3H_2O(l)$

$\Delta H^\circ_{rxn} = 6\Delta H^\circ_f \, CO_2(g) + 3\Delta H^\circ_f \, H_2O(l) - \Delta H^\circ_f \, C_6H_6(l) - 15/2 \, \Delta H^\circ_f \, O_2(g)$

$= 6(-393.5 \text{ kJ}) + 3(-285.83 \text{ kJ}) - 49.0 \text{ kJ} - 15/2 \, (0) = -3267.5 \text{ kJ/mol } C_6H_6$

$\dfrac{-3267.5 \text{ kJ}}{1 \text{ mol } C_6H_6} \times \dfrac{1 \text{ mol } C_6H_6}{78.114 \text{ g } C_6H_6} = 41.830 = 42 \text{ kJ/g } C_6H_6$

5.109 $\Delta E_p = m\,g\,d$. Be careful with units. $1\,J = 1\,kg\text{-}m^2/s^2$

$$200\,lb \times \frac{1\,kg}{2.205\,lb} \times \frac{9.81\,m}{s^2} \times \frac{45\,ft}{time} \times \frac{1\,yd}{3\,ft} \times \frac{1\,m}{1.0936\,yd} \times 20\,times$$

$$= 2.441\times10^5\,kg\text{-}m^2/s^2 = 2.441\times10^5\,J = 2.4\times10^2\,kJ$$

$1\,Cal = 1\,kcal = 4.184\,kJ$

$$2.441\times10^2\,kJ \times \frac{1\,Cal}{4.184\,kJ} = 58.34 = 58\,Cal$$

No, if all work is used to increase the man's potential energy, 20 rounds of stair-climbing will not compensate for one extra order of 245 Cal fries. In fact, more than 58 Cal of work will be required to climb the stairs, because some energy is required to move limbs and some energy will be lost as heat (see Solution 5.96).

Integrative Exercises

5.112   (a) $CH_4(g) + 2O_2(g) \rightarrow CO_2(g) + 2H_2O(l)$

        $\Delta H^\circ = \Delta H_f^\circ\,CO_2(g) + 2\Delta H_f^\circ\,H_2O(l) - \Delta H_f^\circ\,CH_4(g) - 2\Delta H_f^\circ\,O_2(g)$

          $= -393.5\,kJ + 2(-285.83\,kJ) - (-74.8\,kJ) - 2(0) = -890.36 = -890.4\,kJ/mol\,CH_4$

$$\frac{-890.36\,kJ}{mol\,CH_4} \times \frac{1000\,J}{1\,kJ} \times \frac{1\,mol}{6.022\times10^{23}\,molecules\,CH_4} = 1.4785\times10^{-18}$$

$$= 1.479\times10^{-18}\,J/molecule$$

    (b) $1\,eV = 96.485\,kJ/mol$

$$8\,keV \times \frac{1000\,eV}{1\,keV} \times \frac{96.485\,kJ}{eV\text{-}mol} \times \frac{1\,mol}{6.022\times10^{23}} \times \frac{1000\,J}{kJ} = 1.282\times10^{-15} = 1\times10^{-15}\,J/X\text{-ray}$$

The X-ray has approximately 1000 times more energy than is produced by the combustion of 1 molecule of $CH_4(g)$.

5.114   (a) $\Delta H^\circ = \Delta H_f^\circ\,NaNO_3(aq) + \Delta H_f^\circ\,H_2O(l) - \Delta H_f^\circ\,HNO_3(aq) - \Delta H_f^\circ\,NaOH(aq)$

        $\Delta H^\circ = -446.2\,kJ - 285.83\,kJ - (-206.6\,kJ) - (-469.6\,kJ) = -55.8\,kJ$

        $\Delta H^\circ = \Delta H_f^\circ\,NaCl(aq) + \Delta H_f^\circ\,H_2O(l) - \Delta H_f^\circ\,HCl(aq) - \Delta H_f^\circ\,NaOH(aq)$

        $\Delta H^\circ = -407.1\,kJ - 285.83\,kJ - (-167.2\,kJ) - (-469.6\,kJ) = -56.1\,kJ$

        $\Delta H^\circ = \Delta H_f^\circ\,NH_3(aq) + \Delta H_f^\circ\,Na^+(aq) + \Delta H_f^\circ\,H_2O(l) - \Delta H_f^\circ\,NH_4^+(aq) - \Delta H_f^\circ\,NaOH(aq)$

          $= -80.29\,kJ - 240.1\,kJ - 285.83\,kJ - (-132.5\,kJ) - (-469.6\,kJ) = -4.1\,kJ$

    (b) $H^+(aq) + OH^-(aq) \rightarrow H_2O(l)$

(c)   The $\Delta H°$ values for the first two reactions are nearly identical, -55.8 kJ and -56.1 kJ. The spectator ions by definition do not change during the course of a reaction, so $\Delta H°$ is the enthalpy change for the net ionic equation. Since the first two reactions have the same net ionic equation, it is not surprising that they have the same $\Delta H°$.

(d)   Strong acids are more likely than weak acids to donate H^+. The neutralizations of the two strong acids are energetically favorable, while the neutralization of $NH_4^+(aq)$ is significangly less favorable. $NH_4^+(aq)$ is probably a weak acid.

5.116   (a) $AgNO_3(aq) + NaCl(aq) \rightarrow NaNO_3(aq) + AgCl(s)$

net ionic equation: $Ag^+(aq) + Cl^-(aq) \rightarrow AgCl(s)$

$\Delta H° = \Delta H_f° \, AgCl(s) - \Delta H_f° \, Ag^+(aq) - \Delta H_f° Cl^-(aq)$

$\Delta H° = -127.0 \text{ kJ} - (105.90 \text{ kJ}) - (-167.2 \text{ kJ}) = -65.7 \text{ kJ}$

(b) $\Delta H°$ for the complete molecular equation will be the same as $\Delta H°$ for the net ionic equation. $Na^+(aq)$ and $NO_3^-(aq)$ are spectator ions; they appear on both sides of the chemical equation. Since the overall enthalpy change is the enthalpy of the products minus the enthalpy of the reactants, the contributions of the spectator ions cancel.

(c) $\Delta H° = \Delta H_f° \, NaNO_3(aq) + \Delta H_f° \, AgCl(s) - \Delta H_f° \, AgNO_3(aq) - \Delta H_f° \, NaCl(aq)$

$\Delta H_f° \, AgNO_3(aq) = \Delta H_f° \, NaNO_3(aq) + \Delta H_f° \, AgCl(s) - \Delta H_f° \, NaCl(aq) - \Delta H°$

$\Delta H_f° \, AgNO_3(aq) = -446.2 \text{ kJ} + (-127.0 \text{ kJ}) - (-407.1 \text{ kJ}) - (-65.7 \text{ kJ})$

$\Delta H_f° \, AgNO_3(aq) = -100.4 \text{ kJ/mol}$

6 Electronic Structure of Atoms

Visualizing Concepts

6.2   Given: 2450 MHz radiation. Hz = s^{-1}, unit of frequency. M = 1×10^6; 2450×10^6 Hz = 2.45×10^9 Hz = 2.45×10^9 s^{-1}.

    (a)   Find 2.45×10^9 s^{-1} on the frequency axis of Figure 6.4. The wavelength that corresponds to this frequency is approximately $1 \times 10^{-1} = 0.1$ m or 10 cm.

    (b)   No, visible radiation has wavelengths of 4×10^{-7} to 7×10^{-7} m, much shorter than 0.1 m.

    (c)   Energy and wavelength are inversely proportional. Photons of the longer 0.1 m radiation have less energy than visible photons.

    (d)   Radiation of 0.1 m is in the low energy portion of the microwave region. The appliance is probably a microwave oven. (The highest FM radio frequency is 108 MHz, so the device is not an FM radio.)

6.5   (a)   Increase. The rainbow has shorter wavelength blue light on the inside and longer wavelength red light on the outside. (See Figure 6.4.)

    (b)   Decrease. Wavelength and frequency are inversely related. Wavelength increases so frequency decreases going from the inside to the outside of the rainbow.

    (c)   The light from the hydrogen discharge tube is not a continuous spectrum, so not all visible wavelengths will be in our "hydrogen discharge rainbow." Starting with the shortest wavelengths, it will be violet followed by blue-violet and blue-green on the inside. Then there will be a gap, and finally a red band. (See the H spectrum in Figure 6.11.)

6.8   (a)   1

    (b)   p (dumbbell shape, node at the nucleus)

    (c)   The lobes in the contour representation would extend farther along the y axis. A larger principle quantum number (4p vs. 3p) implies a greater average distance from the nucleus for electrons occupying the orbital.

The Wave Nature of Light (section 6.1)

6.11   (a)   Meters (m)   (b)   1/seconds (s^{-1})   (c)   meters/second ($m \cdot s^{-1}$ or m/s)

6.13   (a)   True.

    (b)   False. Ultraviolet light has shorter wavelengths than visible light. [See Solution 6.12(b).]

(c)　　False. X-rays travel at the same speed as microwaves. (X-rays and microwaves are both electromagnetic radiation.)

(d)　　False. Electromagnetic radiation and sound waves travel at different speeds. (Sound is not a form of electromagnetic radiation.)

6.15　　*Analyze/Plan*. Use the electromagnetic spectrum in Figure 6.4 to determine the wavelength of each type of radiation; put them in order from shortest to longest wavelength. *Solve*.

Wavelength of X-rays < ultraviolet < green light < red light < infrared < radio waves

Check. These types of radiation should read from left to right on Figure 6.4

6.17　　*Analyze/Plan*. These questions involve relationships between wavelength, frequency, and the speed of light. Manipulate the equation $\nu = c/\lambda$ to obtain the desired quantities, paying attention to units.　　*Solve*.

(a)　　$\nu = c/\lambda$; $\dfrac{2.998 \times 10^8 \text{ m}}{\text{s}} \times \dfrac{1}{10 \text{ } \mu m} \times \dfrac{1 \text{ } \mu m}{1 \times 10^{-6} \text{ m}} = 3.0 \times 10^{13} \text{ s}^{-1}$

(b)　　$\lambda = c/\nu$; $\dfrac{2.998 \times 10^8 \text{ m}}{\text{s}} \times \dfrac{1 \text{ s}}{5.50 \times 10^{14}} = 5.45 \times 10^{-7} \text{ m}$　(545 nm)

(c)　　The radiation in (b) is in the visible region and is "visible" to humans.

　　　　The 10 μm (1×10^{-5} m) radiation in (a) is in the infrared region and is not visible.

(d)　　$50.0 \text{ } \mu s \times \dfrac{1 \text{ s}}{1 \times 10^6 \text{ } \mu s} \times \dfrac{2.998 \times 10^8 \text{ m}}{\text{s}} = 1.50 \times 10^4 \text{ m}$

Check. Confirm that powers of 10 make sense and units are correct.

6.19　　*Analyze/Plan*. $\nu = c/\lambda$; change nm $\rightarrow$ m.

Solve. $\nu = c/\lambda$; $\dfrac{2.998 \times 10^8 \text{ m}}{1 \text{ s}} \times \dfrac{1}{532 \text{ nm}} \times \dfrac{1 \text{ nm}}{1 \times 10^{-9} \text{ m}} = 5.64 \times 10^{14} \text{ s}^{-1}$

The color is green.

Check. $(3000 \times 10^5 / 500 \times 10^{-9}) = 6 \times 10^{14} \text{ s}^{-1}$; units are correct.

Quantized Energy and Photons (section 6.2)

6.21　　Quantization means that energy changes can only happen in certain allowed increments. If the human growth quantum is one-foot, growth occurs instantaneously in one-foot increments. That is, a child experiences growth spurts of one-foot; her height changes by one-foot increments.

6.23　　*Analyze/Plan*. These questions deal with the relationships between energy, wavelength, and frequency. Use the relationships $E = h\nu = hc/\lambda$ to calculate the desired quantities. Pay attention to units.　　*Solve*.

(a)　　$E = h\nu = 6.626 \times 10^{-34} \text{ J-s} \times 6.75 \times 10^{12} \text{ s}^{-1} = 4.47 \times 10^{-21} \text{ J}$

(b)　　$\Delta E = h\nu = \dfrac{hc}{\lambda} = \dfrac{6.626 \times 10^{-34} \text{ J-s} \times 2.998 \times 10^8 \text{ m/s}}{322 \text{ nm}} \times \dfrac{1 \text{ nm}}{1 \times 10^{-9} \text{ m}} = 6.17 \times 10^{-19} \text{ J}$

(c) $\lambda = hc/\Delta E = \dfrac{6.626 \times 10^{-34} \text{ J-s}}{2.87 \times 10^{-18} \text{ J}} \times \dfrac{2.998 \times 10^8 \text{ m}}{1 \text{ s}} = 6.92 \times 10^{-8} \text{ m} = 69.2 \text{ nm}$

6.25   *Analyze/Plan.* Use $E = hc/\lambda$; pay close attention to units.   *Solve.*

(a) $E = hc/\lambda = 6.626 \times 10^{-34} \text{ J-s} \times \dfrac{2.998 \times 10^8 \text{ m}}{1 \text{ s}} \times \dfrac{1}{3.3 \,\mu\text{m}} \times \dfrac{1 \,\mu\text{m}}{1 \times 10^{-6} \text{ m}}$

$$= 6.0 \times 10^{-20} \text{ J}$$

$E = hc/\lambda = 6.626 \times 10^{-34} \text{ J-s} \times \dfrac{2.998 \times 10^8 \text{ m}}{1 \text{ s}} \times \dfrac{1}{0.154 \text{ nm}} \times \dfrac{1 \text{ nm}}{1 \times 10^{-9} \text{ m}}$

$$= 1.29 \times 10^{-15} \text{ J}$$

Check. $(6.6 \times 3/3.3) \times (10^{-34} \times 10^8/10^{-6}) \approx 6 \times 10^{-20} \text{ J}$

$(6.6 \times 3/0.15) \times (10^{-34} \times 10^8/10^{-9}) \approx 120 \times 10^{-17} \approx 1.2 \times 10^{-15} \text{ J}$

The results are reasonable. We expect the longer wavelength 3.3 μm radiation to have the lower energy.

(b)   The 3.3 μm photon is in the infrared and the 0.154 nm (1.54×10^{-10} m) photon is in the X-ray region; the X-ray photon has the greater energy.

6.27   *Analyze/Plan.* Use $E = hc/\lambda$ to calculate J/photon; Avogadro's number to calculate J/mol; photon/J [the result from part (a)] to calculate photons in 1.00 mJ. Pay attention to units. *Solve.*

(a) $E_{photon} = hc/\lambda = \dfrac{6.626 \times 10^{-34} \text{ J-s}}{325 \times 10^{-9} \text{ m}} \times \dfrac{2.998 \times 10^8 \text{ m}}{\text{s}} = 6.1122 \times 10^{-19}$

$$= 6.11 \times 10^{-19} \text{ J/photon}$$

(b) $\dfrac{6.1122 \times 10^{-19} \text{ J}}{1 \text{ photon}} \times \dfrac{6.022 \times 10^{23} \text{ photons}}{1 \text{ mol}} = 3.68 \times 10^5 \text{ J/mol} = 368 \text{ kJ/mol}$

(c) $\dfrac{1 \text{ photon}}{6.1122 \times 10^{-19} \text{ J}} \times 1.00 \text{ mJ} \times \dfrac{1 \times 10^{-3} \text{ J}}{1 \text{ mJ}} = 1.64 \times 10^{15} \text{ photons}$

Check. Powers of 10 (orders of magnitude) and units are correct.

(d)   If the energy of one 325 nm photon breaks exactly one bond, one mol of photons break 1 mol of bonds. The average bond energy in kJ/mol is the energy of 1 mol of photons (from part b) 368 kJ/mol.

6.29   *Analyze/Plan.* $E = hc/\lambda$ gives J/photon. Use this result with J/s (given) to calculate photons/s. *Solve.*

(a)   The $\sim 1 \times 10^{-6}$ m radiation is infrared but very near the visible edge.

(b) $E_{photon} = hc/\lambda = \dfrac{6.626 \times 10^{-34} \text{ J-s}}{987 \times 10^{-9} \text{ m}} \times \dfrac{2.998 \times 10^8 \text{ m}}{1 \text{ s}} = 2.0126 \times 10^{-19}$

$$= 2.01 \times 10^{-19} \text{ J/photon}$$

$\dfrac{0.52 \text{ J}}{32 \text{ s}} \times \dfrac{1 \text{ photon}}{2.0126 \times 10^{-19} \text{ J}} = 8.1 \times 10^{16} \text{ photons/s}$

Check. $(7 \times 3/1000) \times (10^{-34} \times 10^8/10^{-9}) \approx 21 \times 10^{-20} \approx 2.1 \times 10^{-19} \text{ J/photon}$

$(0.5/30/2) \times (1/10^{-19}) = 0.008 \times 10^{19} = 8 \times 10^{16} \text{ photons/s}$

Units are correct; powers of 10 are reasonable.

6.31   *Analyze/Plan.* Use $E = h\nu$ and $\nu = c/\lambda$. Calculate the desired characteristics of the photons. Assume 1 photon interacts with 1 electron. Compare E_{min} and E_{120} to calculate maximum kinetic energy of the emitted electron.   *Solve.*

(a) $E = h\nu = 6.626 \times 10^{-34}\ \text{J·s} \times 1.09 \times 10^{15}\ \text{s}^{-1} = 7.22 \times 10^{-19}\ \text{J}$

(b) $\lambda = c/\nu = \dfrac{2.998 \times 10^8\ \text{m}}{1\ \text{s}} \times \dfrac{1\ \text{s}}{1.09 \times 10^{15}} = 2.75 \times 10^{-7}\ \text{m} = 275\ \text{nm}$

(c) $E_{120} = hc/\lambda = 6.626 \times 10^{-34}\ \text{J·s} \times \dfrac{2.998 \times 10^8\ \text{m}}{1\ \text{s}} \times \dfrac{1}{120\ \text{nm}} \times \dfrac{1\ \text{nm}}{1 \times 10^{-9}\ \text{m}}$

$$= 1.655 \times 10^{-18} = 1.66 \times 10^{-18}\ \text{J}$$

The excess energy of the 120 nm photon is converted into the kinetic energy of the emitted electron.

$E_k = E_{120} - E_{min} = 16.55 \times 10^{-19}\ \text{J} - 7.22 \times 10^{-19}\ \text{J} = 9.3 \times 10^{-19}\ \text{J/electron}$

Check. E_{120} must be greater than E_{min} in order for the photon to impart kinetic energy to the emitted electron. Our calculations are consistent with this requirement.

Bohr's Model; Matter Waves (sections 6.3 and 6.4)

6.33   When applied to atoms, the notion of quantized energies means that only certain energies can be gained or lost, only certain values of ΔE are allowed. The allowed values of ΔE are represented by the lines in the emission spectra of excited atoms.

6.35   *Analyze/Plan.* An isolated electron is assigned an energy of zero; the closer the electron comes to the nucleus, the more negative its energy. Thus, as an electron moves closer to the nucleus, the energy of the electron decreases and the excess energy is emitted. Conversely, as an electron moves further from the nucleus, the energy of the electron increases and energy must be absorbed.   *Solve.*

(a)   As the principle quantum number decreases, the electron moves toward the nucleus and energy is emitted.

(b)   An increase in the radius of the orbit means the electron moves away from the nucleus; energy is absorbed.

(c)   An isolated electron is assigned an energy of zero. As the electron moves to the $n = 3$ state closer to the H^+ nucleus, its energy becomes more negative (decreases) and energy is emitted.

6.37   *Analyze/Plan.* Equation 6.5: $E = (-2.18 \times 10^{-18}\ \text{J})(1/n^2)$.   *Solve.*

(a) $E_2 = -2.18 \times 10^{-18}\ \text{J}/(2)^2 = -5.45 \times 10^{-19}\ \text{J}$

$E_6 = -2.18 \times 10^{-18}\ \text{J}/(6)^2 = -6.0556 \times 10^{-20} = -0.606 \times 10^{-19}\ \text{J}$

$\Delta E = E_2 - E_6 = (-5.45 \times 10^{-19}\ \text{J}) - (-0.606 \times 10^{-19}\ \text{J})$

$\qquad = -4.844 \times 10^{-19}\ \text{J} = -4.84 \times 10^{-19}\ \text{J}$

$\lambda = hc/\Delta E = \dfrac{6.626 \times 10^{-34}\ \text{J·s}}{4.844 \times 10^{-19}\ \text{J}} \times \dfrac{2.998 \times 10^8\ \text{m}}{\text{s}} = 4.10 \times 10^{-7}\ \text{m} = 410\ \text{nm}$

(b) The visible range is 400–700 nm, so this line is visible; the observed color is violet.

Check. We expect E_6 to be a more positive (or less negative) than E_2, and it is. ΔE is negative, which indicates emission. The orders of magnitude make sense and units are correct.

6.39 (a) Only lines with $n_f = 2$ represent ΔE values and wavelengths that lie in the visible portion of the spectrum. Lines with $n_f = 1$ have larger ΔE values and shorter wavelengths that lie in the ultraviolet. Lines with $n_f > 2$ have smaller ΔE values and lie in the lower energy longer wavelength regions of the electromagnetic spectrum.

(b) *Analyze/Plan.* Use Equation 6.7 to calculate ΔE, then $\lambda = hc/\Delta E$.   *Solve.*

$$n_i = 3, n_f = 2; \quad \Delta E = -2.18 \times 10^{-18} \text{ J} \left[\frac{1}{n_f^2} - \frac{1}{n_i^2} \right] = -2.18 \times 10^{-18} \text{ J} (1/4 - 1/9)$$

$$\lambda = hc/E = \frac{6.626 \times 10^{-34} \text{ J·s} \times 2.998 \times 10^8 \text{ m/s}}{-2.18 \times 10^{-18} \text{ J} (1/4 - 1/9)} = 6.56 \times 10^{-7} \text{ m}$$

This is the red line at 656 nm.

$$n_i = 4, n_f = 2; \quad \lambda = hc/E = \frac{6.626 \times 10^{-34} \text{ J·s} \times 2.998 \times 10^8 \text{ m/s}}{-2.18 \times 10^{-18} \text{ J} (1/4 - 1/16)} = 4.86 \times 10^{-7} \text{ m}$$

This is the blue-green line at 486 nm.

$$n_i = 5, n_f = 2; \quad \lambda = hc/E = \frac{6.626 \times 10^{-34} \text{ J·s} \times 2.998 \times 10^8 \text{ m/s}}{-2.18 \times 10^{-18} \text{ J} (1/4 - 1/25)} = 4.34 \times 10^{-7} \text{ m}$$

This is the blue-violet line at 434 nm.

Check. The calculated wavelengths correspond well to three lines in the H emission spectrum in Figure 6.11, so the results are sensible.

6.41 (a) $93.8 \text{ nm} \times \dfrac{1 \times 10^{-9} \text{ m}}{1 \text{ nm}} = 9.38 \times 10^{-8} \text{ m};$ this line is in the ultraviolet region.

(b) *Analyze/Plan.* Only lines with $n_f = 1$ have a large enough ΔE to lie in the ultraviolet region (see Solutions 6.39 and 6.40). Solve Equation 6.7 for n_i, recalling that ΔE is negative for emission.   *Solve.*

$$\frac{-hc}{\lambda} = -2.18 \times 10^{-18} \text{ J} \left[\frac{1}{n_f^2} - \frac{1}{n_i^2} \right]; \quad \frac{hc}{\lambda(2.18 \times 10^{-18} \text{ J})} = \left[1 - \frac{1}{n_i^2} \right]$$

$$-\frac{1}{n_i^2} = \left[\frac{hc}{\lambda(2.18 \times 10^{-18} \text{ J})} - 1 \right]; \quad \frac{1}{n_i^2} = \left[1 - \frac{hc}{\lambda(2.18 \times 10^{-18} \text{ J})} \right]$$

$$n_i^2 = \left[1 - \frac{hc}{\lambda(2.18 \times 10^{-18} \text{ J})} \right]^{-1}; \quad n_i = \left[1 - \frac{hc}{\lambda(2.18 \times 10^{-18} \text{ J})} \right]^{-1/2}$$

$$n_i = \left(1 - \frac{6.626 \times 10^{-34} \text{ J·s} \times 2.998 \times 10^8 \text{ m/s}}{9.38 \times 10^{-8} \text{ m} \times 2.18 \times 10^{-18} \text{ J}} \right)^{-1/2} = 6 \text{ (n values must be integers)}$$

$n_i = 6, n_f = 1$

Check. From Solution 6.40, we know that $n_i > 4$ for $\lambda = 93.8$ nm. The calculated result is close to 6, so the answer is reasonable.

6.43 *Analyze/Plan.* $\lambda = \dfrac{h}{mv}$; $1 \text{ J} = \dfrac{1 \text{ kg} \cdot \text{m}^2}{\text{s}^2}$;

Change mass to kg and velocity to m/s in each case.   *Solve.*

(a) $\dfrac{50 \text{ km}}{1 \text{ hr}} \times \dfrac{1000 \text{ m}}{1 \text{ km}} \times \dfrac{1 \text{ hr}}{60 \text{ min}} \times \dfrac{1 \text{ min}}{60 \text{ s}} = 13.89 = 14$ m/s

$\lambda = \dfrac{6.626 \times 10^{-34} \text{ kg} \cdot \text{m}^2 \cdot \text{s}}{1 \text{ s}^2} \times \dfrac{1}{85 \text{ kg}} \times \dfrac{1 \text{ s}}{13.89 \text{ m}} = 5.6 \times 10^{-37}$ m

(b) $10.0 \text{ g} \times \dfrac{1 \text{ kg}}{1000 \text{ g}} = 0.0100$ kg

$\lambda = \dfrac{6.626 \times 10^{-34} \text{ kg} \cdot \text{m}^2 \cdot \text{s}}{1 \text{ s}^2} \times \dfrac{1}{0.0100 \text{ kg}} \times \dfrac{1 \text{ s}}{250 \text{ m}} = 2.65 \times 10^{-34}$ m

(c) We need to calculate the mass of a single Li atom in kg.

$\dfrac{6.94 \text{ g Li}}{1 \text{ mol Li}} \times \dfrac{1 \text{ kg}}{1000 \text{ g}} \times \dfrac{1 \text{ mol}}{6.022 \times 10^{23} \text{ Li atoms}} = 1.152 \times 10^{-26} = 1.15 \times 10^{-26}$ kg

$\lambda = \dfrac{6.626 \times 10^{-34} \text{ kg} \cdot \text{m}^2 \cdot \text{s}}{1 \text{ s}^2} \times \dfrac{1}{1.152 \times 10^{-26} \text{ kg}} \times \dfrac{1 \text{ s}}{2.5 \times 10^5 \text{ m}} = 2.3 \times 10^{-13}$ m

(d) Calculate the mass of a single O_3 molecule in kg.

$\dfrac{48.00 \text{ g O}_3}{1 \text{ mol O}_3} \times \dfrac{1 \text{ kg}}{1000 \text{ g}} \times \dfrac{1 \text{ mol}}{6.022 \times 10^{23} \text{ O}_3 \text{ molecules}} = 7.971 \times 10^{-26}$

$= 7.97 \times 10^{-26}$ kg

$\lambda = \dfrac{6.626 \times 10^{-34} \text{ kg} \cdot \text{m}^2 \cdot \text{s}}{1 \text{ s}^2} \times \dfrac{1}{7.971 \times 10^{-26} \text{ kg}} \times \dfrac{1 \text{ s}}{550 \text{ m}}$

$= 1.51 \times 10^{-11}$ m (15 pm)

6.45 *Analyze/Plan.* Use $v = h/m\lambda$; change wavelength to meters and mass of neutron (back inside cover) to kg.   *Solve.*

$\lambda = 0.955 \text{ Å} \times \dfrac{1 \times 10^{-10} \text{ m}}{1 \text{ Å}} = 0.955 \times 10^{-10}$ m; $m = 1.6749 \times 10^{-27}$ kg

$v = \dfrac{6.626 \times 10^{-34} \text{ kg} \cdot \text{m}^2 \cdot \text{s}}{1 \text{ s}^2} \times \dfrac{1}{1.6749 \times 10^{-27} \text{ kg}} \times \dfrac{1}{0.955 \times 10^{-10} \text{ m}} = 4.14 \times 10^3$ m/s

Check. $(6.6/1.6/1) \times (10^{-34}/10^{-27}/10^{-10}) \approx 4 \times 10^3$ m/s

6.47 *Analyze/Plan.* Use $\Delta x \geq h/4\pi \, m \, \Delta v$, paying attention to appropriate units. Note that the uncertainty in speed of the particle (Δv) is important, rather than the speed itself. *Solve.*

(a) $m = 1.50 \text{ mg} \times \dfrac{1 \text{ g}}{1000 \text{ mg}} \times \dfrac{1 \text{ kg}}{1000 \text{ g}} = 1.50 \times 10^{-6}$ kg; $\Delta v = 0.01$ m/s

$\Delta x \geq \dfrac{6.626 \times 10^{-34} \text{ J} \cdot \text{s}}{4\pi(1.50 \times 10^{-6} \text{ kg})(0.01 \text{ m/s})} \geq 3.52 \times 10^{-27} = 4 \times 10^{-27}$ m

(b) $m = 1.673 \times 10^{-24}$ g $= 1.673 \times 10^{-27}$ kg; $\Delta v = 0.01 \times 10^4$ m/s

$$\Delta x \geq = \frac{6.626 \times 10^{-34} \text{ J-s}}{4\pi(1.673 \times 10^{-27} \text{ kg})(0.01 \times 10^4 \text{ m/s})} \geq 3 \times 10^{-10} \text{ m}$$

Check. The more massive particle in (a) has a much smaller uncertainty in position.

Quantum Mechanics and Atomic Orbitals (sections 6.5 and 6.6)

6.49   (a)   The uncertainty principle states that there is a limit to how precisely we can simultaneously know the position and momentum (related to energy) of an electron. The Bohr model states that electrons move about the nucleus in precisely circular orbits of known radius; each permitted orbit has an allowed energy associated with it. Thus, according to the Bohr model, we can know the exact distance of an electron from the nucleus and its energy. This violates the uncertainty principle.

(b)   De Broglie stated that electrons demonstrate the properties of both particles and waves, that each particle has a wave associated with it. A wave function is the mathematical description of the matter wave of an electron.

(c)   Although we cannot predict the exact location of an electron in an allowed energy state, we can determine the likelihood or probability of finding an electron at a particular position (or within a particular volume). This statistical knowledge of electron location is called the *probability density* or electron density and is a function of ψ^2, the square of the wave function ψ.

6.51   (a)   The possible values of l are $(n-1)$ to 0. $n = 4$, $l = 3, 2, 1, 0$

(b)   The possible values of m_l are $-l$ to $+l$. $l = 2$, $m_l = -2, -1, 0, 1, 2$

(c)   Since the value of m_l is less than or equal to the value of l, $m_l = 2$ must have an l-value greater than or equal to 2. In terms of elements that have been observed, the possibilities are 2, 3 and 4.

6.53   (a)   3p: $n = 3$, $l = 1$ (b)   2s: $n = 2$, $l = 0$

(c)   4f: $n = 4$, $l = 3$ (d)   5d: $n = 5$, $l = 2$

6.55   Impossible: (a) 1p, only $l = 0$ is possible for $n = 1$; (d) 2d, for $n = 2$, $l = 1$ or 0, but not 2

6.57

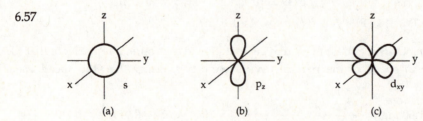

6.59   (a)   The 1s and 2s orbitals of a hydrogen atom have the same overall spherical shape. The 2s orbital has a larger radial extension and one node, while the 1s orbital has continuous electron density. Since the 2s orbital is "larger," there is greater probability of finding an electron further from the nucleus in the 2s orbital.

(b)   A single 2p orbital is directional in that its electron density is concentrated along one of the three Cartesian axes of the atom. The $d_{x^2-y^2}$ orbital has electron density along both the x- and y-axes, while the p_x orbital has density only along the x-axis.

(c)   The average distance of an electron from the nucleus in a 3s orbital is greater than for an electron in a 2s orbital. In general, for the same kind of orbital, the larger the *n* value, the greater the average distance of an electron from the nucleus of the atom.

(d)   1s < 2p < 3d < 4f < 6s. In the hydrogen atom, orbitals with the same *n* value are degenerate and energy increases with increasing *n* value. Thus, the order of increasing energy is given above.

Many-Electron Atoms and Electron Configurations (sections 6.7 – 6.9)

6.61   (a)   In the hydrogen atom, orbitals with the same principle quantum number, *n*, have the same energy; they are degenerate.

(b)   In a many-electron atom, for a given *n*-value, orbital energy increases with increasing *l*-value: s < p < d < f.

6.63   (a)   There are two main pieces of experimental evidence for electron "spin": the Stern-Gerlach experiment and line spectra of multi-electron atoms. The Stern-Gerlach experiment shows that a beam of neutral Ag atoms passed through a nonhomogeneous magnetic field is deflected equally in two directions. This shows that atoms with a single unpaired electron interact differently with the magnetic field and suggests that there are two and only two values for an atom's own magnetic field. Examination of the fine details of emission line-spectra of multi-electron atoms reveals that each line is really a close pair of lines. Both observations can be rationalized if electrons have the property of spin.

(b)   Note that 2 electrons in the same orbital have opposite spin; this is called spin-pairing.

(c)

6.65   *Analyze/Plan.* Each subshell has an l-value associated with it. For a particular l-value, permissible m_l-values are $-l$ to $+l$. Each m_l-value represents an orbital, which can hold two electrons.   *Solve.*

   (a)   6   (b)   10   (c)   2   (d)   14

6.67   (a)   "Valence electrons" are those involved in chemical bonding. They are part (or all) of the outer-shell electrons listed after core electrons in a condensed electron configuration.

   (b)   "Core electrons" are inner shell electrons that have the electron configuration of the nearest noble-gas element.

   (c)   Each box represents an orbital.

   (d)   Each half-arrow in an orbital diagram represents an electron. The direction of the half-arrow represents electron spin.

6.69   *Analyze/Plan.* Follow the logic in Sample Exercise 6.9. *Solve.*

   (a)   Cs: $[Xe]6s^1$ (b)   Ni: $[Ar]4s^2 3d^8$

   (c)   Se: $[Ar]4s^2 3d^{10} 4p^4$ (d)   Cd: $[Kr]5s^2 4d^{10}$

   (e)   U: $[Rn]5f^3 6d^1 7s^2$. (Note the U and several other f-block elements have irregular d- and f-electron orders.)

   (f)   Pb: $[Xe]6s^2 4f^{14} 5d^{10} 6p^2$

6.71   (a)   Be, 0 unpaired electrons   (b)   O, 2 unpaired electrons

   (c)   Cr, 6 unpaired electrons   (d)   Te , 2 unpaired electrons

6.73   (a)   The fifth electron would fill the 2p subshell (same n-value as 2s) before the 3s.

   (b)   The Ne core has filled 2s and 2p subshells. Either the core is [He] or the outer electron configuration should be $3s^2 3p^3$.

   (c)   The 3p subshell would fill before the 3d because it has the lower l-value and the same n-value. (If there were more electrons, 4s would also fill before 3d.)

Additional Exercises

6.75   (a) $\lambda_A = 1.6 \times 10^{-7}$ m $/ 4.5 = 3.56 \times 10^{-8} = 3.6 \times 10^{-8}$ m

   $\lambda_B = 1.6 \times 10^{-7}$ m $/ 2 = 8.0 \times 10^{-8}$ m

   (b) $\nu = c/\lambda$; $\nu_A = \dfrac{2.998 \times 10^8 \text{ m}}{1 \text{ s}} \times \dfrac{1}{3.56 \times 10^{-8} \text{ m}} = 8.4 \times 10^{15}$ s^{-1}

   $\nu_B = \dfrac{2.998 \times 10^8 \text{ m}}{1 \text{ s}} \times \dfrac{1}{8.0 \times 10^{-8} \text{ m}} = 3.7 \times 10^{15}$ s^{-1}

   (c)   A: ultraviolet, B: ultraviolet

6.78   All electromagnetic radiation travels at the same speed, 2.998×10^8 m/s. Change miles to meters and seconds to some appropriate unit of time.

$$746 \times 10^6 \text{ mi} \times \frac{1.6093 \text{ km}}{1 \text{ mi}} \times \frac{1000 \text{ m}}{1 \text{ km}} \times \frac{1 \text{ s}}{2.998 \times 10^8 \text{ m}} \times \frac{1 \text{ min}}{60 \text{ s}} = 66.7 \text{ min}$$

6.82
$$\frac{2.6 \times 10^{-12} \text{ C}}{1 \text{ s}} \times \frac{1 e^-}{1.602 \times 10^{-19} \text{ C}} \times \frac{1 \text{ photon}}{1 e^-} = 1.623 \times 10^7 = 1.6 \times 10^7 \text{ photons/s}$$

$$\frac{E}{\text{photon}} = hc/\lambda = \frac{6.626 \times 10^{-34} \text{ J·s}}{630 \text{ nm}} \times \frac{2.998 \times 10^8 \text{ m}}{1 \text{ s}} \times \frac{1 \text{ nm}}{1 \times 10^{-9} \text{ m}} \times \frac{1.623 \times 10^7 \text{ photon}}{\text{s}}$$

$$= 5.1 \times 10^{-12} \text{ J/s}$$

6.85 (a) Lines with $n_f = 1$ lie in the ultraviolet (see Solution 6.40) and with $n_f = 2$ lie in the visible (see Solution 6.39). Lines with $n_f = 3$ will have smaller ΔE and longer wavelengths and lie in the infrared.

(b) Use Equation 6.7 to calculate ΔE, then $\lambda = hc/\Delta E$.

$$n_i = 4, n_f = 3; \Delta E = -2.18 \times 10^{-18} \text{ J}\left[\frac{1}{n_f^2} - \frac{1}{n_i^2}\right] = -2.18 \times 10^{-18} \text{ J} \,(1/9 - 1/16)$$

$$\lambda = hc/E = \frac{6.626 \times 10^{-34} \text{ J·s} \times 2.998 \times 10^8 \text{ m/s}}{-2.18 \times 10^{-18} \,(1/9 - 1/16)} = 1.87 \times 10^{-6} \text{ m}$$

$$n_i = 5, n_f = 3; \lambda = hc/E = \frac{6.626 \times 10^{-34} \text{ J·s} \times 2.998 \times 10^8 \text{ m/s}}{-2.18 \times 10^{-18} \,(1/9 - 1/25)} = 1.28 \times 10^{-6} \text{ m}$$

$$n_i = 6, n_f = 3; \lambda = hc/E = \frac{6.626 \times 10^{-34} \text{ J·s} \times 2.998 \times 10^8 \text{ m/s}}{-2.18 \times 10^{-18} \,(1/9 - 1/36)} = 1.09 \times 10^{-6} \text{ m}$$

These three wavelengths are all greater than 1 µm or 1×10^{-6} m. They are in the infrared, close to the visible edge (0.7×10^{-6} m).

6.90 (a) l (b) n and l (c) m_s (d) m_l

6.92 (a) The p_z orbital has a nodal plane where $z = 0$. This is the xy plane.

(b) The d_{xy} orbital has four lobes and two nodal planes, the two planes where $x = 0$ and $y = 0$. These are the yz and xz planes.

(c) The $d_{x^2 - y^2}$ has four lobes and two nodal planes, the planes where $x^2 - y^2 = 0$. These are the planes that bisect the x and y axes and contain the z axis.

6.94 If m_s had three allowed values instead of two, each orbital would hold three electrons instead of two. Assuming that the same orbitals are available (that there is no change in the n, l, and m_l values), the number of elements in each of the first four rows would be:

1st row: 1 orbital × 3 = 3 elements

2nd row: 4 orbitals × 3 = 12 elements

3rd row: 4 orbitals × 3 = 12 elements

4th row: 9 orbitals × 3 = 27 elements

The s-block would be 3 columns wide, the p-block 9 columns wide and the d-block 15 columns wide.

Integrative Exercises

6.97   (a)   We know the wavelength of microwave radiation, the volume of coffee to be heated, and the desired temperature change. Assume the density and heat capacity of coffee are the same as pure water. We need to calculate: (i) the total energy required to heat the coffee and (ii) the energy of a single photon in order to find (iii) the number of photons required.

        (i)   From Chapter 5, the heat capacity of liquid water is 4.184 J/g°C.

               To find the mass of 200 mL of coffee at 23°C, use the density of water given in Appendix B.

$$200 \text{ mL} \times \frac{0.997 \text{ g}}{1 \text{ mL}} = 199.4 = 199 \text{ g coffee}$$

$$\frac{4.184 \text{ J}}{1 \text{ g} \,°C} \times 199.4 \text{ g} \times (60°C - 23°C) = 3.087 \times 10^4 \text{ J} = 31 \text{ kJ}$$

        (ii) $E = hc/\lambda = 6.626 \times 10^{-34} \text{ J-s} \times \dfrac{2.998 \times 10^8 \text{ m}}{1 \text{ s}} \times \dfrac{1}{0.112 \text{ m}}$

$$= \frac{1.77 \times 10^{-24} \text{ J}}{1 \text{ photon}}$$

        (iii) $3.087 \times 10^4 \text{ J} \times \dfrac{1 \text{ photon}}{1.774 \times 10^{-24} \text{ J}} = 1.7 \times 10^{28}$ photons

    (The answer has 2 sig figs because the temperature change, 37°C, has 2 sig figs.)

    (b)   1 W = 1 J/s. 900 W = 900 J/s. From part (a), 31 kJ are required to heat the coffee.

$$3.087 \times 10^4 \text{ J} \times \frac{1 \text{ s}}{900 \text{ J}} = 34.30 = 34 \text{ s}$$

6.101   (a)   Bohr's theory was based on the Rutherford "nuclear" model of the atom. That is, Bohr theory assumed a dense positive charge at the center of the atom and a diffuse negative charge (electrons) surrounding it. Bohr's theory then specified the nature of the diffuse negative charge. The prevailing theory before the nuclear model was Thomson's plum pudding or watermelon model, with discrete electrons scattered about a diffuse positive charge cloud. Bohr's theory could not have been based on the Thomson model of the atom.

    (b)   De Broglie's hypothesis is that electrons exhibit both particle and wave properties. Thomson's conclusion that electrons have mass is a particle property, while the nature of cathode rays is a wave property. De Broglie's hypothesis actually rationalizes these two seemingly contradictory observations about the properties of electrons.

7 Periodic Properties of the Elements

Visualizing Concepts

7.3 (a) The bonding atomic radius of A, r_A, is $d_1/2$. The distance d_2 is the sum of the bonding atomic radii of A and X, $r_A + r_X$. Since we know that $r_A = d_1/2$, $d_2 = r_X + d_1/2$, $r_X = d_2 - d_1/2$.

(b) The length of the X-X bond is $2r_X$.

$$2r_X = 2(d_2 - d_1/2) = 2d_2 - d_1.$$

7.6 (a) $X + 2F_2 \rightarrow XF_4$

(b) If X is a nonmetal, XF_4 is a molecular compound. If X is a metal, XF_4 is ionic. For an ionic compound with this formula, X would have a charge of 4+, and a much smaller bonding atomic radius than F^-. X in the diagram has about the same bonding radius as F, so it is likely to be a nonmetal.

Periodic Table; Effective Nuclear Charge (sections 7.1 and 7.2)

7.7 The number of columns in the various blocks of the periodic chart corresponds to the maximum number of electrons that can occupy the various kinds of atomic orbitals: 2 columns on the left for 2 electrons in s orbitals, 10 columns in the transition metals for 10 electrons in d orbitals, 6 columns on the right for 6 electrons in p orbitals, 14-member rows below for 14 electrons in f orbitals. The order of blocks corresponds to the filling order of atomic orbitals, and the row number corresponds to the principal quantum number of the valence electrons of elements in that row, ns, np, $(n-1)$d, $(n-2)$f.

7.9 According to Figure 7.1, the first elements to be discovered, those known since ancient times, were Fe, Cu, Ag, Au, Hg, Sn, Pb, C and S. These elements are present in nature in elemental form; they can be observed directly and their isolation does not require chemical processing. In general, elements are discovered according to their ease of isolation in elemental form. This depends on chemical properties rather than relative abundance.

7.11 (a) *Effective nuclear charge*, Z_{eff}, is a representation of the average electrical field experienced by a single electron. It is the average environment created by the nucleus and the other electrons in the molecule, expressed as a net positive charge at the nucleus. It is approximately the nuclear charge, Z, minus the number of core electrons.

(b) Going from left to right across a period, nuclear charge increases while the number of electrons in the core is constant. This results in an increase in Z_{eff}.

7.13 (a) *Analyze/Plan.* Z_{eff} = Z–S. Find the atomic number, Z, of Na and K. Write their electron configurations and count the number of core electrons. Assume S = number of core electrons.

Solve. Na: Z = 11; [Ne]$3s^1$. In the Ne core there are 10 electrons. Z_{eff} = 11–10 = 1.
K: Z = 19; [Ar]$4s^1$. In the Ar core there are 18 electrons. Z_{eff} = 19–18 = 1.

(b) *Analyze/Plan.* Z_{eff} = Z–S. Write the complete electron configuration for each element to show counting for Slater's rules. S = 0.35 (# of electrons with same *n*) + 0.85 (# of electrons with (*n*–1) + 1 (# of electrons with (*n*–2).

Solve. Na: $1s^2 2s^2 2p^6 3s^1$. S = 0.35(0) + 0.85(8) + 1(2) = 8.8. Z_{eff} = 11 – 8.8 = 2.2

K: $1s^2 2s^2 2p^6 3s^2 3p^6 4s^1$. S = 0.35(0) + 0.85(8) + 1(10) = 16.8. Z_{eff} = 19 – 16.8 = 2.2

(c) For both Na and K, the two values of Z_{eff} are 1.0 and 2.2. The Slater value of 2.2 is closer to the values of 2.51 (Na) and 3.49 (K) obtained from detailed calculations.

(d) Both approximations, 'core electrons 100% effective' and Slater, yield the same value of Z_{eff} for Na and K. Neither approximation accounts for the gradual increase in Z_{eff} moving down a group.

(e) Following the trend from detailed calculations, we predict a Z_{eff} value of approximately 4.5 for Rb.

7.15 Krypton has a larger nuclear charge (Z = 36) than argon (Z = 18). The shielding of electrons in the *n* = 3 shell by the 1s, 2s and 2p core electrons in the two atoms is approximately equal, so the *n* = 3 electrons in Kr experience a greater effective nuclear charge and are thus situated closer to the nucleus.

Atomic and Ionic Radii (section 7.3)

7.17 (a) Atomic radii are determined by measuring distances between nuclei (interatomic distances) in various bonding and nonbonding situations.

(b) Bonding radii are calculated from the internuclear separation of two atoms joined by a chemical bond. Nonbonding radii are calculated from the internuclear separation between two gaseous atoms that collide and move apart, but do not bond.

(c) For a given element, the nonbonding radius is always larger than the bonding radius. When a chemical bond forms, electron clouds of the two atoms interpenetrate, bringing the two nuclei closer together and resulting in a smaller bonding atomic radius (Figure 7.5).

(d) If a free atom reacts to become part of a covalent molecule, the atom gets smaller; its radius changes from nonbonding to bonding.

7.19 (a) The atomic (*metallic*) radius of W is the interatomic W-W distance divided by 2, 2.74 Å/2= 1.37 Å.

(b) Under high pressure, we expect atoms in a pure substance to move closer together. That is, the distance between W atoms will decrease.

7.21 From bonding atomic radii in Figure 7.6, As–I = 1.19 Å + 1.33 Å = 2.52 Å. This is very close to the experimental value of 2.55 Å in AsI_3.

7.23 (a) Atomic radii decrease moving from left to right across a row and (b) increase from top to bottom within a group.

(c) O < Si < Ge < I. The order for the first three atoms is unambiguous according to the trends of increasing atomic radius moving down a column and to the left in a row of the table. Moving from Ge to I, radius increases moving down but decreases moving right. It is usually true that a change in row number, which involves a change in principal quantum number of valence electrons, is a larger effect than a horizontal change across a row. This is verified by values for the atomic radii of Ge and I on Figure 7.6.

7.25 *Plan.* Locate each element on the periodic charge and use trends in radii to predict their order. *Solve.*

(a) Cs > K > Li   (b) Pb > Sn > Si   (c) N > O > F

7.27 (a) False. Cations are smaller than their corresponding neutral atoms. Electrostatic repulsions are reduced by removing an electron from a neutral atom, Z_{eff} increases, and the cation is smaller.

(b) True. [See (a) above.]

(c) False. I^- is bigger than Cl^-. Going down a column, the *n* value of the valence electrons increases and they are farther from the nucleus. Valence electrons also experience greater shielding by core electrons. The greater radial extent of the valence electrons outweighs the increase in Z, and the size of particles with like charge increases.

7.29 The size of the red sphere decreases on reaction, so it loses one or more electrons and becomes a cation. Metals lose electrons when reacting with nonmetals, so the red sphere represents a metal. The size of the blue sphere increases on reaction, so it gains one or more electrons and becomes an anion. Nonmetals gain electrons when reacting with metals, so the blue sphere represents a nonmetal.

7.31 (a) An isoelectronic series is a group of atoms or ions that have the same number of electrons.

(b) Ga^{3+}: Ni; Zr^{4+}: Kr; Mn^{7+}: Ar; I^-: Xe; Pb^{2+}: Hg

7.33 (a) Cl^-: Ar   (b) Sc^{3+}: Ar

(c) Fe^{2+}: $[Ar]3d^6$. Fe^{2+} has 24 electrons. Neutral Cr has 24 electrons, $[Ar]4s^13d^5$. Because transition metals fill the *s* subshell first but also lose *s* electrons first when they form ions, many transition metal ions do not have neutral atoms with the same electron configuration.

(d) Zn^{2+}: $[Ar]3d^{10}$; no neutral atom with same configuration [same reason as (c)].

(e) Sn^{4+}: $[Kr]4d^{10}$; no neutral atom with same electron configuration [same reason as (c)], but Sn^{4+} does have the same configuration as Cd^{2+}.

7.35   (a) K^+ (larger Z) is smaller.

       (b) Cl^- and K^+: $[Ne]3s^23p^6$. 10 core electrons
             Cl^-, Z = 17. Z_{eff} = 17 – 10 = 7
             K^+, Z = 19. Z_{eff} = 19 – 10 = 9

       (c)   Valence electron, n = 3; 7 other n = 3 electrons; eight n = 2 electrons;
             two n = 1 electrons. S = 0.35(7) + 0.85(8) + 1(2) = 11.25
             Cl^-: Z_{eff} = 17 – 11.25 = 5.75. K^+: Z_{eff} = 19 – 11.25 = 7.75

       (d)   For isoelectronic ions (without d electrons), the electron configurations and therefore shielding values (S) are the same. Only the nuclear charge changes. So, as nuclear charge (Z) increases, effective nuclear charge (Z_{eff}) increases and ionic radius decreases.

7.37   (a) $Se < Se^{2-} < Te^{2-}$ (b) $Co^{3+} < Fe^{3+} < Fe^{2+}$ (c) $Ti^{4+} < Sc^{3+} < Ca$ (d) $Be^{2+} < Na^+ < Ne$

Ionization Energies; Electron Affinities (sections 7.4 and 7.5)

7.39 $Al(g) \rightarrow Al^+(g) + 1e^-$; $Al^+(g) \rightarrow Al^{2+}(g) + 1e^-$; $Al^{2+}(g) \rightarrow Al^{3+}(g) + 1e^-$

       The process for the first ionization energy requires the least amount of energy. When an electron is removed from an atom or ion, electrostatic repulsions are reduced, Z_{eff} increases, and the energy required to remove the next electron increases. This rationale is confirmed by the ionization energies listed in Table 7.2.

7.41   (a)   False. Ionization energies are always positive quantities.

       (b)   False. F has a greater first ionization energy than O, because Z_{eff} for F is greater than Z_{eff} for O.

       (c)   True.

7.43   (a)   In general, the smaller the atom, the larger its first ionization energy.

       (b)   According to Figure 7.9, He has the largest and Cs the smallest first ionization energy of the nonradioactive elements.

7.45   *Plan.* Use periodic trends in first ionization energy.   *Solve.*

       (a)   Cl   (b)   Ca   (c)   K   (d)   Ge   (e)   Sn

7.47   *Plan.* Follow the logic of Sample Exercise 7.7.   *Solve.*

       (a) Fe^{2+}: $[Ar]3d^6$ (b) Hg^{2+}: $[Xe]4f^{14}5d^{10}$

       (c) Mn^{2+}: $[Ar]3d^5$ (d) Pt^{2+}: $[Xe]4f^{14}5d^8$

       (e) P^{3-}: $[Ne]3s^23p^6$

7.49   *Plan.* Focus on transition metals, which have d electrons in their outer shell. Use Figure 7.14 to find representative oxidations states for transition metals. Note that, by definition, metals lose electrons to form positive ions.

       Solve. Elements in group 10 and beyond have at least 8 d electrons. Of these, the group 10 metals all form 2+ ions with the electron configuration nd^8. Of the elements in groups 11 and 12, only Au adopts a sufficiently high positive charge to form an ion with the configuration nd^8. (Other possibilities not listed on Figure 7.14 exist.)

Ni^{2+}: $[Ar]3d^8$; Pd^{2+}: $[Kr]4d^8$; Pt^{2+}: $[Xe]4f^{14}5d^8$; Au^{3+}: $[Xe]4f^{14}5d^8$

7.51   *Analyze/Plan.* Consider the definitions of ionization energy, electron affinity and the electron configuration of Ar. *Solve.*

(a)   Argon is a noble gas, with a very stable core electron configuration. This causes the element to resist chemical change. Positive, endothermic, values for ionization energy and electron affinity mean that energy is required to either remove or add electrons. Valence electrons in Ar experience the largest Z_{eff} of any element in the third row, because the nuclear buildup is not accompanied by an increase in screening. This results in a large, positive ionization energy. When an electron is added to Ar, the n = 3 electrons become core electrons which screen the extra electrons so effectively that Ar^- has a higher energy than an Ar atom and a free electron. This results in a large positive electron affinity.

(b)   The units of electron affinity are kJ/mol.

7.53   *Analyze/Plan.* Consider the definitions of ionization energy and electron affinity, along with pertinent electron configurations. *Solve.*

Electron affinity : $Br(g)$ $+1e^-$ $\rightarrow$ $Br^-(g)$
                        $[Ar]4s^2 3d^{10}4p^5$ $[Ar]4s^2 3d^{10}4p^6$

When a Br atom gains an electron, the Br^- ion adopts the stable electron configuration of Kr. Since the electron is added to the same 4p subshell as other outer electrons, it experiences essentially the same attraction for the nucleus. Thus, the energy of the Br^- ion is lower than the total energy of a Br atom and an isolated electron, and electron affinity is negative.

Electron affinity : $Kr(g)$ $+1e^-$ $\rightarrow$ $Kr^-(g)$
                        $[Ar]4s^2 3d^{10}4p^6$ $[Ar]4s^2 3d^{10}4p^6 5s^1$

Energy is required to add an electron to a Kr atom; Kr^- has a higher energy than the isolated Kr atom and free electron. In Kr^- the added electron would have to occupy the higher energy 5s orbital; a 5s electron is farther from the nucleus and effectively shielded by the spherical Kr core and is not stabilized by the nucleus.

7.55   *Analyze/Plan.* Consider the definitions of ionization energy and electron affinity, along with the appropriate electron configurations. *Solve.*

(a)   Ionization energy of Ne: $Ne(g)$ $\rightarrow$ $Ne^+(g)$ + $1e^-$
                                    $[He]2s^2 2p^6$ $[He]2s^2 2p^5$

       Electron affinity of F: $F(g)$ + $1e^-$ $\rightarrow$ $F^-(g)$
                                    $[He]2s^2 2p^5$ $[He]2s^2 2p^6$

(b)   The I_1 of Ne is positive, while E_1 of F is negative. All ionization energies are positive.

(c)   One process is apparently the reverse of the other, with one important difference. The Z (and Z_{eff}) for Ne is greater than Z (and Z_{eff}) for F^-. So we expect $I_1(Ne)$ to be somewhat greater in magnitude and opposite in sign to $E_1(F)$. [Repulsion effects approximately cancel; repulsion decrease upon I_1 causes smaller positive value; repulsion increase upon E_1 causes smaller negative value.]

Properties of Metals and Nonmetals (section 7.6)

7.57   The smaller the first ionization energy of an element, the greater the metallic character of that element.

7.59   *Analyze/Plan.* Use Figure 7.12, "Metals, metalloids and nonmetals", and Figure 7.14, "Representative oxidation states of the elements", to inform our discussion.

Solve. Agree. An element that commonly forms a cation is a metal. The only exception to this statement shown on Figure 7.14 is antimony, Sb, a metalloid which commonly forms cations. Although Sb is a metalloid, it is far down (in the fifth row) on the chart and likely to have significant metallic character.

7.61   *Analyze/Plan.* Ionic compounds are formed by combining a metal and a nonmetal; molecular compounds are formed by two or more nonmetals. *Solve.*

Ionic: SnO_2, Al_2O_3, Li_2O, Fe_2O_3;   molecular: CO_2, H_2O

Sn, Al, Li and Fe are metals; C and H are nonmetals.

7.63   (a)   When dissolved in water, an "acidic oxide" produces an acidic (pH < 7) solution. A "basic oxide" dissolved in water produces a basic (pH > 7) solution.

(b)   Oxides of nonmetals are acidic. Example: $SO_3(g) + H_2O(l) \rightarrow H_2SO_4(aq)$. Oxides of metals are basic. Example: CaO (quick lime). $CaO(s) + H_2O(l) \rightarrow Ca(OH)_2(aq)$.

7.65   *Analyze/Plan.* Cl_2O_7 is a molecular compound formed by two nonmetallic elements. More specifically, it is a nonmetallic oxide and acidic.   *Solve.*

(a)   Dichlorineheptoxide

(b)   Elemental chlorine and oxygen are diatomic gases.

$2Cl_2(g) + 7O_2(g) \rightarrow 2Cl_2O_7(l)$

(c)   Most nonmetallic oxides we have seen, such as CO_2 and SO_3, are gases. However, oxides with more atoms, such as $P_2O_3(l)$ and $P_2O_5(s)$, exist in other states. A boiling point of 81°C is not totally unexpected for a large molecule like Cl_2O_7.

(d) Cl_2O_7 is an acidic oxide, so it will be more reactive to base, OH^-.

$Cl_2O_7(l) + 2OH^-(aq) \rightarrow 2ClO_4^-(aq) + H_2O(l)$

(e)   The oxidation state of Cl in Cl_2O_7 is +7. In this oxidation state, the electron configuration of Cl is $[He]2s^22p^6$ or [Ne].

7.67   (a) $BaO(s) + H_2O(l) \rightarrow Ba(OH)_2(aq)$

(b) $FeO(s) + 2HClO_4(aq) \rightarrow Fe(ClO_4)_2(aq) + H_2O(l)$

(c) $SO_3(g) + H_2O(l) \rightarrow H_2SO_4(aq)$

(d) $CO_2(g) + 2NaOH(aq) \rightarrow Na_2CO_3(aq) + H_2O(l)$

Group Trends in Metals and Nonmetals (sections 7.7 and 7.8)

7.69   Yes, the reactivity of a metal correlates with its first ionization energy. Since metals lose electrons when they form ions, the less energy required for this process, the more reactive the metal. However, we usually observe reactivity of metals in the solid state and ionization energy is a gas phase property, so there are differences between the two properties.

7.71   (a)   Ca and Mg are both metals; they tend to lose electrons and form cations when they react. Ca is more reactive because it has a lower ionization energy than Mg. The Ca valence electrons in the 4s orbital are less tightly held because they are farther from the nucleus than the 3s valence electrons of Mg.

(b)   K and Ca are both metals; they tend to lose electrons and form cations when they react. K is more reactive because it has the lower first ionization energy. The 4s valence electron in K is less tightly held because it has the same n value as the valence electrons of Ca and experiences smaller Z and Z_{eff}.

7.73   (a) $2K(s) + Cl_2(g) \rightarrow 2KCl(s)$

(b) $SrO(s) + H_2O(l) \rightarrow Sr(OH)_2(aq)$

(c) $4Li(s) + O_2(g) \rightarrow 2Li_2O(s)$

(d) $2Na(s) + S(l) \rightarrow Na_2S(s)$

7.75   (a)   The reactions of the alkali metals with hydrogen and with a halogen are redox reactions. In both classes of reaction, the alkali metal loses electrons and is oxidized. Both hydrogen and the halogen gain electrons and are reduced. The product is an ionic solid, where either hydride ion, H^-, or a halide ion, X^-, is the anion and the alkali metal is the cation.

(b) $Ca(s) + F_2(g) \rightarrow CaF_2(s)$ $Ca(s) + H_2(g) \rightarrow CaH_2(s)$

Both products are ionic solids containing Ca^{2+} and the corresponding anion in a 1:2 ratio.

7.77

	Br	**Cl**
(a)	$[Ar]4s^2 4p^5$	$[Ne]3s^2 3p^5$
(b)	–1	–1
(c)	1140 kJ/mol	1251 kJ/mol
(d)	reacts slowly to form HBr+HOBr	reacts slowly to form HCl+HOCl
(e)	–325 kJ/mol	–349 kJ/mol
(f)	1.14 Å	0.99 Å

The $n = 4$ valence electrons in Br are farther from the nucleus and less tightly held than the $n = 3$ valence electrons in Cl. Therefore, the ionization energy of Cl is greater, the electron affinity is more negative and the atomic radius is smaller.

7.79 (a) The term "inert" was dropped because it no longer described all the group 8A elements.

(b) In the 1960s, scientists discovered that Xe would react with substances such as F_2 and PtF_6 that have a strong tendency to remove electrons. Thus, Xe could not be categorized as an "inert" gas.

(c) The group is now called the noble gases.

7.81 (a) $2O_3(g) \rightarrow 3O_2(g)$

(b) $Xe(g) + F_2(g) \rightarrow XeF_2(g)$

$Xe(g) + 2F_2(g) \rightarrow XeF_4(s)$

$Xe(g) + 3F_2(g) \rightarrow XeF_6(s)$

(c) $S(s) + H_2(g) \rightarrow H_2S(g)$

(d) $2F_2(g) + 2H_2O(l) \rightarrow 4HF(aq) + O_2(g)$

Additional Exercises

7.83 Up to Z = 82, there are three instances where atomic weights are reversed relative to atomic numbers: Ar and K; Co and Ni; Te and I.

In each case, the most abundant isotope of the element with the larger atomic number (Z) has one more proton, but fewer neutrons than the element with the smaller atomic number. The smaller number of neutrons causes the element with the larger Z to have a smaller than expected atomic weight.

7.85 (a) P: $[Ne]3s^23p^3$. $Z_{eff} = Z - S = 15 - 10 = 5$.

(b) Four other $n = 3$ electrons, eight $n = 2$ electrons, two $n = 1$ electron. $S = 0.35(4) + 0.85(8) + 1(2) = 10.2$. $Z_{eff} = Z - S = 15 - 10.2 = 4.8$. The approximation in (a) and Slater's rules give very similar values for Z_{eff}.

(c) The 3s electrons penetrate the [Ne] core electrons (by analogy to Figure 7.3) and experience less shielding than the 3p electrons. That is, S is greater for 3p electrons, owing to the penetration of the 3s electrons, so Z – S (3p) is less than Z – S (3s).

(d) The 3p electrons are the outermost electrons; they experience a smaller Z_{eff} than 3s electrons and thus a smaller attraction for the nucleus, given equal n-values. The first electron lost is a 3p electron. Each 3p orbital holds one electron, so there is no preference as to which 3p electron will be lost.

7.88 (a) The estimated distances in the table below are the sum of the radii of the group 5A elements and H from Figure 7.6.

bonded atoms	estimated distance	measured distance
P – H	1.43	1.419
As – H	1.56	1.519
Sb – H	1.75	1.707

In general, the estimated distances are a bit longer than the measured distances. This probably shows a systematic bias in either the estimated radii or in the method of obtaining the measured values. (Recall that the radii in Figure 7.6 come from measuring many different molecules for each element, not just the bonds listed in this exercise.)

(b) The principal quantum number of the outer electrons and thus the average distance of these electrons from the nucleus increases from P (n = 3) to As (n = 4) to Sb (n = 5). This causes the systematic increase in M – H distance.

7.92 (a) $2Sr(s) + O_2(g) \rightarrow 2SrO(s)$

(b) Assume that the corners of the cube are at the centers of the outermost O^{2-} ions, and that the edges each pass through the center of one Sr^{2+} ion. The length of an edge is then $r(O^{2-}) + 2r(Sr^{2+}) + r(O^{2-}) = 2r(O^{2-}) + 2r(Sr^{2+}) = 2(1.32 \text{ Å}) + 2(1.26 \text{ Å}) = 5.16 \text{ Å}$.

(c) Density is the ratio of mass to volume.

$$d = \frac{\text{mass SrO in cube}}{\text{vol cube}} = \frac{\text{\# SrO units} \times \text{mass of SrO}}{\text{vol cube}}$$

Calculate the mass of 1 SrO unit in grams and the volume of the cube in cm^3; solve for number of SrO units.

$$\frac{103.62 \text{ g SrO}}{\text{mol}} \times \frac{1 \text{ mol SrO}}{6.022 \times 10^{23} \text{ SrO units}} = 1.7207 \times 10^{-22} = 1.721 \times 10^{-22} \text{ g/SrO unit}$$

$$V = (5.16)^3 \text{ Å}^3 \times \frac{(1 \times 10^{-8})^3 \text{ cm}^3}{\text{Å}^3} = 1.3739 \times 10^{-22} = 1.37 \times 10^{-22} \text{ cm}^3$$

$$d = \frac{\text{\# of SrO units} \times 1.7207 \times 10^{-22} \text{ g/SrO unit}}{1.3739 \times 10^{-22} \text{ cm}^3} = 5.10 \text{ g/cm}^3$$

$$\text{\# of SrO units} = 5.10 \text{ g/cm}^3 \times \frac{1.3739 \times 10^{-22} \text{ cm}^3}{1.7207 \times 10^{-22} \text{ g/SrO unit}} = 4.07 \text{ units}$$

Since the number of formula units must be an integer, there are four SrO formula units in the cube. Using average values for ionic radii to estimate the edge length probably leads to the small discrepancy.

7.95 (a) O: $[He]2s^22p^4$

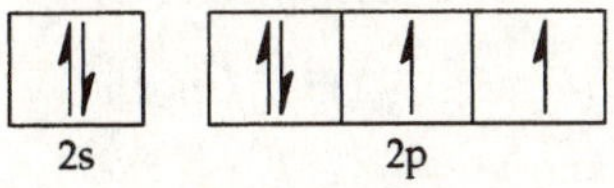

O^{2-}: $[He]2s^22p^6 = [Ne]$

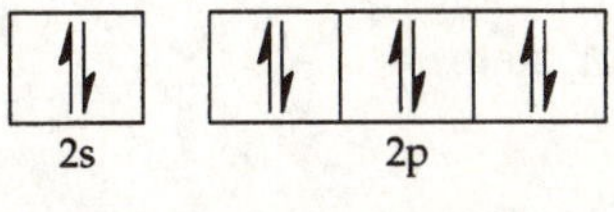

(b) O^{3-}: $[Ne]3s^1$ The third electron would be added to the 3s orbital, which is farther from the nucleus and more strongly shielded by the [Ne] core. The overall attraction of this 3s electron for the O nucleus is not large enough for O^{3-} to be a stable particle.

7.98   (a)   For both H and the alkali metals, the added electron will complete an ns subshell (1s for H and ns for the alkali metals) so shielding and repulsion effects will be similar. For the halogens, the electron is added to an np subshell, so the energy change is likely to be quite different.

    (b)   True. Only He has a smaller estimated "bonding" atomic radius, and no known compounds of He exist. The electron configuration of H is $1s^1$. The single 1s electron experiences no repulsion from other electrons and feels the full unshielded nuclear charge. It is held very close to the nucleus. The outer electrons of all other elements that form compounds are shielded by a spherical inner core of electrons and are less strongly attracted to the nucleus, resulting in larger bonding atomic radii.

    (c)   Ionization is the process of removing an electron from an atom. For the alkali metals, the ns electron being removed is effectively shielded by the core electrons, so ionization energies are low. For the halogens, a significant increase in nuclear charge occurs as the np orbitals fill, and this is not offset by an increase in shielding. The relatively large effective nuclear charge experienced by np electrons of the halogens is similar to the unshielded nuclear charge experienced by the H 1s electron. Both H and the halogens have large ionization energies.

    (d)   Ionization energy of hydride: $H^-(g) \rightarrow H(g) + 1e^-$

    (e)   Electron affinity of hydroden: $H(g) + 1e^- \rightarrow H^-(g)$

    The two processes in parts (d) and (e) are the exact reverse of one another. The value for the ionization energy of hydride is equal in magnitude but opposite in sign to the electron affinity of hydrogen.

7.103   *Plan.* According to the periodic table on the inside cover of your text, element 116 is in group 6A, so element 117 will be in group 7A, the halogens. Write the electron configuration and use information from Figures 7.6, 7.9, 7.11, 7.14 and Table 7.7 along with periodic trends to estimate values for properties. Remember that element 117 is two rows below iodine, and that the increase in Z and Z_{eff} that accompanies filling of the f orbitals will decrease the size of the changes in ionization energy, electron affinity and atomic size. *Solve.*

Electron configuration:	$[Rn]7s^2 5f^{14} 6d^{10} 7p^5$
First ionization energy:	805 kJ/mol
Electron affinity:	–235 kJ/mol
Atomic size:	1.65 Å
Common oxidation state:	–1

Integrative Exercises

7.106   (a)   Li: $[He]2s^1$. Assume that the [He] core is 100% effective at shielding the 2s valence electron $Z_{eff} = Z - S \approx 3 - 2 = 1+$.

(b) The first ionization energy represents loss of the 2s electron.

ΔE = energy of free electron (n = ∞) – energy of electron in ground state (n = 2)

$\Delta E = I_1 = [-2.18 \times 10^{-18} \text{ J } (Z^2/\infty^2)] - [-2/18 \times 10^{-18} \text{ J } (Z^2/2^2)]$

$\Delta E = I_1 = 0 + 2.18 \times 10^{-18} \text{ J } (Z^2/2^2)$

For Li, which is not a one-electron particle, let $Z = Z_{eff}$.

$\Delta E \approx 2.18 \times 10^{-18} \text{ J } (+1^2/4) \approx 5.45 \times 10^{-19} \text{ J/atom}$

(c) Change the result from part (b) to kJ/mol so it can be compared to the value in

Table 7.4. $5.45 \times 10^{-19} \dfrac{\text{J}}{\text{atom}} \times \dfrac{6.022 \times 10^{23} \text{ atom}}{\text{mol}} \times \dfrac{1 \text{ kJ}}{1000 \text{ J}} = 328 \text{ kJ/mol}$

The value in Table 7.4 is 520 kJ/mol. This means that our estimate for Z_{eff} was a lower limit, that the [He] core electrons do not perfectly shield the 2s electron from the nuclear charge.

(d) From Table 7.4, I_1 = 520 kJ/mol.

$\dfrac{520 \text{ kJ}}{\text{mol}} \times \dfrac{1000 \text{ J}}{\text{kJ}} \times \dfrac{1 \text{ mol}}{6.022 \times 10^{23} \text{ atoms}} = 8.6350 \times 10^{-19} \text{ J/atom}$

Use the relationship for I_1 and Z_{eff} developed in part (b).

$Z_{eff}^2 = \dfrac{4(8.6350 \times 10^{-19} \text{ J})}{2.18 \times 10^{-18} \text{ J}} = 1.5844 = 1.58; \ Z_{eff} = 1.26$

This value, Z_{eff} = 1.26, based on the experimental ionization energy, is greater than our estimate from part (a), which is consistent with the explanation in part (c).

7.108 (a) The X-ray source had an energy of 1253.6 eV. Change eV to J/photon and use the relationship $\lambda = hc/E$ to find wavelength.

$1253.6 \text{ eV} \times \dfrac{96.485 \text{ kJ}}{1 \text{ eV} \cdot \text{mol}} \times \dfrac{1 \text{mol}}{6.0221 \times 10^{23} \text{ photons}} \times \dfrac{1000 \text{ J}}{1 \text{ kJ}} = 2.0085 \times 10^{-16} \text{ J/photon}$

$\lambda = \dfrac{6.6261 \times 10^{-34} \text{ J} \cdot \text{s} \times 2.9979 \times 10^{8} \text{ m/s}}{2.0085 \times 10^{-16} \text{ J}} = 9.8902 \times 10^{-10} \text{ m} = 0.98902 \text{ nm} = 9.8902 \text{ Å}$

(b) Express energies of Hg 4f and O 1s electrons in terms of kJ/mol for comparison with data from Figure 7.9 of the text.

Hg 4f: $105 \text{ eV} \times \dfrac{96.485 \text{ kJ}}{1 \text{ eV} \cdot \text{mol}} = 10{,}131 = 1.01 \times 10^4 \text{ kJ/mol}$

O 1s: $531 \text{ eV} \times \dfrac{96.485 \text{ kJ}}{1 \text{ eV} \cdot \text{mol}} = 51{,}234 = 5.12 \times 10^4 \text{ kJ/mol}$

By definition, the first ionization energy is the minimum energy required to remove the first electron from an atom. This first electron is the highest energy valence electron in the neutral atom. We expect the energies of valence electrons to be higher than those of core electrons, and first ionization energies to be less than the energy required to remove a lower energy core electron.

For Hg, the first ionization energy is 1007 kJ/mol, while the XPS energy of the 4f electron is 10,100 kJ/mol. The energy required to remove a 4f core electron is 10 times the energy required to remove a 6s valence electron.

For O, the first ionization energy is 1314 kJ/mol, while the XPS energy of a 1s electron is 51,200 kJ/mol. The energy required to remove a 1s core electron is 50 times that required to remove a 2p valence electron.

(c) Hg^{2+}: $[Xe]4f^{14}5d^{10}$; valence electrons are 5d

O^{2-} : $[He]2s^22p^6$ or [Ne]; valence electrons are 2p

(d) (Recall that Slater's rules are best applied to elements that do not contain d aor f electrons.)
Hg^{2+} 5d valence: $n = 5$ (5s+5p+5d) has $18\ e^-$; subtract one for the electron under consideration $= 17\ e^-$ with the same n value as the one under consideration.
$n = 4$ (4s+4p+4d+4f) has $32\ e^-$ with $(n-1)$; $n = 3, 2, 1$ have $(18+8+2) = 28\ e^-$ with $(n-2)$ or less.

$S\ (5d) = 0.35(17) + 0.85(32) + 1.0(28) = 61.15.$ $Z_{eff} = Z - S = 80 - 61.15 = 18.85$

Hg^{2+} 4f: The 4f electrons are not shielded by electrons with $n > 4$. $n = 4$ has $(32 - 1) = 31\ e^-$; $n = 3$ has $18\ e^-$ with $(n-1)$; $n = 2, 1$ have $10\ e^-$ with $(n-2)$ or less.

$S\ (4f) = 0.35(31) + 0.85(18) + 1.0(10) = 36.15.$ $Z_{eff} = Z - S = 80 - 36.15 = 43.85$

O^{2-} 2p valence: $n = 2$ has $(8-1) = 7\ e^-$; $n = 1$ has $2\ e^-$.

$S = 0.35(7) + 0.85(2) = 4.15;$ $Z_{eff} = 8 - 4.15 = 3.85$

7.110 (a) Mg_3N_2

(b) $Mg_3N_2(s) + 3H_2O(l) \rightarrow 3MgO(s) + 2NH_3(g)$

The driving force is the production of $NH_3(g)$.

(c) After the second heating, all the Mg is converted to MgO.

Calculate the initial mass Mg.

$0.486\ g\ MgO \times \dfrac{24.305\ g\ Mg}{40.305\ g\ MgO} = 0.293\ g\ Mg$

$x = g\ Mg$ converted to MgO; $y = g\ Mg$ converted to Mg_3N_2; $x = 0.293 - y$

$g\ MgO = x\left(\dfrac{40.305\ g\ MgO}{24.305\ g\ Mg}\right);\ g\ Mg_3N_2 = y\left(\dfrac{100.929\ g\ Mg_3N_2}{72.915\ g\ Mg}\right)$

g MgO + g Mg_3N_2 = 0.470

$$(0.293 - y)\left(\frac{40.305}{24.305}\right) + y\left(\frac{100.929}{72.915}\right) = 0.470$$

$$(0.293 - y)(1.6583) + y(1.3842) = 0.470$$

$$-1.6583\, y + 1.3842\, y = 0.470 - 0.48588$$

$$-0.2741\, y = -0.01588 = -0.016$$

$$y = 0.05794 = 0.058 \text{ g Mg in } Mg_3N_2$$

$$\text{g } Mg_3N_2 = 0.05794 \text{ g Mg} \times \frac{100.929 \text{ g } Mg_3N_2}{72.915 \text{ g Mg}} = 0.0802 = 0.080 \text{ g } Mg_3N_2$$

$$\text{mass \% } Mg_3N_2 = \frac{0.0802 \text{ g } Mg_3N_2}{0.470 \text{ g } (MgO + Mg_3N_2)} \times 100 = 17\%$$

(The final mass % has two sig figs because the mass of Mg obtained from solving simultaneous equations has two sig figs.)

(d) $3Mg(s) + 2NH_3(g) \rightarrow Mg_3N_2(s) + 3H_2(g)$

$$6.3 \text{ g Mg} \times \frac{1 \text{ mol Mg}}{24.305 \text{ g Mg}} = 0.2592 = 0.26 \text{ mol Mg}$$

$$2.57 \text{ g } NH_3 \times \frac{1 \text{ mol } NH_3}{17.031 \text{ g } NH_3} = 0.1509 = 0.15 \text{ mol } NH_3$$

$$0.2592 \text{ mol Mg} \times \frac{2 \text{ mol } NH_3}{3 \text{ mol Mg}} = 0.1728 = 0.17 \text{ mol } NH_3$$

0.26 mol Mg requires more than the available NH_3 so NH_3 is the limiting reactant.

$$0.1509 \text{ mol } NH_3 \times \frac{3 \text{ mol } H_2}{2 \text{ mol } NH_3} \times \frac{2.016 \text{ g } H_2}{\text{mol } H_2} = 0.4563 = 0.46 \text{ g } H_2$$

(e) $\Delta H^{\circ}_{rxn} = \Delta H^{\circ}_f\, Mg_3N_2(s) + 3\Delta H^{\circ}_f\, H_2(g) - 3\Delta H^{\circ}_f\, Mg(s) - 2\Delta H^{\circ}_f\, NH_3(g)$

$$= -461.08 \text{ kJ} + 3(0) - 3(0) - 2(-46.19) = -368.70 \text{ kJ}$$

8 Basic Concepts of Chemical Bonding

Visualizing Concepts

8.1   *Analyze/Plan.* Count the number of electrons in the Lewis symbol. This corresponds to the 'A'-group number of the family.   *Solve.*

(a)   Group 4A or 14

(b)   Group 2A or 2

(c)   Group 5A or 15

(These are the appropriate groups in the s and p blocks, where Lewis symbols are most useful.)

8.4   *Analyze/Plan.* Count the valence electrons in the orbital diagram, take ion charge into account, and find the element with this orbital electron count on the periodic table. Write the complete electron configuration for the ion.   *Solve.*

(a)   This ion has six 4d electrons. Transition metals, or d-block elements, have valence electrons in d-orbitals. Transition metal ions first lose electrons from the 5s orbital, then from 4d if required by the charge. This 2+ ion has lost two electrons from 5s, none from 4d. The transition metal with six 4d-electrons is ruthenium, Ru.

(b)   The electron configuration of Ru is $[Kr]5s^2 4d^6$. (The configuration of Ru^{2+} is $[Kr]4d^6$).

8.7   *Analyze/Plan.* Since there are no unshared pairs in the molecule, we use single bonds to H to complete the octet of each C atom. For the same pair of bonded atoms, the greater the bond order, the shorter and stronger the bond.   *Solve.*

(a)   Moving from left to right along the molecule, the first C needs two H atoms, the second needs one, the third needs none, and the fourth needs one. The complete molecule is:

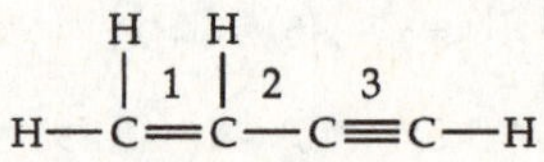

(b)   In order of increasing bond length: 3 < 1 < 2

(c)   In order of increasing bond enthalpy (strength): 2 < 1 < 3

Lewis Symbols (section 8.1)

8.9   (a)   Valence electrons are those that take part in chemical bonding, those in the outermost electron shell of the atom. This usually means the electrons beyond the core noble-gas configuration of the atom, although it is sometimes only the outer shell electrons.

(b)   N : [He] $2s^2 2p^3$

            |_____| A nitrogen atom has 5 valence electrons.

       Valence electrons

(c) $1s^2 2s^2 2p^6$ $3s^2 3p^2$ The atom (Si) has 4 valence electrons.

         |_____| |_____|

        [Ne]   valence electrons

8.11   Si: $1s^2 2s^2 2p^6 3s^2 3p^2$. The 3s and 3p electrons are valence electrons; the 1s, 2s and 2p electrons are nonvalence or core electrons. Valence electrons are involved in chemical bonding, while nonvalence or core electrons are not.

8.13   (a) $\cdot \dot{A}l \cdot$ (b) $: \ddot{B}r :$ (c) $: \ddot{A}r :$ (d) $\dot{S}r$

Ionic Bonding

8.15 $Mg \cdot + \cdot \ddot{O} : \longrightarrow Mg^{2+} + \left[: \ddot{\ddot{O}} : \right]^{2-}$

8.17   (a) AlF_3 (b) K_2S (c) Y_2O_3 (d) Mg_3N_2

8.19   (a) Sr^{2+}: $[Ar]4s^2 3d^{10} 4p^6 = [Kr]$, noble-gas configuration

       (b) Ti^{2+}: $[Ar]3d^2$

       (c) Se^{2-}: $[Ar]4s^2 3d^{10} 4p^6 = [Kr]$, noble-gas configuration

       (d) Ni^{2+}: $[Ar]3d^8$

       (e) Br^-: $[Ar]4s^2 3d^{10} 4p^6 = [Kr]$, noble-gas configuration

       (f) Mn^{3+}: $[Ar]3d^4$

8.21   (a)   *Lattice energy* is the energy required to totally separate one mole of solid ionic compound into its gaseous ions.

       (b)   The magnitude of the lattice energy depends on the magnitudes of the charges of the two ions, their radii, and the arrangement of ions in the lattice. The main factor is the charges, because the radii of ions do not vary over a wide range.

8.23   KF, 808 kJ/mol; CaO, 3414 kJ/mol; ScN, 7547 kJ/mol

      The sizes of the ions vary as follows: $Sc^{3+} < Ca^{2+} < K^+$ and $F^- < O^{2-} < N^{3-}$. Therefore, the inter-ionic distances are similar. According to Coulomb's law for compounds with similar ionic separations, the lattice energies should be related as the product of the charges of the ions. The lattice energies above are approximately related as (1)(1): (2)(2): (3)(3) or 1:4:9. Slight variations are due to the small differences in ionic separations.

8.25   Since the ionic charges are the same in the two compounds, the K–Br and Cs–Cl separations must be approximately equal. Since the radii are related as $Cs^+ > K^+$ and $Br^- > Cl^-$, the difference between Cs^+ and K^+ must be approximately equal to the difference between Br^- and Cl^-. This is somewhat surprising, since K^+ and Cs^+ are two rows apart and Cl^- and Br^- are only one row apart.

8.27   Equation 8.4 predicts that as the oppositely charged ions approach each other, the energy of interaction will be large and negative. This more than compensates for the energy required to form Ca^{2+} and O^{2-} from the neutral atoms (see Figure 8.5 for the formation of NaCl).

8.29 $RbCl(s) \rightarrow Rb^+(g) + Cl^-(g)$ ΔH (lattice energy) = ?

By analogy to NaCl, Figure 8.5, the lattice energy is

$$\Delta H_{latt} = -\Delta H_f^{\circ} \, RbCl(s) + \Delta H_f^{\circ} \, Rb(g) + \Delta H_f^{\circ} \, Cl(g) + I_1(Rb) + E(Cl)$$
$$= -(-430.5 \text{ kJ}) + 85.8 \text{ kJ} + 121.7 \text{ kJ} + 403 \text{ kJ} + (-349 \text{ kJ}) = +692 \text{ kJ/mol}$$

This value is smaller than that for NaCl (+788 kJ/mol) because Rb^+ has a larger ionic radius than Na^+. This means that the value of d in the denominator of Equation 8.4 is larger for RbCl, and the potential energy of the electrostatic attraction is smaller.

Covalent Bonding, Electronegativity, and Bond Polarity
(sections 8.3 and 8.4)

8.31   (a)   A *covalent bond* is the bond formed when two atoms share one or more pairs of electrons.

(b)   Any simple compound whose component atoms are nonmetals, such as H_2, SO_2, and CCl_4, are molecular and have covalent bonds between atoms.

(c)   Covalent because it is a gas even below room temperature.

8.33   *Analyze/Plan.* Follow the logic in Sample Exercise 8.3.   *Solve.*

Check. Each pair of shared electrons in $SiCl_4$ is shown as a line; each atom is surrounded by an octet of electrons.

8.35   (a) $:\ddot{O}\!\!=\!\!\ddot{O}:$

(b)   A double bond is required because there are not enough electrons to satisfy the octet rule with single bonds and unshared pairs.

(c)   The greater the number of shared electron pairs between two atoms, the shorter the distance between the atoms. If O_2 has a double bond, the O–O distance will be shorter than the O–O single bond distance.

8.37   (a)   *Electronegativity* is the ability of an atom in a molecule (a bonded atom) to attract electrons to itself.

(b)   The range of electronegativities on the Pauling scale is 0.7–4.0.

(c)   Fluorine, F, is the most electronegative element.

(d)   Cesium, Cs, is the least electronegative element that is not radioactive.

8.39   *Plan.* Electronegativity increases going up and to the right in the periodic table. *Solve.*

(a)   Mg   (b)   S   (c)   C   (d)   As

Check. The electronegativity values in Figure 8.7 confirm these selections.

8.41   The bonds in (a), (c) and (d) are polar because the atoms involved differ in electronegativity. The more electronegative element in each polar bond is:

      (a) F   (c) O   (d) I

8.43   (a)   *Analyze/Plan.* Q is the charge at either end of the dipole. $Q = \mu/r$. The values for HBr are $\mu = 0.82$ D and $r = 1.41$ Å. Change Å to m and use the definition of debyes and the charge of an electron to calculate effective charge in units of e. *Solve.*

$$Q = \frac{\mu}{r} = \frac{0.82\,\text{D}}{1.41\,\text{Å}} \times \frac{1\,\text{Å}}{1 \times 10^{-10}\,\text{m}} \times \frac{3.34 \times 10^{-30}\,\text{C·m}}{1\,\text{D}} \times \frac{1\,e}{1.60 \times 10^{-19}\,\text{C}} = 0.12\,e$$

      (b)   From Sample Exercise 8.5, the effective charges on H and Cl in the HCl molecule are $+0.178\,e$ and $-0.178\,e$, respectively. From part (a), the effective charges on H and Br are $+0.12\,e$ and $-0.12\,e$. HBr has a smaller dipole moment and longer bond length than HCl; these properties both contribute to the smaller charge separation in HBr.

8.45   *Analyze/Plan.* Generally, compounds formed by a metal and a nonmetal are described as ionic, while compounds formed from two or more nonmetals are covalent. However, substances with metals in a high oxidation state often have properties of molecular compounds. In this exercise we know that one substance in each pair is molecular and one is ionic; we may need to distinguish by comparison. *Solve.*

      (a) SiF_4, metalloid and nonmetal, molecular, silicon tetrafluoride.

             LaF_3, metal and nonmetal, ionic, lanthanum(III) fluoride

      (b) $FeCl_2$, metal and nonmetal, ionic, iron(II) chloride

             $ReCl_6$, metal in high oxidation state, Re(VI), molecular, rhenium hexachloride

      (c) $PbCl_4$, metal and nonmetal, Pb(IV) is relatively high oxidation state, molecular (by contrast with RbCl, which is definitely ionic), lead tetrachloride

             RbCl, metal and nonmetal, ionic, rubidium chloride

Lewis Structures; Resonance Structures (sections 8.5 and 8.6)

8.47   *Analyze.* Counting the **correct number of valence electrons** is the foundation of every Lewis structure.   *Plan/Solve.*

      (a)   Count valence electrons: $4 + (4 \times 1) = 8$ e$^-$, 4 e$^-$ pairs. Follow the procedure in Sample Exercise 8.6.

$$
\begin{array}{c}
\text{H} \\
| \\
\text{H--Si--H} \\
| \\
\text{H}
\end{array}
$$

      (b)   Valence electrons: $4 + 6 = 10$ e$^-$, 5 e$^-$ pairs

$$:\text{C} \equiv \text{O}:$$

      (c)   Valence electrons: $[6 + (2 \times 7)] = 20$ e$^-$, 10 e$^-$ pairs

$$:\ddot{\text{F}} - \ddot{\text{S}} - \ddot{\text{F}}:$$

i.   Place the S atom in the middle and connect each F atom with a single bond; this requires 2 e⁻ pairs.

ii. Complete the octets of the F atoms with nonbonded pairs of electrons; this requires an additional 6 e⁻ pairs.

iii. The remaining 2 e⁻ pairs complete the octet of the central S atom.

(d) (Draw the structure that obeys the octet rule, for now.) 32 valence e⁻, 16 e⁻ pairs

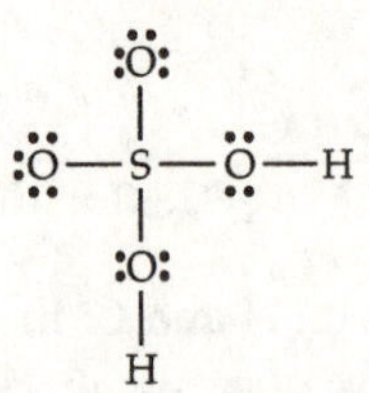

(e) Follow Sample Exercise 8.8. 20 valence e⁻, 10 e⁻ pairs

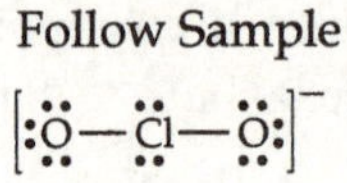

(f) 14 valence e⁻, 7 e⁻ pairs

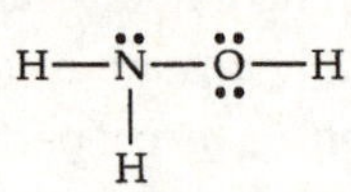

Check. In each molecule, bonding e⁻ pairs are shown as lines, and each atom is surrounded by an octet of electrons (duet for H).

8.49 (a) *Formal charge* is the charge on each atom in a molecule, assuming all atoms have the same electronegativity.

(b) Formal charges are not actual charges. They are a bookkeeping system that assumes perfect covalency, one extreme for the possible electron distribution in a molecule.

(c) The other extreme is represented by oxidation numbers, a bookkeeping system which assumes that the more electronegative element holds all electrons in a bond. The true electron distribution is some composite of the two extremes.

8.51 *Analyze/Plan.* Draw the correct Lewis structure: count valence electrons in each atom, total valence electrons and electron pairs in the molecule or ion; connect bonded atoms with a line, place the remaining e⁻ pairs as needed, in nonbonded pairs or multiple bonds, so that each atom is surrounded by an octet (or duet for H). Calculate formal charges: assign electrons to individual atoms [nonbonding e⁻ + 1/2 (bonding e⁻)]; formal charge = valence electrons − assigned electrons. Assign oxidation numbers, assuming that the more electronegative element holds all electrons in a bond.

Solve. Formal charges are shown near the atoms, oxidation numbers (ox. #) are listed below the structures.

(a) 16 e⁻, 8 e⁻ pairs

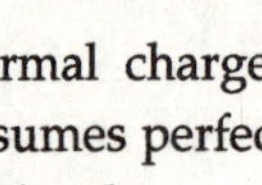

ox. #: O, −2; C, +4; S, −2

(b) 26 valence e⁻, 13 e⁻ pairs

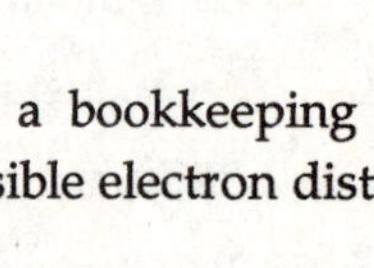

ox #: S, +4; Cl, −1; O, −2

(c)   26 valence e^-, 13 e^- pairs   (d)   20 valence e^-, 10 e^- pairs

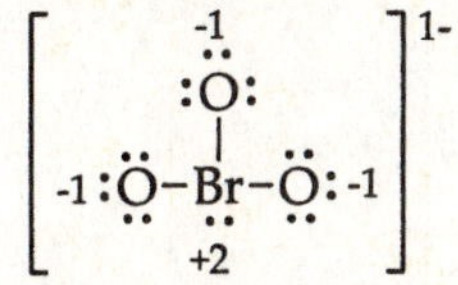

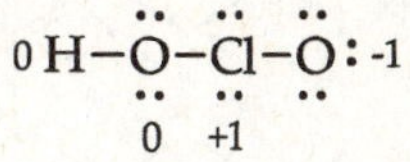

$$\begin{bmatrix} \overset{-1}{\ddot{O}} \\ | \\ {}_{-1}\ddot{O}-\underset{+2}{Br}-\ddot{O}{:}_{-1} \end{bmatrix}^{1-}$$

$${}_{0}H-\overset{0}{\underset{..}{\ddot{O}}}-\overset{+1}{\underset{..}{Cl}}-\ddot{O}{:}_{-1}$$

ox. #: Br, +5; O, –2   ox. #: Cl, +3; H, +1; O, –2

Check. Each atom is surrounded by an octet (or duet) and the sum of the formal charges and oxidation numbers is the charge on the particle.

8.53   (a)   *Plan.* Count valence electrons, draw all possible correct Lewis structures, taking note of alternate placements for multiple bonds.   *Solve.*

18 e^-, 9 e^- pairs

$$\begin{bmatrix} \ddot{O}=\ddot{N}-\ddot{O}{:} \end{bmatrix}^- \longleftrightarrow \begin{bmatrix} {:}\ddot{O}-\ddot{N}=\ddot{O} \end{bmatrix}^-$$

Check. The octet rule is satisfied.

(b)   *Plan.* Isoelectronic species have the same number of valence electrons and the same electron configuration.   *Solve.*

A single O atom has 6 valence electrons, so the neutral ozone molecule O_3 is isoelectronic with NO_2^-.

$$\ddot{O}=\ddot{O}-\ddot{O}{:} \longleftrightarrow {:}\ddot{O}-\ddot{O}=\ddot{O}$$

Check. The octet rule is satisfied.

(c)   Since each N–O bond has partial double bond character, the N–O bond length in NO_2^- should be shorter than N–O single bonds but longer than N=O double bonds.

8.55   *Plan/Solve.* The Lewis structures are as follows:

5 e^- pairs   8 e^- pairs

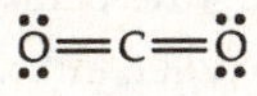

$${:}C{\equiv}O{:}$$ $$\ddot{O}=C=\ddot{O}$$

12 e^- pairs

$$\begin{bmatrix} {:}\ddot{O}{:} \\ | \\ {:}\ddot{O} \quad C \quad \ddot{O}{:} \end{bmatrix}^{2-} \begin{bmatrix} {:}\ddot{O}{:} \\ | \\ {:}\ddot{O} \quad C \quad \ddot{O}{:} \end{bmatrix}^{2-} \begin{bmatrix} {:}O{:} \\ \| \\ {:}\ddot{O} \quad C \quad \ddot{O}{:} \end{bmatrix}^{2-}$$

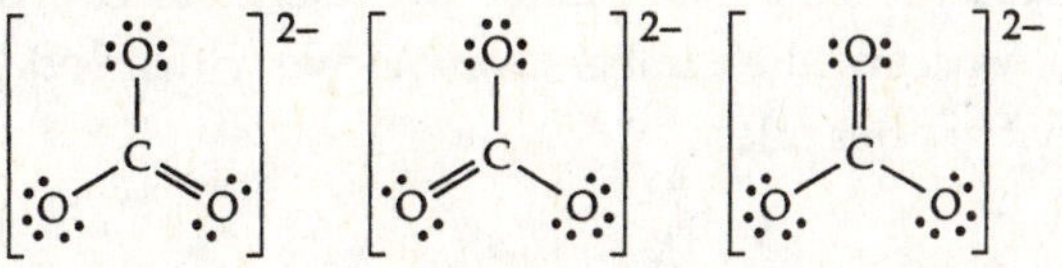

The more pairs of electrons shared by two atoms, the shorter the bond between the atoms. The average number of electron pairs shared by C and O in the three species is 3 for CO, 2 for CO_2, and 1.33 for CO_3^{2-}. This is also the order of increasing bond length: $CO < CO_2 < CO_3^{2-}$.

8.57   (a)   Two equally valid Lewis structures can be drawn for benzene.

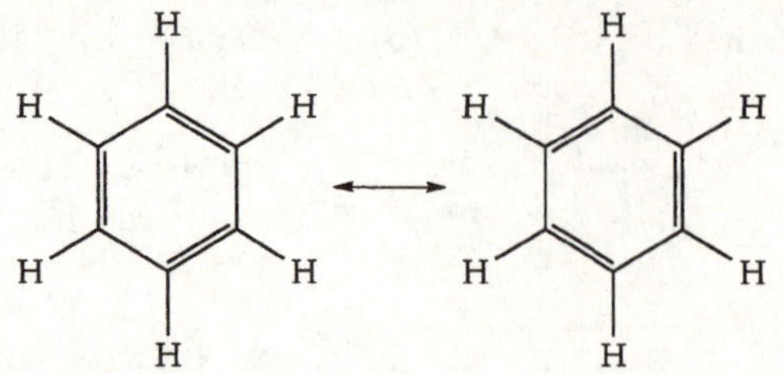

Each structure consists of alternating single and double C–C bonds; a particular bond is single in one structure and double in the other. The concept of resonance dictates that the true description of bonding is some hybrid or blend of the two Lewis structures. The most obvious blend of these two resonance structures is a molecule with six equivalent C–C bonds, each with some but not total double-bond character. If the molecule has six equivalent C–C bonds, the lengths of these bonds should be equal.

(b)   The resonance model described in (a) has six equivalent C–C bonds, each with some double bond character. That is, more than one pair but less than two pairs of electrons is involved in each C–C bond. This model predicts a uniform C–C bond length that is shorter than a single bond but longer than a double bond.

Exceptions to the Octet Rule (section 8.7)

8.59   (a)   The *octet rule* states that atoms will gain, lose, or share electrons until they are surrounded by eight valence electrons.

(b)   The octet rule applies to both ionic and covalent compounds, with some exceptions. In ionic compounds, the cation has lost electrons to achieve an octet and the anion has gained electrons to achieve an octet. For example, in $MgCl_2$, Mg loses 2 e^- to become Mg^{2+} with the electron configuration of Ne. Each Cl atom gains one electron to form Cl^- with the electron configuration of Ar. In covalent compounds, such as CCl_4, atoms share electrons in order to surround themselves with an octet.

8.61   No chlorine oxide will obey the octet rule. Oxygen has six valence electrons, an even number. Any number of oxygen atoms in the molecule will result in an even number of valence electrons from oxygen. Chlorine has seven valence electrons, an odd number. For neutral chlorine oxide molecules, the total number of valence electrons will be an (odd + even) sum, which is always an odd number. A molecule with an odd number of valence electrons cannot obey the octet rule.

8.63   (a)   26 e^-, 13 e^- pairs

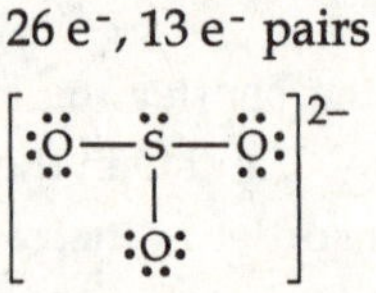

$$\left[:\ddot{O}—\ddot{S}—\ddot{O}: \atop \quad \underset{:\ddot{O}:}{|} \right]^{2-}$$

Other resonance structures with one, two, or three double bonds can be drawn. While a structure with three double bonds minimizes formal charges, all structures with double bonds violate the octet rule. The octet rule vs formal charge debate is ongoing.

96

(b) $6\ e^-$, $3\ e^-$ pairs

$$H—Al—H$$
$$\overset{|}{H}$$

6 electrons around Al; impossible to satisfy octet rule with only 6 valence electrons.

(c) $16\ e^-$, $8\ e^-$ pairs

$$\left[:N\equiv N—\ddot{\underset{..}{N}}:\right]^{-} \longleftrightarrow \left[:\ddot{\underset{..}{N}}—N\equiv N:\right]^{-} \longleftrightarrow \left[:\ddot{N}=N=\ddot{N}:\right]^{-}$$

3 resonance structures; all obey octet rule.

(d) $20\ e^-$, $10\ e^-$ pairs

$$:\ddot{\underset{..}{C}}l:$$
$$:\ddot{\underset{..}{C}}l—\overset{|}{\underset{|}{C}}—H$$
$$\overset{}{H}$$

Obeys octet rule.

(e) $40\ e^-$, $20\ e^-$ pairs

$$:\ddot{F}:$$
$$:\ddot{F}—Sb\overset{\ddot{F}:}{\underset{\ddot{F}:}{}}$$
$$:\ddot{F}:$$

Does not obey octet rule; $10\ e^-$ around central Sb

Does not obey the octet rule.

8.65   (a) $16\ e^-$, $8\ e^-$ pairs

$$:\ddot{\underset{..}{C}}l—Be—\ddot{\underset{..}{C}}l:$$

This structure violates the octet rule; Be has only $4\ e^-$ around it.

(b) $\ddot{\underset{..}{C}}l=Be=\ddot{\underset{..}{C}}l \longleftrightarrow :\ddot{\underset{..}{C}}l—Be\equiv Cl: \longleftrightarrow :Cl\equiv Be—\ddot{\underset{..}{C}}l:$

(c)   The formal charges on each of the atoms in the four resonance structures are:

$$:\ddot{\underset{..}{C}}l—Be—\ddot{\underset{..}{C}}l: \qquad \ddot{\underset{..}{C}}l=Be=\ddot{\underset{..}{C}}l \qquad :\ddot{\underset{..}{C}}l—Be\equiv Cl: \qquad :Cl\equiv Be—\ddot{\underset{..}{C}}l:$$
$$\quad 0 \qquad 0 \qquad 0 \qquad\quad +1 \quad -2 \quad +1 \qquad 0 \quad -2 \quad +2 \qquad +2 \quad -2 \quad 0$$

Formal charges are minimized on the structure that violates the octet rule; this form is probably dominant.

8.67 $26\ e^-$, $13\ e^-$ pairs

$$\left[\begin{array}{c} \overset{0}{}\ \overset{+1}{}\ \overset{-1}{} \\ H—\ddot{\underset{..}{O}}—\overset{}{S}—\ddot{\underset{..}{O}}: \\ \underset{|}{} \\ :\ddot{\underset{..}{O}}: \\ -1 \end{array}\right]^{1-} \qquad \left[\begin{array}{c} \overset{0}{}\ \overset{0}{}\ \overset{0}{} \\ H—\ddot{\underset{..}{O}}—S=\ddot{\underset{..}{O}} \\ \underset{|}{} \\ :\ddot{\underset{..}{O}}: \\ -1 \end{array}\right]^{1-} \qquad \left[\begin{array}{c} \overset{0}{}\ \overset{-1}{}\ \overset{0}{} \\ H—\ddot{\underset{..}{O}}—S=\ddot{\underset{..}{O}} \\ \underset{\|}{} \\ :\ddot{\underset{..}{O}}: \\ 0 \end{array}\right]^{1-}$$

Three resonance structures for HSO_3^- are shown above. Because we are dealing with an ion with a 1– charge, the sum of the formal charges of the atoms will be –1. That is, no correct Lewis structure will have all atoms with zero formal charge. The structure with no double bonds obeys the octet rule for all atoms, but does not lead to minimum formal charge. The structures with one and two double bonds both minimize formal charge but do not obey the octet rule. Of these two, the structure with one double bond is preferred because the formal charge is localized on the more electronegative oxygen atom.

Bond Enthalpies (section 8.8)

8.69   *Analyze*. Given: structural formulas. Find: enthalpy of reaction.

Plan. Count the number and kinds of bonds that are broken and formed by the reaction. Use bond enthalpies from Table 8.4 and Equation 8.12 to calculate the overall enthalpy of reaction, ΔH.   *Solve*.

(a) $\Delta H = 2D(O-H) + D(O-O) + 4D(C-H) + D(C=C)$

        $-2D(O-H) - 2D(O-C) - 4D(C-H) - D(C-C)$

    $\Delta H = D(O-O) + D(C=C) - 2D(O-C) - D(C-C)$

        $= 146 + 614 - 2(358) - 348 = -304 \text{ kJ}$

(b) $\Delta H = 5D(C-H) + D(C \equiv N) + D(C=C) - 5D(C-H) - D(C \equiv N) - 2D(C-C)$

        $= D(C=C) - 2D(C-C) = 614 - 2(348) = -82 \text{ kJ}$

(c) $\Delta H = 6D(N-Cl) - 3D(Cl-Cl) - D(N \equiv N)$

        $= 6(200) - 3(242) - 941 = -467 \text{ kJ}$

8.71   *Plan*. Draw structural formulas so bonds can be visualized. Then use Table 8.4 and Equation 8.12.   *Solve*.

(a) 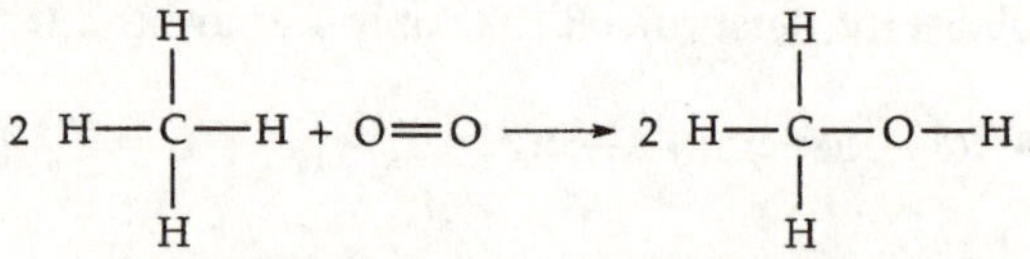
$$2 \underset{\underset{H}{|}}{\overset{\overset{H}{|}}{H-C-H}} + O{=}O \longrightarrow 2 \underset{\underset{H}{|}}{\overset{\overset{H}{|}}{H-C-O-H}}$$

    $\Delta H = 8D(C-H) + D(O=O) - 6D(C-H) - 2D(C-O) - 2D(O-H)$

        $= 2D(C-H) + D(O=O) - 2D(C-O) - 2D(O-H)$

        $= 2(413) + (495) - 2(358) - 2(463) = -321 \text{ kJ}$

(b) $H-H \quad + \quad Br-Br \quad \rightarrow \quad 2 \text{ H-Br}$

    $\Delta H = D(H-H) + D(Br-Br) - 2D(H-Br)$

        $= (436) + (193) - 2(366) = -103 \text{ kJ}$

(c) $2 \text{ H-O-O-H} \rightarrow 2 \text{ H-O-H} + O{=}O$

    $\Delta H = 4D(O-H) + 2D(O-O) - 4D(O-H) - D(O=O)$

    $\Delta H = 2D(O-O) - D(O=O) = 2(146) - (495) = -203 \text{ kJ}$

8.73   *Plan.* Draw structural formulas so bonds can be visualized. Then use Table 8.4 and
        Equation 8.12.   *Solve.*

(a) $:N\equiv N: + 3\,H—H \longrightarrow 2\,H—\overset{..}{\underset{\underset{H}{|}}{N}}—H$

$$\Delta H = D(N \equiv N) + 3D(H–H) - 6(N–H) = 941\ \text{kJ} + 3(436\ \text{kJ}) - 6(391\ \text{kJ})$$

$$= -97\ \text{kJ}/2\ \text{mol NH}_3;\ \text{exothermic}$$

(b)   *Plan.* Use Equation 5.31 to calculate ΔH_{rxn} from ΔH_f° values.

$\Delta H_{rxn}^\circ = \Sigma n\,\Delta H_f^\circ\ (\text{products}) - \Sigma n\,\Delta H_f^\circ\ (\text{reactants}).\quad \Delta H_f^\circ\ \text{NH}_3(g) = -46.19\ \text{kJ}.$
Solve.

$$\Delta H_{rxn}^\circ = 2\,\Delta H_f^\circ\ \text{NH}_3(g) - 3\,\Delta H_f^\circ\ \text{H}_2(g) - \Delta H_f^\circ\ \text{N}_2(g)$$

$$\Delta H_{rxn}^\circ = 2(-46.19) - 3(0) - 0 = -92.38\ \text{kJ}/2\ \text{mol NH}_3$$

The ΔH calculated from bond enthalpies is slightly more exothermic (more
negative) than that obtained using ΔH_f° values.

8.75   The average Ti–Cl bond enthalpy is just the average of the four values listed.
        430 kJ/mol.

Additional Exercises

8.77   Six nonradioactive elements in the periodic table have Lewis symbols with single dots.
        Yes, they are in the same family, assuming H is placed with the alkali metals, as it is on
        the inside cover of the text. This is because the Lewis symbol represents the number of
        valence electrons of an element, and all elements in the same family have the same
        number of valence electrons. By definition of a family, all elements with the same Lewis
        symbol must be in the same family.

8.81   The charge on M is likely to be 3+. According to Table 8.2, the lattice energy for an ionic
        compound with the general formula MX and a charge of 2+ on the metal will be in the
        range of $3\text{-}4 \times 10^3$ kJ/mol. The charge on M must be greater than 2+. ScN, where the
        charge on Sc is 3+, has a lattice energy of 7547 kJ/mol. It is reasonable to conclude that
        the charge on M is 3+, and the M–X distance is greater than the Sc–N distance.

8.85   (a)   B–O. The most polar bond will be formed by the two elements with the greatest
              difference in electronegativity. Since electronegativity increases moving right and
              up on the periodic table, the possibilities are B–O and Te–O. These two bonds are
              likely to have similar electronegativitiy differences (3 columns apart vs. 3 rows
              apart). Values from Figure 8.6 confirm the similarity, and show that B–O is
              slightly more polar.

        (b)   Te–I. Both are in the fifth row of the periodic table and have the two largest
              covalent radii among this group of elements.

(c) TeI_2. Te needs to participate in two covalent bonds to satisfy the octet rule, and each I atom needs to participate in one bond, so by forming a TeI_2 molecule, the octet rule can be satisfied for all three atoms.

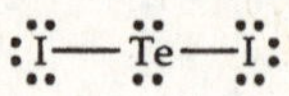

$$:\ddot{I}—\ddot{Te}—\ddot{I}:$$

(d) B_2O_3. Although this is probably not a purely ionic compound, it can be understood in terms of gaining and losing electrons to achieve a noble-gas configuration. If each B atom were to lose 3 e^- and each O atom were to gain 2 e^-, charge balance and the octet rule would be satisfied.

P_2O_3. Each P atom needs to share 3 e^- and each O atom 2 e^- to achieve an octet. Although the correct number of electrons seem to be available, a correct Lewis structure is difficult to imagine. In fact, phosphorus (III) oxide exists as P_4O_6 rather than P_2O_3 (Chapter 22).

8.90   Formal charge (FC) = # valence e^- – (# nonbonding e^- + 1/2 # bonding e^-)

(a)   18 e^-, 9 e^- pairs

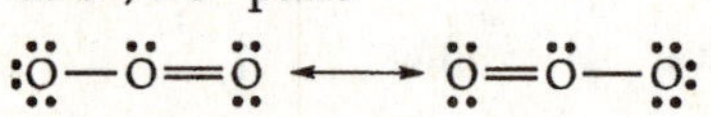

$$:\ddot{O}—\ddot{O}=\ddot{O} \longleftrightarrow \ddot{O}=\ddot{O}—\ddot{O}:$$

FC for the central O = 6 – [2 + 1/2 (6)] = +1

(b)   48 e^-, 24 e^- pairs

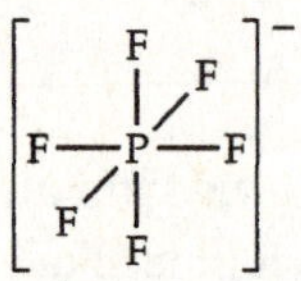

FC for P = 5 – [0 + 1/2 (12)] = –1

The three nonbonded pairs on each F have been omitted.

(c)   17 e^-; 8 e^- pairs, 1 odd e^-

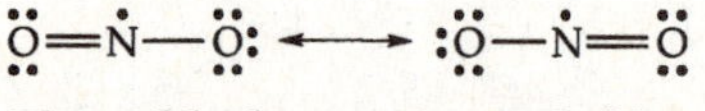

$$\ddot{O}=\dot{N}—\ddot{O}: \longleftrightarrow :\ddot{O}—\dot{N}=\ddot{O}$$

The odd electron is probably on N because it is less electronegative than O. Assuming the odd electron is on N, FC for N = 5 – [1 + 1/2 (6)] = +1. If the odd electron is on O, FC for N = 5 – [2 + 1/2 (6)] = 0.

(d)   28 e^-, 14 e^- pairs   (e)   32 e^-, 16 e^- pairs

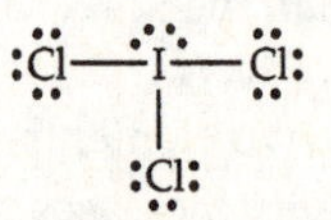

FC for I = 7 – [4 + 1/2 (6)] = 0   FC for Cl = 7 – [0 + 1/2 (8)] = +3

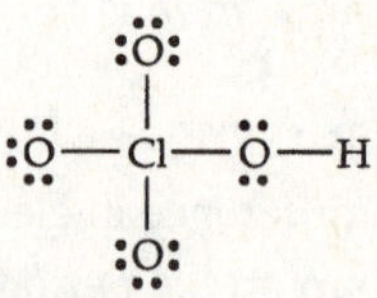

8.95   An experimentally determined molecular structure will reveal bond lengths and angles of the B–A=B molecule. If resonance structures are important, the two B–A bond lengths will be identical, or nearly so. If the molecule features one single and one double bond, the lengths will be significantly different. (While bond angles often reveal bonding details, in this case bond lengths are telling.)

8.98   (a)

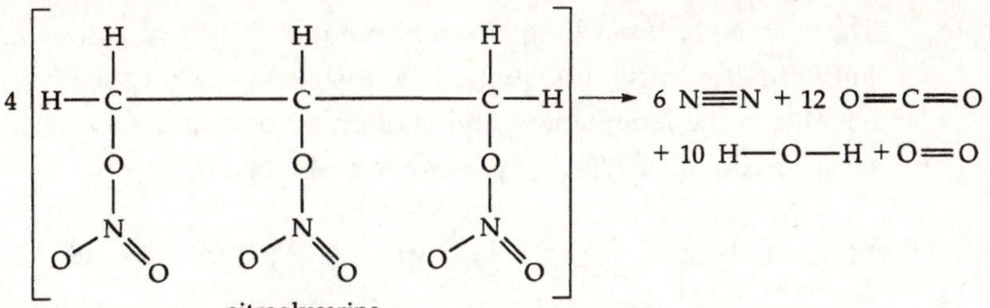

nitroglycerine

$$\Delta H = 20D(C-H) + 8D(C-C) + 12D(C-O) + 24D(O-N) + 12D(N=O)$$
$$- [6D(N \equiv N) + 24D(C=O) + 20D(H-O) + D(O=O)]$$

$$\Delta H = 20(413) + 8(348) + 12(358) + 24(201) + 12(607)$$
$$- [6(941) + 24(799) + 20(463) + 495]$$
$$= -7129 \text{ kJ}$$

$$1.00 \text{ g } C_3H_5N_3O_9 \times \frac{1 \text{ mol } C_3H_5N_3O_9}{227.1 \text{ g } C_3H_5N_3O_9} \times \frac{-7129 \text{ kJ}}{4 \text{ mol } C_3H_5N_3O_9} = 7.85 \text{ kJ/g } C_3H_5N_3O_9$$

(b) $4C_7H_5N_3O_6(s) \rightarrow 6N_2(g) + 7CO_2(g) + 10H_2O(g) + 21C(s)$

Integrative Exercises

8.101   (a) Ti^{2+} : $[Ar]3d^2$; Ca : $[Ar]4s^2$. Yes. The two valence electrons in Ti^{2+} and Ca are in different principle quantum levels and different subshells.

(b)   According to the Aufbau Principle, valence electrons will occupy the lowest energy empty orbital. Thus, in Ca the 4s is lower in energy than the 3d, while in Ti^{2+}, the 3d is lower in energy than the 4s.

(c)   No. Since there is only one 4s orbital, the two valence electrons in Ca are paired. There are five degenerate 3d orbitals, so the two valence electrons in Ti^{2+} are unpaired. Ca has no unpaired electrons, Ti^{2+} has two.

8.107   (a)   Assume 100 g.

$$62.04 \text{ g Ba} \times \frac{1 \text{ mol}}{137.33 \text{ g Ba}} = 0.4518 \text{ mol Ba}; 0.4518 / 0.4518 = 1.0$$

$$37.96 \text{ g N} \times \frac{1 \text{ mol}}{14.007 \text{ g N}} = 2.710 \text{ mol N}; 2.710 / 0.4518 = 6.0$$

The empirical formula is BaN_6. Ba has an ionic charge of 2+, so there must be two 1– azide ions to balance the charge. The formula of each azide ion is N_3^-.

(b)   16 e⁻, 8 e⁻ pairs

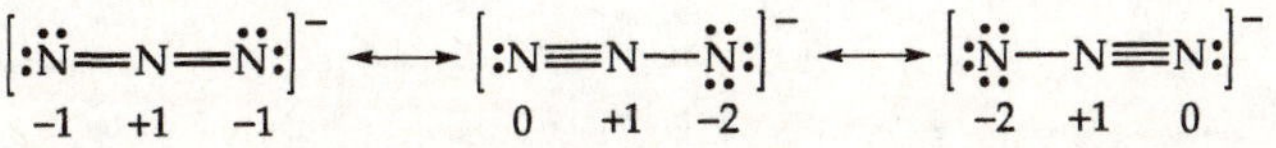

(c)   The structure with two double bonds minimizes formal charges and is probably the main contributor.

(d)   The two N–N bond lengths will be equal. The two minor contributors would individually cause unequal N–N distances, but collectively they contribute equally to the lengthening and shortening of each bond. The N–N distance will be approximately 1.24 Å, the average N=N distance.

8.112   (a) $Br_2(l) \rightarrow 2Br(g)$ $\Delta H° = 2\Delta H_f° \, Br(g) = 2(111.8) \, kJ = 223.6 \, kJ$

(b) $CCl_4(l) \rightarrow C(g) + 4Cl(g)$

$\Delta H° = \Delta H_f° \, C(g) + 4\Delta H_f° \, Cl(g) - \Delta H_f° \, CCl_4(l)$

$\qquad = 718.4 \, kJ + 4(121.7) \, kJ - (-139.3) \, kJ = 1344.5$

$\dfrac{1344.5 \, kJ}{4 \, C-Cl \text{ bonds}} = 336.1 \, kJ$

(c) $H_2O_2(l) \rightarrow 2H(g) + 2O(g)$

$\underline{2H(g) + 2O(g) \rightarrow 2OH(g)}$

$\qquad H_2O_2(l) \rightarrow 2OH(g)$

$D(O-O)(l) = 2\Delta H_f° \, H(g) + 2\Delta H_f° \, O(g) - \Delta H_f° \, H_2O_2(l) - 2D(O-H)(g)$

$\qquad\qquad = 2(217.94) \, kJ + 2(247.5) \, kJ - (-187.8) \, kJ - 2(463) \, kJ$

$\qquad\qquad = 193 \, kJ$

(d)   The data are listed below.

bond	D gas kJ/mol	D liquid kJ/mol
Br–Br	193	223.6
C–Cl	328	336.1
O–O	146	192.7

Breaking bonds in the liquid requires more energy than breaking bonds in the gas phase. For simple molecules, bond dissociation from the liquid phase can be thought of in two steps:

       molecule (l) → molecule (g)

       molecule (g) → atoms (g)

The first step is evaporation or vaporization of the liquid and the second is bond dissociation in the gas phase. Average bond enthalpy in the liquid phase is then the sum of the enthalpy of vaporization for the molecule and the gas phase bond dissociation enthalpies, divided by the number of bonds dissociated. This is greater than the gas phase bond dissociation enthalpy owing to the contribution from the enthalpy of vaporization.

9 Molecular Geometry and Bonding Theories

Visualizing Concepts

9.1   Removing an atom from the equatorial plane of trigonal bipyramid in Figure 9.3 creates a seesaw shape. It might appear that you could also obtain a seesaw by removing two atoms from the square plane of the octahedron. However, one of the B–A–B angles in the seesaw is 120°, so it must be derived from a trigonal bipyramid.

9.3   *Analyze/Plan.*   Visualize the molecular geometry and the electron domain geometries that could produce it. Confirm your choices with Tables 9.2 and 9.3. On Table 9.3, note that octahedral electron domain geometry results in only 3 possible molecular geometries: octahedral, square pyramidal and square planar (not T-shaped, bent or linear).   *Solve.*

   (a)   2. Molecular geometry: linear. Possible electron domain geometries: linear, trigonal bipyramidal

   (b)   1. Molecular geometry, T-shaped.   Possible electron domain geometries: trigonal bipyramidal

   (c)   1. Molecular geometry, octahedral. Possible electron domain geometries: octahedral

   (d)   1. Molecular geometry, square-pyramidal. Possible electron domain geometries: octahedral

   (e)   1. Molecular geometry, square planar. Possible electron domain geometries: octahedral

   (f)   1. Molecular geometry, triangular pyramid. Possible electron domain geometries: trigonal bipyramidal. This is an unusual molecular geometry which is not listed in Table 9.3. It could occur if the equatorial substituents on the trigonal bipyramid were extremely bulky, causing the nonbonding electron pair to occupy an axial position.

9.5   (a)   The reference point for zero energy on the diagram corresponds to a state where the two Cl atoms are separate and not interacting. This corresponds to an infinite Cl–Cl distance beyond the right extreme of the horizontal axis. The point near the left side of the plot where the curve intersects the x-axis at E = 0 has no special meaning.

   (b)   According to the valence-bond model, as atoms approach, their valence atomic orbitals overlap, allowing two electrons of opposite spin to mutually occupy space between the two nuclei. Energy decreases as atom separation decreases because the valence electrons of one atom come close enough to the other atom to be stabilized by both nuclei instead of just one nucleus.

(c)   The Cl–Cl distance at the energy minimum on the plot is the Cl–Cl bond length.

(d)   At interatomic separations shorter than the bond distance, the two nuclei begin to repel each other, increasing the overall energy of the system.

(e)   Energy is shown on the y-axis of the plot. The minimum energy for the two atoms represents the stabilization obtained by bringing two Cl atoms together at the optimum (bond) distance. The y-coordinate of the minimum point on the plot corresponds to the Cl–Cl bond energy, or bond strength.

9.6 SiF_4 32 e⁻, 16 e⁻ pr PF_3 26 e⁻, 13 e⁻ pr SF_2 20 e⁻, 10 e⁻ pr

        ~109°   ~107°   ~105°

In all three molecules, the electron domain geometry is tetrahedral and the approximate bond angle is 109°. But, SiF_4 has 0 nonbonding electron pairs, PF_3 has 1 nonbonding pair, and SF_2 has 2 nonbonding pairs. If nonbonding electron pairs occupy more space than bonding pairs, we expect the bond angles to decrease in the series, perhaps 109°, 107°, 105°.

9.9   *Analyze/Plan.* σ molecular orbitals (MOs) are symmetric about the internuclear axis, π MOs are not. Bonding MOs have most of their electron density in the area between the nuclei, antibonding MOs have a node between the nuclei.

(a)   (i)   Two s atomic orbitals (electron density at each nucleus).

        (ii)   Two p atomic orbitals overlapping end-to-end (node near each nucleus).

        (iii)   Two p atomic orbitals overlapping side-to-side (node near each nucleus).

(b)   (i)   σ-type (symmetric about the internuclear axis, s orbitals can produce only σ overlap).

        (ii)   σ-type (symmetric about internuclear axis)

        (iii)   π-type (not symmetric about internuclear axis, side-to-side overlap)

(c)   (i)   antibonding (node between nuclei)

        (ii)   bonding (concentration of electron density between nuclei)

        (iii)   antibonding (node between nuclei)

(d)   (i)   The nodal plane is between the atom centers, perpendicular to the interatomic axis and equidistant from each atom.

        (ii)   There are two nodal planes; both are perpendicular to the interatomic axis. One is left of the left atom and the second is right of the right atom.

        (iii)   There are two nodal planes; one is between the atom centers, perpendicular to the interatomic axis and equidistant from each atom. The second contains the interatomic axis and is perpendicular to the first.

Molecular Shapes; the VSEPR Model (sections 9.1 and 9.2)

9.11   (a)   Yes. The stated shape, linear, defines the bond angle (180°) and the A–B bond length tells the size.

      (b)   No. Atom A could have 0 or 3 nonbonding electron pairs, depending on the total number of electron domains around atom A. Only molecules with 2 or 5 total electron domains about the central atom can possibly result in linear molecular geometry. Refer to Tables 9.2 and 9.3.

9.13   A molecule with tetrahedral molecular geometry has an atom at each vertex of the tetrahedron. A trigonal pyramidal molecule has one vertex of the tetrahedron occupied by a nonbonding electron pair rather than an atom. That is, a trigonal pyramid is a tetrahedron with one vacant vertex.

9.15   (a)   An *electron domain* is a region in a molecule where electrons are most likely to be found.

      (b)   Each balloon in Figure 9.5 occupies a volume of space. The best arrangement is one where each balloon has its "own" space, where they are as far apart as possible and repulsions are minimized. Electron domains are negatively charged regions, so they also adopt an arrangement where repulsions are minimized.

9.17   (a)   The number of electron domains in a molecule or ion is the number of bonds (double and triple bonds count as one domain) **plus** the number of nonbonding (lone) electron pairs.

      (b)   A *bonding electron domain* is a region between two bonded atoms that contains one or more pairs of bonding electrons. A *nonbonding electron domain* is localized on a single atom and contains one pair of nonbonding electrons (a lone pair).

9.19   *Analyze/Plan.* Draw the Lewis structure of each molecule and take note of nonbonding (lone) electron pairs about the central atom. *Solve.*

      (a) SiH_4, 8 valence e^-, 4 e^- pr, 0 nonbonding pairs, no effect on molecular shape

$$\begin{array}{c} H \\ | \\ H-Si-H \\ | \\ H \end{array}$$

      (b) PF_3, 26 valence e^-, 13 e^- pr, 1 nonbonding pair on P, influences molecular shape

$$\begin{array}{c} :\ddot{F}-\ddot{P}-\ddot{F}: \\ | \\ :\ddot{F}: \end{array}$$

      (c)   HBr, 8 valence e^-, 4 e^- pr, 3 nonbonding pairs on Br, no effect on molecular shape because Br is not "central"

$$H-\ddot{Br}:$$

      (d)   HCN, 10 valence e^-, 5 e^- pr, 0 nonbonding pairs on C, no effect on molecular shape

$$H-C\equiv N:$$

(e) SO_2, 18 valence e^-, 9 e^- pr, 1 nonbonding pair on S, influences molecular shape

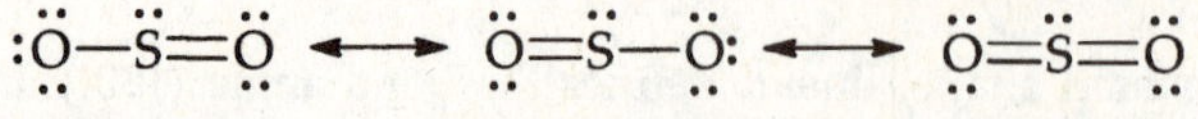

$$:\ddot{O}-\ddot{S}=\ddot{O} \longleftrightarrow \ddot{O}=\ddot{S}-\ddot{O}: \longleftrightarrow \ddot{O}=\ddot{S}=\ddot{O}$$

9.21   *Analyze/Plan.* Draw the Lewis structure of each molecule and count the number of nonbonding (lone) electron pairs. Note that the question asks 'in the molecule' rather than just around the central atom. *Solve.*

(a) $(CH_3)_2S$, 20 valence e^-, 10 e^- pr, 2 nonbonding pairs

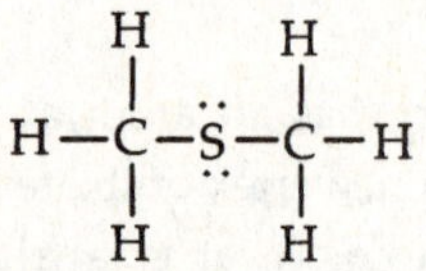

$$\begin{array}{ccc}
H & & H \\
| & & | \\
H-C-\ddot{S}-C-H \\
| & & | \\
H & & H
\end{array}$$

(b)   HCN, 10 valence e^-, 5 e^- pr, 1 nonbonding pair

$$H-C\equiv N:$$

(c) H_2C_2, 10 valence e^-, 5 e^- pr, 0 nonbonding pairs

$$H-C\equiv C-H$$

(d) CH_3F, 14 valence e^-, 7 e^- pr, 3 nonbonding pairs

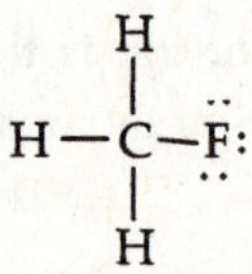

$$\begin{array}{c}
H \\
| \\
H-C-\ddot{F}: \\
| \\
H
\end{array}$$

9.23   The electron-domain geometry indicated by VSEPR describes the arrangement of all bonding and nonbonding electron domains. The molecular geometry describes just the atomic positions. H_2O has the Lewis structure given below; there are four electron domains around oxygen so the electron-domain geometry is tetrahedral, but the molecular geometry of the three atoms is bent.

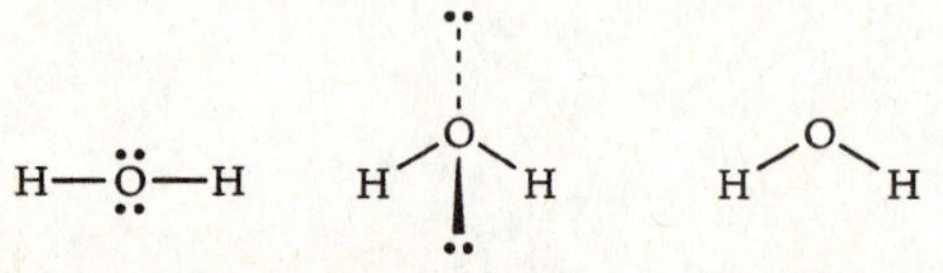

$$H-\ddot{O}-H \qquad \qquad H\overset{\displaystyle O}{\diagdown}H \qquad \qquad H\overset{\displaystyle O}{\diagup\diagdown}H$$

Lewis structure   electron-domain   molecular geometry
                  geometry

We make this distinction because all electron domains must be considered when describing the atomic arrangement and bond angles in a molecule but the molecular geometry or shape is a description of just the atomic positions.

9.25   *Analyze/Plan.* See Tables 9.2 and 9.3. *Solve.*

| | electron-domain | molecular |
| Lewis structure | geometry | geometry |

(a)

            tetrahedral   tetrahedral

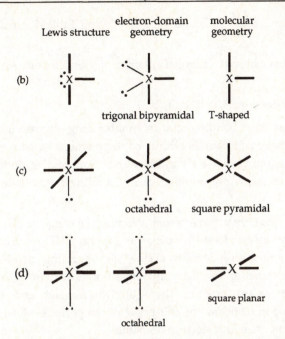

	Lewis structure	electron-domain geometry	molecular geometry
(b)		trigonal bipyramidal	T-shaped
(c)		octahedral	square pyramidal
(d)		octahedral	square planar

9.27　　*Analyze/Plan.* Follow the logic in Sample Exercise 9.1. *Solve.*

bent (b), linear (l), octahedral (oh), seesaw (ss), square pyramidal (sp), square planar (spl), tetrahedral (td), trigonal bipyramidal (tbp), trigonal planar (tr), trigonal pyramidal (tp), T-shaped (T)

	Molecule or ion	Valence electrons	Lewis structure	Electron-domain geometry	Molecular geometry
(a)	HCN	10	:N≡C—H　　N≡C—H	l	l
(b)	SO_3^{2-}	26	*$[:\ddot{O}-\ddot{S}-\ddot{O}:]^{2-}$　$\left[\begin{array}{c}S\\O\ \ O\end{array}\right]^{2-}$	td	tp
(c)	SF_4	34	:F̈—S̈—F̈:　　$\begin{array}{c}F\\F-S:\\F\ \ F\end{array}$	tbp	ss
(d)	PF_6^-	48	$\left[:\ddot{F}-\ddot{P}-\ddot{F}:\right]^-$　$\left[\begin{array}{c}F\ \ F\\F-P-F\\F\ \ F\end{array}\right]^-$	oh	oh
(e)	NH_3Cl^+	14	$\left[\begin{array}{c}:\ddot{Cl}:\\H-N-H\\H\end{array}\right]^+$　$\left[\begin{array}{c}Cl\\N\\H\ \ H\ \ H\end{array}\right]^+$	td	td
(f)	N_3^-	16	*$[:\ddot{N}-N≡N:]^-$　$[N-N≡N]^-$	l	l

*More than one resonance structure is possible. All equivalent resonance structures predict the same molecular geometry.

9.29   *Analyze/Plan.* Work backwards from molecular geometry, using Tables 9.2 and 9.3. *Solve.*

(a)   Electron-domain geometries: i, trigonal planar; ii, tetrahedral; iii, trigonal bipyramidal

(b)   nonbonding electron domains: i, 0; ii, 1; iii, 2

(c)   N and P. Shape ii has three bonding and one nonbonding electron domains. Li and Al would form ionic compounds with F, so there would be no nonbonding electron domains. Assuming that F always has three nonbonding domains, BF_3 and ClF_3 would have the wrong number of nonbonding domains to produce shape ii.

(d)   Cl (also Br and I, since they have seven valence electrons). This T-shaped molecular geometry arises from a trigonal bipyramidal electron-domain geometry with two nonbonding domains (Table 9.3). Assuming each F atom has three nonbonding domains and forms only single bonds with A, A must have seven valence electrons to produce these electron-domain and molecular geometries. It must be in or below the third row of the periodic table, so that it can accommodate more than four electron domains.

9.31   *Analyze/Plan.* Follow the logic in Sample Exercise 9.3.   *Solve.*

(a)   1 – 109°, 2 – 109°   (b)   3 – 109°, 4 – 109°

(c)   5 – 180°   (d)   6 – 120°, 7 – 109°, 8 – 109°

9.33   *Analyze/Plan.* Given the formula of each molecule or ion, draw the correct Lewis structure and use principles of VSEPR to answer the question.   *Solve.*

The three nonbonded electron pairs on each F atom have been omitted for clarity.

The two molecules with trigonal bipyramidal electron domain geometry, PF_5 and SF_4, have more than one F–A–F bond angle.

9.35   *Analyze.* Given: molecular formulas. Find: explain features of molecular geometries.

Plan. Draw the correct Lewis structures for the molecules and use VSEPR to predict and explain observed molecular geometry.   *Solve.*

(a) BrF_4^- 36 e⁻, 18 e⁻ pr BF_4^- 32 e⁻, 16 e⁻ pr

$$\left[\begin{array}{c} :\ddot{F}: \\ | \\ :\ddot{F}\!-\!\underset{..}{Br}\!-\!\ddot{F}: \\ | \\ :\ddot{F}: \end{array}\right]^{-} \qquad \left[\begin{array}{c} :\ddot{F}: \\ | \\ :\ddot{F}\!-\!B\!-\!\ddot{F}: \\ | \\ :\ddot{F}: \end{array}\right]^{-}$$

6 e⁻ pairs around Br   4 e⁻ pairs around B,

octahedral e⁻ domain geometry   tetrahedral e⁻ domain geometry

square planar molecular geometry   tetrahedral molecular geometry

The fundamental feature that determines molecular geometry is the number of electron domains around the central atom, and the number of these that are bonding domains. Although BrF_4^- and BF_4^- are both of the form AX_4^-, the central atoms and thus the number of valence electrons in the two ions are different. This leads to different numbers of e^- domains about the two central atoms. Even though both ions have four bonding electron domains, the six total domains around Br require octahedral domain geometry and square planar molecular geometry, while the four total domains about B lead to tetrahedral domain and molecular geometry.

(b) H_2X, $8\,e^-$, $4\,e^-$ pr

$$H \overset{\displaystyle \cdot\cdot}{\underset{\displaystyle \cdot\cdot}{\;\;X\;\;}} H$$

All molecules in the series have tetrahedral electron domain geometry and bent molecular structure. To a first approximation, the H–X–H angles will be 109°. Any variation will be due to differences in repulsion among the nonbonding and bonding electron domains. The less electronegative the central atom, the larger the nonbonding electron domain, the greater the effect of repulsive forces on adjacent bonding domains. The less electronegative the central atom, the larger the deviation from ideal tetrahedral angles. The angles will vary as $H_2O > H_2S > H_2Se$.

Shapes and Polarity of Polyatomic Molecules (section 9.3)

9.37   A bond dipole is the asymmetric charge distribution between two bonded atoms with unequal electronegativities. A molecular dipole moment is the three-dimensional sum of all the bond dipoles in a molecule. (A molecular dipole moment is a measurable physical property; a bond dipole is not measurable, unless the molecule is diatomic.)

9.39   *Analyze/plan.* Follow the logic in Sample Exercise 9.4. *Solve.*

(a) SCl_2, $20\,e^-$, $10\,e^-$ pr

$$:\overset{\cdot\cdot}{\underset{\cdot\cdot}{Cl}}—\overset{\cdot\cdot}{\underset{\cdot\cdot}{S}}—\overset{\cdot\cdot}{\underset{\cdot\cdot}{Cl}}:$$

tetrahedral e^- domain geometry
bent molecular geometry

S and Cl have different electronegativities; the S–Cl bonds are polar. The bond dipoles are not opposite each other, so the molecule is polar. The dipole moment vector bisects the Cl–S–Cl bond angle. (A more difficult question is which end of the dipole moment vector is negative. The resultant of the two bond dipoles has its negative end toward the Cl atoms. However, the partial negative charge due to the lone pairs on S points opposite to the negative end of the resultant. A reasonable guess is that the negative end of the dipole moment vector is in the direction of the lone pairs.)

(b) $BeCl_2$, 16 e⁻, 8 e⁻ pr

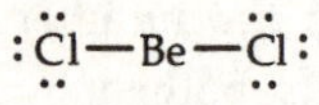

$:\ddot{C}l—Be—\ddot{C}l:$

linear electron domain and molecular geometry

(Resonance structures with Be=Cl can be drawn, but electronegativity arguments predict that most electron density will reside on Cl and that the structure above is the main resonance contributor.)   Be and Cl have very different electronegativities, so the Be–Cl bonds are polar. The individual bond dipoles are equal and opposite, so the net molecular dipole moment is zero.

9.41   (a)   In Exercise 9.29, molecules ii and iii will have nonzero dipole moments. Molecule i has zero nonbonding electron pairs on A, and the 3 A–F dipoles are oriented so that the sum of their vectors is zero (the bond dipoles cancel). Molecules ii and iii have nonbonding electron pairs on A and their bond dipoles do not cancel. A nonbonding electron pair (or pairs) on a central atom almost guarantees at least a small molecular dipole moment, because no bond dipole exactly cancels a nonbonding pair. (Exceptions are molecular geometries with nonbonding electron domains 180° apart.)

(b) AF_4 molecules will have a zero dipole moment if the 4 A–F bond dipoles are arranged (symmetrically) so that they cancel, and any nonbonding pairs are arranged so that they cancel. Therefore, in Exercise 9.30, molecules i and ii have zero dipole moments and are nonpolar.

9.43   *Analyze/Plan.* Given molecular formulas, draw correct Lewis structures, determine molecular structure and polarity. *Solve.*

(a)   Polar, $\Delta EN > 0$

I–F

(b)   Nonpolar, the molecule is linear and the bond dipoles cancel.

$S=C=S$

(c)   Nonpolar, in a symmetrical trigonal planar structure, the bond dipoles cancel.

O
‖
S
O   O

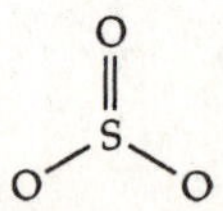

(d)   Polar, although the bond dipoles are essentially zero, there is an unequal charge distribution due to the nonbonded electron pair on P.

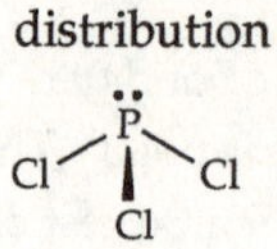

$\ddot{P}$
Cl   Cl
|
Cl

(e)   Nonpolar, symmetrical octahedron

F
F | F
S
F | F
F

(f)   Polar, square pyramidal molecular geometry, bond dipoles do not cancel.

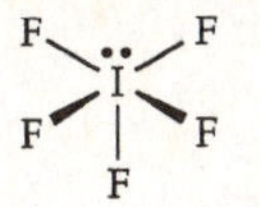

9.45   *Analyze/Plan.* Given molecular formulas, draw correct Lewis structures, analyze molecular structure and determine polarity.   *Solve.*

(a) $C_2H_2Cl_2$, each isomer has 24 e^-, 12 e^- pr. Lewis structures:

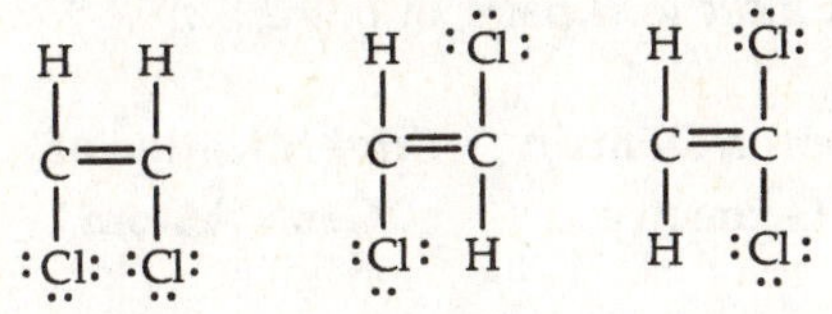

Molecular geometries:

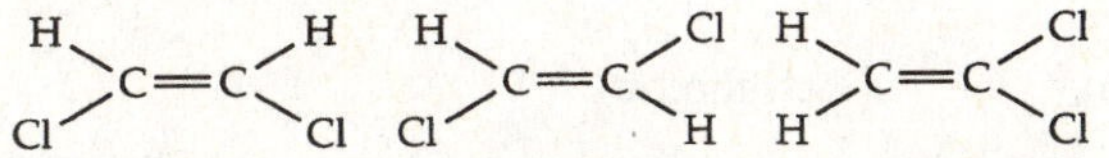

(b)   All three isomers are planar. The molecules on the left and right are polar because the C–Cl bond dipoles do not point in opposite directions. In the middle isomer, the C–Cl bonds and dipoles are pointing in opposite directions (as are the C–H bonds), the molecule is nonpolar and has a measured dipole moment of zero.

(c) C_2H_3Cl (lone pairs on Cl omitted for clarity)

There are four possible placements for Cl:

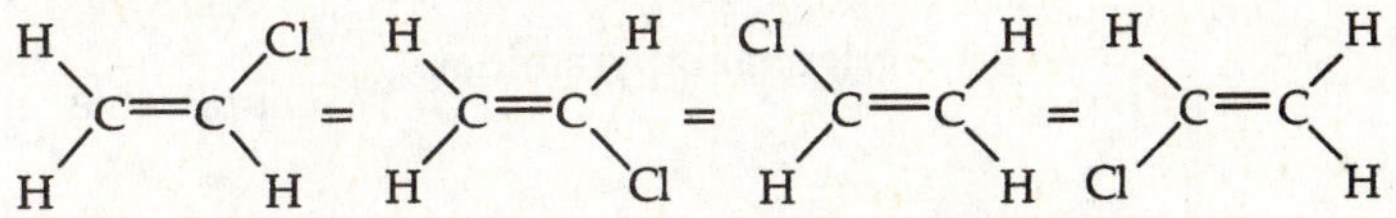

By rotating each of these structures in various directions, it becomes clear that the four structures are equivalent; C_2H_3Cl has only one isomer. Because C_2H_3Cl has only one C–Cl bond, the bond dipoles do not cancel, and the molecule has a dipole moment.

Orbital Overlap; Hybrid Orbitals (sections 9.4 and 9.5)

9.47   (a)   *Orbital overlap* occurs when a valence atomic orbital on one atom shares the same region of space with a valence atomic orbital on an adjacent atom.

(b)   A chemical bond is a concentration of electron density between the nuclei of two atoms. This concentration can take place because orbitals on the two atoms overlap.

9.49   (a)   4 valence e^-, 2 e^- pairs

H—Mg—H

2 bonding e^- domains, linear e^- domain and molecular geometry

(b)   The linear electron domain geometry in MgH_2 requires sp hybridization.

(c)

9.51   *Analyze/Plan.* For entries where the molecule is listed, follow the logic in Sample
Exercises 9.4 and 9.5. For entries where no molecule is listed, decide electron domain
geometry from hybridization (or vice-versa). If the molecule is nonpolar, the terminal
atoms will be identical. If the molecule is polar, the terminal atoms will be different, or
the central atom will have one or more lone pairs, or both. *Solve.*

Molecule	Molecular Structures	Electron Domain Geometry	Hybrdization of CentralAtom	Dipole Moment Yes or No
CO_2	O=C=O	linear	sp	no
NH_3	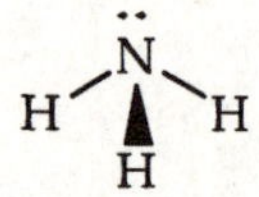	tetrahedral	sp³	yes
CH_4	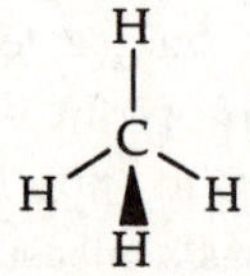	tetrahedral	sp³	no
BH_3	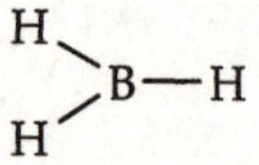	trigonal planar	sp²	no
SF_4	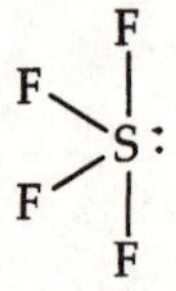	trigonal bipyramidal	not applicable	yes
SF_6	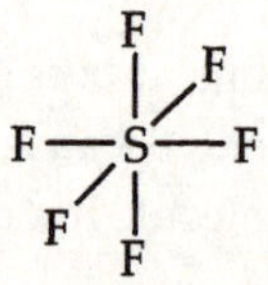	octahedral	not applicable	no
H_2CO	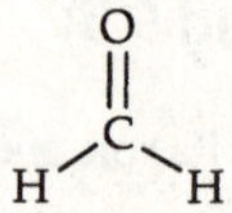	trigonal planar	sp²	yes
PF_5	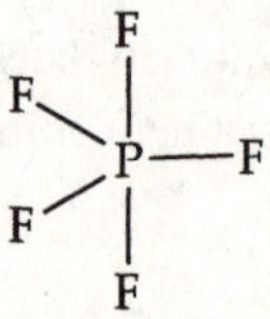	trigonal bipyramidal	not applicable	no
XeF_2	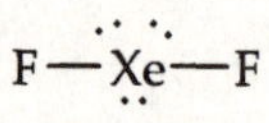 F—Xe—F	trigonal bipyramidal	not applicable	no

112

9.53   (a)   B: $[He]2s^2 2p^1$

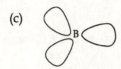

2s   2p $\xrightarrow{promote}$ 2s   2p $\xrightarrow{hybridize}$ sp^2 p

(b)   The hybrid orbitals are called sp^2.

(c)

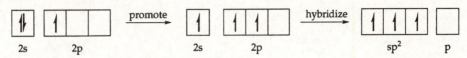

(d)   A single 2p orbital is unhybridized. It lies perpendicular to the trigonal plane of the sp^2 hybrid orbitals.

9.55   *Analyze/Plan.* Given the molecular (or ionic) formula, draw the correct Lewis structure and determine the electron domain geometry, which determines hybridization. *Solve.*

(a)   24 e⁻, 12 e⁻ pairs

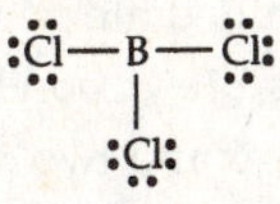

:Cl—B—Cl:
       |
      :Cl:

3 e⁻ pairs around B, trigonal planar e⁻ domain geometry, sp^2 hybridization

(b)   32 e⁻, 16 e⁻ pairs

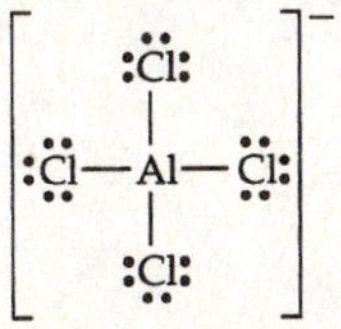

$$\begin{bmatrix} & :\ddot{C}l: & \\ & | & \\ :\ddot{C}l{-}Al{-}\ddot{C}l: & \\ & | & \\ & :\ddot{C}l: & \end{bmatrix}^{-}$$

4 e⁻ domains around Al, tetrahedral e⁻ domain geometry, sp^3 hybridization

(c)   16 e⁻, 8 e⁻ pairs

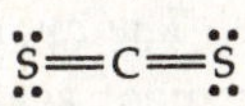

S̈=C=S̈

2 e⁻ domains around C, linear e⁻ domain geometry, sp hybridization

(d)   8 e⁻, 4 e⁻ pairs

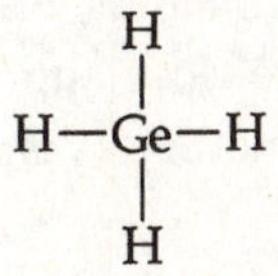

        H
        |
H—Ge—H
        |
        H

4 e⁻ pairs around Ge, tetrahedral e⁻ domain geometry, sp^3 hybridization

9.57   (a)   No hybrid orbitals discussed in this chapter have angles of 90°; p atomic orbitals are perpendicular to each other.

(b)   Angles of 109.5° are characteristic of sp^3 hybrid orbitals.

(c)   Angles of 120° can be formed by sp^2 hybrids.

Multiple Bonds (section 9.6)

9.59   (a) 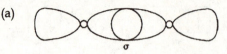   (b)

(c)   A σ bond is generally stronger than a π bond, because there is more extensive orbital overlap.

(d)   Two *s* orbitals cannot form a π bond. A π bond has no electron density along the internuclear axis. Overlap of *s* orbitals results in electron density along the internuclear axis. (Another way to say this is that *s* orbitals have the wrong symmetry to form a π bond.)

9.61   *Analyze/Plan.* Draw the correct Lewis structures, count electron domains and decide hybridization. Molecules with π bonds that require all bonded atoms to be in the same plane are planar. For bond-type counting, single bonds are σ bonds, double bonds consist of one σ and one π bond, triple bonds consist of one σ and two π bonds. *Solve.*

(a)

$$H-\underset{\underset{H}{|}}{\overset{\overset{H}{|}}{C}}-\underset{\underset{H}{|}}{\overset{\overset{H}{|}}{C}}-H \qquad \underset{H}{\overset{H}{>}}C=C\underset{H}{\overset{H}{<}} \qquad H-C\equiv C-H$$

(b) sp^3 sp^2 sp

(c)   nonplanar   planar   planar

(d) $7\,\sigma, 0\,\pi$ $5\,\sigma, 1\,\pi$ $3\,\sigma, 2\,\pi$

(e)   Since Si has the same valence electron configuration as C, the structures of the Si analogs would be the same as the C-based molecules. The hybridization at Si is then the same as the hybridization at C. This argument assumes that the Si analogs exist.

Silicon, which lies in the row below C, has a larger bonding atomic radius and larger atomic orbitals than C. In the analogous compounds, Si–Si distances will be greater than C–C distances. This means that the close approach of Si atoms required to form strong, stable π bonds in Si_2H_4 and Si_2H_2 is not possible and these Si analogs do not readily form.

9.63   *Analyze/Plan.* Single bonds are σ bonds, double bonds consist of 1 σ and 1 π bond. Each bond is formed by a pair of valence electrons. *Solve.*

(a) C_3H_6 has $3(4) + 6(1) = 18$ valence electrons

(b)   8 pairs or 16 total valence electrons form σ bonds

(c)   1 pair or 2 total valence electrons form π bonds

(d)   no valence electrons are nonbonding

(e)   The left and central C atoms are sp^2 hybridized; the right C atom is sp^3 hybridized.

9.65   *Analyze/Plan.* Given the correct Lewis structure, analyze the electron domain geometry at each central atom. This determines the hybridization and bond angles at that atom. *Solve.*

(a)   ~109° bond angles about the left most C, sp^3; ~120° bond angles about the right-hand C, sp^2

(b)   The doubly bonded O can be viewed as sp^2, the other as sp^3; the nitrogen is sp^3 with approximately 109° bond angles.

(c)   9 σ bonds, 1 π bond

9.67   (a)   In a localized π bond, the electron density is concentrated strictly between the two atoms forming the bond. In a delocalized π bond, parallel p orbitals on more than two adjacent atoms overlap and the electron density is spread over all the atoms that contribute p orbitals to the network. There are still two regions of overlap, above and below the σ framework of the molecule.

(b)   The existence of more than one resonance form is a good indication that a molecule will have delocalized π bonding.

(c)

The existence of more than one resonance form for NO_2 indicates that the π bond is delocalized. From an orbital perspective, the electron-domain geometry around N is trigonal planar, so the hybridization at N is sp^2. This leaves a p orbital on N and one on each O atom perpendicular to the trigonal plane of the molecule, in the correct orientation for delocalized π overlap. Physically, the two N–O bond lengths are equal, indicating that the two N–O bonds are equivalent, rather than one longer single bond and one shorter double bond.

9.69   *Analyze/Plan.* Count valence e^- and e^- pairs in each molecule. Complete the Lewis structure by placing nonbonding electron pairs. Analyze the electron domain geometry at each central atom; visualize and describe the molecular structure. *Solve.*

(a)   26 e^-, 13 e^- pairs

$$H—C≡C—C≡C—C≡N:$$

The molecule is linear. Each C atom has 2 bonding e^- domains, linear geometry and sp hybridization. This requires that all atoms not only lie in the same plane, but in a line.

(b)   34 e^-, 17 e^- pairs

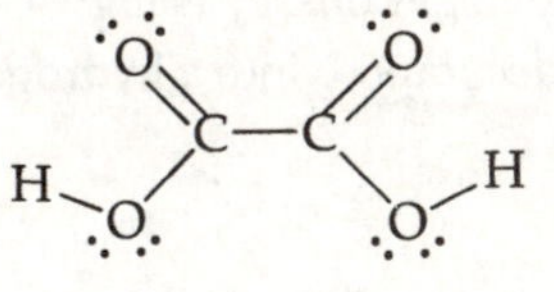

The two central C atoms each have 3 bonding e⁻ domains, trigonal planar geometry and sp^2 hybridization. Each O–C–O group is planar, while the terminal H atoms can rotate out of these planes. In principle, there is free rotation about the C–C σ bond, but delocalization of the π electrons is possible if the two planes are coincident. It is possible to put all 8 atoms in the same plane.

(c)　12 e⁻, 6 e⁻ pairs

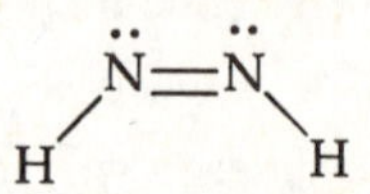

$$\overset{\displaystyle \cdot\cdot \quad \cdot\cdot}{\ddot{N}=\ddot{N}}$$

H　　　　　H

The molecule is planar. Each N atom has 3 bonding e⁻ domains, trigonal planar geometry and sp^2 hybridization. Since the N atoms share a π bond, the planes must be coincident and all 4 atoms are required to lie in this plane.

Molecular Orbitals and Second Row Diatomic Molecules
(sections 9.7 and 9.8)

9.71　(a)　Hybrid orbitals are mixtures (linear combinations) of atomic orbitals from a single atom; the hybrid orbitals remain localized on that atom. Molecular orbitals are combinations of atomic orbitals from two or more atoms. They are associated with the entire molecule, not a single atom.

(b)　Each MO, like each AO or hybrid, can hold a maximum of two electrons.

(c)　Antibonding MOs can have electrons in them.

9.73　(a)

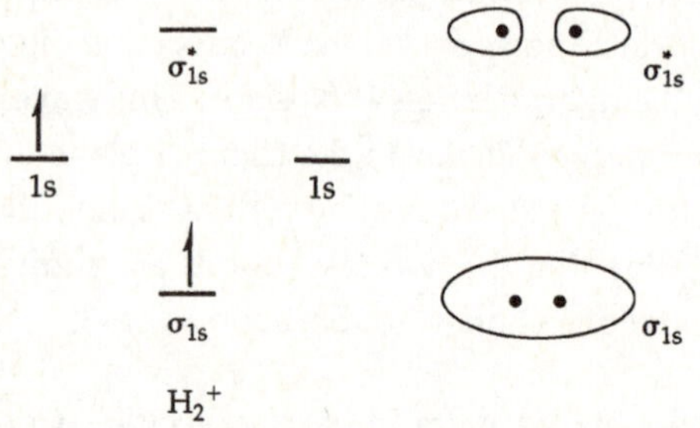

$$\sigma_{1s}^*$$

$$\uparrow \quad\quad\quad \\ 1s \quad\quad\quad 1s$$

$$\uparrow \\ \sigma_{1s}$$

$$H_2^+$$

$$\sigma_{1s}^*$$

$$\sigma_{1s}$$

(b)　There is one electron in H_2^+.

(c)

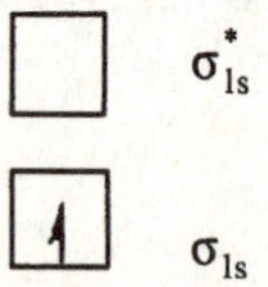

☐　σ_{1s}^*

⤈　σ_{1s}

(d)　Bond order = 1/2 (1-0) = 1/2

(e)　Fall apart. The stability of H_2^+ is due to the lower energy state of the σ bonding molecular orbital relative to the energy of a H 1s atomic orbital. If the single electron in H_2^+ is excited to the σ^*_{1s} orbital, its energy is higher than the energy of an H 1s atomic orbital and H_2^+ will decompose into a hydrogen atom and a hydrogen ion.

$$H_2^+ \overset{h\nu}{\rightarrow} H + H^+.$$

9.75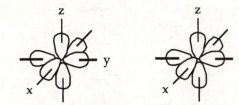

(a) One. With three mutually perpendicular p orbitals on each atom, only one set can be oriented for end-to-end, sigma overlap.

(b) Two. The $2p$ orbitals on each atom not involved in σ bonding can be aligned for side-to-side π overlap.

(c) Three, 1 σ^* and 2 π^*. There are a total of 6 p orbitals on the two atoms. When combining AOs to form MOs, total number of orbitals is conserved. If 3 of the 6 MOs are bonding MOs, as described in (a) and (b), then the remaining 3 MOs must be antibonding. They will have the same symmetry as the bonding MOs, 1 σ^* and 2 π^*.

9.77 (a) When comparing the same two bonded atoms, the greater the bond order, the shorter the bond length and the greater the bond energy. That is, bond order and bond energy are directly related, while bond order and bond length are inversely related. When comparing different bonded nuclei, there are no simple relationships (see Solution 8.100).

(b)
Be_2, 4 e$^-$		Be_2^+, 3 e$^-$	
⇅	σ^*_{2s}	↑	σ^*_{2s}
⇅	σ_{2s}	⇅	σ_{2s}
BO = 1/2(2-2) = 0		BO = 1/2(2-1) = 0.5	

Be_2 has a bond order of zero and is not energetically favored over isolated Be atoms; it is not expected to exist. Be_2^+ has a bond order of 0.5 and is slightly lower in energy than isolated Be atoms. It will probably exist under special experimental conditions, but be unstable.

9.79 (a), (b) Substances with no unpaired electrons are weakly repelled by a magnetic field. This property is called *diamagnetism*.

(c) O_2^{2-}, Be_2^{2+} [see Figure 9.43 and Solution 9.77(b)]

9.81

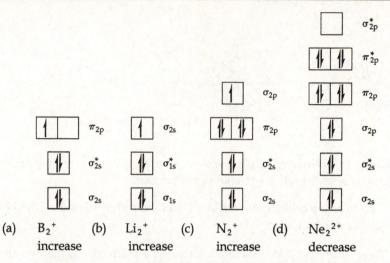

(a)	(b)	(c)	(d)
B_2^+	Li_2^+	N_2^+	Ne_2^{2+}
increase	increase	increase	decrease

Addition of an electron increases bond order if it occupies a bonding MO and decreases stability if it occupies an antibonding MO.

9.83   *Analyze/Plan.* Determine the number of "valence" (non-core) electrons in each molecule or ion. Use the homonuclear diatomic MO diagram from Figure 9.43 (shown below) to calculate bond order and magnetic properties of each species. The electronegativity difference between heteroatomics increases the energy difference between the 2s AO on one atom and the 2p AO on the other, rendering the "no interaction" MO diagram in Figure 9.43 appropriate.   *Solve.*

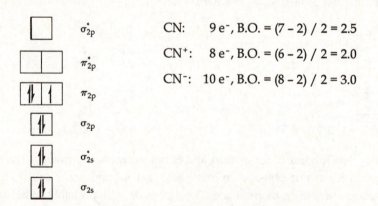

CN: $9\,e^-$, B.O. $= (7 - 2) / 2 = 2.5$

CN^+: $8\,e^-$, B.O. $= (6 - 2) / 2 = 2.0$

CN^-: $10\,e^-$, B.O. $= (8 - 2) / 2 = 3.0$

(a) CN^- has the highest bond order and therefore the strongest C–N bond.

(b)   CN and CN^+. CN has an odd number of valence electrons, so it must have an unpaired electron. The electron configuration for CN is shown in the diagram. Removing one electron from the π_{2p} MOs to form CN^+ produces an ion with two unpaired electrons. Adding one electron to the π_{2p} MOs of CN to form CN^- produces an ion with all electrons paired.

(c) NO^+ is isoelectronic with N_2, and NO^- is isoelectronic with O_2.

9.85   (a) $3s, 3p_x, 3p_y, 3p_z$ (b) π_{3p}

(c)   Two. Note that there are two degenerate π_{3p} bonding molecular orbitals; each holds two electrons. A total of 4 electrons can be designated as π_{3p}, but no single molecular orbital can hold more than two electrons.

(d)   If the MO diagram for P_2 is similar to that of N_2, P_2 will have no unpaired electrons and be diamagnetic.

Additional Exercises

9.89

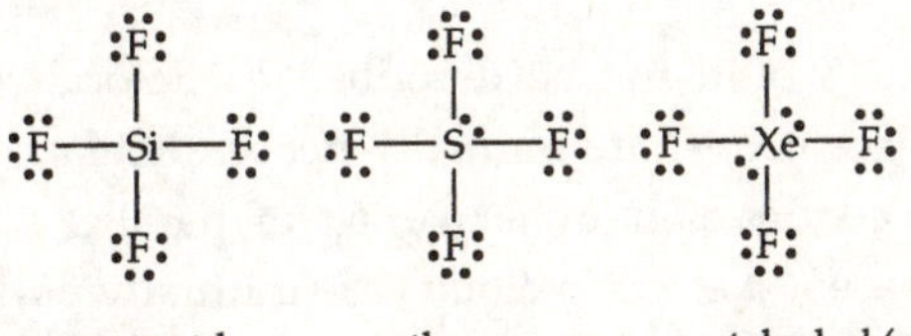

e⁻ domain geometry td   tbp   octahedral (oh)

molecular shape td   seesaw (ss)   square planar (s)

Although there are four bonding electron domains in each molecule, the number of nonbonding domains is different in each case. The bond angles and thus the molecular shape are influenced by the total number of electron domains.

9.92   (a) CO_2, 16 valence e⁻   (b) $(CN)_2$, 18 valence e⁻

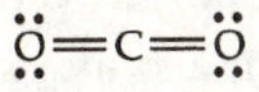

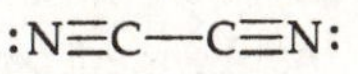

2σ 2π $3\sigma, 4\pi$

(c) H_2CO, 12 valence e⁻   (d)   HCOOH, 18 valence e⁻

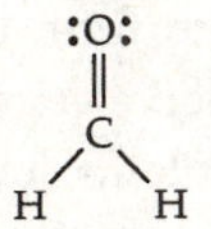

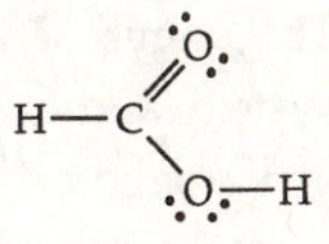

$3\sigma, 1\pi$ $4\sigma, 1\pi$

9.94

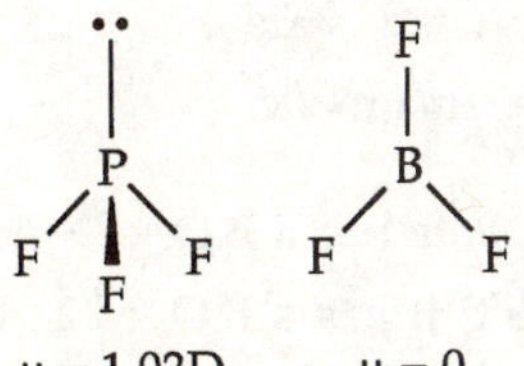

$\mu = 1.03D$ $\mu = 0$

BF_3 is a trigonal planar molecule with the central B atom symmetrically surrounded by the three F atoms (Figure 9.12). The individual B–F bond dipoles cancel, and the molecule has a net dipole moment of zero. PF_3 has tetrahedral electron-domain geometry with one of the positions in the tetrahedron occupied by a nonbonding electron pair. The individual P–F bond dipoles do not cancel and the presence of a nonbonding electron pair ensures an asymmetrical electron distribution; the molecule is polar.

9.99

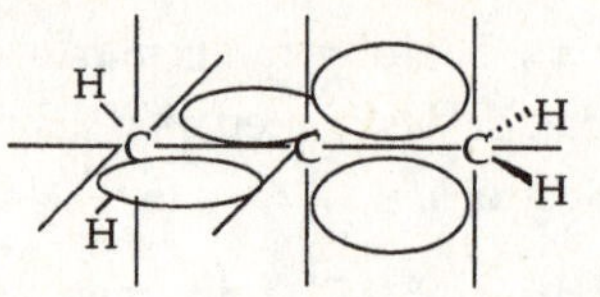

(a) The molecule is not planar. The CH_2 planes at each end are twisted 90° from one another.

(b) Allene has no dipole moment.

(c) The bonding in allene would not be described as delocalized. The π electron clouds of the two adjacent C=C are mutually perpendicular. The mechanism for delocalization of π electrons is mutual overlap of parallel p atomic orbitals on adjacent atoms. If adjacent π electron clouds are mutually perpendicular, there is no overlap and no delocalization of π electrons.

9.101 (a)

To accommodate the π bonding by all 3 O atoms indicated in the resonance structures above, all O atoms are sp^2 hybridized.

(b) For the first resonance structure, both sigma bonds are formed by overlap of sp^2 hybrid orbitals, the π bond is formed by overlap of atomic p orbitals, one of the nonbonded pairs on the right terminal O atom is in a p atomic orbital, and the remaining five nonbonded pairs are in sp^2 hybrid orbitals.

(c) Only unhybridized p atomic orbitals can be used to form a delocalized π system.

(d) The unhybridized p orbital on each O atom is used to form the delocalized π system, and in both resonance structures one nonbonded electron pair resides in a p atomic orbital. The delocalized π system then contains four electrons, two from the π bond and two from the nonbonded pair in the p orbital.

9.104

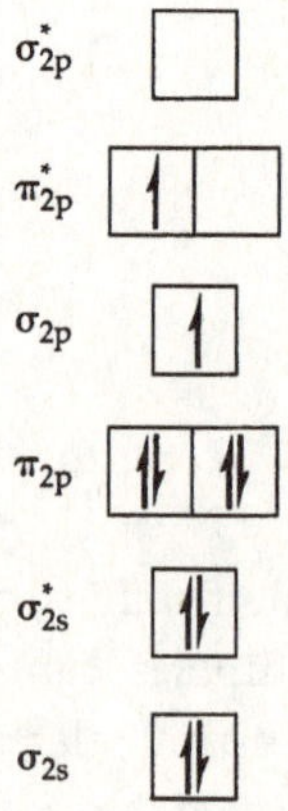

(a) N_2 in the first excited state has two unpaired electrons and is paramagnetic.

(b) N_2 in the ground state has a B.O. of 3; in the first excited state (at left) it has a B.O. of 2. Owing to the reduction in bond order, N_2 in the first excited state has a weaker (and longer) N–N bond than N_2 in the ground state.

Integrative Exercises

9.110   (a) $2SF_4(g) + O_2(g) \rightarrow 2OSF_4(g)$

(b) $40\ e^-,\ 20\ e^-$ pairs

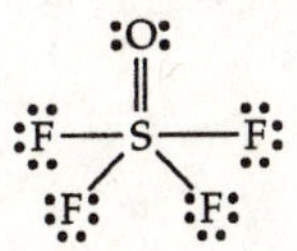

There must be a double bond drawn between O and S in order for their formal charges to be zero.

(c) $\Delta H = 8D(S–F) + D(O=O) – 8D(S–F) – 2D(S=O)$

$\Delta H = D(O=O) – 2D(S=O) = 495 – 2(523) = –551$ kJ, exothermic

(d)   trigonal bipyramidal electron-domain geometry

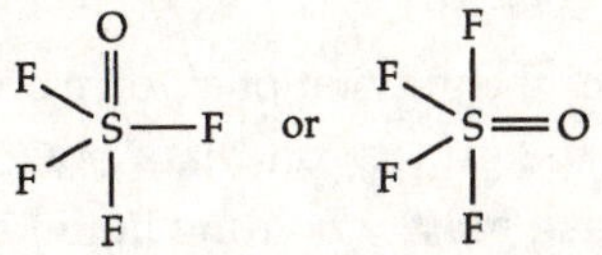

(e)   Because F is more electronegative than O, S–F bonding domains are smaller than the S=O domain. The structure that minimizes S=O repulsions is more likely. That is, the structure with fewer 90° F– S=O angles and more 120° F– S=O angles is favored. The structure on the right, with O in the equatorial position is more likely. Note that a double bond involving an atom with an expanded octet of electrons, such as the S=O in this molecule, does not have the same geometric implications as a double bond to a first row element.

10 Gases

Visualizing Concepts

10.1   It would be much easier to drink from a straw on Mars. When a straw is placed in a glass of liquid, the liquid level in the straw equals the liquid level in the glass. The atmosphere exerts equal pressure inside and outside the straw. When we drink through a straw, we withdraw air, thereby reducing the pressure on the liquid inside. If only 0.007 atm is exerted on the liquid in the glass, a very small reduction in pressure inside the straw will cause the liquid to rise.

Another approach is to consider the gravitational force on Mars. Since the pull of gravity causes atmospheric pressure, the gravity on Mars must be much smaller than that on Earth. With a very small Martian gravity holding liquid in a glass, it would be very easy to raise the liquid through a straw.

10.4   (a)   At constant pressure and temperature, the container volume is directly proportional to the number of particles present (Avogadro's Law). As the reaction proceeds, 3 gas molecules are converted to 2 gas molecules, so the container volume decreases. If the reaction goes to completion, the final volume would be 2/3 of the initial volume.

   (b)   At constant volume and temperature, pressure is directly proportional to the number of particles . Since the number of molecules decreases as the reaction proceeds, the pressure also decreases. At completion, the final pressure would be 2/3 the initial pressure.

10.7   (a)   Partial pressure depends on the number of particles of each gas present. Red has the fewest particles, then yellow, then blue. $P_{red} < P_{yellow} < P_{blue}$

   (b) $P_{gas} = \chi_{gas} P_t$. Calculate the mole fraction, χ_{gas} = [mol gas / total moles] or [particles gas / total particles]. This is true because Avogadro's number is a counting number, and mole ratios are also particle ratios.

   χ_{red} = 2 red atoms / 10 total atoms = 0.2; P_{red} = 0.2(1.40 atm) = 0.28 atm

   χ_{yellow} = 3 yellow atoms / 10 total atoms = 0.3; P_{yellow} = 0.3(1.40 atm) = 0.42 atm

   χ_{blue} = 5 blue atoms / 10 total atoms = 0.5; P_{blue} = 0.5(1.40 atm) = 0.70 atm

   Check. (0.28 atm + 0.42 atm + 0.70 atm) = 1.40 atm. The sum of the calculated partial pressures equals the given total pressure.

10.10   (a)   Total pressure is directly related to total number of particles (or total mol particles). P(ii) < P(i) = P(iii)

122

(b) Partial pressure of He is directly related to number of He atoms (yellow) or mol He atoms. $P_{He}(iii) < P_{He}(ii) < P_{He}(i)$

(c) Density is total mass of gas per unit volume. We can use the atomic or molar masses of He (4) and N_2 (28), as relative masses of the particles.

$$mass(i) = 5(4) + 2(28) = 76$$

$$mass(ii) = 3(4) + 1(28) = 40$$

$$mass(iii) = 2(4) + 5(28) = 148$$

Since the container volumes are equal, d(ii) < d(i) < d(iii).

(d) At the same temperature, all gases have the same averageg kinetic energy. The average kinetic energies of the particles in the three containers are equal.

Gas Characteristics; Pressure (sections 10.1 and 10.2)

10.13 In the gas phase, molecules are far apart and in constant motion. In the liquid phase, molecules maintain contact with each other while they move.

(a) A gas is much less dense than a liquid because most of the volume of a gas is empty space.

(b) A gas is much more compressible because of the distance between molecules.

(c) Gaseous molecules are so far apart that there is no barrier to mixing, regardless of the identity of the molecule. All mixtures of gases are homogeneous. Liquid molecules are touching. In order to mix, they must displace one another. Similar molecules displace each other and form homogeneous mixtures. Very dissimilar molecules form heterogeneous mixtures.

(d) Gas molecules are far apart and constantly moving, so a gas expands to the shape and volume of its container. In a liquid sample, molecules are moving but in close contact with each other; there is very little empty space in a liquid. A liquid takes the shape of its container because the molecules are mobile, but maintains its own volume because the molecules are touching.

10.15 *Analyze.* Given: mass, area. Find: pressure. *Plan.* $P=F/A = m \times a/A$; use this relationship, paying attention to units. *Solve.*

(a) $1\,Pa = \dfrac{1\,N}{m^2} = \dfrac{1\,kg \cdot m}{s^2} \times \dfrac{1}{m^2} = \dfrac{1\,kg}{m \cdot s^2}$ Change mass to kg and area to m^2.

$P = \dfrac{m \times a}{A} = \dfrac{130\,lb}{0.50\,in^2} \times \dfrac{9.81\,m}{1\,s^2} \times \dfrac{0.454\,kg}{1\,lb} \times \dfrac{39.4^2\,in^2}{1\,m^2} = 1.798 \times 10^6 \dfrac{kg}{m \cdot s^2}$

$$= 1.8 \times 10^6\,Pa = 1.8 \times 10^3\,kPa$$

Check. $[1.30 \times 10 \times 0.5 \times (40)^2/0.5] \approx (130 \times 16,000) \approx 2.0 \times 10^6\,Pa \approx 2.0 \times 10^3\,kPa$. The units are correct.

(b) 1 atm = 101.325 kPa

$$1.798 \times 10^3\,kPa \times \dfrac{1\,atm}{101.325\,kPa} = 1.774 \times 10^1 = 18\,atm$$

(c) $1 \text{ atm} = 14.70 \text{ lb/in}^2$

$$17.74 \text{ atm} \times \frac{14.70 \text{ lb/in}^2}{1 \text{ atm}} = 260.8 = 2.6 \times 10^2 \text{ lb/in}^2$$

10.17   *Analyze.* Given: 760 mm column of Hg, densities of Hg and H_2O. Find: height of a column of H_2O at same pressure.

Plan. We must develop a relationship between pressure, height of a column of liquid, and density of the liquid. Relationships that might prove useful: $P = F/A$; $F = m \times a$; $m = d \times V$(density)(volume); $V = A \times$ height   *Solve.*

$$P = \frac{F}{A} = \frac{m \times a}{A} = \frac{d \times V \times a}{A} = \frac{d \times A \times h \times a}{A} = d \times h \times a$$

(a) $P_{Hg} = P_{H_2O}$; Using the relationship derived above: $(d \times h \times a)_{H_2O} = (d \times h \times a)_{Hg}$

Since a, the acceleration due to gravity, is equal in both liquids,
$(d \times h)_{H_2O} = (d \times h)_{Hg}$

$1.00 \text{ g/mL} \times h_{H_2O} = 13.6 \text{ g/mL} \times 760 \text{ mm}$

$$h_{H_2O} = \frac{13.6 \text{ g/mL} \times 760 \text{ mm}}{1.00 \text{ g/mL}} = 1.034 \times 10^4 = 1.03 \times 10^4 \text{ mm} = 10.3 \text{ m}$$

(b)   Pressure due to H_2O:

$1 \text{ atm} = 1.034 \times 10^4 \text{ mm } H_2O$ (from part (a))

$$39 \text{ ft } H_2O \times \frac{12 \text{ in}}{1 \text{ ft}} \times \frac{2.54 \text{ cm}}{1 \text{ in}} \times \frac{10 \text{ mm}}{1 \text{ cm}} \times \frac{1 \text{ atm}}{1.034 \times 10^4 \text{ mm}} = 1.150 = 1.2 \text{ atm}$$

$P_{total} = P_{atm} + P_{H_2O} = 0.97 \text{ atm} + 1.150 \text{ atm} = 2.120 = 2.1 \text{ atm}$

10.19   (a)   The tube can have any cross-sectional area. (The height of the Hg column in a barometer is independent of the cross-sectional area. See the expression for pressure derived in Solution 10.17.)

(b)   At equilibrium, the force of gravity per unit area acting on the mercury column is not equal to the force of gravity per unit area acting on the atmosphere. (F = ma; the acceleration due to gravity is equal for the two substances, but the mass of Hg for a given cross-sectional area is different than the mass of air for this same area.)

(c)   The column of mercury is held up by the pressure of the atmosphere applied to the exterior pool of mercury.

(d)   If you took the mercury barometer with you on a trip from the beach to high mountains, the height of the mercury column would decrease with elevation. (Atmospheric pressure decreases as elevation increases.)

10.21   *Analyze/Plan.* Follow the logic in Sample Exercise 10.1.   *Solve.*

(a) $265 \text{ torr} \times \dfrac{1 \text{ atm}}{760 \text{ torr}} = 0.349 \text{ atm}$

(b) $265 \text{ torr} \times \dfrac{1 \text{ mm Hg}}{1 \text{ torr}} = 265 \text{ mm Hg}$

(c) $265 \text{ torr} \times \dfrac{1.01325 \times 10^5 \text{ Pa}}{760 \text{ torr}} = 3.53 \times 10^4 \text{ Pa}$

(d) $265 \text{ torr} \times \dfrac{1.01325 \times 10^5 \text{ Pa}}{760 \text{ torr}} \times \dfrac{1 \text{ bar}}{1 \times 10^5 \text{ Pa}} = 0.353 \text{ bar}$

(e) $265 \text{ torr} \times \dfrac{1 \text{ atm}}{760 \text{ torr}} \times \dfrac{14.70 \text{ psi}}{1 \text{ atm}} = 5.13 \text{ psi}$

10.23   *Analyze/Plan.* Follow the logic in Sample Exercise 10.1.   *Solve.*

(a) $30.45 \text{ in Hg} \times \dfrac{25.4 \text{ mm}}{1 \text{ in}} \times \dfrac{1 \text{ torr}}{1 \text{ mm Hg}} = 773.4 \text{ torr}$

[The result has 4 sig figs because 25.4 mm/in is considered to be an exact number.]

(b) $30.45 \text{ in Hg} = 773.4 \text{ torr}; \quad 773.4 \text{ torr} \times \dfrac{1 \text{ atm}}{760 \text{ torr}} = 1.018 \text{ atm}$

(c)   The pressure in Chicago is greater than **standard atmospheric pressure**, 760 torr or 1 atm, so it makes sense to classify this weather system as a "high pressure system."

10.25   *Analyze/Plan.* Follow the logic in Sample Exercise 10.2.   *Solve.*

(i)   The Hg level is lower in the open end than the closed end, so the gas pressure is less than atmospheric pressure.

$P_{gas} = 0.995 \text{ atm} - \left(52 \text{ cm} \times \dfrac{1 \text{ atm}}{76.0 \text{ cm}} \right) = 0.31 \text{ atm}$

(ii)   The Hg level is higher in the open end, so the gas pressure is greater than atmospheric pressure.

$P_{gas} = 0.995 \text{ atm} + \left(67 \text{ cm Hg} \times \dfrac{1 \text{ atm}}{76.0 \text{ cm Hg}} \right) = 1.8766 = 1.88 \text{ atm}$

(iii)   This is a closed-end manometer, so $P_{gas} = h$.

$P_{gas} = 10.3 \text{ cm} \times \dfrac{1 \text{ atm}}{76.0 \text{ cm}} = 0.136 \text{ atm}$

The Gas Laws (section 10.3)

10.27   *Analyze/Plan.* Given certain changes in gas conditions, predict the effect on other conditions. Consider the gas law relationships in section 10.3.   *Solve.*

(a)   P and V are inversely proportional at constant T. If the volume decreases by a factor of 4, the pressure increases by a factor of 4.

(b)   P and T are directly proportional at constant V. If T decreases by a factor of 2, P also decreases by a factor of 2.

(c)   P and n are directly proportional at constant V and T. If n decreases by a factor of 4, P also decreases by a factor of 4.

10.29   (a)   Avogadro's hypothesis states that equal volumes of gases at the same temperature and pressure contain equal numbers of molecules. Since molecules react in the ratios of small whole numbers, it follows that the volumes of reacting gases (at the same temperature and pressure) are in the ratios of small whole numbers.

         (b)   Since the two gases are at the same temperature and pressure, the ratio of the numbers of atoms is the same as the ratio of volumes. There are 1.5 times as many Xe atoms as Ne atoms.

         (c)   Yes. By definition, one mole of an ideal gas (or any other substance) contains Avogadro's number of particles. At a given temperature and pressure, equal numbers of particles occupy the same volume, so one mole of an ideal gas will always occupy the same volume at the given temperature and pressure.

The Ideal-Gas Equation (section 10.4)

(In *Solutions to Exercises*, the symbol for molar mass is MM.)

10.31   (a)   An ideal gas exhibits pressure, volume, and temperature relationships which are described by the equation $PV = nRT$. (An ideal gas obeys the ideal-gas equation.)

         (b) $V = \text{constant}/P$;   Boyle's Law
                 $V = \text{constant} \times T$;   Charles Law
                 $V = \text{constant} \times n$;   Avogadro's Law

                 Collecting all the equalities, $V = \dfrac{\text{constant} \times T \times n}{P}$

                 If we call the constant R, and multiply both sides by P,
                 $PV = RTn$ or $PV = nRT$

         (c) $PV = nRT$; P in atmospheres, V in liters, n in moles, T in kelvins

         (d) $R = \dfrac{0.08206\,\text{L-atm}}{\text{mol-K}} \times \dfrac{1.01325 \times 10^5\,\text{Pa}}{1\,\text{atm}} \times \dfrac{1\,\text{bar}}{1 \times 10^5\,\text{Pa}} = \dfrac{0.08315\,\text{L-bar}}{\text{mol-K}}$

10.33   *Analyze/Plan.* $PV = nRT$. At constant volume and temperature, P is directly proportional to n.

         Solve. For samples with equal masses of gas, the gas with MM = 30 will have twice as many moles of particles and twice the pressure. Thus, flask A contains the gas with MM = 30 and flask B contains the gas with MM = 60.

10.35   *Analyze/Plan.* Follow the strategy for calculations involving many variables given in section 10.4.   *Solve.*

$$T = \frac{PV}{nR} = 2.00\,\text{atm} \times \frac{1.00\,\text{L}}{0.500\,\text{mol}} \times \frac{\text{mol-K}}{0.08206\,\text{L-atm}} = 48.7\,\text{K}$$

$$K = 27°C + 273 = 300\,\text{K}$$

$$n = \frac{PV}{RT} = 0.300\,\text{atm} \times \frac{0.250\,\text{L}}{300\,\text{K}} \times \frac{\text{mol-K}}{0.08206\,\text{-atm}} = 3.05 \times 10^{-3}\,\text{mol}$$

$$650\,\text{torr} \times \frac{1\,\text{atm}}{760\,\text{torr}} = 0.85526 = 0.855\,\text{atm}$$

$$V = \frac{nRT}{P} = 0.333\,mol \times \frac{350\,K}{0.85526\,atm} \times \frac{0.08206\,L\text{-}atm}{mol\text{-}K} = 11.2\,L$$

$585\,mL = 0.585\,L$

$$P = \frac{nRT}{V} = 0.250\,mol \times \frac{295\,K}{0.585\,L} \times \frac{0.08206\,L\text{-}atm}{mol\text{-}K} = 10.3\,atm$$

P	V	N	T
2.00 atm	1.00 L	0.500 mol	**48.7 K**
0.300 atm	0.250 L	**3.05 × 10⁻³ mol**	27°C
650 torr	**11.2 L**	0.333 mol	350 K
10.3 atm	585 mL	0.250 mol	295 K

10.37   *Analyze/Plan.* Follow the strategy for calculations involving many variables.   *Solve.*

$n = g/MM;\ PV = nRT;\ PV = gRT/MM;\ g = MM \times PV/RT$

$P = 1.0\,atm,\ T = 23°C = 296\,K,\ V = 1.75 \times 10^5\,ft^3$. Change ft^3 to L, then calculate grams (or kg).

$$1.75 \times 10^5\,ft^3 \times \frac{(12)^3\,in^3}{ft^3} \times \frac{(2.54)^3\,cm^3}{in^3} \times \frac{1\,L}{1 \times 10^3\,cm^3} = 4.9554 \times 10^6 = 4.96 \times 10^6\,L$$

$$g = \frac{4.003\,g\,He}{1\,mol\,He} \times \frac{mol\text{-}K}{0.08206\,L\text{-}atm} \times \frac{1.0\,atm \times 4.955 \times 10^6\,L}{296\,K} = 8.2 \times 10^5\,g = 820\ kg\ He$$

10.39   *Analyze/Plan.* Follow the strategy for calculations involving many variables.   *Solve.*

(a) $V = 2.25\,L;\ T = 273 + 37°C = 310\,K;\ P = 735\,torr \times \dfrac{1\,atm}{760\,torr} = 0.96710 = 0.967\,atm$

$PV = nRT,\ n = PV/RT$, number of molecules (#) $= n \times 6.022 \times 10^{23}$

$$\# = \frac{0.9671\,atm \times 2.25\,L}{310\,K} \times \frac{mol\text{-}K}{0.08206\,L\text{-}atm} \times \frac{6.022 \times 10^{23}\,molecules}{mol}$$
$$= 5.15 \times 10^{22}\,molecules$$

(b) $V = 5.0 \times 10^3\,L;\ T = 273 + 0°C = 273\,K;\ P = 1.00\,atm;\ MM = 28.98\,g/mol$

$$PV = \frac{g}{MM}RT; \quad g = \frac{MM \times PV}{RT}$$

$$g = \frac{28.98\,g\,air}{1\,mol\,air} \times \frac{mol\text{-}K}{0.08206\,L\text{-}atm} \times \frac{1.00\,atm \times 5.0 \times 10^3\,L}{273\,K}$$
$$= 6,468\,g = 6.5 \times 10^3\,g\ air = 6.5\ kg\ air$$

10.41   *Analyze/Plan.* Follow the strategy for calculations involving many variables.   *Solve.*

(a) $P = \dfrac{nRT}{V}; n = 0.29\,kg\,O_2 \times \dfrac{1000\,g}{1\,kg} \times \dfrac{1\,mol\,O_2}{32.00\,g\,O_2} = 9.0625 = 9.1\,mol;\ V = 2.3\,L;$

$T = 273 + 9°C = 282\,K$

$$P = \frac{9.0625\,mol}{2.3\,L} \times \frac{0.08206\,L\text{-}atm}{mol\text{-}K} \times 282\ K = 91\,atm$$

(b) $V = \dfrac{nRT}{P}; = \dfrac{9.0625\,mol}{0.95\,atm} \times \dfrac{0.08206\,L\text{-}atm}{mol\text{-}K} \times 299\,K = 2.3 \times 10^2\,L$

10.43   *Analyze/Plan.* Follow the strategy for calculations involving many variables.   *Solve.*

$V = 8.70\,L, T = 24°C = 297\,K, P = 895\,torr \times \dfrac{1\,atm}{760\,torr} = 1.1776 = 1.18\,atm$

(a) $g = \dfrac{MM \times PV}{RT}; g = \dfrac{70.91\,g\,Cl_2}{1\,mol\,Cl_2} \times \dfrac{mol\text{-}K}{0.08206\,L\text{-}atm} \times \dfrac{1.1776\,atm}{297\,K} \times 8.70\,L$

$= 29.8\,g\,Cl_2$

(b) $V_2 = \dfrac{P_1V_1T_2}{T_1P_2} = \dfrac{895\,torr \times 8.70\,L \times 273\,K}{297\,K \times 760\,torr} = 9.42\,L$

(c) $T_2 = \dfrac{P_2V_2T_1}{P_1V_1} = \dfrac{876\,torr \times 15.00\,L \times 297\,K}{895\,torr \times 8.70\,L} = 501\,K$

(d) $P_2 = \dfrac{P_1V_1T_2}{V_2T_1} = \dfrac{895\,torr \times 8.70\,L \times 331\,K}{5.00\,L \times 297\,K} = 1.73 \times 10^3\,torr = 2.28\,atm$

10.45   *Analyze.* Given: mass of cockroach, rate of O_2 consumption, temperature, percent O_2 in air, volume of air. Find: mol O_2 consumed per hour; mol O_2 in 1 quart of air; mol O_2 consumed in 48 hr.

(a)   *Plan/Solve.* V of O_2 consumed = rate of consumption $\times$ mass $\times$ time. n = PV/RT.

$5.2\,g \times 1\,hr \times \dfrac{0.8\,mL\,O_2}{1\,g-hr} = 4.16 = 4\,mL\,O_2$ consumed

(b) $n = \dfrac{PV}{RT} = 1\,atm \times \dfrac{mol\text{-}K}{0.08206\,L\text{-}atm} \times \dfrac{0.00416\,L}{297\,K} = 1.71 \times 10^{-4} = 2 \times 10^{-4}\,mol\,O_2$

(c)   *Plan/Solve.* qt air $\rightarrow$ L air $\rightarrow$ L O_2 available. mol O_2 available = PV/RT.
mol O_2/hr (from part (a)) $\rightarrow$ total mol O_2 consumed. Compare O_2 available and O_2 consumed.

$1\,qt\,air \times \dfrac{0.946\,L}{1\,qt} \times 0.21\,O_2$ in air $= 0.199\,L\,O_2$ available

$n = 1\,atm \times \dfrac{mol\text{-}K}{0.08206\,L\text{-}atm} \times \dfrac{0.199\,L}{297\,K} = 8.17 \times 10^{-3} = 8 \times 10^{-3}\,mol\,O_2$ available

roach uses $\dfrac{1.71 \times 10^{-4}\,mol}{1\,hr} \times 48\,hr = 8.21 \times 10^{-3} = 8 \times 10^{-3}\,mol\,O_2$ consumed

Not only does the roach use 20% of the available O_2, it needs all the O_2 in the jar.

10.47   (a)   *Analyze.* Given: 119 tons Hg, 1 atm, 298 K. Find: vol Hg(g). *Plan.* Change tons to grams; use V = gRT/MM×P to calculate volume.   *Solve.*

$119\,tons \times \dfrac{2000\,lb}{1\,ton} \times \dfrac{453.59\,g}{1\,lb} = 1.07954 \times 10^8 = 1.08 \times 10^8\,g\,Hg$

$V = \dfrac{1.07954 \times 10^8\,g\,Hg \times 1\,mol\,Hg}{200.59\,g\,Hg} \times \dfrac{0.08206\,L\text{-}atm}{mol\text{-}K} \times \dfrac{298\,K}{1\,atm}$

$= 1.3161 \times 10^7 = 1.32 \times 10^7\,L$

(b)   *Analyze.* Given: total vol atmospheric gases (air) = 51×10^{12} m³; 245 ppb Hg(g) by volume. Find: mol Hg(g) in atmospheres. *Plan.* Change m³ to L; use definition of ppb to get L Hg(g); use n = PV/RT to calculate mol Hg. Assume STP.   *Solve.*

$$51 \times 10^{12} \text{ m}^3 \times \frac{10^3 \text{ dm}^3}{1 \text{ m}^3} \times \frac{1 \text{ L}}{1 \text{ dm}^3} = 5.1 \times 10^{16} \text{ L air}$$

$$5.1 \times 10^{16} \text{ L air} \times \frac{245 \text{ L Hg(g)}}{1 \times 10^9 \text{ L air}} = 1.2495 \times 10^{10} = 1.2 \times 10^{10} \text{ L Hg(g)}$$

$$n = \frac{1 \text{ atm} \times 1.2495 \times 10^{10} \text{ L Hg(g)}}{273 \text{ K}} \times \frac{\text{mol-K}}{0.08206 \text{ L-atm}} = 5.5775 \times 10^8 = 5.6 \times 10^8 \text{ mol Hg(g)}$$

Check. Note that calculating amount of Hg from volume Hg(g) depends on the the assumed temperature of the atmosphere.

Further Applications of the Ideal-Gas Equation (section 10.5)

10.49 Cl_2(g) is the most dense at 1.00 atm and 298 K. Gas density is directly proportional to molar mass and pressure, and inversely proportional to temperature (Equation [10.10]). For gas samples at the same conditions, molar mass determines density. Of the three gases listed, Cl_2 has the largest molar mass.

10.51   (c) Because the helium atoms are of lower mass than the average air molecule, the helium gas is less dense than air. The balloon thus weighs less than the air displaced by its volume.

10.53   *Analyze/Plan.* Conditions (P, V, T) and amounts of gases are given. Rearrange the relationship PV × MM = gRT to obtain the desired of quantity, paying attention (as always!) to units.   *Solve.*

(a) $d = \dfrac{MM \times P}{RT}$; MM = 46.0 g/mol; P = 0.970 atm, T = 35°C = 308 K

$$d = \frac{46.0 \text{ g } NO_2}{1 \text{ mol}} \times \frac{\text{mol-K}}{0.08206 \text{ L-atm}} \times \frac{0.970 \text{ atm}}{308 \text{ K}} = 1.77 \text{ g/L}$$

(b) $MM = \dfrac{gRT}{PV} = \dfrac{2.50 \text{ g}}{0.875 \text{ L}} \times \dfrac{0.08206 \text{ L-atm}}{\text{mol-K}} \times \dfrac{308 \text{ K}}{685 \text{ torr}} \times \dfrac{760 \text{ torr}}{1 \text{ atm}} = 80.1 \text{ g/mol}$

10.55   *Analyze/Plan.* Given: mass, conditions (P, V, T) of unknown gas. Find: molar mass. MM = gRT/PV. *Solve.*

$$MM = \frac{gRT}{PV} = \frac{1.012 \text{ g}}{0.354 \text{ L}} \times \frac{0.08206 \text{ L-atm}}{\text{mol-K}} \times \frac{372 \text{ K}}{742 \text{ torr}} \times \frac{760 \text{ torr}}{1 \text{ atm}} = 89.4 \text{ g/mol}$$

10.57   *Analyze/Plan.* Follow the logic in Sample Exercise 10.9.   *Solve.*

$$\text{mol } O_2 = \frac{PV}{RT} = 3.5 \times 10^{-6} \text{ torr} \times \frac{1 \text{ atm}}{760 \text{ torr}} \times \frac{\text{mol-K}}{0.08206 \text{ L-atm}} \times \frac{0.452 \text{ L}}{300 \text{ K}} = 8.456 \times 10^{-11}$$

$$= 8.5 \times 10^{-11} \text{ mol } O_2$$

$$8.456 \times 10^{-11} \text{ mol } O_2 \times \frac{2 \text{ mol Mg}}{1 \text{ mol } O_2} \times \frac{24.3 \text{ g Mg}}{1 \text{ mol Mg}} = 4.1 \times 10^{-9} \text{ g Mg (4.1 ng Mg)}$$

10.59   (a)   *Analyze/Plan.* g glucose → mol glucose → mol CO_2 → V CO_2 *Solve.*

$$24.5\,g \times \frac{1\,mol\,glucose}{180.1\,g} \times \frac{6\,mol\,CO_2}{1\,mol\,glucose} = 0.8162 = 0.816\,mol\,CO_2$$

$$V = \frac{nRT}{P} = 0.8162\,mol \times \frac{0.08206\,L\text{-}atm}{mol\text{-}K} \times \frac{310\,K}{0.970\,atm} = 21.4\,L\,CO_2$$

(b)   *Analyze/Plan.* g glucose → mol glucose → mol O_2 → V O_2 *Solve.*

$$50.0\,g \times \frac{1\,mol\,glucose}{180.1\,g} \times \frac{6\,mol\,O_2}{1\,mol\,glucose} = 1.6657 = 1.67\,mol\,O_2$$

$$V = \frac{nRT}{P} = 1.6657\,mol \times \frac{0.08206\,L\text{-}atm}{mol\text{-}K} \times \frac{298\,K}{1\,atm} = 40.7\,L\,O_2$$

10.61   *Analyze/Plan.* The gas sample is a mixture of $H_2(g)$ and $H_2O(g)$. Find the partial pressure of $H_2(g)$ and then the moles of $H_2(g)$ and Zn(s). *Solve.*

$$P_t = 738\,torr = P_{H_2} + P_{H_2O}$$

From Appendix B, the vapor pressure of H_2O at 24°C = 22.38 torr

$$P_{H_2} = (738\,torr - 22.38\,torr) \times \frac{1\,atm}{760\,torr} = 0.9416 = 0.942\,atm$$

$$n_{H_2} = \frac{P_{H_2}V}{RT} = 0.9416\,atm \times \frac{mol\text{-}K}{0.08206\,L\text{-}atm} \times \frac{0.159\,L}{297\,K} = 0.006143 = 0.00614\,mol\,H_2$$

$$0.006143\,mol\,H_2 \times \frac{1\,mol\,Zn}{1\,mol\,H_2} \times \frac{65.39\,g\,Zn}{1\,mol\,Zn} = 0.402\,g\,Zn$$

Partial Pressures (section 10.6)

10.63   (a)   When the stopcock is opened, the volume occupied by $N_2(g)$ increases from 2.0 L to 5.0 L. At constant T, $P_1V_1 = P_2V_2$. 1.0 atm × 2.0 L = P_2 × 5.0 L; P_2 = 0.40 atm

(b)   When the gases mix, the volume of $O_2(g)$ increases from 3.0 L to 5.0 L. At constant T, $P_1V_1 = P_2V_2$. 2.0 atm × 3.0 L = P_2 × 5.0 L; P_2 = 1.2 atm

(c) $P_t = P_{N_2} + P_{O_2} = 0.40\,atm + 1.2\,atm = 1.6\,atm$

10.65   *Analyze.* Given: amount, V, T of three gases. Find: P of each gas, total P.

Plan. P = nRT/V; $P_t = P_1 + P_2 + P_3 + \cdots$ *Solve.*

(a) $P_{He} = \dfrac{nRT}{V} = 0.765\,mol \times \dfrac{0.08206\,L\text{-}atm}{mol\text{-}K} \times \dfrac{298\,K}{10.00\,L} = 1.871 = 1.87\,atm$

$P_{Ne} = \dfrac{nRT}{V} = 0.330\,mol \times \dfrac{0.08206\,L\text{-}atm}{mol\text{-}K} \times \dfrac{298\,K}{10.00\,L} = 0.8070 = 0.807\,atm$

$P_{Ar} = \dfrac{nRT}{V} = 0.110\,mol \times \dfrac{0.08206\,L\text{-}atm}{mol\text{-}K} \times \dfrac{298\,K}{10.00\,L} = 0.2690 = 0.269\,atm$

(b) $P_t = 1.871\,atm + 0.8070\,atm + 0.2690\,atm = 2.9470 = 2.95\,atm$

10.67   *Analyze.* Given 390 ppm CO_2 in the atmosphere; 390 L CO_2 in 10^6 total L air. Find: mole fraction CO_2 in the atmosphere. *Plan.* Avogadro's law deals with the relationship between volume and moles of a gas.

Solve. Avogadro's law states that volume of a gas at constant temperature and pressure is directly proportional to moles of the gas. Using volume fraction to express concentration assumes that the 390 L CO_2 and 10^6 total L air are at the same temperature and pressure. That is, 390 L is the volume that the number of moles of CO_2 present in 10^6 L air would occupy at atmospheric temperature and pressure. The mole fraction of CO_2 in the atmosphere is then just the volume fraction from the concentration by volume.

$$\chi_{CO_2} = \frac{390 \text{ L } CO_2}{10^6 \text{ L air}} = 0.00039$$

10.69　*Analyze.* Given: mass CO_2 at V, T; pressure of air at same V, T. Find: partial pressure of CO_2 at these conditions, total pressure of gases at V, T.

Plan. $g\,CO_2 \to mol\,CO_2 \to P_{CO_2}$ (via P = nRT/V); $P_t = P_{CO_2} + P_{air}$ 　　　*Solve.*

$$5.50 \text{ g } CO_2 \times \frac{1 \text{ mol } CO_2}{44.01 \text{ g } CO_2} = 0.12497 = 0.125 \text{ mol } CO_2; T = 273 + 24°C = 297 \text{ K}$$

$$P_{CO_2} = 0.12497 \text{ mol} \times \frac{297 \text{ K}}{10.0 \text{ L}} \times \frac{0.08206 \text{ L-atm}}{\text{mol-K}} = 0.30458 = 0.305 \text{ atm}$$

$$P_{air} = 705 \text{ torr} \times \frac{1 \text{ atm}}{760 \text{ torr}} = 0.92763 = 0.928 \text{ atm}$$

$$P_t = P_{CO_2} + P_{air} = 0.30458 + 0.92763 = 1.23221 = 1.232 \text{ atm}$$

(Result has 3 decimal places and 4 sig figs.)

10.71　*Analyze/Plan.* The partial pressure of each component is equal to the mole fraction of that gas times the total pressure of the mixture. Find the mole fraction of each component and then its partial pressure.　　　*Solve.*

$n_t = 0.75 \text{ mol } N_2 + 0.30 \text{ mol } O_2 + 0.15 \text{ mol } CO_2 = 1.20 \text{ mol}$

$$\chi_{N_2} = \frac{0.75}{1.20} = 0.625 = 0.63; P_{N_2} = 0.625 \times 2.15 \text{ atm} = 1.344 = 1.3 \text{ atm}$$

$$\chi_{O_2} = \frac{0.30}{1.20} = 0.250 = 0.25; P_{O_2} = 0.250 \times 2.15 \text{ atm} = 0.5375 = 0.54 \text{ atm}$$

$$\chi_{CO_2} = \frac{0.15}{1.20} = 0.125 = 0.13; P_{CO_2} = 0.125 \times 2.15 \text{ atm} = 0.2688 = 0.27 \text{ atm}$$

Check. $P_T = 1.344 + 0.5375 + 0.2688 = 2.1503 = 2.15$ atm. The sum of the partial pressures agrees with the stated total pressure.

10.73　*Analyze/Plan.* Mole fraction = pressure fraction. Find the desired mole fraction of O_2 and change to mole percent.　　　*Solve.*

$$\chi_{O_2} = \frac{P_{O_2}}{P_t} = \frac{0.21 \text{ atm}}{8.38 \text{ atm}} = 0.025; \text{ mole \%} = 0.025 \times 100 = 2.5\%$$

10.75　*Analyze/Plan.* $N_2(g)$ and $O_2(g)$ undergo changes of conditions and are mixed. Calculate the new pressure of each gas and add them to obtain the total pressure of the mixture.

$P_2 = P_1 V_1 T_2 / V_2 T_1; P_T = P_{N_2} + P_{O_2}.$　　　*Solve.*

$$P_{N_2} = \frac{P_1 V_1 T_2}{V_2 T_1} = \frac{5.25 \, \text{atm} \times 1.00 \, \text{L} \times 293 \, \text{K}}{12.5 \, \text{L} \times 299 \, \text{K}} = 0.41157 = 0.412 \, \text{atm}$$

$$P_{O_2} = \frac{P_1 V_1 T_2}{V_2 T_1} = \frac{5.25 \, \text{atm} \times 5.00 \, \text{L} \times 293 \, \text{K}}{12.5 \, \text{L} \times 299 \, \text{K}} = 2.05786 = 2.06 \, \text{atm}$$

$$P_t = 0.41157 \, \text{atm} + 2.05786 \, \text{atm} = 2.46943 = 2.47 \, \text{atm}$$

Kinetic-Molecular Theory of Gases; Effusion and Diffusion
(sections 10.7 and 10.8)

10.77   (a)   Increase in temperature at constant volume, decrease in volume, increase in pressure

(b)   Decrease in temperature

(c)   Increase in volume, decrease in pressure

(d)   Increase in temperature

10.79   The fact that gases are readily compressible supports the assumption that most of the volume of a gas sample is empty space.

10.81   Average speed is the numerical mean speed, the sum of the speeds of all particles divided by the total number of particles. The root mean square (rms) speed is the speed of a molecule with the same kinetic energy as the average kinetic energy of the sample. For a given gas sample at a fixed temperature, rms speed is larger (greater) than average speed, but the difference between the two is small.

10.83   *Analyze/Plan.* Apply the concepts of the Kinetic-Molecular Theory (KMT) to the situation where a gas is heated at constant volume. Determine how the quantities in (a)-(d) are affected by this change. *Solve.*

(a)   Average kinetic energy is proportional to temperature (K), so average kinetic energy of the molecules increases.

(b)   The average kinetic energy of a gas is $1/2 \, m u_{rms}^2$. Molecular mass doesn't change as T increases; average kinetic energy increases so rms speed (u) increases. (Also, $u_{rms} = (3RT/MM)^{1/2}$, so u_{rms} is directly related to T.)

(c)   As T and thus rms molecular speed increase, molecular momentum (mu) increases and the strength of an average impact with the container wall increases.

(d)   As T and rms molecular speed increase, the molecules collide more frequently with the container walls, and the total number of collisions per second increases.

10.85   (a)   *Plan.* The larger the molar mass, the slower the average speed (at constant temperature).

Solve. In order of increasing speed (and decreasing molar mass):
$HBr < NF_3 < SO_2 < CO < Ne$

(b)   *Plan.* Follow the logic of Sample Exercise 10.14. *Solve.*

$$u_{rms} = \sqrt{\frac{3RT}{MM}} = \left(\frac{3 \times 8.314\,kg\text{-}m^2/s^2\text{-}mol\text{-}K \times 298\,K}{71.0 \times 10^{-3}\,kg/mol} \right)^{1/2} = 324\,m/s$$

(c)   *Plan.* Use Equation 10.23 to calculate the most probable speed, u_{mp}. MM of O_3 = 48.0 g/mol; T = 270 K.   *Solve.*

$$u_{mp} = \sqrt{\frac{2RT}{MM}} = \left(\frac{2 \times 8.314\,kg\text{-}m^2/s^2\text{-}mol\text{-}K \times 270\,K}{48.0 \times 10^{-3}\,kg/mol} \right)^{1/2} = 306\,m/s$$

10.87   *Effusion* is the escape of gas molecules through a tiny hole. *Diffusion* is the distribution of a gas throughout space or throughout another substance. On a macroscopic scale, effusion is like hundreds of students leaving an auditorium through one small door, while diffusion is analogous to these students making their way around the quad, which may or may not already contain other students.

10.89   *Plan.* The heavier the molecule, the slower the rate of effusion. Thus, the order for increasing rate of effusion is in the order of decreasing mass.   *Solve.*

rate $^2H^{37}Cl$ < rate $^1H^{37}Cl$ < rate $^2H^{35}Cl$ < rate $^1H^{35}Cl$

10.91   *Analyze.* Given: relative effusion rates of two gases at same temperature. Find: molecular formula of one of the gases. *Plan.* Use Graham's law to calculate the formula weight of arsenic (III) sulfide, and thus the molecular formula.   *Solve.*

$$\frac{rate\,(sulfide)}{rate\,(Ar)} = \left[\frac{39.9}{MM\,(sulfide)} \right]^{1/2} = 0.28$$

MM (sulfide) = 39.9 / 0.28^2 = 509 g/mol (two significant figures)
The empirical formula of arsenic(III) sulfide is As_2S_3, which has a formula mass of 246.1. Twice this is 490 g/mol, close to the value estimated from the effusion experiment. Thus, the formula of the gas phase molecule is As_4S_6.

Nonideal-Gas Behavior (section 10.9)

10.93   (a)   Nonideal gas behavior is observed at very high pressures and/or low temperatures.

(b)   The real volumes of gas molecules and attractive intermolecular forces between molecules cause gases to behave nonideally.

(c)   The ratio PV/RT is equal to the number of moles of particles in an ideal-gas sample; this number should be a constant for all pressure, volume, and temperature conditions. If the value of this ratio changes with increasing pressure, the gas sample is not behaving ideally. That is, the gas is not behaving according to the ideal-gas equation.

Negative deviations indicate fewer "effective" particles in the sample, a result of attractive forces among particles. Positive deviations indicate more "effective" particles, a result of the real volume occupied by the particles.

10.95   *Plan.* The constants *a* and *b* are part of the correction terms in the van der Waals equation. The smaller the values of *a* and *b*, the smaller the corrections and the more ideal the gas. *Solve.*

Ar (a = 1.34, b = 0.0322) will behave more like an ideal gas than CO_2 (a = 3.59, b = 0.0427) at high pressures.

10.97   *Analyze/Plan.* Follow the logic in Sample Exercise 10.16. Use the ideal-gas equation to calculate pressure in (a), the van der Waals equation in (b). n = 1.00 mol, V = 5.00 L, T = 25°C = 298 K; a = 6.49 L²-atm/mol², b = 0.0562 L/mol.

(a) $P = \dfrac{nRT}{V} = 1.00 \text{ mol} \times \dfrac{298 \text{ K}}{5.00 \text{ L}} \times \dfrac{0.08206 \text{ L - atm}}{\text{mol - K}} = 4.89 \text{ atm}$

(b) $P = \dfrac{nRT}{V - nb} - \dfrac{n^2 a}{V^2}$;

$$P = \dfrac{(1.00 \text{ mol})(298 \text{ K})(0.08206 \text{ L - atm/mol - K})}{5.00 \text{ L} - (1.00 \text{ mol})(0.0562 \text{ L/mol})} - \dfrac{(1.00 \text{ mol})^2 (6.49 \text{ L}^2 \text{ - atm / mol}^2)}{(5.00 \text{ L})^2}$$

P = 4.9463 atm – 0.2596 atm = 4.6868 = 4.69 atm

(c)   From Sample Exercise 10.16, the difference at 22.41 L between the ideal and van der Waals results is (1.00 – 0.990) = 0.010 atm. At 5.00 L, the difference is (4.89 – 4.69) = 0.20 atm. The effects of both molecular attractions, the *a* correction, and molecular volume, the *b* correction, increase with decreasing volume. For the *a* correction, V^2 appears in the denominator, so the correction increases exponentially as V decreases. For the *b* correction, nb is a larger portion of the total volume as V decreases. That is, 0.0562 L is 1.1% of 5.0 L, but only 0.25% of 22.41 L. Qualitatively, molecular attractions are more important as the amount of free space decreases and the number of molecular collisions increase. Molecular volume is a larger part of the total volume as the container volume decreases.

10.99   *Analyze.* Given the *b* value of Xe, 0.0510 L/mol, calculate the radius of a Xe atom.

Plan. Use Avogadro's number to chang L/mol to L/atom. Use the volume formula, $V = 4/3 \pi r^3$ and units conversion to obtain the radius in A. 1 L = 1 dm³.   *Solve.*

$$\dfrac{0.0510 \text{ L}}{1 \text{ mol Xe}} \times \dfrac{1 \text{ mol Xe}}{6.022 \times 10^{23} \text{ Xe atoms}} \times \dfrac{1 \text{ dm}^3}{1 \text{ L}} = 8.4689 \times 10^{-26} = 8.47 \times 10^{-26} \text{ dm}^3$$

$V = 4/3 \pi r^3$; $r^3 = 3V/4\pi$; $r = (3V/4\pi)^{1/3}$

$$r = \left(\dfrac{3 \times 8.4689 \times 10^{-26} \text{ dm}^3}{4 \times 3.14159} \right)^{1/3} = 2.7243 \times 10^{-9} = 2.72 \times 10^{-9} \text{ dm}$$

$$2.72 \times 10^{-9} \text{ dm} \times \dfrac{1 \text{ m}}{10 \text{ dm}} \times \dfrac{1 \text{ Å}}{1 \times 10^{-10} \text{ m}} = 2.72 \text{ Å}$$

The calculated value is the nonbonding radius. From Figure 7.6 in section 7.3, the bonding atomic radius of Xe is 1.30 Å. We expect the nonbonding radius of an atom to be larger than the bonding radius, but our calculated value is more than twice as large.

Additional Exercises

10.101 $P_1V_1 = P_2V_2$; $V_2 = P_1V_1/P_2$

$$V_2 = \frac{3.0\,\text{atm} \times 1.0\,\text{mm}^3}{730\,\text{torr}} \times \frac{760\,\text{torr}}{1\,\text{atm}} = 3.1\,\text{mm}^3$$

10.105   (a) $n = \dfrac{PV}{RT} = 3.00\,\text{atm} \times \dfrac{\text{mol-K}}{0.08206\,\text{L-atm}} \times \dfrac{110\,\text{L}}{300\,\text{K}} = 13.4\,\text{mol}\;C_3H_8(g)$

(b) $\dfrac{0.590\,\text{g }C_3H_8\,(l)}{1\,\text{mL}} \times 110 \times 10^3\,\text{mL} \times \dfrac{1\,\text{mol }C_3H_8}{44.094\,\text{g}} = 1.47 \times 10^3\,\text{mol }C_3H_8\,(l)$

(c)   Using C_3H_8 in a 110 L container as an example, the ratio of moles liquid to moles gas that can be stored in a certain volume is $\dfrac{1.47 \times 10^3\,\text{mol liquid}}{13.4\,\text{mol gas}} = 110$.

A container with a fixed volume holds many more moles (molecules) of $C_3H_8(l)$ because in the liquid phase the molecules are touching. In the gas phase, the molecules are far apart (statement 2, section 10.7), and many fewer molecules will fit in the container.

10.108   It is simplest to calculate the partial pressure of each gas as it expands into the total volume, then sum the partial pressures.

$P_2 = P_1V_1/V_2$; $P_{N_2} = 265\,\text{torr}\,(1.0\,\text{L}/2.5\,\text{L}) = 106 = 1.1 \times 10^2\,\text{torr}$

$P_{Ne} = 800\,\text{torr}\,(1.0\,\text{L}/2.5\,\text{L}) = 320 = 3.2 \times 10^2\,\text{torr}$; $P_{H_2} = 532\,\text{torr}\,(0.5\,\text{L}/2.5\,\text{L})$

$$= 106 = 1.1 \times 10^2\,\text{torr}$$

$P_t = P_{N_2} + P_{Ne} + P_{H_2} = (106 + 320 + 106)\,\text{torr} = 532 = 5.3 \times 10^2\,\text{torr}$

10.111   V and T are the same for He and O_2.

$P_{He}V = n_{He}RT$, $P_{He}/n_{He} = RT/V$; $P_{O_2}/n_{O_2} = RT/V$

$\dfrac{P_{He}}{n_{He}} = \dfrac{P_{O_2}}{n_{O_2}} = n_{O_2} \times \dfrac{P_{O_2} \times n_{He}}{P_{He}}$; $n_{He} = 1.42\,\text{g He} \times \dfrac{1\,\text{mol He}}{4.003\,\text{g He}} = 0.3547 = 0.355\,\text{mol He}$

$n_{O_2} = \dfrac{158\,\text{torr}}{42.5\,\text{torr}} \times 0.355\,\text{mol} = 1.3188 = 1.32\,\text{mol }O_2$; $1.3188\,\text{mol }O_2 \times \dfrac{32.00\,\text{g }O_2}{1\,\text{mol }O_2} = 42.2\,\text{g }O_2$

10.115 $u_{rms} = (3RT/MM)^{1/2}$; $u_{rms2} = 2\,u_{rms1}$; $T_1 = -33° = 240\,\text{K}$

$u_{rms1} = (3RT_1/MM)^{1/2}$; $u_{rms1}{}^2 = 3(240)R/MM = 720R/MM$

$u_{rms1} = (720R/MM)^{1/2}$; $u_{rms2} = 2u_{rms1} = (2)(720R/MM)^{1/2}$

$(2)(720R/MM)^{1/2} = (3RT_2/MM)^{1/2}$

$(2)^2(720R/MM) = 3RT_2/MM$; $(2)^2(720) = 3T_2$

$T_2 = (4)(720)/3 = 960\,\text{K} = 687°C$

Increasing the rms speed (u) by a factor of 2 requires heating to 960 K (or 687°C), increasing the temperature by a factor of 4.

10.120   (a) $120.00 \text{ kg N}_2(g) \times \dfrac{1000 \text{ g}}{1 \text{ kg}} \times \dfrac{1 \text{ mol N}_2}{28.0135 \text{ g N}} = 4283.6 \text{ mol N}_2$

$P = \dfrac{nRT}{V} = 4283.6 \text{ mol} \times \dfrac{0.08206 \text{ L-atm}}{\text{mol-K}} \times \dfrac{553 \text{ K}}{1100.0 \text{ L}} = 176.72 = 177 \text{ atm}$

(b)   According to Equation [10.26], $P = \dfrac{nRT}{V - nb} - \dfrac{n^2 a}{V^2}$

$P = \dfrac{(4283.6 \text{ mol})(0.08206 \text{ L-atm/mol-K})(553 \text{ K})}{1100.0 \text{ L} - (4283.6 \text{ mol})(0.0391 \text{ L/mol})} - \dfrac{(4283.6 \text{ mol})^2 (1.39 \text{ L}^2\text{-atm/mol}^2)}{(1100.0 \text{ L})^2}$

$P = \dfrac{194{,}388 \text{ L-atm}}{1100.0 \text{ L} - 167.5 \text{ L}} - 21.1 \text{ atm} = 208.5 \text{ atm} - 21.1 \text{ atm} = 187.4 \text{ atm}$

(c)   The pressure corrected for the real volume of the N_2 molecules is 208.5 atm, 31.8 atm higher than the ideal pressure of 176.7 atm. The 21.1 atm correction for intermolecular forces reduces the calculated pressure somewhat, but the "real" pressure is still higher than the ideal pressure. The correction for the real volume of molecules dominates. Even though the value of b is small, the number of moles of N_2 is large enough so that the molecular volume correction is larger than the attractive forces correction.

Integrative Exercises

10.123   (a)   Get g C from mL CO_2; get g H from mL H_2O. Also calculate mol C and H, to use in part (b). Get g N by subtraction. Calculate % composition.
n = PV/RT. At STP, P = 1 atm, T = 273 K. (STP implies an infinite number of sig figs)

$n_{CO_2} = 0.08316 \text{ L} \times \dfrac{1 \text{ atm}}{273 \text{ K}} \times \dfrac{\text{mol-K}}{0.08206 \text{ L-atm}} = 0.003712 \text{ mol CO}_2$

$0.003712 \text{ mol CO}_2 \times \dfrac{1 \text{ mol C}}{1 \text{ mol CO}_2} \times \dfrac{12.0107 \text{ g C}}{\text{mol C}} = 0.044585 = 0.04458 \text{ g CO}_2$

$n_{H_2O} = 0.07330 \text{ L} \times \dfrac{1 \text{ atm}}{273 \text{ K}} \times \dfrac{\text{mol-K}}{0.08206 \text{ L-atm}} = 3.2720 \times 10^{-3}$

$= 3.272 \times 10^{-3} \text{ mol H}_2O$

$3.2720 \times 10^{-3} \text{ mol H}_2O \times \dfrac{2 \text{ mol H}}{1 \text{ mol H}_2O} \times \dfrac{1.00794 \text{ g H}}{\text{mol H}} = 6.5959 \times 10^{-3}$

$= 6.596 \times 10^{-3} \text{ g H}$

$\text{mass \% X} = \dfrac{\text{mass X}}{\text{sample mass}} \times 100; \quad \text{sample mass} = 100.0 \text{ mg} = 0.1000 \text{ g}$

$\text{\% C} = \dfrac{0.044585 \text{ g}}{0.1000 \text{ g}} \times 100 = 44.585 = 44.58\% \text{ C}$

$\text{\% H} = \dfrac{6.5959 \times 10^{-3} \text{ g H}}{0.1000 \text{ g}} \times 100 = 6.5959 = 6.596\% \text{ H}$

$\text{\% Cl} = \dfrac{0.01644 \text{ g Cl}}{0.1000 \text{ g}} \times 100 = 16.44\% \text{ Cl}$

% N = 100 − 44.58 − 6.596 − 16.44 = 32.38% N

(b)   0.003712 mol C; $2(3.272 \times 10^{-3}) = 6.544 \times 10^{-3}$ mol H

$$0.01644 \text{ g Cl} \times \frac{1 \text{ mol C}}{35.453 \text{ g Cl}} = 4.637 \times 10^{-4} \text{ mol Cl}$$

0.1000 g sample $\times$ 0.3238 mass fraction N = 0.03238 g N

$$0.03238 \text{ g N} \times \frac{1 \text{ mol N}}{14.0067 \text{ g N}} = 0.0023118 = 0.002312 \text{ mol N}$$

Divide by the smallest number of mol to find the simplest ratio of moles.

$$\frac{0.003712 \text{ mol C}}{4.637 \times 10^{-4}} = 8.005 \text{ C}$$

$$\frac{6.544 \times 10^{-3} \text{ mol H}}{4.637 \times 10^{-4}} = 14.11 \text{ H}$$

$$\frac{4.637 \times 10^{-4} \text{ mol Cl}}{4.637 \times 10^{-4}} = 1.000 \text{ Cl}$$

$$\frac{0.002312 \text{ mol N}}{4.637 \times 10^{-4}} = 4.985 \text{ N}$$

If we assume 14.11 is "close" to 14 (a reasonable assumption), the empirical formula is $C_8H_{14}N_5Cl$.

(c)   Molar mass of the compound is required in order to determine molecular formula when the empirical formula is known.

10.128   (a) ft^3 $CH_4 \to L$ $CH_4 \to$ mol $CH_4 \to$ mol $CH_3OH \to$ g $CH_3OH \to L$ CH_3OH

$$10.7 \times 10^9 \text{ ft}^3 \text{ CH}_4 \times \frac{1 \text{ yd}^3}{3^3 \text{ ft}^3} \times \frac{1 \text{ m}^3}{(1.0936)^3 \text{ yd}^3} \times \frac{1 \text{ L}}{1 \times 10^{-3} \text{ m}^3} = 3.03001 \times 10^{11}$$

$$= 3.03 \times 10^{11} \text{ L CH}_4$$

$$n = \frac{PV}{RT} = \frac{3.03 \times 10^{11} \text{ L} \times 1.00 \text{ atm}}{298 \text{ K}} \times \frac{\text{mol-K}}{0.08206 \text{ L-atm}} = 1.2391 \times 10^{10}$$

$$= 1.24 \times 10^{10} \text{ mol CH}_4$$

1 mol CH_4 = 1 mol CH_3OH

$$1.2391 \times 10^{10} \text{ mol CH}_3\text{OH} \times \frac{32.04 \text{ g CH}_3\text{OH}}{\text{mol CH}_3\text{OH}} \times \frac{1 \text{ mL CH}_3\text{OH}}{0.791 \text{g}} \times \frac{1 \text{ L}}{1000 \text{ mL}}$$

$$= 5.0189 \times 10^8 = 5.02 \times 10^8 \text{ L CH}_3\text{OH}$$

(b) $CH_4(g) + 2O_2(g) \to CO_2(g) + 2H_2O(l)$ $\Delta H^\circ = -890.4$ kJ

$\Delta H^\circ = \Delta H_f^\circ CO_2(g) + 2\Delta H_f^\circ H_2O(l) - \Delta H_f^\circ CH_4(g) - 2\Delta H_f^\circ O_2(g)$

$\Delta H^\circ = -393.5 \text{ kJ} + 2(-285.83 \text{ kJ}) - (-74.8 \text{ kJ}) - 0 = -890.4$ kJ

$$1.2391 \times 10^{10} \text{ mol CH}_4 \times \frac{-890.4 \text{ kJ}}{1 \text{ mol CH}_4} = -1.10 \times 10^{13} \text{ kJ}$$

$$CH_3OH(l) + 3/2\, O_2(g) \rightarrow CO_2(g) + 2H_2O(l) \quad \Delta H^\circ = -726.6 \text{ kJ}$$

$$\Delta H^\circ = \Delta H_f^\circ\, CO_2(g) + 2\Delta H_f^\circ\, H_2O(l) - \Delta H_f^\circ\, CH_3OH(l) - 3/2\, \Delta H_f^\circ\, O_2(g)$$

$$= -393.5 \text{ kJ} + 2(-285.83 \text{ kJ}) - (-238.6 \text{ kJ}) - 0 = -726.6 \text{ kJ}$$

$$1.2391 \times 10^{10} \text{ mol CH}_3\text{OH} \times \frac{-726.6 \text{ kJ}}{1 \text{ mol CH}_3\text{OH}} = -9.00 \times 10^{12} \text{ kJ}$$

(c)   Assume a volume of 1.00 L of each liquid.

$$1.00\, L\, CH_4(l) \times \frac{466 \text{ g}}{1\,L} \times \frac{1 \text{ mol}}{16.04 \text{ g}} \times \frac{-890.4 \text{ kJ}}{\text{mol CH}_4} = -2.59 \times 10^4 \text{ kJ/L CH}_4$$

$$1.00\, L\, CH_3OH \times \frac{791 \text{ g}}{1\,L} \times \frac{1 \text{ mol}}{32.04 \text{ g}} \times \frac{-726.6 \text{ kJ}}{\text{mol CH}_3\text{OH}} = -1.79 \times 10^4 \text{ kJ/L CH}_3\text{OH}$$

Clearly $CH_4(l)$ has the higher enthalpy of combustion per unit volume.

11 Liquids and Intermolecular Forces

Visualizing Concepts

In this chapter we will use the temperature units °C and K interchangeably when designating specific heats and *changes* in temperature.

11.1   The diagram best describes a **liquid**. In the diagram, the particles are close together, mostly touching but there is no regular arrangement or order. This rules out a gaseous sample, where the particles are far apart, and a crystalline solid, which has a regular repeating structure in all three directions.

11.4   When heat is added to a liquid, the temperature of the liquid rises. If enough heat is added to reach the boiling point, any excess heat is used to vaporize the liquid. If heat is still available when all the liquid is converted to gas, the temperature of the gas rises.

Use the specific heat of $CH_4(l)$ to calculate the amount of heat required to raise the temperature of 32.0 g of $CH_4(l)$ from –170 °C to –161.5 °C. If this is less than 42 kJ, use ΔH_{vap} to calculate the energy required to vaporize the liquid, and so on, until exactly 42.0 kJ has been used to increase the temperature and/or change the state of CH_4.

Heat the liquid to its boiling point: $\Delta T = [-161.5 \, °C - (-170 \, °C)] = 8.5 \, °C = 8.5 \, K$

$$\frac{3.48 \, J}{g \cdot K} \times 32.0 \, g \, CH_4 \times 8.5 \, °C = 946.56 = 9.5 \times 10^2 \, J = 0.95 \, kJ$$

Heating $CH_4(l)$ to its boiling point requires only 0.95 kJ. We have added 42 kJ, so there is definitely enough heat to vaporize the liquid. ΔH_{vap} for $CH_4(l)$ is 8.20 kJ/mol. The 32.0 g sample is 2.00 mol $CH_4(l)$, so the energy required to vaporize the sample at –161.5 °C is (2 × 8.20 kJ/mol =) 16.4 kJ. The energy used to heat the sample to –161.5 °C and vaporize it at this temperature is (0.947 kJ + 16.4 kJ) = 17.347 = 17.3 kJ. We have (42.0 kJ – 17.347 kJ) = 24.653 = 24.7 kJ left to heat the gas.

$$\Delta T = 24.65 \, kJ \times \frac{g \cdot K}{2.22 \, J} \times \frac{1000 \, J}{kJ} \times \frac{1}{32.0 \, g \, CH_4} = 346.99 = 347 \, °C$$

The final temperature of the methane gas, $CH_4(g)$, is then
(–161.5 °C + 346.99 °C) = 185.49 = 185 °C.

11.5   (a)   385 mm Hg. Find 30 °C on the horizontal axis, and follow a vertical line from this point to its intersection with the red vapor pressure curve. Follow a horizontal line from the intersection to the vertical axis and read the vapor pressure.

    (b)   22 °C. Reverse the procedure outlined in part (a). Find 300 torr on the vertical axis, follow it to the curve and down to the value on the horizontal axis.

(c)   47 °C. The normal boiling point of a liquid is the temperature at which its vapor pressure is 1 atm, or 760 mm Hg. A vapor pressure of 1 atm is very near the top of this diagram, at approximately 47 °C.

11.6   The stronger the intermolecular forces, the greater the average kinetic energy required to escape these forces, and the higher the boiling point. Propanol, $CH_3CH_2CH_2OH$, has hydrogen bonding, by virtue of its –OH group, so it has the higher boiling point. Van der Waals forces in the two liquids are similar because molar masses are the same for both molecules.

11.7   (a)   360 K, the normal boiling point; 260 K, normal freezing point. The left-most line is the freezing/melting curve, the right-most line is the condensation/boiling curve. The normal boiling and freezing points are the temperatures of boiling and freezing at 1 atm pressure.

         (b)   The material is solid in the left-most green (or pale blue) zone, liquid in the blue zone, and gas in the tan zone. (i) gas (ii) solid (iii) liquid

         (c)   The triple point, where all three phases are in equilibrium, is the point where the three lines on the phase diagram meet. For this substance, the triple point is approximately 185 K at 0.45 atm.

Molecular Comparisons of Gases, Liquids, and Solids (section 11.1)

11.9   (a)   solid < liquid < gas

         (b)   gas < liquid < solid

         (c)   Matter in the gaseous state is most easily compressed, because particles are far apart and there is much empty space.

11.11   By observation (or a quick WebElements© search), we know that Ar is a gas, CCl_4 is a liquid and Si is a solid at standard conditions. (Si is a substance in which the particles are held together by chemical bonds, like the ones listed in Table 11.2.) The order of increasing boiling point is then   Ar < CCl_4 < Si.

11.13   (a)   At standard temperature and pressure, the molar volumes of Cl_2 and NH_3 are nearly the same because they are both gases. In the gas phase, molecules are far apart and most of the volume occupied by the substance is empty space. Differences in molecular characteristics such as weight, shape and dipole moment have little bearing on the molar volume of a gas.

             The ideal gas law states that one mole of any gas at STP will occupy a fixed volume. The slight difference in molar volumes of the two gases is predicted by the van der Waals correction, which quantifies deviation from ideal behavior.

         (b)   On cooling to 160 K, both compounds condense from the gas phase to the solid state. Condensation, as the word implies, eliminates most of the empty space between molecules, so we expect a significant decrease in the molar volume.

         (c) $\dfrac{1\,cm^3}{2.02\,g\,Cl_2} \times \dfrac{70.096\,g\,Cl_2}{1\,mol\,Cl_2} = 35.1\,cm^3/mol\,Cl_2 = 0.0351\ L/mol\,Cl_2$

$$\frac{1\,cm^3}{0.84\,g\,NH_3} \times \frac{17.031\,g\,NH_3}{1\,mol\,NH_3} = 20.3\,cm^3/mol\,NH_3 = 0.0203\ L/mol\,NH_3$$

(d) Solid state molar volumes are not as similar as those in the gaseous state. In the solid state, most of the empty space is gone, so molecular characteristics do influence molar volumes. Cl_2 is heavier than NH_3 and the Cl–Cl bond distance is almost double the N–H bond distance (Figure 7.6). Intermolecular attractive forces among polar NH_3 molecules bind them more tightly than forces among nonpolar Cl_2 molecules. These factors all contribute to a molar volume for $Cl_2(s)$ that is almost twice that of $NH_3(s)$.

(e) Like solids, liquids are condensed phases. That is, there is little empty space between molecules in the liquid state. We expect the molar volumes of the liquids to be closer to those in the solid state than those in the gaseous state.

Intermolecular Forces (section 11.2)

11.15 (a) London dispersion forces

(b) dipole-dipole and London dispersion forces

(c) dipole-dipole or in certain cases hydrogen bonding

11.17 (a) SO_2 is a polar covalent molecule, so dipole-dipole and London dispersion forces must be overcome to convert the liquid to a gas.

(b) CH_3COOH is a polar covalent molecule that experiences London dispersion, dipole-dipole, and hydrogen-bonding (O–H bonds) forces. All of these forces must be overcome to convert the liquid to a gas.

(c) H_2Se is a polar covalent molecule that experiences London dispersion and dipole-dipole forces, so these must be overcome to change the liquid into a gas. (H–Se bonds do not lead to hydrogen-bonding interactions.)

11.19 (a) *Polarizability* is the ease with which the charge distribution (electron cloud) in a molecule can be distorted to produce an instantaneous dipole.

(b) Sb is most polarizable because its valence electrons are farthest from the nucleus and least tightly held.

(c) Polarizability increases as molecular size (and thus molecular weight) increases. In order of increasing polarizability: $CH_4 < SiH_4 < SiCl_4 < GeCl_4 < GeBr_4$

(d) The magnitude of London-dispersion forces and thus the boiling points of molecules increase as polarizability increases. The order of increasing boiling points is the order of increasing polarizability:
$CH_4 < SiH_4 < SiCl_4 < GeCl_4 < GeBr_4$

11.21 *Analyze/Plan.* For molecules with similar structures, the strength of dispersion forces increases with molecular size (molecular weight and number of electrons in the molecule).

Solve: (a) H_2S (b) CO_2 (c) GeH_4

11.23 Both hydrocarbons experience dispersion forces. Rod-like butane molecules can contact each other over the length of the molecule, while spherical 2-methylpropane molecules can only touch tangentially. The larger contact surface of butane facilitates stronger forces and produces a higher boiling point.

11.25   (a)   A molecule must contain H atoms bound to either N, O or F atoms in order to participate in hydrogen bonding with like molecules.

       (b)   **CH_3NH_2** and **CH_3OH** have N–H and O–H bonds, respectively; they will form hydrogen bonds with other molecules of the same kind. (CH_3F has C–F and C–H bonds, but no H–F bonds.)

11.27   (a)   Replacing a hydroxyl hydrogen with a CH_3 group eliminates hydrogen bonding in that part of the molecule. This reduces the strength of intermolecular forces and leads to a (much) lower boiling point.

       (b) $CH_3OCH_2CH_2OCH_3$ is a larger, more polarizable molecule with stronger London-dispersion forces and thus a higher boiling point.

11.29

Physical Property	H_2O	H_2S
Normal Boiling Point, °C	100.00	−60.7
Normal Melting Point, °C	0.00	−85.5

       (a)   Based on its much higher normal melting and boiling point, H_2O has stronger intermolecular forces. H_2O, with H bound to O, has hydrogen bonding. H_2S, with H bound to S, has dipole-dipole forces. (The electronegativities of H and S, 2.1 and 2.5, respectively, are similar. The H–S bond dipoles in H_2S are not large, but S does have two nonbonded electron pairs. The molecule has medium polarity.)

       (b) H_2S is probably a typical compound, where there is less empty space in the more ordered solid and the solid is denser than the liquid. For H_2O, maximizing the number of hydrogen bonds to each molecule in the solid requires more empty space than in the liquid. The solid is less dense than the liquid and ice floats.

       (c)   Specific heat is the energy required to raise the temperature of one gram of the substance one degree Celsius. Raising the temperature of a substance increases average kinetic energy and molecular motion. Hydrogen bonding in water is such a strong attractive interaction that the energy required to disrupt it and increase molecular motion is large.

11.31 SO_4^{2-} has a greater negative charge than BF_4^-, so ion-ion electrostatic attractions are greater in sulfate salts. These strong forces limit the ion mobility required for the formation of an ionic liquid. (This is called an electronic effect.)

Select Properties of Liquids (section 11.3)

11.33   (a)   Surface tension and viscosity are the result of intermolecular attractive forces or cohesive forces among molecules in a liquid sample. As temperature increases, the number of molecules with sufficient kinetic energy to overcome these attractive forces increases, and viscosity and surface tension decrease.

       (b)   Surface tension and viscosity are both directly related to the strength of intermolecular attractive forces. The same attractive forces that cause surface molecules to be difficult to separate cause molecules elsewhere in the sample to resist movement relative to one another. Liquids with high surface tension have intermolecular attractive forces sufficient to produce a high viscosity as well.

11.35   (a) $CHBr_3$ has a higher molar mass, is more polarizable, and has stronger dispersion forces, so the surface tension is greater.

(b)   As temperature increases, the viscosity of the oil decreases because the average kinetic energy of the molecules increases [Solution 11.33(a)].

(c)   Adhesive forces between polar water and nonpolar car wax are weak, so the large surface tension of water draws the liquid into the shape with the smallest surface area, a sphere.

(d)   Surface tension due to dispersion forces in oil is not great. Adhesive forces between nonpolar oil and nonpolar car wax are similar to cohesive forces in oil, so the oil drops spread out on the waxed hood.

11.37   (a)   The three molecules have similar structures and all experience hydrogen bonding, dipole-dipole and dispersion forces. The main difference in the series is the increase in the number of carbon atoms in the alkyl chain, with a corresponding increase in chain length, molecular weight, and strength of dispersion forces. The boiling points, surface tension, and viscosities all increase because the strength of dispersion forces increases.

(b)   Ethylene glycol has an –OH group at both ends of the molecule. This greatly increases the possibilities for hydrogen bonding, so the overall intermolecular attractive forces are greater and the viscosity of ethylene glycol is much greater.

(c)   Water has the highest surface tension but lowest viscosity because it is the smallest molecule in the series. Since water molecules are small, they approach each other closely and form many strong hydrogen bonds. There is no hydrocarbon chain to disrupt hydrogen bond formation or to inhibit their attraction to molecules in the interior of the drop. Water molecules at the surface of a drop are missing a few hydrogen bonds and are strongly pulled into the center of the drop, resulting in high surface tension. The absence of an alkyl chain also means the molecules can move around each other easily, resulting in the low viscosity.

Phase Changes (Section 11.4)

11.39   (a)   melting, endothermic

(b)   evaporation or vaporization, endothermic

(c)   deposition, exothermic

(d)   condensation, exothermic

11.41   The heat energy required to increase the kinetic energy of molecules enough to melt the solid does not produce a large separation of molecules. The specific order is disrupted, but the molecules remain close together. On the other hand, when a liquid is vaporized, the intermolecular forces which maintain close molecular contacts must be overcome. Because molecules are being separated, the energy requirement is higher than for melting.

11.43 *Analyze.* The heat required to vaporize 60 g of H_2O equals the heat lost by the cooled water.

Plan. Using the enthalpy of vaporization, calculate the heat required to vaporize 60 g of H_2O in this temperature range. Using the specific heat capacity of water, calculate the mass of water than can be cooled 15 °C if this much heat is lost.

Solve. Evaporation of 60 g of water requires:

$$60\,\text{g } H_2O \times \frac{2.4\,\text{kJ}}{1\,\text{g } H_2O} = 1.44 \times 10^2 \text{ kJ} = 1.4 \times 10^5 \text{ J}$$

Cooling a certain amount of water by 15 °C:

$$1.44 \times 10^5 \text{ J} \times \frac{1\,\text{g} \cdot \text{K}}{4.184\,\text{J}} \times \frac{1}{15\,°\text{C}} = 2294 = 2.3 \times 10^3 \text{ g } H_2O$$

Check. The units are correct. A surprisingly large mass of water (2300 g ≈ 2.3 L) can be cooled by this method.

11.45 *Analyze/Plan.* Follow the logic in Sample Exercise 11.3. *Solve.* Physical data for ethanol, C_2H_5OH, is: mp = –114°C; ΔH_{fus} = 5.02 kJ/mol; $C_{s(solid)}$ = 0.97 J/g-K; bp = 78°C; ΔH_{vap} = 38.56 kJ/mol; $C_{s(liquid)}$ = 2.3 J/g-K. *Solve.*

(a) Heat the liquid from 35 °C to 78 °C, ΔT = 43 °C = 43 K.

$$42.0\,\text{g } C_2H_5OH \times \frac{2.3\,\text{J}}{\text{g} \cdot \text{K}} \times 43\,\text{K} \times \frac{1\,\text{kJ}}{1000\,\text{J}} = 4.1538 = 4.2 \text{ kJ}$$

Vaporize (boil) the liquid at 78 °C, using ΔH_{vap}.

$$42.0\,\text{g } C_2H_5OH \times \frac{1\,\text{mol } C_2H_5OH}{46.07\,\text{g}} \times \frac{38.56\,\text{kJ}}{\text{mol}} = 35.1535 = 35.2 \text{ kJ}$$

Total energy required is 4.1538 kJ + 35.1535 kJ = 39.3073 = 39.3 kJ.

(b) Heat the solid from –155 °C to –114 °C, ΔT = 41 °C = 41 K.

$$42.0\,\text{g } C_2H_5OH \times \frac{0.97\,\text{J}}{\text{g} \cdot \text{K}} \times 41\,\text{K} \times \frac{1\,\text{kJ}}{1000\,\text{J}} = 1.6703 = 1.7 \text{ kJ}$$

Melt the solid at –114 °C, using ΔH_{fus}.

$$42.0\,\text{g } C_2H_5OH \times \frac{1\,\text{mol } C_2H_5OH}{46.07\,\text{g}} \times \frac{5.02\,\text{kJ}}{\text{mol}} = 4.5765 = 4.58 \text{ kJ}$$

Heat the liquid from –114°C to 78°C, ΔT = 192°C = 192 K.

$$42.0\,\text{g } C_2H_5OH \times \frac{2.3\,\text{J}}{\text{g} \cdot \text{K}} \times 192\,\text{K} \times \frac{1\,\text{kJ}}{1000\,\text{J}} = 18.5472 = 19 \text{ kJ}$$

From (a), vaporizing (boiling) 42.0 g of C_2H_5OH liquid at 78°C requires 35.1535 kJ = 35.2 kJ.

Total energy required = 1.6703 kJ + 4.5765 kJ + 18.5472 kJ + 35.1535 kJ = 59.9476

$$= 60 \text{ kJ.}$$

Check. The relative energies of the various steps are reasonable; vaporization is the largest. The sum has no decimal places because (19 kJ) has no decimal places.

11.47 (a) The critical pressure is the pressure required to cause liquefaction at the critical temperature.

(b) The critical temperature is the highest temperature at which a gas can be liquefied, regardless of pressure. As the force of attraction between molecules increases, the critical temperature of the compound increases.

(c) The temperature of $N_2(l)$ is 77 K. All of the gases in Table 11.5 have critical temperatures higher than 77 K, so all of them can be liquefied at this temperature, given sufficient pressure.

Vapor Pressure (section 11.5)

11.49 (a) No effect.

(b) No effect.

(c) Vapor pressure decreases with increasing intermolecular attractive forces because fewer molecules have sufficient kinetic energy to overcome attractive forces and escape to the vapor phase.

(d) Vapor pressure increases with increasing temperature because average kinetic energies of molecules increases.

(e) Vapor pressure decreases with increasing density. Density, the ratio of mass to volume occupied, increases as molecular weight increases. The strength of attractive dispersion forces also increases, so fewer molecules have sufficient kinetic energy to escape to the vapor phase and vapor pressure decreases.

11.51 (a) *Analyze/Plan.* Given the molecular formulae of several substances, determine the kind of intermolecular forces present, and rank the strength of these forces. The weaker the forces, the more volatile the substance. *Solve*.

$CBr_4 < CHBr_3 < CH_2Br_2 < CH_2Cl_2 < CH_3Cl < CH_4$

The weaker the intermolecular forces, the higher the vapor pressure, the more volatile the compound. The order of increasing volatility is the order of decreasing strength of intermolecular forces. By analogy to attractive forces in HCl (Section 11.2), the trend will be dominated by dispersion forces, even though four of the molecules ($CHBr_3$, CH_2Br_2, CH_2Cl_2 and CH_3Cl) are polar. Thus, the order of increasing volatility is the order of decreasing molar mass and decreasing strength of dispersion forces.

(b) $CH_4 < CH_3Cl < CH_2Cl_2 < CH_2Br_2 < CHBr_3 < CBr_4$

Boiling point increases as the strength of intermolecular forces increases, so the order of boiling points is the order of increasing strength of forces. This is the order of decreasing volatility and the reverse of the order in part (a).

11.53 (a) The water in the two pans is at the same temperature, the boiling point of water at the atmospheric pressure of the room. During a phase change, the temperature of a system is constant. All energy gained from the surroundings is used to accomplish the transition, in this case to vaporize the liquid water. The pan of water that is boiling vigorously is gaining more energy and the liquid is being vaporized more quickly than in the other pan, but the temperature of the phase change is the same.

(b)   Vapor pressure does not depend on either volume or surface area of the liquid. As long as the containers are at the same temperature, the vapor pressures of water in the two containers are the same.

11.55   *Analyze/Plan.* Follow the logic in Sample Exercise 11.4. The boiling point is the temperature at which the vapor pressure of a liquid equals atmospheric pressure. *Solve.*

(a)   The boiling point of ethanol at 200 torr is ~48 °C.

(b)   The vapor pressure of ethanol at 60 °C is approximately 340 torr. Thus, at 60 °C ethyl alcohol would boil at an external pressure of 340 torr.

(c)   The boiling point of diethyl ether at 400 torr is ~17 °C.

(d)   40 °C is above the normal boiling point of diethyl ether, so the pressure at which 40 °C is the boiling point is greater than 760 torr. According to Figure 11.25, a boiling point of 40 °C requires an external pressure of 1000 torr. (At these conditions, the vapor pressure of diethyl ether is 1000 torr.)

Phase Diagrams (section 11.6)

11.57   (a)   The *critical point* is the temperature and pressure beyond which the gas and liquid phases are indistinguishable.

(b)   The gas/liquid line ends at the critical point because at conditions beyond the critical temperature and pressure, there is no distinction between gas and liquid. In experimental terms, a gas cannot be liquefied at temperatures higher than the critical temperature, regardless of pressure.

11.59   (a)   The water vapor would deposit to form a solid at a pressure of around 4 torr. At higher pressure, perhaps 5 atm or so, the solid would melt to form liquid water. This occurs because the melting point of ice, which is 0 °C at 1 atm, decreases with increasing pressure.

(b)   In thinking about this exercise, keep in mind that the **total** pressure is being maintained at a constant 0.50 atm. That pressure is composed of water vapor pressure and some other pressure, which could come from an inert gas. At 100 °C and 0.50 atm, water is in the vapor phase. As it cools, the water vapor will condense to the liquid at the temperature where the vapor pressure of liquid water is 0.50 atm. From Appendix B, we see that condensation occurs at approximately 82 °C. Further cooling of the liquid water results in freezing to the solid at approximately 0 °C. The freezing point of water increases with decreasing pressure, so at 0.50 atm, the freezing temperature is very slightly above 0 °C.

11.61   *Analyze/Plan.* Follow the logic in Sample Exercise 11.5, using the phase diagram for neon.   *Solve.*

(a)   The normal melting point is the temperature where solid becomes liquid at 1 atm pressure. Following a horizontal line at 1 atm to the solid-liquid line, the normal melting point is approximately 24 K.

     (b)   Neon sublimes, changes directly from solid to gas, at pressures less than the triple point pressure, 0.43 atm.

     (c)   Room temperature is 298 K, in the region where neon is a supercritical fluid. Neon cannot be liquefied at any temperature above the critical temperature, 44 K, regardless of pressure.

11.63   *Analyze/Plan.* Follow the logic in Sample Exercise 11.5, using the phase diagram for methane in Figure 11.30.   *Solve.*

     (a)   According to Sample Exercise 11.5, the triple point of methane (CH_4) is approximately (–180 °C, 0.1 atm). The solid-liquid line in the phase diagram is essentially vertical in the pressure range 0.1-100 atm. This means that conditions at the surface of Titan (–178 °C, 1.6 atm) are very close to the solid-liquid line. Methane on the surface of titan is likely to exist in both solid and liquid forms.

     (b)   Methane is a liquid at –178 °C and 1.6 atm. Moving upward through Titan's atmosphere at a constant temperature of –178 °C, pressure decreases. At a pressure slightly greater than 0.1 atm, we expect to see vaporization to gaseous methane. If we begin with solid methane at 1.6 atm and a temperature slightly below –180 °C, we expect sublimation to gaseous methane at a pressure slightly less than 0.1 atm.

Liquid Crystals (section 11.7)

11.65   In a nematic liquid crystalline phase, molecules are aligned along their long axes, but the molecular ends are not aligned. In an ordinary liquid, molecules have no orderly arrangement; they are randomly oriented. Both an ordinary liquid and a nematic liquid crystal phase are fluids; molecules are free to move relative to one another. In an ordinary liquid, molecules can move in any direction. In a nematic phase, molecules are free to translate in all dimensions. Molecules cannot tumble or rotate out of the molecular plane, or the order of the nematic phase is lost and the sample becomes an ordinary liquid.

11.67   The presence of polar groups or nonbonded electron pairs leads to relatively strong dipole-dipole interactions between molecules. These are a significant part of the orienting forces necessary for liquid crystal formation.

11.69   Because order is maintained in at least one dimension, the molecules in a liquid-crystalline phase are not totally free to change orientation. This makes the liquid-crystalline phase more resistant to flow, more viscous, than the isotropic liquid.

11.71   As the temperature of a substance increases, the average kinetic energy of the molecules increases. More molecules have sufficient kinetic energy to overcome intermolecular attractive forces, so overall ordering of the molecules decreases as temperature increases. Melting provides kinetic energy sufficient to disrupt alignment in one dimension in the solid, producing a smectic phase with ordering in two dimensions. Additional heating of the smectic phase provides kinetic energy sufficient to disrupt alignment in another dimension, producing a nematic phase with one-dimensional order.

Additional Exercises

11.73 (a) decrease   (b) increase   (c) increase   (d) increase

       (e) increase   (f) increase   (g) increase

11.77 When a halogen atom (Cl or Br) is substituted for H in benzene, the molecule becomes polar. These molecules experience dispersion forces similar to those in benzene plus dipole-dipole forces, so they have higher boiling points than benzene. C_6H_5Br has a higher molar mass and is more polarizable than C_6H_5Cl, so it has the higher boiling point. C_6H_5OH experiences hydrogen bonding, the strongest force between neutral molecules, so it has the highest boiling point.

11.82 (a) Sweat, or salt water, on the surface of the body vaporizes to establish its typical vapor pressure at atmospheric pressure. Since the atmosphere is a totally open system, typical vapor pressure is never reached, and the sweat evaporates continuously. Evaporation is an endothermic process. The heat required to vaporized sweat is absorbed from your body, helping to keep it cool.

       (b) The vacuum pump reduces the pressure of the atmosphere (air + water vapor) above the water. Eventually, atmospheric pressure equals the vapor pressure of water and the water boils. Boiling is an endothermic process, and the temperature drops if the system is not able to absorb heat from the surroundings fast enough. As the temperature of the water decreases, the water freezes. (On a molecular level, the evaporation of water removes the molecules with the highest kinetic energies from the liquid. This decrease in average kinetic energy is what we experience as a temperature decrease.)

11.86 In a liquid crystal display (Figure 11.36), the molecules must be oriented to rotate polarized light by 90°. Light is then reflected back through both horizontal and vertical polarizers, producing a bright display. When voltage is applied, the molecules align with the voltage and light cannot pass through the horizontal polarizer, producing a dark spot on the display. At low Antarctic temperatures, the liquid crystalline phase is closer to its freezing point. The molecules have less kinetic energy due to temperature and the applied voltage may not be sufficient to overcome orienting forces among the molecules. If some or all of the molecules do not rotate when the voltage is applied, the display will not function properly.

Integrative Exercises

11.90

It is useful to draw the structural formulas because intermolecular forces are determined by the size and shape (structure) of molecules.

(a)   *Molar mass*: compounds (i) and (ii) have similar rod-like structures; (ii) has a longer rod. The longer chain leads to greater molar mass, stronger London-dispersion forces and higher heat of vaporization.

(b)   *Molecular shape*: compounds (iii) and (v) have the same chemical formula and molar mass but different molecular shapes (they are structural isomers). The more rod-like shape of (v) leads to more contact between molecules, stronger dispersion forces and higher heat of vaporization.

(c)   *Molecular polarity*: rod-like hydrocarbons (i) and (ii) are essentially nonpolar, owing to free rotation about C–C σ bonds, while (iv) is quite polar, owing to the C=O group. (iv) has a smaller molar mass than (ii) but a larger heat of vaporization, which must be due to the presence of dipole-dipole forces in (iv). [Note that (iii) and (iv), with similar shape and molecular polarity, have very similar heats of vaporization.]

(d)   *Hydrogen-bonding interactions*: molecules (v) and (vi) have similar structures, but (vi) has hydrogen bonding and (v) does not. Even though molar mass and thus dispersion forces are larger for (v), (vi) has the higher heat of vaporization. This must be due to hydrogen bonding interactions.

11.93 $P = \dfrac{nRT}{V} = \dfrac{g\,RT}{M\,V}$; T = 273.15 + 26.0 °C = 299.15 = 299.2 K; V = 5.00 L

$g\,C_6H_6(g) = 7.2146 - 5.1493 = 2.0653\ g\,C_6H_6(g)$

$$P\,(vapor) = \frac{2.0653\ g}{78.11\ g/mol} \times \frac{299.15\ K}{5.00\ L} \times \frac{0.08206\ L\text{-}atm}{mol\text{-}K} \times \frac{760\ torr}{1\ atm} = 98.660 = 98.7\ torr$$

12 Solids and Modern Materials

Visualizing Concepts

12.1   When choosing a unit cell, remember that the environment of each lattice point must be identical and that unit cells must *tile* to generate the complete two-dimensional lattice. For a given structure, there are often several ways to draw a unit cell. We will select the unit cell with higher symmetry (more 90° or 120° angles) and smaller area ($a \times b$).

Two-dimensional structure	(i)	(ii)
(a) unit cell		

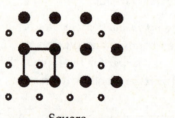 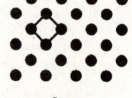

(b) lattice	Square	Square
(c) cell contents	One black, one white	One black

12.3   (a)   There is a Re atom (gray sphere) at each corner of the unit cell: ($8 \times 1/8 = 1$). There is an O atom (red sphere) in the middle of each cell edge: ($12 \times 1/4 = 3$). There are 1 Re and 3 O atoms per unit cell, for an empirical formula of ReO_3.

(b)   Each cell edge goes through two half Re atoms (at the corners) and one full O atom (centered on the edge). The length of an edge, a, is then

$a = 2r_{Re} + 2r_O = 2(0.70 \text{ Å}) + 2(1.26 \text{ Å}) = 3.92 \text{ Å}$

(c)   The density of a crystalline solid is the mass of the unit cell contents divided by the unit cell volume. There is one ReO_3 unit in each primitive cubic unit cell. The unit cell volume is a^3, $(3.92 \text{ Å})^3$.

$$\frac{1 \, ReO_3 \text{ unit}}{(3.92 \text{ Å})^3} \times \frac{1 \text{Å}^3}{(10^{-8} \text{ cm})^3} \times \frac{234.205 \text{ g } ReO_3}{6.022 \times 10^{23} \, ReO_3 \text{ units}} = 6.46 \text{ g/cm}^3$$

Check. The units of density are correct.

12.4   (a)   Band A is the valence band.

(b)   Band B is the conduction band.

(c)   Band A (the valence band) consists of bonding molecular orbitals (MOs).

(d)   This is the electronic structure of a p-type doped semiconductor. The electronic structure shows a few empty MOs or "positive" holes in the valence band. This fits the description of a p-type doped semiconductor.

(e)   The dopant is Ga. Of the three elements listed, only Ga has fewer valance electrons than Ge, the requirement for a p-type dopant.

Classification of Solids (section 12.1)

12.7   In molecular solids, relatively weak intermolecular forces (hydrogen bonding, dipole-dipole, dispersion) bind the molecules in the lattice, so relatively little energy is required to disrupt these forces. In covalent-network solids, covalent bonds join atoms into an extended network. Melting or deforming a covalent-network solid means breaking these covalent bonds, which requires a large amount of energy.

12.9   (a)   hydrogen bonding, dipole-dipole forces, London dispersion forces

         (b)   covalent chemical bonds (mainly)

         (c)   ionic bonds, the electrostatic attraction between anions and cations (mainly)

         (d)   metallic bonds

12.11   (a)   ionic   (b)   metallic

         (c)   covalent-network   (Based on melting point, it is definitely a network solid. Transition metals in high oxidation states often form bonds with nonmetals that have significant covalent character. It could also be characterized as ionic with some covalent character to the bonds.)

         (d)   molecular   (e)   molecular   (f)   molecular

12.13   Because of its relatively high melting point and properties as a conducting solution, the solid must be ionic.

Structures of Solids (section 12.2)

12.15   *Analyze/Plan.* Crystalline solids have a regular repeat in all three directions. Amorphous solids have no regular repeating structure. Draw diagrams that reflect these definitions. *Solve.*

a)

b)

                       crystalline   amorphous

12.17   *Analyze.* Given two two-dimensional structures, draw and describe the unit cells and lattice vectors. *Plan.* When choosing a unit cell, the environment of each lattice point must be identical and the unit cells must *tile* to generate the complete two-dimensional lattice. For a given structure, there are often several ways to draw a unit cell. The radii of A and B are equal. *Solve.*

Two-dimensional   (i)   (ii)
      structure

(a) unit cell

	(b) γ, a, b	$\gamma = 90°$, $a = b$	$\gamma = 120°$, $a = b$
	(c) lattice type	square	hexagonal

12.19　*Plan.* Refer to Figure 12.6 to find geometric characteristics of the seven three dimensional primitive lattices.

(a)　Orthorhombic, $a \neq b \neq c$, $\alpha = \beta = \gamma = 90°$. A rectangular base has unequal side lengths and an angle of 90°. The third lattice vector is perpendicular to the first two and a different length, so all angles are 90° and all vector lengths are unequal.

(b)　Tetragonal, $a = b \neq c$, $\alpha = \beta = \gamma = 90°$. The only difference between the two lattices is that the second has a square (rather than rectangular) base. This means that two of the edge lengths are equal but the third is not, and all angles remain 90°.

12.21　*Plan.* Refer to Figure 12.6 to find geometric characteristics of the seven three dimensional primitive lattices.

If no lattice vectors are perpendicular to each other, none of the unit cell angles (α, β, γ) are 90°. This is characteristic of two of the three dimensional primitive lattices: triclinic and rhombohedral.

12.23　*Analyze.* Given an element with a body centered cubic lattice, find the minimum number of atoms in a unit cell. *Plan.* Refer to Figures 12.11 and 12.12 to visualize an element (all the same kind of atoms) with a body centered cubic lattice. *Solve.*

A body centered cubic lattice is composed of body centered cubic unit cells. A unit cell contains the minimum number of atoms when it has atoms only at the lattice points. A body centered cubic unit cell like this is shown in Figure 12.12(b). There is one atom totally inside the cell (1×1) and one at each corner ($8 \times 1/8$) for a total of 2 (metal) atoms in the unit cell. (Only metallic elements have body-centered cubic lattices and unit cells.)

12.25　*Analyze.* Given a diagram of the unit cell dimensions and contents of nickel arsenide, determine what kind of lattice this crystal possess, and the empirical formula of the compound. *Plan.* Refer to Figure 12.6 to find geometric characteristics of the seven three dimensional primitive lattices. Decide where atoms of the two elements are located in the unit cell and use Table 12.1 to help determine the empirical formula.

(a)　$a = b = 3.57$ Å. $c = 5.10$ Å $\neq a$ or b. $\alpha = \beta = 90°$, $\gamma = 120°$. This unit cell is hexagonal. There are no atoms in the exact middle of the cell or on the face centers, so it is a primitive hexagonal unit cell. Nickel arsenide has a primitive hexagonal unit cell and crystal lattice.

(b)　There are Ni atoms at each corner of the cell ($8 \times 1/8$) and centered on four of the unit cell edges ($4 \times 1/4$) for a total of 2 Ni atoms. There are 2 As atoms totally inside the cell. The unit cell contains 2 Ni and 2 As atoms; the empirical formula is NiAs.

Metallic Solids (section 12.3)

12.27   *Analyze/Plan.* Consider trends in atomic mass and volume of the elements listed to explain the variation in density. *Solve.*

Moving left to right in the period, atomic mass and Z_{eff} increase. The increase in Z_{eff} leads to smaller bonding atomic radii and thus atomic volume. Mass increases, volume decreases, and density increases in the series.

The variation in densities reflects shorter metal-metal bond distances. These shorter distances suggest that the extent of metal-metal bonding increases in the series. This is consistent with greater occupancy of the bonding band (Section 12.4) as the number of valence electrons increases up to 6. The strength of metal-metal bonds in the series is probably the most important factor influencing the increase in density.

12.29   *Analyze.* Give diagrams of three structure types, find which is most densely packed and which is least densely packed. *Plan.* Assume that the same element packs in each of the three structures, so that atomic mass and volume are constant. Then we are analyzing the packing efficiency or relative amount of empty space in each structure. *Solve.*

Structure type A has a face centered cubic unit cell with metal atoms only at the lattice points; this corresponds to a cubic close packed structure. Structure type B has a body centered cubic unit cell with metal atoms at the lattice points; this is also a body centered cubic structure. Structure type C has a hexagonal unit cell with two atoms totally inside the cell. Building up many unit cells into a lattice (Figure 12.14) leads to a hexagonal close packed structure.

(a)   In both cubic and hexagonal close packed structures, any individual atom has twelve nearest neighbor atoms. Both structures are "closest" packed and have equal amounts of empty space. Structure types A and C have equally dense packing and are more densely packed than structure type B.

(b)   Structure type B, which is not close packed, has the least dense atom packing.

12.31   *Analyze.* Given the cubic unit cell edge length and arrangement of Ir atoms, calculate the atomic radius and the density of the metal. *Plan.* There is space between the atoms along the unit cell edge, but they touch along the face diagonal. Use the geometry of the right equilateral triangle to calculate the atomic radius. From the definition of density and paying attention to units, calculate the density of Ir(s). *Solve.*

(a)   The length of the face diagonal of a face-centered cubic unit cell is four times the radius of the atom and $\sqrt{2}$ times the unit cell dimension or edge length, *a* for cubic unit cells.

$$4\,r = \sqrt{2}\,a; \, r = \sqrt{2}\,a/4 = \frac{\sqrt{2} \times 3.833\ \text{Å}}{4} = 1.3552 = 1.355\ \text{Å}$$

(b)   The density of iridium is the mass of the unit cell contents divided by the unit cell volume. There are 4 Ir atoms in a face-centered cubic unit cell.

$$\rho = \frac{4\ \text{Ir atoms}}{(3.833 \times 10^{-8}\ \text{cm})^3} \times \frac{192.22\ \text{g Ir}}{6.022 \times 10^{23}\ \text{Ir atoms}} = 22.67\ \text{g/cm}^3$$

Check. The units of density are correct. Note that Ir is quite dense.

12.33   *Analyze.* Given the structure of aluminum metal and the atomic radius of and Al atom, find the number of Al atoms in each unit cell and the coordination number of each Al atom. Calculate (estimate) the length of the unit cell edge and the density of aluminum metal.

Plan. Use Figure 12.14(a) to count the number of Al atoms in one unit cell and visualize the coordination number of each Al atom. There is space between the atoms along the unit cell edge, but they touch along the face diagonal. Use the geometry of the right equilateral triangle and the atomic radius to calculate the unit cell edge length. From the definition of density and paying attention to units, calculate the density of aluminum metal. *Solve.*

(a)   8 corners × 1/8 atom/corner + 6 faces × ½ atom/face = 4 atoms

(b)   Each aluminum atom is in contact with 12 nearest neighbors, 6 in one plane, 3 above that plane, and 3 below. Its coordination number is thus 12.

(c)   The length of the face diagonal of a face-centered cubic unit cell is four times the radius of the metal and $\sqrt{2}$ times the unit cell dimension (usually designated a for cubic cells).

$$4 \times 1.43\,\text{Å} = \sqrt{2} \times a; \quad a = \frac{4 \times 1.43\,\text{Å}}{\sqrt{2}} = 4.0447 = 4.04\,\text{Å} = 4.04 \times 10^{-8}\,\text{cm}$$

(d)   The density of the metal is the mass of the unit cell contents divided by the volume of the unit cell.

$$\text{density} = \frac{4\,\text{Al atoms}}{(4.0447 \times 10^{-8}\,\text{cm})^3} \times \frac{26.98\,\text{g Al}}{6.022 \times 10^{23}\,\text{Al atoms}} = 2.71\,\text{g/cm}^3$$

12.35   *Analyze/Plan.* Use information in Section 12.3 to define *alloy*, and compare the various types of alloys.   *Solve.*

An *alloy* contains atoms of more than one element and has the properties of a metal. *Solution alloys* are homogeneous mixtures with different kinds of atoms dispersed randomly and uniformly. In *heterogeneous alloys* the components (elements or compounds) are not evenly dispersed and their properties depend not only on composition but methods of preparation. In an *intermetallic compound* the component elements have interacted to form a compound substance, for example, Cu_3As. As with more familiar compounds, these are homogeneous and have definite composition and properties.

12.37   *Analyze/Plan.* Consider the descriptions of various alloy types in Section 12.3. *Solve.*

(a) $Fe_{0.97}Si_{0.03}$; interstitial alloy. The radii of Fe and Si are substantially different, so Si could fit in "holes" in the Fe lattice. Also, the small amount of Si relative to Fe is characteristic of an interstitial alloy.

(b) $Fe_{0.60}Ni_{0.40}$, substitutional alloy. The two metals have very similar atomic radii and are present in similar amounts.

(c) $SmCo_5$, intermetallic compound. The two elements are present in stoichiometric amounts.

12.39　(a)　True

　　　(b)　False.　Interstitial alloys form between elements with very different bonding atomic radii.

　　　(c)　False.　Non-metallic elements are typically found in interstitial alloys.

12.41　*Analyze*.　Given the color of a gold alloy, find the other element(s) in the alloy and the type of alloy formed.　*Plan*.　Refer to 'Chemistry Put to Work: ALLOYS OF GOLD'.　*Solve*.

　　　(a)　White gold, nickel or palladium, substitutional alloy

　　　(b)　Rose gold, copper, substitutional alloy

　　　(c)　Blue gold, indium, intermetallic compound

　　　(d)　Green gold, silver, substitutional alloy

Metallic Bonding (section 12.4)

12.43　*Analyze/Plan*.　Apply the description of the electron-sea model of metallic bonding given in Section 12.4 to the conductivity of metals.　*Solve*.

In the electron-sea model for metallic bonding, valence electrons move about the three-dimensional metallic lattice, while the metal atoms maintain regular lattice positions.

Under the influence of an applied potential the electrons can move throughout the structure, giving rise to high electrical conductivity. The mobility of the electrons facilitates the transfer of kinetic energy and leads to high thermal conductivity.

12.45　*Plan*.　By analogy to Figure 12.23, the most bonding, lowest energy MOs have the fewest nodes.　As energy increases, the number of nodes increases.　When constructing an MO diagram from AOs, total number of orbitals is conserved.　The MO diagram for a linear chain of six Li atoms will have six MOs, starting with zero nodes and maximum overlap, and ending with five nodes and minimum overlap.　*Solve*.

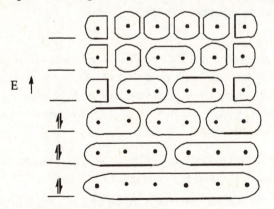

　　　(a)　Six AOs require six MOs.

　　　(b)　Zero nodes in lowest energy orbital

　　　(c)　Five nodes in highest energy orbital

　　　(d)　Two nodes in the HOMO

　　　(e)　Three nodes in the LUMO

12.47   *Analyze/Plan.* Consider the definition of ductility, as well as the discussion of metallic bonding in Section 12.4.   *Solve.*

Ductility is the property related to the ease with which a solid can be drawn into a wire. Basically, the softer the solid the more ductile it is. The more rigid the solid, the less ductile it is. For metals, ductility decreases as the strength of metal-metal bonding increases, producing a stiffer lattice less susceptible to distortion.

(a)   Ag is more ductile. Mo, with 6 valence electrons, has a filled bonding band, strong metal-metal interactions, and a rigid lattice. This predicts high hardness and low ductility. Ag, with 12 valence electrons, has filled antibonding as well as bonding bands. Bonding is weaker than in Mo, and Ag is more ductile.

(b)   Zn is more ductile. Si is a covalent-network solid with all valence electrons localized in bonds between Si atoms. Covalent-network substances are high-melting, hard, and not particularly ductile.

12.49   Moving across the fifth period from Y to Mo, the melting points of the metals increase. The number of valence electrons also increases, from 3 for Y to 6 for Mo. More valence electrons (up to 6) mean increased occupancy of the bonding molecular orbital band, and increased strength of metallic bonding. Melting requires that atoms are moving relative to each other. Stronger metallic bonding requires more energy to break bonds and mobilize atoms, resulting in higher melting points from Y to Mo.

Ionic and Molecular Solids (sections 12.5 and 12.6)

12.51   (a)   Sr: Sr atoms occupy the 8 corners of the cube.

                8 corners × 1/8 sphere/corner = 1 Sr atom

        O: O atoms occupy the centers of the 6 faces of the cube.

                6 faces × 1/2 atom/face = 3 O atoms

        Ti: There is 1 Ti atom at the body center of the cube.

        Formula: $SrTiO_3$

(b)   Twelve. Each Sr atom occupies one corner of 8 unit cells. Sr is coordinated to 3 oxygen positions in each unit cell for a total of 24 oxygen positions. However, each O position is in the center of a cell face, with half-occupancy in each cell. 24 oxygen positions × ½ occupancy = 12 oxygen atoms. Each Ti atom is coordinated to 12 oxygen atoms.

12.53   *Analyze/Plan.* Use the structure of NaCl in Figure 12.26(b) to estimate the unit cell edge length based on ionic radii of Na^+ and F^- from Figure 7.7. *Solve.*

(a)   According to Figure 12.26(b), the length of the unit cell edge in this structure is $2r_{Na^+} + 2r_{F^-}$. $a = 2(1.16\ \text{Å}) + 2(1.19\ \text{Å}) = 4.70\ \text{Å}$

(b)   The density of NaF is the mass of the unit cell contents divided by the volume of the unit cell. By analogy to the NaCl structure in Figure 12.26(b)and 12.27, there are 8 fluoride ions (green spheres) on the corners of the unit cell, and 6 fluoride ions in the middle of the faces. The number of fluoride ions per unit cell is then

8(1/8) + 6(1/2) = 4.　There is 1 sodium ion (purple sphere almost hidden in figure) completely inside the unit cell and 12 sodium ions (purple spheres) along the unit cell edges. The number of sodium ions is then 1 + 12(1/4) = 4. This result satisfies charge balance requirements.

4 Na$^+$, 4 F$^-$, 4 NaF formula units. The mass of 1 NaF formula unit is 41.99 g/6.022 × 10^{23} NaF units.

$$d = \frac{4 \text{ NaF units}}{(4.70 \text{ Å})^3} \times \frac{41.99 \text{ g}}{6.022 \times 10^{23} \text{ NaF units}} \times \left(\frac{1 \text{ Å}}{1 \times 10^{-8} \text{ cm}}\right)^3 = 2.69 \text{ g/cm}^3$$

Check. The value for the density of NaF(s) reported in the *CRC Handbook of Chemistry and Physics*, 74th Ed., is 2.56 g/cm³. The calculated density based on the estimated value of *a* is within 5% of the reported value.

12.55　*Analyze.* Given the atomic arrangement and length of the unit cell side, calculate the density of HgS and HgSe. Qualitatively and quantitatively compare the densities of the two solids.　*Plan.* Calculate the mass and volume of a single unit cell and then use them to find density. The unit cell volume is the cube of edge length.　*Solve.* The unit cell edge is designated *a*.

(a)　According to Figure 12.26(c) and 12.27, sulfide ions (yellow spheres) occupy the corners and faces of the unit cell (in a face-centered cubic arrangement), for a total of 6(1/2) + 8(1/8) = 4 S^{2-} ions per unit cell. There are 4 mercury ions (gray spheres) totally in the interior of the cell. This means there are 4 HgS units in a unit cell with the zinc blende structure.

$$\text{density} = \frac{4 \text{ HgS units}}{(5.852 \text{ Å})^3} \times \frac{232.655 \text{ g}}{6.022 \times 10^{23} \text{ HgS units}} \times \left(\frac{1 \text{ Å}}{1 \times 10^{-8} \text{ cm}}\right)^3 = 7.711 \text{ g/cm}^3$$

(b)　We expect Se^{2-} to have a larger ionic radius than S^{2-}, since Se is below S in the chalcogen family and both ions have the same charge. Thus, HgSe will occupy a larger volume and the unit cell edge will be longer.

(c)　For HgSe, also with the zinc blende structure:

$$\text{density} = \frac{4 \text{ HgSe units}}{(6.085 \text{ Å})^3} \times \frac{279.55 \text{ g HgSe}}{6.022 \times 10^{23} \text{ HgSe units}} \times \left(\frac{1 \text{ Å}}{1 \times 10^{-8} \text{ cm}}\right)^3 = 8.241 \text{ g/cm}^3$$

Even though HgSe has a larger unit cell volume than HgS, it also has a larger molar mass. The mass of Se is more than twice that of S, while the radius of Se^{2-} is only slightly larger than that of S^{2-} (Figure 7.7). The greater mass of Se accounts for the greater density of HgSe.

12.57　*Analyze.* Given that CuI, CsI and NaI uniquely adopt one of the structure types pictured in Figure 12.26, match the ionic compound with its structure. Use ionic radii to inform your decision.　*Plan.* Note the relative cation/anion ratio in the three structures in Figure 12.26. Match these ratios with those of CuI, CsI and NaI.　*Solve.*

(a)　In the CsCl structure, the anion and cation have about the same radius; in the NaCl structure, the anion is somewhat larger than the cation; in the ZnS structure the anion is much larger than the cation.

In the three compounds given, Cs^+ (r = 1.81 Å) and I^- (r = 2.06 Å) have the most similar radii; CsI will adopt the CsCl-type structure. The radii of Na^+ (r = 1.16 Å) and I^- (r = 2.06 Å) are somewhat different; NaI will adopt the NaCl-type structure. The radii of Cu^+ (r = 0.74 Å) and I^- (r = 2.06 Å) are very different; CuI has the ZnS-type structure.

(b)   As cation size decreases, coordination number of the anion decreases. In CsI, I^- has a coordination number of eight ; in NaI, I^- has a coordination number of six,; in CuI, I^- has a coordination number of four.

12.59   *Analyze.* Given three magnesium compounds in which the coordination number (CN) of Mg^{2+} is six, determine the coordination number of the anion. *Plan.* Use Equation 12.1 with the cation/anion ratio of each compound and the Mg^{2+} coordination number to calculate the anion coordination number in each compound. *Solve.*

(a)   MgS: 1 cation, 1 anion, cation CN = 6

$$\frac{\text{\# of cations per formula unit}}{\text{\# of anions per formula unit}} = \frac{\text{anion coordination number}}{\text{cation coordination number}}$$

$$\text{anion CN} = \frac{\text{cation CN} \times \text{\# of cations per formula unit}}{\text{\# of anions per formula unit}} = \frac{6 \times 1}{1} = 6$$

(b) MgF_2: 1 cation, 2 anions, cation CN = 6

$$\text{anion CN} = \frac{\text{cation CN} \times \text{\# of cations per formula unit}}{\text{\# of anions per formula unit}} = \frac{6 \times 1}{2} = 3$$

(c)   MgO: 1 cation, 1 anion, cation CN = 6, anion CN = 6. The cation/anion ratio and cation CN are the same as in part (a), so the anion CN is the same (6).

12.61   (a)   False. Although both molecular solids and covalent-network solids have covalent bonds, the melting points of molecular solids are much lower because intermolecular forces among their molecules are much weaker than covalent bonds among atoms in a covalent-network solid.

(b)   True. The statement is true if there are no significant differences in polarity or molar mass of the molecules being compared.

Covalent-Network Solids (section 12.7)

12.63   (a)   Ionic solids are much more likely to dissolve in water. Polar water molecules can disrupt ionic bonds to surround and separate ions and form a solution, but they cannot break the covalent bonds of a covalent-network solid.

(b)   Covalent-network solids can become electricity conducting via chemical substitution. Most semiconductors are covalent-network solids, and doping or chemical substitution changes their electrical properties.

12.65   *Analyze/Plan.* Follow the logic in Sample Exercise 12.3. *Solve.*

(a)   CdS. Both semiconductors contain Cd; S and Te are in the same family, and S is higher.

(b)   GaN. Ga is in the same family and higher than In; N is in the same family and higher than P.

(c)   GaAs. Both semiconductors contain As; Ga and In are in the same family and Ga is higher.

12.67   *Analyze.* Given: GaAs. Find: dopant to make n-type semiconductor.

Plan. An n-type semiconductor has extra negative charges. If the dopant replaces a few Ga atoms, it should have more valence electrons than Ga, Group 3A.

Solve. The obvious choice is a Group 4A element, either Ge or Si. Ge would be closer to Ga in bonding atomic radius (Figure 7.6).

12.69   (a)   *Analyze.* Given: 1.1 eV. Find: wavelength in meters that corresponds to the energy 1.1 eV. *Plan.* Use dimensional analysis to find wavelength.

Solve. 1 eV = 1.602×10^{-19} J (inside-back cover of text); $\lambda = hc/E$

$$\lambda = 6.626 \times 10^{-34} \text{ J-s} \times \frac{3.00 \times 10^8 \text{ m}}{\text{s}} \times \frac{1}{1.1 \text{ eV}} \times \frac{1 \text{ eV}}{1.602 \times 10^{-19} \text{ J}} = 1.128 \times 10^{-6}$$
$$= 1.1 \times 10^{-6} \text{ m}$$

(b)   Si can absorb energies ≥ 1.1 eV, or wavelengths $\leq 1.1 \times 10^{-6}$ m. The range of wavelengths in the solar spectrum at sea level is 3×10^{-6} to 2×10^{-7} m, or 30×10^{-7} to 2×10^{-7} m, a span of 28×10^{-7} m. Si can absorb 11×10^{-7} to 2×10^{-7} m, a span of

9×10^{-7} m. This represents $\dfrac{9 \times 10^{-7}}{28 \times 10^{-7}} \times 100 = 32\%$ of the wavelengths in the

solar spectrum. According to the diagram, these wavelengths represent much more than 32% of the total flux. Silicon does absorb a significant "portion" of the visible light that comes from the sun, whether that "portion" is defined by wavelength range or flux.

12.71   *Plan/Solve.* Follow the logic in Solution 12.69.

$$\lambda = hc/E = 6.626 \times 10^{-34} \text{ J-s} \times \frac{3.00 \times 10^8 \text{ m}}{\text{s}} \times \frac{1}{2.2 \text{ eV}} \times \frac{1 \text{ eV}}{1.602 \times 10^{-19} \text{ J}} = 5.640 \times 10^{-7}$$
$$= 5.6 \times 10^{-7} \text{ m} = 560 \text{ nm}$$

12.73   *Analyze/Plan.* From Table 12.4, E_g for GaAs (x = 0) is 1.43 eV and for GaP (x = 1) is 2.26 eV. If E_g varies linearly with x, the band gap for x = 0.5 should be approximately the average of the two extreme values.

Solve. (1.43 + 2.26)/2 = 1.845 = 1.85 eV.

$$\lambda = hc/E = \frac{6.626 \times 10^{-34} \text{ J-s} \times 3.00 \times 10^8 \text{ m}}{\text{s}} \times \frac{1}{1.845 \text{ eV}} \times \frac{1 \text{ eV}}{1.602 \times 10^{-19} \text{ J}}$$
$$= 6.72 \times 10^{-7} \text{ m} = 672 \text{ nm}$$

Polymeric Solids (section 12.8)

12.75 Monomers are small molecules with low molecular mass that are joined together to form polymers. They are the repeating units of a polymer. Three (of the many) monomers mentioned in this chapter are

$$CH_2=CH \qquad CH_2=CH \qquad \begin{array}{c} CH_3 \qquad H \\ C=C \\ CH_2 \qquad CH_2 \end{array}$$
$$\quad\quad\ |$$
$$\quad\ CH_3$$

  propylene   styrene   isoprene
  (propene)   (phenyl ethene)   (2-methyl-1,3-butadiene)

12.77 *Analyze.* Given two types of reactant molecules, we are asked to write a condensation reaction with an ester product. *Plan.* A condensation reaction occurs when two smaller molecules combine to form a larger molecule and a small molecule, often water. Consider the structures of the two reactants and how they could combine to join the larger fragments and split water. *Solve.*

A carboxylic acid contains the $-\overset{\overset{\displaystyle O}{\|}}{C}-OH$ functional group; an alcohol contains the

–OH functional group. These can be arranged to form the $-\overset{\overset{\displaystyle O}{\|}}{C}-O-C$ ester

functional group and H_2O. Condensation reaction to form an ester:

$$CH_3-\overset{\overset{\displaystyle O}{\|}}{C}-\boxed{O-H + H}-O-CH_2-CH_3 \longrightarrow CH_3-\overset{\overset{\displaystyle O}{\|}}{C}-O-CH_2CH_3 + H_2O$$

   acetic acid   ethanol   ethyl acetate

If a dicarboxylic acid (two –COOH groups, usually at opposite ends of the molecule) and a dialcohol (two –OH groups, usually at opposite ends of the molecule) are combined, there is the potential for propagation of the polymer chain at both ends of both monomers. Polyethylene terephthalate (Table 12.5) is an example of a polyester formed from the monomers ethylene glycol and terephthalic acid.

12.79 *Analyze/Plan.* Decide whether the given polymer is an addition or condensation polymer. Select the smallest repeat unit and deconstruct it into the monomer(s) with the specific functional group(s) that would form the stated polymer.   *Solve.*

(a)
$$\begin{array}{c} H \qquad\qquad H \\ C=C \\ H \qquad\qquad Cl \end{array}$$

vinyl chloride (chloroethylene or chloroethene)

(b)
$$H_2N \overset{CH_2}{\diagdown} \overset{CH_2}{\diagup}\diagdown \overset{CH_2}{\diagup}\diagdown \overset{CH_2}{\diagup} NH_2 \quad +$$
$$\quad CH_2 \quad\quad CH_2 \quad\quad CH_2$$

        hexanediamine

$$HO \overset{\diagdown}{\underset{\underset{\displaystyle O}{\|}}{C}} \overset{CH_2}{\diagup}\diagdown \overset{CH_2}{\diagup}\diagdown \overset{\overset{\displaystyle O}{\|}}{C} OH$$
$$\qquad\quad CH_2 \quad\quad CH_2$$

           adipic acid

(c)

ethylene glycol   terephthalic acid

12.81 *Plan/Solve.* When nylon polymers are made, H_2O is produced as the C–N bonds are formed. Reversing this process (adding H_2O across the C–N bond), we see that the monomers used to produce Nomex™ are:

HOOC   COOH H_2N NH_2

and

12.83 Most of a polymer backbone is composed of σ bonds. The geometry around individual atoms is tetrahedral with bond angles of 109°, so the polymer is not flat, and there is relatively free rotation around the σ bonds. The flexibility of the molecular chains causes flexibility of the bulk material. Flexibility is enhanced by molecular features that inhibit order, such as branching, and diminished by features that encourage order, such as cross-linking or delocalized π electron density.

Cross-linking is the formation of chemical bonds between polymer chains. It reduces flexibility of the molecular chains and increases the hardness of the material. Cross-linked polymers are less chemically reactive because of the links.

12.85 No. The function of the material (polymer) determines whether high molecular mass and high degree of crystallinity are desirable properties. If the material will be formed into containers or pipes, the rigidity and structural strength associated with high molecular mass are required. If the polymer will be used as a flexible wrapping or as a garment material, high molecular mass and rigidity are undesirable properties.

Nanomaterials (section 12.9)

12.87 Continuous energy bands of molecular orbitals require a large number of atoms contributing a large number of atomic orbitals to the molecular orbital scheme. If a solid has dimensions 1–10 nm, nanoscale dimensions, there may not be enough contributing atomic orbitals to produce continuous energy bands of molecular orbitals.

12.89 (a) False. As particle size decreases, the band gap increases. The smaller the particle, the fewer AOs that contribute to the MO scheme, the more localized the bonding and the larger the band gap.

     (b) False. The wavelength of emitted light corresponds to the energy of the band gap. As particle size decreases, band gap increases and wavelength decreases ($E = hc/\lambda$).

12.91 *Analyze.* Given: Au, 4 atoms per unit cell, 4.08 Å cell edge, volume of sphere $= 4/3\,\pi\,r^3$. Find: Au atoms in 20 nm diameter sphere.

Plan. Relate the number of Au atoms in the volume of 1 cubic unit cell to the number of Au atoms in a 20 nm diameter sphere. Change units to Å (you could just as well have chosen nm as the common unit), calculate the volumes of the unit cell and sphere, and use a ratio to calculate atoms in the sphere.

Solve. vol of unit cell = $(4.08 \text{ Å})^3$ = 67.9173 = 67.9 Å^3

$$20 \text{ nm diameter} = 10 \text{ nm radius}; 10 \text{ nm} \times \frac{1 \times 10^{-9} \text{ m}}{1 \text{ nm}} \times \frac{1 \text{ Å}}{1 \times 10^{-10} \text{ m}} = 100 \text{ Å radius}$$

(Note that 1 nm = 10 Å.)

vol. of sphere = $4/3 \times 3.14159 \times (100 \text{ Å})^3 = 4.18879 \times 10^6 = 4.19 \times 10^6$ Å^3

$$\frac{4 \text{ Au atoms}}{67.9173 \text{ Å}^3} = \frac{x \text{ Au atoms}}{4.18879 \times 10^6 \text{ Å}^3}; x = 2.46699 \times 10^5 = 2.47 \times 10^5 \text{ Au atoms}$$

Additional Exercises

12.94　Qualitatively from Figure 12.12, a face-centered cubic structure has a greater portion of its volume occupied by metal atoms and less "empty" space than a body-centered structure. The face-centered structure will have the greater density.

Quantitatively, use the atomic radius of iron, 1.25 Å from Figure 7.6, to estimate the unit cell edge length in each of the structures, then calculate the estimated density of both structures. Recall that a face centered cubic structure has 4 atoms per unit cell, and a body centered structure has 2.

Face centered cubic:　$4 r_{Fe} = \sqrt{2}\, a; \quad a = 4 r_{Fe} / \sqrt{2} = \dfrac{4 \times 1.25 \text{ Å}}{\sqrt{2}} = 3.5355 = 3.54 \text{ Å}$

$$\rho = \frac{4 \text{ Fe atoms}}{(3.5355 \times 10^{-8} \text{ cm})^3} \times \frac{55.845 \text{ g Fe}}{6.022 \times 10^{23} \text{ Fe atoms}} = 8.3934 = 8.39 \text{ g/cm}^3$$

Body centered cubic:　$4 r_{Fe} = \sqrt{3}\, a; \quad a = 4 r_{Fe} / \sqrt{3} = \dfrac{4 \times 1.25 \text{ Å}}{\sqrt{3}} = 2.8868 = 2.89 \text{ Å}$

$$\rho = \frac{2 \text{ Fe atoms}}{(2.8868 \times 10^{-8} \text{ cm})^3} \times \frac{55.845 \text{ g Ca}}{6.022 \times 10^{23} \text{ Ca atoms}} = 7.7098 = 7.71 \text{ g/cm}^3$$

The face centered cubic structure has the greater density.

12.98　(a)　CsCl, primitive cubic lattice (Figure 12.26)

　　　　(b)　Au, face-centered cubic lattice (Figure 12.18)

　　　　(c)　NaCl, face-centered cubic lattice (Figure 12.26)

　　　　(d)　Po, primitive cubic lattice, rare for metals (Section 12.3)

　　　　(e)　ZnS, face-centered cubic lattice (Figure 12.26)

12.99　*Analyze/Plan*. Recall the diamond [Figure 12.30(a)] and close-packed metallic structures (Figure 12.14 and others) described in this chapter. Use these structures to draw conclusions about Sn–Sn distance and electrical conductivity in the two allotropes. *Solve*.

White tin, with a characteristic metallic structure, is expected to be more metallic in character. The white allotropic form has the properties of a metal, including high electrical conductivity, because the valence electrons are shared with 12 nearest neighbors rather than being localized in four bonds to nearest neighbors as in gray tin. Gray tin, on the other hand, has the diamond structure characteristic of other Group IV semiconductors.

The Sn–Sn distance should be longer in white tin; there are only four valence electrons from each atom, and 12 nearest neighbors. The **average** tin–tin bond order can, therefore, be only about 1/3, whereas in gray tin the bond order is one. Gray tin, with the higher bond order, has a shorter Sn–Sn distance, 2.81 Å. The bond length in white tin, with the lower bond order, is 3.02 Å.

12.103　　

Teflon™ is formed by addition polymerization.

12.105　X-ray *diffraction* is the phenomenon that enables us to measure inter-atomic distances in crystals. Diffraction is most efficient when the wavelength of light is similar to the size of the object (e.g., the slit) doing the diffracting. Inter-atomic distances are on the order of 1-10 Å, and the wavelengths of X-rays are also in this range. Visible light has wavelengths of 400-700 nm, or 4000-7000 Å, too long to be diffracted efficiently by atoms (electrons) in crystals.

12.107　Germanium is in the same family but below Si on the periodic chart. This means that Ge will probably have bonding characteristics and crystal structure similar to those of Si. Since Ge has a larger bonding atomic radius than Si, we expect a larger unit cell and *d*-spacing for Ge. In Bragg's law, $n\lambda = 2d \sin\theta$, *d* and $\sin\theta$ are inversely proportional. That is, the larger the *d*-spacing, the smaller the value of $\sin\theta$ and θ. In a diffraction experiment, we expect a Ge crystal to diffract X-rays at a smaller θ-angle than a Si crystal, assuming the X-rays have the same wavelength.

12.109　(a)　n-Type doping will weaken the bonds in a semiconductor. For example, Si, with four valence electrons per atom, has a valence band (made from bonding MOs) which is completely full, and a conduction band (made from antibonding MOs) which is totally empty. Doping a few atoms with more than four valence electrons places a few electrons in the antibonding conduction band, reduces bond order and weakens the bonds in the semiconductor.

　　　　(b)　p-Type doping will also weaken the bonds in a semiconductor. Again consider silicon. Doping a few atoms with less than four valence electrons places fewer electrons in the bonding conduction band, reduces bond order and weakens the bonds in the semiconductor.

Integrative Exercises

12.112 Refer to Section 12.7 and Figure 12.30.

(a)   In diamond, each C atom is bound to 4 other C atoms. According to VSEPR, the geometry around a central atom with 4 bonding electron pairs is tetrahedral and the C–C–C bond angles are $109°$.

(b)   In graphite, each C atom is bound to 3 other C atoms in a trigonal planar arrangement. The C–C–C bond angles are $120°$.

12.113 (a)

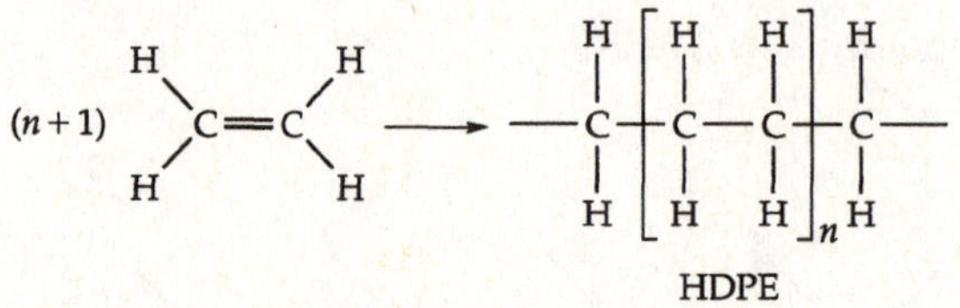

HDPE

$$\Delta H = D(C{=}C) - 2D(C{-}C) = 614 - 2(348) = -82 \text{ kJ/mol } C_2H_4$$

(b) $(n+1)\ HOOC{-}(CH_2)_6{-}COOH + (n+1)\ H_2N{-}(CH_2)_6{-}NH_2 \longrightarrow$

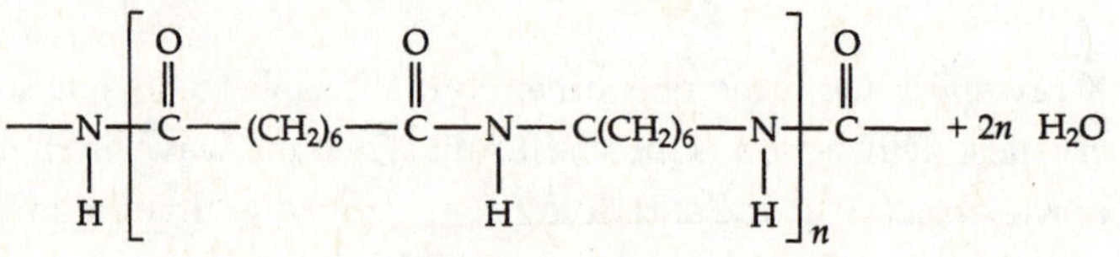

Nylon 6,6

$$\Delta H = 2D(C{-}O) + 2D(N{-}H) - 2D(C{-}N) - 2D(H{-}O)$$

$$\Delta H = 2(358) + 2(391) - 2(293) - 2(463) = -14 \text{ kJ/mol}$$

(This is –14 kJ/mol of either reactant.)

(c) $(n+1)\ HOOC{-}\langle\bigcirc\rangle{-}COOH + (n+1)\ HO{-}CH_2{-}CH_2{-}OH \longrightarrow$

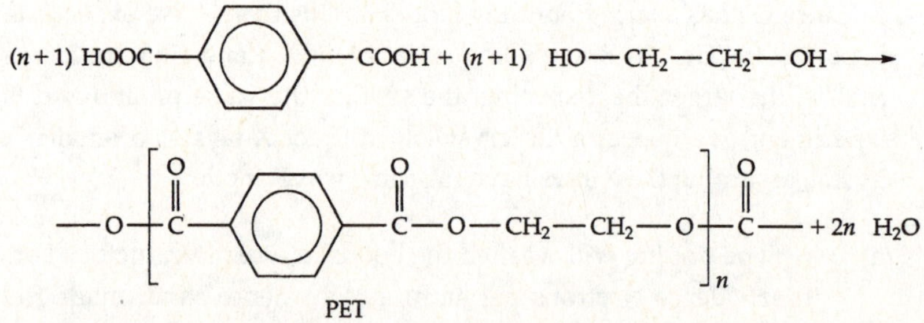

PET

$$\Delta H = 2D(C{-}O) + 2D(O{-}H) - 2D(C{-}O) - 2D(O{-}H) = 0 \text{ kJ}$$

13 Properties of Solutions

Visualizing Concepts

13.1   (a) < (b) < (c). In Section 13.1, *entropy* is qualitatively defined as randomness or dispersal in space. In container (a) the two kinds of particles are not mixed and the particles are close together, so (a) has the least entropy. In container (b), the particles occupy approximately the same volume as container (a) but the two kinds of particles are homogeneously mixed, so the degree of dispersal and randomness is greater than in (a). In container (c) the two kinds of particles are homogeneously mixed and they occupy a larger volume than in (b), so (c) has the greatest entropy.

13.3   Lattice energy is the main component of ΔH_{solute}, the enthalpy required to separate solute particles. The greater the lattice energy of the ionic solid, the more endothermic the contribution from $\Delta H_{solute} + \Delta H_{solvent}$ and the more endothermic the overall dissolving process. If ΔH_{soln} is prohibitively endothermic, the substance is not very soluble. The greater the lattice energy of an ionic solid, the less soluble it is in water.

13.7   Vitamin B_6 is likely to be largely water soluble. The three –OH groups and the —$\ddot{N}$— can enter into many hydrogen bonding interactions with water. The relatively small molecular size indicates that dispersion forces will not play a large role in intermolecular interactions and the hydrogen bonding will dominate. Vitamin E is likely to be largely fat soluble. The long, rod-like hydrocarbon chain will lead to stronger dispersion forces among vitamin E and the mostly nonpolar fats. Although vitamin E has one –OH and one —$\ddot{O}$— group, the long hydrocarbon chain prevents water from surrounding and separating the vitamin E molecules, reducing its water-solubility.

13.9   (a)   Yes, the *molarity* changes with a change in temperature. Molarity is defined as moles solute per unit volume of solution. If solution volume is different, molarity is different.

      (b)   No, *molality* does not change with change in temperature. Molality is defined as moles solute per kilogram of solvent. Even though the volume of solution has changed due to increased kinetic energy, the mass of solute and solvent have not changed, and the molality stays the same.

The Solution Process (section 13.1)

13.13 If the magnitude of the enthalpy released due to solute-solvent attractive forces ($-\Delta H_{mix}$) is at least as large as the magnitude of the enthalpy required to separate the solute particles ($+\Delta H_{solute}$), the overall enthalpy of solution (ΔH_{soln}) will be either slightly endothermic (owing to $+\Delta H_{solvent}$) or exothermic. Even if ΔH_{soln} is slightly endothermic, the increase in disorder due to mixing will cause a significant amount of solute to dissolve. If the magnitude of ΔH_{mix} is small relative to the magnitude of ΔH_{solute}, ΔH_{soln} will be large and endothermic (energetically unfavorable) and not much solute will dissolve.

13.15   *Analyze/Plan.* Decide whether the solute and solvent in question are ionic, polar covalent, or nonpolar covalent. Draw Lewis structures as needed. Then state the appropriate type of solute-solvent interaction. *Solve.*

     (a) CCl_4, nonpolar; benzene, nonpolar; dispersion forces

     (b)   methanol, polar with hydrogen bonding; water, polar with hydrogen bonding; hydrogen bonding

     (c)   KBr, ionic; water, polar; ion-dipole forces

     (d)   HCl, polar; CH_3CN, polar; dipole-dipole forces

13.17   Very soluble. In order for ΔH_{soln} to be negative (exothermic), ΔH_{mix} must have a greater magnitude than ($\Delta H_{solute} + \Delta H_{solvent}$). The collective attractive interactions formed upon mixing must be greater than the interactions being disrupted in the pure solvent and solute. The entropy of mixing always encourages solubility. In this case, the enthalpy of the system decreases and the entropy increases, so the ionic compound dissolves.

13.19   (a)   Lattice energy is the amount of energy required to completely separate a mole of solid ionic compound into its gaseous ions (Section 8.2). For ionic solutes, this corresponds to ΔH_{solute} (solute-solute interactions) in Equation [13.1].

     (b)   In Equation [13.1], ΔH_{mix} is always exothermic. Formation of attractive interactions, no matter how weak, always lowers the energy of the system, relative to the energy of the isolated particles.

13.21   (a) ΔH_{soln} is determined by the relative magnitudes of the "old" solute-solute (ΔH_{solute}) and solvent-solvent ($\Delta H_{solvent}$) interactions and the new solute-solvent interactions (ΔH_{mix}); $\Delta H_{soln} = \Delta H_{solute} + \Delta H_{solvent} + \Delta H_{mix}$. Since the solute and solvent in this case experience very similar London dispersion forces, the energy required to separate them individually and the energy released when they are mixed are approximately equal. $\Delta H_{solute} + \Delta H_{solvent} \approx -\Delta H_{mix}$. Thus, ΔH_{soln} is nearly zero.

     (b)   Mixing hexane and heptane produces a homogeneous solution from two pure substances, and the randomness of the system increases. Since no strong intermolecular forces prevent the molecules from mixing, they do so spontaneously due to the increase in disorder.

Saturated Solutions; Factors Affecting Solubility (sections 13.1 and 13.2)

13.23   (a)   Supersaturated

     (b)   Add a seed crystal. Supersaturated solutions exist because not enough solute molecules are properly aligned for crystallization to occur. A seed crystal provides a nucleus of already aligned molecules, so that ordering of the dissolved particles is more facile.

13.25   *Analyze/Plan.* On Figure 13.18, find the solubility curve for the appropriate solute. Find the intersection of 40°C and 40 g solute on the graph. If this point is below the solubility curve, more solute can dissolve and the solution is unsaturated. If the intersection is on or above the curve, the solution is saturated. *Solve.*

     (a)   unsaturated   (b)   saturated   (c)   saturated   (d)   unsaturated

13.27   The liquids water and glycerol form homogenous mixtures (solutions), regardless of the relative amounts of the two components. Glycerol has an –OH group on each C atom in the molecule. This structure facilitates strong hydrogen bonding similar to that in water. Like dissolves like and the two liquids are miscible in all proportions.

13.29   *Analyze/Plan.* Evaluate molecules in the four common laboratory solvents for strength of intermolecular interactions with nonpolar solutes. *Solve.* Toluene, $C_6H_5CH_3$, is the best solvent for nonpolar solutes. Without polar groups or nonbonding electron pairs, it forms only dispersion interactions with itself and other molecules. The enthalpy of solution, ΔH_{soln}, is essentially zero (as in Solution 13.21) and solution occurs because of the favorable entropy of mixing.

13.31   (a)   Dispersion interactions among nonpolar $CH_3(CH_2)_{16}$ –chains dominate the properties of stearic acid. It is more soluble in nonpolar CCl_4 than polar (hydrogen bonding) water, despite the presence of the –COOH group.

     (b)

cyclohexane   dioxane

Dioxane can act as a hydrogen bond acceptor, so it will be more soluble than cyclohexane in water.

13.33   *Analyze/Plan.* Hexane is a nonpolar hydrocarbon that experiences dispersion forces with other nonpolar molecules. Solutes that primarily experience dispersion forces will be more soluble in hexane. *Solve.*

     (a) CCl_4 is more soluble because dispersion forces among nonpolar CCl_4 molecules are similar to dispersion forces in hexane. Ionic bonds in $CaCl_2$ are unlikely to be broken by weak solute-solvent interactions. For $CaCl_2$, ΔH_{solute} is large, relative to ΔH_{mix}.

     (b)   Benzene, C_6H_6, is also a nonpolar hydrocarbon and will be more soluble in hexane. Glycerol experiences hydrogen bonding with itself; these solute-solute interactions are less likely to be overcome by weak solute-solvent interactions.

     (c)   Octanoic acid, $CH_3(CH_2)_6COOH$, will be more soluble than acetic acid CH_3COOH. Both solutes experience hydrogen bonding by –COOH groups, but octanoic acid has a long, rod-like hydrocarbon chain with dispersion forces similar to those in hexane, facilitating solubility in hexane.

13.35 (a) Carbonated beverages are stored with a partial pressure of $CO_2(g)$ greater than 1 atm above the liquid. A sealed container is required to maintain this CO_2 pressure.

(b) Since the solubility of gases increases with decreasing temperature, more $CO_2(g)$ will remain dissolved in the beverage if it is kept cool.

13.37 *Analyze/Plan.* Follow the logic in Sample Exercise 13.3. *Solve.*

$$S_{He} = 3.7 \times 10^{-4} \ M/atm \times 1.5 \ atm = 5.6 \times 10^{-4} \ M$$

$$S_{N_2} = 6.0 \times 10^{-4} \ M/atm \times 1.5 \ atm = 9.0 \times 10^{-4} \ M$$

Concentrations of Solutions (section 13.4)

13.39 *Analyze/Plan.* Follow the logic in Sample Exercise 13.3.   *Solve.*

(a) $\text{mass \%} = \dfrac{\text{mass solute}}{\text{total mass solution}} \times 100 = \dfrac{10.6 \ g \ Na_2SO_4}{10.6 \ g \ Na_2SO_4 + 483 \ g \ H_2O} \times 100 = 2.15\%$

(b) $\text{ppm} = \dfrac{\text{mass solute}}{\text{total mass solution}} \times 10^6; \dfrac{2.86 \ g \ Ag}{1 \ ton \ ore} \times \dfrac{1 \ ton}{2000 \ lb} \times \dfrac{1 \ lb}{453.6 \ g} \times 10^6 = 3.15 \ ppm$

13.41 *Analyze/Plan.* Given masses of CH_3OH and H_2O, calculate moles of each component.

(a) Mole fraction $CH_3OH = (\text{mol } CH_3OH)/(\text{total mol})$

(b) mass % $CH_3OH = [(g \ CH_3OH)/(\text{total mass})] \times 100$

(c) molality $CH_3OH = (\text{mol } CH_3OH)/(kg \ H_2O)$.   *Solve.*

(a) $14.6 \ g \ CH_3OH \times \dfrac{1 \ mol \ CH_3OH}{32.04 \ g \ CH_3OH} = 0.4557 = 0.456 \ mol \ CH_3OH$

$184 \ g \ H_2O \times \dfrac{1 \ mol \ H_2O}{18.02 \ g \ H_2O} = 10.211 = 10.2 \ mol \ H_2O$

$\chi_{CH_3OH} = \dfrac{0.4557}{0.4557 + 10.211} = 0.04272 = 0.0427$

(b) $\text{mass \% } CH_3OH = \dfrac{14.6 \ g \ CH_3OH}{14.6 \ g \ CH_3OH + 184 \ g \ H_2O} \times 100 = 7.35\% \ CH_3OH$

(c) $m = \dfrac{0.4557 \ mol \ CH_3OH}{0.184 \ kg \ H_2O} = 2.477 = 2.48 \ m \ CH_3OH$

13.43 *Analyze/Plan.* Given mass solute and volume solution, calculate mol solute, then molarity = mol solute/L solution. Or, for dilution, $M_c \times L_c = M_d \times L_d$.   *Solve.*

(a) $M = \dfrac{\text{mol solute}}{\text{L soln}}; \dfrac{0.540 \ g \ Mg(NO_3)_2}{0.2500 \ L \ soln} \times \dfrac{1 \ mol \ Mg(NO_3)_2}{148.3 \ g \ Mg(NO_3)_2} = 1.46 \times 10^{-2} \ M \ Mg(NO_3)_2$

(b) $\dfrac{22.4 \ g \ LiClO_4 \cdot 3H_2O}{0.125 \ L \ soln} \times \dfrac{1 \ mol \ LiClO_4 \cdot 3H_2O}{160.4 \ g \ LiClO_4 \cdot 3H_2O} = 1.12 \ M \ LiClO_4 \cdot 3H_2O$

(c) $M_c \times L_c = M_d \times L_d; 3.50 \ M \ HNO_3 \times 0.0250 \ L = ?M \ HNO_3 \times 0.250 \ L$
250 mL of 0.350 M HNO_3

13.45   *Analyze/Plan.* Follow the logic in Sample Exercise 13.4.   *Solve.*

(a) $m = \dfrac{\text{mol solute}}{\text{kg solvent}}; \dfrac{8.66\,\text{g C}_6\text{H}_6}{23.6\,\text{g CCl}_4} \times \dfrac{1\,\text{mol C}_6\text{H}_6}{78.11\,\text{g C}_6\text{H}_6} \times \dfrac{1000\,\text{g CCl}_4}{1\,\text{kg CCl}_4} = 4.70\,m\,\text{C}_6\text{H}_6$

(b)   The density of H_2O = 0.997 g/mL = 0.997 kg/L.

$$\dfrac{4.80\,\text{g NaCl}}{0.350\,\text{L H}_2\text{O}} \times \dfrac{1\,\text{mol NaCl}}{58.44\,\text{g NaCl}} \times \dfrac{1\,\text{L H}_2\text{O}}{0.997\,\text{kg H}_2\text{O}} = 0.235\,m\,\text{NaCl}$$

13.47   *Analyze/Plan.* Assume 1 L of solution. Density gives the total mass of 1 L of solution. The g H_2SO_4/L are also given in the problem. Mass % = (mass solute/total mass solution) × 100. Calculate mass solvent from mass solution and mass solute. Calculate moles solute and solvent and use the appropriate definitions to calculate mole fraction, molality, and molarity.   *Solve.*

(a) $\dfrac{571.6\,\text{g H}_2\text{SO}_4}{1\,\text{L soln}} \times \dfrac{1\,\text{L soln}}{1329\,\text{g soln}} = 0.430098\,\text{g H}_2\text{SO}_4/\text{g soln}$

mass percent is thus 0.4301 × 100 = 43.01% H_2SO_4

(b)   In a liter of solution there are 1329 − 571.6 = 757.4 = 757 g H_2O.

$\dfrac{571.6\,\text{g H}_2\text{SO}_4}{98.09\,\text{g/mol}} = 5.827\,\text{mol H}_2\text{SO}_4 ; \dfrac{757.4\,\text{g H}_2\text{O}}{18.02\,\text{g/mol}} = 42.03 = 42.0\,\text{mol H}_2\text{O}$

$\chi_{\text{H}_2\text{SO}_4} = \dfrac{5.827}{42.03 + 5.827} = 0.122$

(The result has 3 sig figs because (g H_2O) resulting from subtraction is limited to 3 sig figs.)

(c)   molality $= \dfrac{5.827\,\text{mol H}_2\text{SO}_4}{0.7574\,\text{kg H}_2\text{O}} = 7.693 = 7.69\,m\,\text{H}_2\text{SO}_4$

(d)   molarity $= \dfrac{5.827\,\text{mol H}_2\text{SO}_4}{1\,\text{L soln}} = 5.827\,M\,\text{H}_2\text{SO}_4$

13.49   *Analyze/Plan.* Given: 98.7 mL of CH_3CN(l), 0.786 g/mL; 22.5 mL CH_3OH, 0.791 g/mL. Use the density and volume of each component to calculate mass and then moles of each component. Use the definitions to calculate mole fraction, molality, and molarity.   *Solve.*

(a) $\text{mol CH}_3\text{CN} = \dfrac{0.786\,\text{g}}{1\,\text{mL}} \times 98.7\,\text{mL} \times \dfrac{1\,\text{mol CH}_3\text{CN}}{41.05\,\text{g CH}_3\text{CN}} = 1.8898 = 1.89\,\text{mol}$

$\text{mol CH}_3\text{OH} = \dfrac{0.791\,\text{g}}{1\,\text{mL}} \times 22.5\,\text{mL} \times \dfrac{1\,\text{mol CH}_3\text{OH}}{32.04\,\text{g CH}_3\text{OH}} = 0.5555 = 0.556\,\text{mol}$

$\chi_{\text{CH}_3\text{OH}} = \dfrac{0.5555\,\text{mol CH}_3\text{OH}}{1.8898\,\text{mol CH}_3\text{CN} + 0.5555\,\text{mol CH}_3\text{OH}} = 0.227$

(b)   Assuming CH_3OH is the solute and CH_3CN is the solvent,

$98.7\,\text{mL CH}_3\text{CN} \times \dfrac{0.786\,\text{g}}{1\,\text{mL}} \times \dfrac{1\,\text{kg}}{1000\,\text{g}} = 0.07758 = 0.0776\,\text{kg CH}_3\text{CN}$

$m_{\text{CH}_3\text{OH}} = \dfrac{0.5555\,\text{mol CH}_3\text{OH}}{0.07758\,\text{kg CH}_3\text{CN}} = 7.1604 = 7.16\,m\,\text{CH}_3\text{OH}$

(c)   The total volume of the solution is 121.2 mL, assuming volumes are additive.

$$M = \frac{0.5555 \text{ mol CH}_3\text{OH}}{0.1212 \text{ L solution}} = 4.58 \, M \text{ CH}_3\text{OH}$$

13.51   *Analyze/Plan.* Given concentration and volume of solution use definitions of the appropriate concentration units to calculate amount of solute; change amount to moles if needed.   *Solve.*

(a) $\text{mol} = M \times \text{L}; \dfrac{0.250 \text{ mol SrBr}_2}{1 \text{ L soln}} \times 0.600 \text{ L} = 0.150 \text{ mol SrBr}_2$

(b)   Assume that for dilute aqueous solutions, the mass of the solvent is the mass of solution. Use proportions to get mol KCl.

$$\frac{0.180 \text{ mol KCl}}{1 \text{ kg H}_2\text{O}} = \frac{x \text{ mol KCl}}{0.0864 \text{ kg H}_2\text{O}}; x = 1.56 \times 10^{-2} \text{ mol KCl}$$

(c)   Use proportions to get mass of glucose, then change to mol glucose.

$$\frac{6.45 \text{ g C}_6\text{H}_{12}\text{O}_6}{100 \text{ g soln}} = \frac{x \text{ g C}_6\text{H}_{12}\text{O}_6}{124.0 \text{ g soln}}; x = 8.00 \text{ g C}_6\text{H}_{12}\text{O}_6$$

$$8.00 \text{ g C}_6\text{H}_{12}\text{O}_6 \times \frac{1 \text{ mol C}_6\text{H}_{12}\text{O}_6}{180.2 \text{ g C}_6\text{H}_{12}\text{O}_6} = 4.44 \times 10^{-2} \text{ mol C}_6\text{H}_{12}\text{O}_6$$

13.53   *Analyze/Plan.* When preparing solution, we must know amount of solute and solvent. Use the appropriate concentration definition to calculate amount of solute. If this amount is in moles, use molar mass to get grams; use mass in grams directly. Amount of solvent can be expressed as total volume or mass of solution. Combine mass solute and solvent to produce the required amount (mass or volume) of solution.   *Solve.*

(a) $\text{mol} = M \times \text{L}; \dfrac{1.50 \times 10^{-2} \text{ mol KBr}}{1 \text{ L soln}} \times 0.75 \text{ L} \times \dfrac{119.0 \text{ g KBr}}{1 \text{ mol KBr}} = 1.3 \text{ g KBr}$

Weigh out 1.3 g KBr, dissolve in water, dilute with stirring to 0.75 L (750 mL).

(b)   Mass of solution is required, but density is not specified. Use molality to calculate mass fraction, and then the masses of solute and solvent needed for 125 g of solution.

$$\frac{0.180 \text{ mol KBr}}{1000 \text{ g H}_2\text{O}} \times \frac{119.0 \text{ g KBr}}{1 \text{ mol KBr}} = 21.42 = 21.4 \text{ g KBr/kg H}_2\text{O Thus,}$$

$$\text{mass fraction} = \frac{21.42 \text{ g KBr}}{1000 + 21.42} = 0.02097 = 0.0210$$

In 125 g of the 0.180 *m* solution, there are

$$(125 \text{ g soln}) \times \frac{0.02097 \text{ g KBr}}{1 \text{ g soln}} = 2.621 = 2.62 \text{ g KBr}$$

Weigh out 2.62 g KBr, dissolve it in 125 − 2.62 = 122.38 = 122 g H$_2$O to make exactly 125 g of 0.180 *m* solution.

(c)   Using solution density, calculate the total mass of 1.85 L of solution, and from the mass % of KBr, the mass of KBr required.

$$1.85 \text{ L soln} \times \frac{1000 \text{ mL}}{1 \text{ L}} \times \frac{1.10 \text{ g soln}}{1 \text{ mL}} = 2035 = 2.04 \times 10^3 \text{ g soln}$$

$$0.120 \,(2035 \text{ g soln}) = 244.2 = 244 \text{ g KBr}$$

Dissolve 244 g KBr in water, dilute with stirring to 1.85 L.

(d) Calculate moles KBr needed to precipitate 16.0 g AgBr. AgNO₃ is present in excess.

$$16.0 \text{ g AgBr} \times \frac{1 \text{ mol AgBr}}{187.8 \text{ g AgBr}} \times \frac{1 \text{ mol KBr}}{1 \text{ mol AgBr}} = 0.08520 = 0.0852 \text{ mol KBr}$$

$$0.0852 \text{ mol KBr} \times \frac{1 \text{ L soln}}{0.150 \text{ mol KBr}} = 0.568 \text{ L soln}$$

Weigh out 0.0852 mol KBr (10.1 g KBr), dissolve it in a small amount of water, and dilute to 0.568 L.

13.55 *Analyze/Plan.* Assume a solution volume of 1.00 L. Calculate the mass of 1.00 L of solution and the mass of HNO₃ in 1.00 L of solution. Mass % = (mass solute/mass solution) × 100. *Solve.*

$$1.00 \text{ L} \times \frac{1000 \text{ mL}}{1 \text{ L}} \times \frac{1.42 \text{ g soln}}{\text{mL soln}} = 1.42 \times 10^3 \text{ g soln}$$

$$16 \, M = \frac{16 \text{ mol HNO}_3}{1 \text{ L soln}} \times \frac{63.02 \text{ g HNO}_3}{1 \text{ mol HNO}_3} = 1008 = 1.0 \times 10^3 \text{ g HNO}_3$$

$$\text{mass \%} = \frac{1008 \text{ g HNO}_3}{1.42 \times 10^3 \text{ g soln}} \times 100 = 71\% \text{ HNO}_3$$

13.57 *Analyze.* Given: 80.0% Cu, 20.0% Zn by mass; density = 8750 kg/m³. Find: (a) *m* of Zn (b) *M* of Zn

(a) *Plan.* In the brass alloy, Zn is the solute (lesser component) and Cu is the solvent (greater component). *m* = mol Zn/kg Cu. 1 m³ brass alloy weighs 8750 kg. 80.0% is Cu, 20.0% is Zn. Change g Zn → mol Zn and solve for *m*. *Solve.*

$$8750 \text{ kg brass} \times \frac{80 \text{ g Cu}}{100 \text{ g brass}} = 7.00 \times 10^3 \text{ kg Cu}$$

$$8750 \text{ kg brass} - 7000 \text{ kg Cu} = 1750 \text{ kg Zn}$$

$$1750 \text{ kg Zn} \times \frac{1000 \text{ g}}{\text{kg}} \times \frac{1 \text{ mol Zn}}{65.39 \text{ g Zn}} = 26{,}762.5 = 2.68 \times 10^4 \text{ mol Zn}$$

$$m = \frac{2.676 \times 10^4 \text{ mol Zn}}{7000 \text{ kg Cu}} = 3.82 \, m \text{ Zn}$$

(b) *Plan.* *M* = mol Zn/L brass. Use mol Zn from part (a). Change 1 m³ → L brass and calculate *M*. *Solve.*

$$1 \text{ m}^3 \times \frac{(10)^3 \text{ dm}^3}{\text{m}^3} \times \frac{1 \text{ L}}{1 \text{ dm}^3} = 1000 \text{ L}$$

$$M = \frac{2.676 \times 10^4 \text{ mol Zn}}{1000 \text{ L brass}} = 26.76 = 26.8 \, M \text{ Zn}$$

13.59   *Analyze.* Given: 4.6% CO_2 by volume (in air), 1 atm total pressure. Find: partial pressure and molarity of CO_2 in air.

     Plan. 4.6% CO_2 by volume means 4.6 mL of CO_2 could be isolated from 100 mL of air, at the same temperature and pressure. According to Avogadro's Law, equal volumes of gases at the same temperature and pressure contain equal numbers of moles. By inference, the volume ratio of CO_2 to air, 4.6/100 or 0.046, is also the mole ratio.

     Solve. $P_{CO_2} = \chi_{CO_2} \times P_t = 0.046\,(1\,atm) = 0.046\ atm$

     $M = mol\ CO_2/L\ air = n/V. \qquad PV = nRT, \quad M = n/V = P/RT$

$$M_{CO_2} = \frac{P_{CO_2}}{RT} = \frac{0.046\,atm}{310\,K} \times \frac{mol\text{-}K}{0.08206\,L\text{-}atm} = 1.8 \times 10^{-3}\,M$$

Colligative Properties (section 13.5)

13.61   freezing point depression, $\Delta T_f = K_f(m)$; boiling point elevation, $\Delta T_b = K_b(m)$;

     osmotic pressure, $\Pi = M\,RT$; vapor pressure lowering, $P_A = \chi_A P_A^\circ$

13.63   The vapor pressure over the sucrose solution is higher than the vapor pressure over the glucose solution. Since sucrose has a greater molar mass, 10 g of sucrose contains fewer particles than 10 g of glucose. The solution that contains fewer particles, the sucrose solution, will have the higher vapor pressure.

13.65   (a)   *Analyze/Plan.* H_2O vapor pressure will be determined by the mole fraction of H_2O in the solution. The vapor pressure of pure H_2O at 338 K (65°C) = 187.5 torr. *Solve.*

$$\frac{22.5\,g\,C_{12}H_{22}O_{11}}{342.3\,g/mol} = 0.06573 = 0.0657\ mol; \quad \frac{200.0\,g\,H_2O}{18.02\,g/mol} = 11.09878 = 11.10\ mol$$

$$P_{H_2O} = \chi_{H_2O}\,P_{H_2O}^\circ = \frac{11.09878\ mol\ H_2O}{11.09878 + 0.06573} \times 187.5\ torr = 186.4\ torr$$

     (b)   *Analyze/Plan.* For this problem, it will be convenient to express Raoult's law in terms of the lowering of the vapor pressure of the solvent, ΔP_A.

         $\Delta P_A = P_A^\circ - \chi_A P_A^\circ = P_A^\circ (1 - \chi_A)$. $1 - \chi_A = \chi_B$, the mole fraction of the *solute* particles

         $\Delta P_A = \chi_B P_A^\circ$; the vapor pressure of the solvent (A) is lowered according to the mole fraction of solute (B) particles present.   *Solve.*

         P_{H_2O} at $40°C = 55.3$ torr; $\dfrac{340\,g\,H_2O}{18.02\,g/mol} = 18.868 = 18.9\ mol\ H_2O$

$$\chi_{C_3H_8O_2} = \frac{2.88\,torr}{55.3\,torr} = \frac{y\ mol\ C_3H_8O_2}{y\ mol\ C_3H_8O_2 + 18.868\ mol\ H_2O} = 0.05208 = 0.0521$$

         $0.05208 = \dfrac{y}{y + 18.868}; 0.05208\,y + 0.98263 = y;\ 0.94792\,y = 0.98263,$

         $y = 1.0366 = 1.04\ mol\ C_3H_8O_2$

         This result has 3 sig figs because (0.340 kg water) has 3 sig figs.

$$1.0366 \text{ mol } C_3H_8O_2 \times \frac{76.09 \text{ g } C_3H_8O_2}{\text{mol } C_3H_8O_2} = 78.88 = 78.9 \text{ g } C_3H_8O_2$$

13.67 *Analyze/Plan.* At 63.5°C, $P^\circ_{H_2O} = 175$ torr, $P^\circ_{Eth} = 400$ torr. Let G = the mass of H_2O and/or C_2H_5OH. *Solve.*

(a) $$\chi_{Eth} = \frac{\dfrac{G}{46.07 \text{ g } C_2H_5OH}}{\dfrac{G}{46.07 \text{ g } C_2H_5OH} + \dfrac{G}{18.02 \text{ g } H_2O}}$$

Multiplying top and bottom of the right side of the equation by 1/G gives:

$$\chi_{Eth} = \frac{1/46.07}{1/46.07 + 1/18.02} = \frac{0.02171}{0.02171 + 0.05549} = 0.2812$$

(b) $$P_t = P_{Eth} + P_{H_2O}; \quad P_{Eth} = \chi_{Eth} \times P^\circ_{Eth}; \quad P_{H_2O} = \chi_{H_2O} P^\circ_{H_2O}$$

$$\chi_{Eth} = 0.2812, \quad P_{Eth} = 0.2812 (400 \text{ torr}) = 112.48 = 112 \text{ torr}$$

$$\chi_{H_2O} = 1 - 0.2812 = 0.7188; \quad P_{H_2O} = 0.7188(175 \text{ torr}) = 125.8 = 126 \text{ torr}$$

$$P_t = 112.5 \text{ torr} + 125.8 \text{ torr} = 238.3 = 238 \text{ torr}$$

(c) $$\chi_{Eth} \text{ in vapor} = \frac{P_{Eth}}{P_{total}} = \frac{112.5 \text{ torr}}{238.3 \text{ torr}} = 0.4721 = 0.472$$

13.69 (a) Because NaCl is a soluble ionic compound and a strong electrolyte, there are 2 mol dissolved particles for every 1 mol of NaCl solute. $C_6H_{12}O_6$ is a molecular solute, so there is 1 mol of dissolved particles per mol solute. Boiling point elevation is directly related to total moles of dissolved particles; 0.10 *m* NaCl has more dissolved particles so its boiling point is higher than 0.10 *m* $C_6H_{12}O_6$.

(b) *Analyze/Plan.* $\Delta T = K_b \, m$; K_b for H_2O is 0.51 °C/*m* (Table 13.3) *Solve.*

$$0.10 \text{ m NaCl}: \Delta T = \frac{0.51°C}{m} \times 0.20 \text{ m} = 0.102 \text{ °C}; T_b = 100.0 + 0.102 = 100.1°C$$

$$0.10 \text{ m } C_6H_{12}O_6: \Delta T = \frac{0.51°C}{m} \times 0.10 \text{ m} = 0.051 \text{ °C}; T_b = 100.0 + 0.051 = 100.1°C$$

Check. Because K_b for H_2O is so small, there is little real difference in the boiling points of the two solutions.

(c) In solutions of strong electrolytes like NaCl, electrostatic attractions between ions lead to ion pairing. Ion pairing reduces the effective number of particles in solution, decreasing the **change** in boiling point. The actual boiling point is then lower than the calculated boiling point for a 0.1 *M* solution.

13.71 *Analyze/Plan.* Follow the logic in Sample Exercise 13.9. *Solve.*

The more nonvolatile solute particles, the higher the boiling point of the solution. Since LiBr and $Zn(NO_3)_2$ are electrolytes, the particle concentrations in these solutions are 0.10 *m* and 0.15 *m*, respectively (although ion-ion attractive forces may decrease the effective concentrations some-what). Thus, the order of increasing boiling points is:

0.050 *m* LiBr < 0.120 *m* glucose < 0.050 *m* $Zn(NO_3)_2$

13.73   *Analyze/Plan.* $\Delta T = K(m)$; first, calculate the **molality** of each solution. *Solve.*

(a)   0.22 m

(b) $2.45 \text{ mol CHCl}_3 \times \dfrac{119.4 \text{ g CHCl}_3}{\text{mol CHCl}_3} = 292.53 \text{ g} = 0.293 \text{ kg};$

$\dfrac{0.240 \text{ mol C}_{10}\text{H}_8}{0.29253 \text{ kg CHCl}_3} = 0.8204 = 0.820 \ m$

(c) $1.50 \text{ g NaCl} \times \dfrac{1 \text{ mol NaCl}}{58.44 \text{ g NaCl}} \times \dfrac{2 \text{ mol particles}}{1 \text{ mol NaCl}} = 0.05133 = 0.0513 \text{ mol particles}$

$m = \dfrac{0.05133 \text{ mol NaCl}}{0.250 \text{ kg H}_2\text{O}} = 0.20534 = 0.205 \ m$

(d) $2.04 \text{ g KBr} \times \dfrac{1 \text{ mol KBr}}{119.0 \text{ g KBr}} \times \dfrac{2 \text{ mol particles}}{1 \text{ mol KBr}} = 0.03429 = 0.0343 \text{ mol particles}$

$4.82 \text{ g C}_6\text{H}_{12}\text{O}_6 \times \dfrac{1 \text{ mol C}_6\text{H}_{12}\text{O}_6}{180.2 \text{ g C}_6\text{H}_{12}\text{O}_6} = 0.02675 = 0.0268 \text{ mol particles}$

$m = \dfrac{(0.03429 + 0.02675) \text{ mol particles}}{0.188 \text{ kg H}_2\text{O}} = 0.32465 = 0.325 \ m$

Solve. Then, f.p. $= T_f - K_f(m)$; b.p. $= T_b + K_b(m)$; T in °C

	m	T_f	$-K_f(m)$	f.p.	T_b	$+K_b(m)$	b.p.
(a)	0.22	−114.6	−1.99(0.22) = −0.44	−115.0	78.4	1.22(0.22) = 0.27	78.7
(b)	0.820	−63.5	−4.68(0.820) = −3.84	−67.3	61.2	3.63(0.820) = 2.98	64.2
(c)	0.205	0.0	−1.86(0.205) = −0.381	−0.4	100.0	0.51(0.205) = 0.10	100.1
(d)	0.325	0.0	−1.86(0.325) = −0.605	−0.6	100.0	0.51(0.325) = 0.17	100.2

13.75   *Analyze.* Given freezing point of solution and mass of solvent, calculate mass of solute.

Plan. Reverse the logic in Sample Exercise 13.8. Use $\Delta T_f = K_f(m)$ to calculate the required molality, and then apply the definition of molality to calculate moles and grams of $C_2H_6O_2$.

Solve. f.p. of solution $= -5.00$ °C; f.p. of solvent (H_2O) $= 0.0$ °C

$\Delta T_f = 5.00\,°C = K_f(m); \ 5.00\,°C = 1.86\,°C/m(m)$

$m = \dfrac{5.00\ °C}{1.86\ °C/m} = 2.688 = 2.69 \ m \ C_2H_6O_2$

$m = \dfrac{\text{mol C}_2\text{H}_6\text{O}_2}{\text{kg H}_2\text{O}} = C_2H_6O_2 = m \times \text{kg H}_2\text{O}$

$2.688 \ m \ C_2H_6O_2 \times 1.00 \text{ kg H}_2\text{O} = 2.688 = 2.69 \text{ mol C}_2H_6O_2$

$2.688 \ m \ C_2H_6O_2 \times \dfrac{62.07 \text{ g C}_2\text{H}_6\text{O}_2}{1 \text{ mol}} = 166.84 = 167 \text{ g C}_2H_6O_2$

13.77   *Analyze/Plan.* $\Pi = M \text{RT}$; T $= 25$ °C $+ 273 = 298$ K; $M = \text{mol C}_9\text{H}_8\text{O}_4/\text{L soln}$ *Solve.*

$M = \dfrac{44.2 \text{ mg C}_9\text{H}_8\text{O}_4}{0.358 \text{ L}} \times \dfrac{1 \text{ g}}{1000 \text{ mg}} \times \dfrac{1 \text{ mol C}_9\text{H}_8\text{O}_4}{180.2 \text{ g C}_9\text{H}_8\text{O}_4} = 6.851 \times 10^{-4} = 6.85 \times 10^{-4} \ M$

$$\Pi = \frac{6.851 \times 10^{-4} \text{ mol}}{L} \times \frac{0.08206 \text{ L - atm}}{\text{mol - K}} \times 298 \text{ K} = 0.01675 = 0.0168 \text{ atm} = 12.7 \text{ torr}$$

13.79   *Analyze/Plan.* Follow the logic in Sample Exercise 13.11 to calculate the molar mass of adrenaline based on the boiling point data. Use the structure to obtain the molecular formula and molar mass. Compare the two values.   *Solve.*

$$\Delta T_b = K_b \, m; \quad m = \frac{\Delta T_b}{K_b} = \frac{+0.49}{5.02} = 0.0976 = 0.098 \, m \text{ adrenaline}$$

$$m = \frac{\text{mol adrenaline}}{\text{kg CCl}_4} = \frac{\text{g adrenaline}}{\text{MM adrenaline} \times \text{kg CCl}_4}$$

$$\text{MM adrenaline} = \frac{\text{g adrenaline}}{m \times \text{kg CCl}_4} = \frac{0.64 \text{ g adrenaline}}{0.0976 \, m \times 0.0360 \text{ kg CCl}_4} = 1.8 \times 10^2 \text{ g/mol adrenaline}$$

Check. The molecular formula is $C_9H_{13}NO_3$, MM = 183 g/mol. The values agree to 2 sig figs, the precision of the experimental value.

13.81   *Anayze/Plan.* Follow the logic in Sample Exercise 13.12.   *Solve.*

$$\Pi = MRT; \quad M = \frac{\Pi}{RT}; \quad T = 25 \, ^\circ C + 273 = 298 \text{ K}$$

$$M = 0.953 \text{ torr} \times \frac{1 \text{ atm}}{760 \text{ torr}} \times \frac{\text{mol - K}}{0.08206 \text{ L - atm}} \times \frac{1}{298 \text{ K}} = 5.128 \times 10^{-5} = 5.13 \times 10^{-5} \, M$$

$$\text{mol} = M \times L = 5.128 \times 10^{-5} \times 0.210 \text{ L} = 1.077 \times 10^{-5} = 1.08 \times 10^{-5} \text{ mol lysozyme}$$

$$\text{MM} = \frac{g}{\text{mol}} = \frac{0.150 \text{ g}}{1.077 \times 10^{-5} \text{ mol}} = 1.39 \times 10^4 \text{ g/mol lysozyme}$$

13.83   (a)   *Analyze/Plan.* $i = \Pi$ (measured) / Π (calculated for a nonelectrolyte);

Π (calculated) = M RT. *Solve.*

$$\Pi \text{ (calculated)} = \frac{0.010 \text{ mol}}{L} \times \frac{0.08206 \text{ L - atm}}{\text{mol - K}} \times 298 \text{ K} = 0.2445 = 0.24 \text{ atm}$$

$i = 0.674$ atm/0.2445 atm = 2.756 = 2.8

(b)   The van't Hoff factor is the effective number of particles per mole of solute. The closer the measured i value is to a theoretical integer value, the more ideal the solution. Ion-pairing and other interparticle attractive forces reduce the effective number of particles in solution and reduce the measured value of i. The more concentrated the solution, the greater the ion-pairing and the smaller the measured value of i.

Colloids (section 13.6)

13.85   (a)   In the gaseous state, the particles are far apart and intermolecular attractive forces are small. When two gases combine, all terms in Equation [13.1] are essentially zero and the mixture is always homogeneous.

(b)   The outline of a light beam passing through a colloid is visible, whereas light passing through a true solution is invisible unless collected on a screen. This is the Tyndall effect. To determine whether Faraday's (or anyone's) apparently homogeneous dispersion is a true solution or a colloid, shine a beam of light on it and see if the light is scattered.

13.87   (a)   hydrophobic   (b)   hydrophilic   (c)   hydrophobic

(d)   hydrophobic (but stabilized by adsorbed charges)

13.89   Proteins form hydrophilic colloids because they carry charges on their surface (Figure 13.30). When electrolytes are added to a suspension of proteins, the dissolved ions form ion pairs with the protein surface charges, effectively neutralizing them. The protein's capacity for ion-dipole interactions with water is diminished and the colloid separates into a protein layer and a water layer.

Additional Exercises

13.91   The outer periphery of the BHT molecule is mostly hydrocarbon-like groups, such as $-CH_3$. The one $-OH$ group is rather buried inside, and probably does little to enhance solubility in water. Thus, BHT is more likely to be soluble in the nonpolar hydrocarbon hexane, C_6H_{14}, than in polar water.

13.94   (a) $S_{Rn} = kP_{Rn}; k = S_{Rn}/P_{Rn} = 7.27 \times 10^{-3}\ M/1\ atm = 7.27 \times 10^{-3}\ mol/L\text{-atm}$

(b) $P_{Rn} = \chi_{Rn}P_{total}; P_{Rn} = 3.5 \times 10^{-6}\ (32\ atm) = 1.12 \times 10^{-4} = 1.1 \times 10^{-4}\ atm$

$$S_{Rn} = k\,P_{Rn}; S_{Rn} = \frac{7.27 \times 10^{-3}\ mol}{L\text{-atm}} \times 1.12 \times 10^{-4}\ atm = 8.1 \times 10^{-7}\ M$$

13.98   (a) $\dfrac{1.80\ mol\ LiBr}{1\ L\ soln} \times \dfrac{86.85\ g\ LiBr}{1\ mol\ LiBr} = 156.3 = 156\ g\ LiBr$

1 L soln = 826 g soln; g CH_3CN = 826 − 156.3 = 669.7 = 670 g CH_3CN

$$m\ LiBr = \frac{1.80\ mol\ LiBr}{0.6697\ kg\ CH_3CN} = 2.69\ m$$

(b) $\dfrac{669.7\ g\ CH_3CN}{41.05\ g/mol} = 16.31 = 16.3\ mol\ CH_3CN; \chi_{LiBr} = \dfrac{1.80}{1.80 + 16.31} = 0.0994$

(c) $mass\ \% = \dfrac{669.7\ g\ CH_3CN}{826\ g\ soln} \times 100 = 81.1\%\ CH_3CN$

13.100   *Analyze.* Given vapor pressure of both pure water and the aqueous solution and moles H_2O find moles of solute in the solution.

Plan. Use vapor pressure lowering, $P_A = \chi_A P_A^o$, to calculate χ_A, mole fraction solvent, and then use the definition of mole fraction to calculate moles solute particles. Because NaCl is a strong electrolyte, there is one mole NaCl for every two moles solute particles. *Solve.*

$\chi_{H_2O} = P_{soln}/P_{H_2O} = 25.7/31.8 = 0.80818 = 0.808$

$\chi_{H_2O} = \dfrac{mol\ H_2O}{mol\ ions + mol\ H_2O}; \quad 0.80818 = \dfrac{0.115}{(mol\ ions + 0.115)}$

0.80818 (0.115 + mol ions) = 0.115;   0.80818 (mol ions) = 0.115 − 0.092940

mol ions = 0.02206/0.80818 = 0.02730 = 0.0273

mol NaCl = mol ions/2 = 0.02730/2 = 0.01365 = 0.0137 mol NaCl

13.103   (a)   0.100 m K_2SO_4 is 0.300 m in particles. H_2O is the solvent.

$\Delta T_f = K_f m = -1.86(0.300) = -0.558$; $T_f = 0.0 - 0.558 = -0.558°C = -0.6°C$

(b) ΔT_f (nonelectrolyte) $= -1.86(0.100) = -0.186$; $T_f = 0.0 - 0.186 = -0.186°C = -0.2°C$

T_f (measured) $= i \times T_f$ (nonelectrolyte)

From Table 13.4, i for 0.100 m K_2SO_4 = 2.32

T_f (measured) $= 2.32(-0.186°C) = -0.432°C = -0.4°C$

Integrative Exercises

13.106   Since these are very dilute solutions, assume that the density of the solution ≈ the density of H_2O ≈ 1.0 g/mL at 25°C. Then, 100 g solution = 100 g H_2O = 0.100 kg H_2O.

(a) CF_4 : $\dfrac{0.0015 \text{ g } CF_4}{0.100 \text{ kg } H_2O} \times \dfrac{1 \text{ mol } CF_4}{88.00 \text{ g } CF_4} = 1.7 \times 10^{-4} \ m$

$CClF_3$: $\dfrac{0.009 \text{ g } CClF_3}{0.100 \text{ kg } H_2O} \times \dfrac{1 \text{ mol } CClF_3}{104.46 \text{ g } CClF_3} = 8.6 \times 10^{-4} \ m = 9 \times 10^{-4} \ m$

CCl_2F_2 : $\dfrac{0.028 \text{ g } CCl_2F_2}{0.100 \text{ kg } H_2O} \times \dfrac{1 \text{ mol } CCl_2F_2}{120.9 \text{ g } CCl_2F_2} = 2.3 \times 10^{-3} \ m$

$CHClF_2$: $\dfrac{0.30 \text{ g } CHClF_2}{0.100 \text{ kg } H_2O} \times \dfrac{1 \text{ mol } CHClF_2}{86.47 \text{ g } CHClF_2} = 3.5 \times 10^{-2} \ m$

(b) $m = \dfrac{\text{mol solute}}{\text{kg solvent}}$; $M = \dfrac{\text{mol solute}}{\text{L solution}}$

Molality and molarity are numerically similar when kilograms solvent and liters solution are nearly equal. This is true when solutions are dilute, so that the density of the solution is essentially the density of the solvent, and when the density of the solvent is nearly 1 g/mL. That is, for dilute aqueous solutions such as the ones in this problem, $M \approx m$.

(c)   Water is a polar solvent; the solubility of solutes increases as their polarity increases. All the fluorocarbons listed have tetrahedral molecular structures. CF_4, a symmetrical tetrahedron, is nonpolar and has the lowest solubility. As more different atoms are bound to the central carbon, the electron density distribution in the molecule becomes less symmetrical and the molecular polarity increases. The most polar fluorocarbon, $CHClF_2$, has the greatest solubility in H_2O. It may act as a weak hydrogen bond acceptor for water.

(d) $S_g = k\,P_g$. Assume $M = m$ for $CHClF_2$. $P_g = 1$ atm

$k = \dfrac{S_g}{P_g} = \dfrac{M}{P_g}$; $k = \dfrac{3.5 \times 10^{-2} \ M}{1.0 \text{ atm}} = 3.5 \times 10^{-2} \text{ mol/L - atm}$

This value is greater than the Henry's law constant for $N_2(g)$, because $N_2(g)$ is nonpolar and of lower molecular mass than $CHClF_2$. In fact, the Henry's law constant for nonpolar CF_4, 1.7×10^{-4} mol/L-atm is similar to the value for N_2, 6.8×10^{-4} mol L-atm.

13.109   (a)

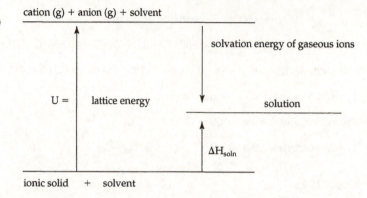

cation (g) + anion (g) + solvent

solvation energy of gaseous ions

U =   lattice energy   solution

ΔH_{soln}

ionic solid   +   solvent

(b)   If the lattice energy (U) of the ionic solid (ion-ion forces) is too large relative to the solvation energy of the gaseous ions (ion-dipole forces), ΔH_{soln} will be too large and positive (endothermic) for solution to occur. This is the case for solutes like NaBr. Lattice energy is inversely related to the distance between ions, so salts with large cations like $(CH_3)_4N^+$ have smaller lattice energies than salts with simple cations like Na^+. The smaller lattice energy of $(CH_4)_3NBr$ causes it to be more soluble in nonaqueous polar solvents. Also, the $-CH_3$ groups in the large cation are capable of dispersion interactions with the $-CH_3$ (or other nonpolar groups) of the solvent molecules. This produces a more negative solvation energy for the salts with large cations.

Overall, for salts with larger cations, U is smaller (less positive), the solvation energy of the gaseous ions is more negative, and ΔH_{soln} is less endothermic. These salts are more soluble in polar nonaqueous solvents.

13.112   The resulting solution is very dilute, so assume ideal behavior. Assume the amount of water consumed in the reaction is negligible. Ignore the solubility of $H_2(g)$ in the solution (see Solution 3.110).

$$1.0 \text{ mm}^3 \times \frac{0.535 \text{ g}}{\text{cm}^3} \times \frac{1^3 \text{ cm}^3}{10^3 \text{ mm}^3} = 5.35 \times 10^{-4} = 5.4 \times 10^{-4} \text{ g Li}$$

$$5.35 \times 10^{-4} \text{ g Li} \times \frac{1 \text{ mol Li}}{6.941 \text{ g Li}} = 7.708 \times 10^{-5} = 7.7 \times 10^{-5} \text{ mol Li}$$

mol Li = mol LiOH; 2 mol ions per mol LiOH

7.708×10^{-5} mol Li = 7.708×10^{-5} mol LiOH = 1.542×10^{-4} mol ions = 1.5×10^{-4} mol ions

$$m = \frac{1.542 \times 10^{-4} \text{ mol ions}}{0.500 \text{ L } H_2O} \times \frac{1 \text{ L}}{1000 \text{ mL}} \times \frac{1 \text{ mL } H_2O}{0.997 \text{ g } H_2O} \times \frac{1000 \text{ g}}{1 \text{ kg}} = 3.092 \times 10^{-4} = 3.1 \times 10^{-4} \text{ m}$$

$\Delta T_f = K_f \, m = -1.86(3.092 \times 10^{-4}) = -5.8 \times 10^{-4} \text{ °C}$; $T_f = 0.00000 - 0.00058 = -0.00058 \text{ °C}$

The freezing point of the LiOH(aq) solution is essentially zero.

14 Chemical Kinetics

Visualizing Concepts

14.1   *Analyze/Plan.* Given three sets of initial conditions for the same chemical reaction, decide which set will react fastest. Consider the four factors that affect reaction rate.

Solve. All vessels are at the same temperature and no catalyst is present, so these two factors are not pertinent. The initial conditions do differ in physical state and concentration. Vessel 2 has the greatest concentration of reactants A and B, and they are in solution; vessel 2 has the fastest reaction. Vessel 1 has fewer reactants and they are in the solid state. The number of reactant collisions is much less than for mobile reactants in solution. Vessel 3 has reactants in solution, but the concentrations are four times less than in Vessel 2.

14.3   Chemical equation (d), B $\rightarrow$ 2A, is consistent with the data. The concentration of A increases with time, and concentration B decreases with time, so B must be a reactant and A must be a product. The ending concentration of A is approximately twice as large as the starting concentration of B, so mole ratio of A:B is 2:1. The reaction is B $\rightarrow$ 2A.

14.9   The reaction profile has a single high point (peak), so the reaction occurs in a single step. This step is necessarily the rate-determining step.

   (1)   Total potential energy of the reactants

   (2) E_a, activation energy of the reaction. This is the difference in energy between the potential energy of the activated complex (transition state) and the potential energy of the reactants.

   (3) ΔE, net energy change for the reaction. This is the difference in energy between the products and reactants. (Under appropriate conditions, this could also be ΔH.) For this reaction, the energy of products is lower than the energy of reactants, and the reaction releases energy to the surroundings.

   (4)   Total potential energy of the products.

14.12   (a) $NO_2 + F_2 \rightarrow NO_2F + F$

                $NO_2 + F \rightarrow NO_2F$

   (b) $2NO_2 + F_2 \rightarrow 2NO_2F$

   (c)   F is the intermediate, because it is produced and then consumed during the reaction.

   (d)   rate = $k[NO_2][F_2]$

179

14.16

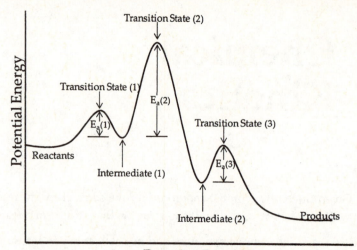

The reaction is exothermic because the energy of products is lower than the energy of reactants. The two intermediates are formed at different rates because $E_a(1) \neq E_a(2)$. In order to have two intermediates, the mechanism must have at least three steps.

Reaction Rates (sections 14.1 and 14.2)

14.17　(a)　*Reaction rate* is the change in the amount of products or reactants in a given amount of time; it is the speed of a chemical reaction.

(b)　Rates depend on concentration of reactants, physical state (or surface area) of reactants, temperature and presence of catalyst.

(c)　No, the rate of disappearance of reactants is not necessarily the same as the rate of appearance of products. The stoichiometry of the reaction (mole ratios of reactants and products) must be known in order to relate rate of disappearance of reactants to rate of appearance of products.

14.19　*Analyze/Plan.* Given mol A at a series of times in minutes, calculate mol B produced, molarity of A at each time, change in M of A at each 10 min interval, and ΔM A/s. For this reaction, mol B produced equals mol A consumed. M of A or [A] = mol A/0.100 L. The average rate of disappearance of A for each 10 minute interval is

$$-\frac{\Delta[A]}{s} = -\frac{[A]_1 - [A]_0}{10 \text{ min}} \times \frac{1 \text{ min}}{60 \text{ s}}$$

Solve.

Time (min)	Mol A	(a) Mol B	[A]	$\Delta[A]$	(b) Rate $-(\Delta[A]/s)$
0	0.065	0.000	0.65		
10	0.051	0.014	0.51	−0.14	2.3×10^{-4}
20	0.042	0.023	0.42	−0.09	2×10^{-4}
30	0.036	0.029	0.36	−0.06	1×10^{-4}
40	0.031	0.034	0.31	−0.05	0.8×10^{-4}

(c) $\dfrac{\Delta M_B}{\Delta t} = \dfrac{(0.029 - 0.014) \text{ mol}/0.100 \text{ L}}{(30 - 10) \text{ min}} \times \dfrac{1 \text{ min}}{60 \text{ s}} = 1.25 \times 10^{-4} = 1.3 \times 10^{-4} \ M/s$

14.21   (a)   *Analyze/Plan.* Follow the logic in Sample Exercises 14.1 and 14.2.   *Solve.*

Time (sec)	Time Interval (sec)	Concentration (M)	ΔM	Rate (M/s)
0		0.0165		
2,000	2,000	0.0110	−0.0055	28×10^{-7}
5,000	3,000	0.00591	−0.0051	17×10^{-7}
8,000	3,000	0.00314	−0.00277	9.23×10^{-7}
12,000	4,000	0.00137	−0.00177	4.43×10^{-7}
15,000	3,000	0.00074	−0.00063	2.1×10^{-7}

(b) $\dfrac{\Delta M_B}{\Delta t} = \dfrac{(0.0165 - 0.00074) \ M}{(15,000 - 0) \ s} = 1.0507 \times 10^{-6} = 1.05 \times 10^{-6} \ M/s$

(c)   From the slopes of the lines in the figure at right, the rates are:

at 5000 s,

$12 \times 10^{-7} \ M/s$;

at 8000 s,

$5.8 \times 10^{-7} \ M/s$

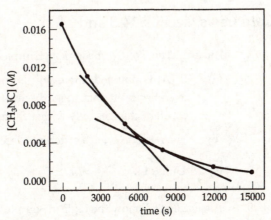

14.23   *Analyze/Plan.* Follow the logic in Sample Exercise 14.3.   *Solve.*

(a) $-\Delta[H_2O_2]/\Delta t = \Delta[H_2]/\Delta t = \Delta[O_2]/\Delta t$

(b) $-\Delta[N_2O]/2\Delta t = \Delta[N_2]/2\Delta t = \Delta[O_2]/\Delta t$

       $-\Delta[N_2O]/\Delta t = \Delta[N_2]/\Delta t = 2\Delta[O_2]/\Delta t$

(c) $-\Delta[N_2]/\Delta t = \Delta[NH_3]/2\Delta t; \ -\Delta[H_2]/3\Delta t = \Delta[NH_3]/2\Delta t$

       $-2\Delta[N_2]/\Delta t = \Delta[NH_3]/\Delta t; \ -\Delta[H_2]/\Delta t = 3\Delta[NH_3]/2\Delta t$

(d) $-\Delta[\ C_2H_5NH_2]/\Delta t = \Delta[\ C_2H_4]/\Delta t = \Delta[NH_3]/\Delta t$

14.24   (a)   rate $= -\Delta[H_2O]/2\Delta t = \Delta[H_2]/2\Delta t = \Delta[O_2]/\Delta t$

(b)   rate $= -\Delta[SO_2]/2\Delta t = -\Delta[O_2]/\Delta t = \Delta[SO_3]/2\Delta t$

(c)   rate $= -\Delta[NO]/2\Delta t = -\Delta[H_2]/2\Delta t = \Delta[N_2]/\Delta t = \Delta[H_2O]/2\Delta t$

(d)   rate $= -\Delta[N_2]/\Delta t = -\Delta[H_2]/2\Delta t = \Delta[N_2H_4]/\Delta t$

14.25   *Analyze/Plan.* Use Equation [14.4] to relate the rate of disappearance of reactants to the rate of appearance of products. Use this relationship to calculate desired quantities. *Solve.*

(a) $\Delta[H_2O]/2\Delta t = -\Delta[H_2]/2\Delta t = -\Delta[O_2]/\Delta t$

H_2 is burning, $-\Delta[H_2]/\Delta t = 0.48$ mol/s

O_2 is consumed, $-\Delta[O_2]/\Delta t = -\Delta[H_2]/2\Delta t = 0.48$ mol/s/2 = 0.24 mol/s

H_2O is produced, $+\Delta[H_2O]/\Delta t = -\Delta[H_2]/\Delta t = 0.48$ mol/s

(b)   The change in total pressure is the sum of the changes of each partial pressure. NO and Cl_2 are disappearing and NOCl is appearing.

$-\Delta P_{NO}/\Delta t = -56$ torr/min

$-\Delta P_{Cl_2}/\Delta t = \Delta P_{NO}/2\Delta t = -28$ torr/min

$+\Delta P_{NOCl}/\Delta t = -\Delta P_{NO}/\Delta t = +56$ torr/min

$\Delta P_T/\Delta t = -56$ torr/min $- 28$ torr/min $+ 56$ torr/min $= -28$ torr/min

Rate Laws (section 14.3 and 14.2)

14.27   *Analyze/Plan.* Follow the logic in Sample Exercises 14.4 and 14.5. *Solve.*

(a)   If [A] is doubled, there will be no change in the rate or the rate constant. The overall rate is unchanged because [A] does not appear in the rate law; the rate constant changes only with a change in temperature.

(b)   The reaction is zero order in A, second order in B and second order overall.

(c)   Units of $k = \dfrac{M/s}{M^2} = M^{-1}\,s^{-1}$

14.29   *Analyze/Plan.* Follow the logic in Sample Exercise 14.4.   *Solve.*

(a)   rate $= k[N_2O_5] = 4.82 \times 10^{-3}\,s^{-1}\,[N_2O_5]$

(b)   rate $= 4.82 \times 10^{-3}\,s^{-1}\,(0.0240\,M) = 1.16 \times 10^{-4}\,M/s$

(c)   rate $= 4.82 \times 10^{-3}\,s^{-1}\,(0.0480\,M) = 2.31 \times 10^{-4}\,M/s$

When the concentration of N_2O_5 doubles, the rate of the reaction doubles.

(d)   rate $= 4.82 \times 10^{-3}\,s^{-1}\,(0.0120\,M) = 5.78 \times 10^{-5}\,M/s$

When the concentration of N_2O_5 is halved, the rate of the reaction is halved.

14.31   *Analyze/Plan.* Write the rate law and rearrange to solve for k. Use the given data to calculate k, including units.   *Solve.*

(a, b) rate $= k[CH_3Br][OH^-]$; $k = \dfrac{\text{rate}}{[CH_3Br][OH^-]}$

at 298 K, $k = \dfrac{0.0432\,M/s}{(5.0 \times 10^{-3}\,M)(0.050\,M)} = 1.7 \times 10^2\,M^{-1}s^{-1}$

(c)   Since the rate law is first order in $[OH^-]$, if $[OH^-]$ is tripled, the rate triples.

(d) If [OH⁻] and [CH₃Br] both triple, the rate increases by a factor of (3)(3) = 9.

14.33 *Analyze/Plan.* Follow the logic in Sample Exercise 14.6.   *Solve.*

(a) From the data given, when [OCl⁻] doubles, rate doubles. When [I⁻] doubles, rate doubles. The reaction is first order in both [OCl⁻] and [I⁻]. rate = k[OCl⁻][I⁻]

(b) Using the first set of data:

$$k = \frac{rate}{[OCl^-][I^-]} = \frac{1.36 \times 10^{-4} \; M/s}{(1.5 \times 10^{-3} \; M)(1.5 \times 10^{-3} \; M)} = 60.444 = 60 \; M^{-1} \, s^{-1}$$

(c) $rate = \dfrac{60.444}{M \cdot s}(2.0 \times 10^{-3} \; M)(5.0 \times 10^{-4} \; M) = 6.0444 \times 10^{-5} = 6.0 \times 10^{-5} \, M/s$

14.35 *Analyze/Plan.* Follow the logic in Sample Exercise 14.6 to deduce the rate law. Rearrange the rate law to solve for k and deduce units. Calculate a k value for each set of concentrations and then average the three values.   *Solve.*

(a) Doubling [NH₃] while holding [BF₃] constant doubles the rate (experiments 1 and 2). Doubling [BF₃] while holding [NH₃] constant doubles the rate (experiments 4 and 5).

Thus, the reaction is first order in both BF₃ and NH₃; rate = k[BF₃][NH₃].

(b) The reaction is second order overall.

(c) From experiment 1: $k = \dfrac{0.2130 \; M/s}{(0.250 \; M)(0.250 \; M)} = 3.41 \; M^{-1} \, s^{-1}$

(Any of the five sets of initial concentrations and rates could be used to calculate the rate constant k. The average of these 5 values is $k_{avg} = 3.408 = 3.41 \; M^{-1}s^{-1}$)

(d) rate = 3.408 $M^{-1}s^{-1}$(0.100 M)(0.500 M) = 0.1704 = 0.170 M/s

14.37 *Analyze/Plan.* Follow the logic in Sample Exercise 4.6 to deduce the rate law. Rearrange the rate law to solve for k and deduce units. Calculate a k value for each set of concentrations and then average the three values.   *Solve.*

(a) Increasing [NO] by a factor of 2.5 while holding [Br₂] constant (experiments 1 and 2) increases the rate by a factor 6.25 or (2.5)². Increasing [Br₂] by a factor of 2.5 while holding [NO] constant increases the rate by a factor of 2.5. The rate law for the appearance of NOBr is: rate = Δ[NOBr]/Δt = k[NO]²[Br₂].

(b) From experiment 1: $k_1 = \dfrac{24 \; M/s}{(0.10 \; M)^2 \, (0.20 \; M)} = 1.20 \times 10^4 = 1.2 \times 10^4 \; M^{-2} \, s^{-1}$

$k_2 = 150/(0.25)^2(0.20) = 1.20 \times 10^4 = 1.2 \times 10^4 \, M^{-2} \, s^{-1}$

$k_3 = 60/(0.10)^2(0.50) = 1.20 \times 10^4 = 1.2 \times 10^4 \, M^{-2} \, s^{-1}$

$k_4 = 735/(0.35)^2(0.50) = 1.2 \times 10^4 = 1.2 \times 10^4 \, M^{-2} \, s^{-1}$

$k_{avg} = (1.2 \times 10^4 + 1.2 \times 10^4 + 1.2 \times 10^4 + 1.2 \times 10^4)/4 = 1.2 \times 10^4 \, M^{-2} \, s^{-1}$

(c) Use the reaction stoichiometry and Equation [14.4] to relate the designated rates. Δ[NOBr]/2Δt = –Δ[Br₂]/Δt; the rate of disappearance of Br₂ is half the rate of appearance of NOBr.

(d) Note that the data are given in terms of appearance of NOBr.

$$\frac{-\Delta[Br_2]}{\Delta t} = \frac{k[NO]^2[Br_2]}{2} = \frac{1.2 \times 10^4}{2 \, M^2 \, s} \times (0.075 \, M)^2 \times (0.250 \, M) = 8.4 \, M/s$$

Change of Concentration with Time (section 14.4)

14.39 (a) $[A]_0$ is the molar concentration of reactant A at time zero, the initial concentration of A. $[A]_t$ is the molar concentration of reactant A at time t. $t_{1/2}$ is the time required to reduce $[A]_0$ by a factor of 2, the time when $[A]_t = [A]_0/2$. k is the rate constant for a particular reaction. k is independent of reactant concentration but varies with reaction temperature.

(b) A graph of ln[A] vs time yields a straight line for a first-order reaction.

(c) On graph of ln[A] vs time, the rate constant, k, is the (–slope) of the straight line.

14.41 For a reaction A → B that is zero-order in A, rate = k. A plot of [A] vs time will be linear, with a slope of –k. (A plot of [B] vs time will also be linear, with a slope of +k.)

14.43 *Analyze/Plan.* The half-life of a first-order reaction depends only on the rate constant, $t_{1/2} = 0.693/k$. Use this relationship to calculate k for a given $t_{1/2}$, and, at a different temperature, $t_{1/2}$ given k. *Solve.*

(a) $t_{1/2} = 2.3 \times 10^5$ s; $t_{1/2} = 0.693/k$, $k = 0.693/t_{1/2}$

$k = 0.693/2.3 \times 10^5$ s $= 3.0 \times 10^{-6}$ s^{-1}

(b) $k = 2.2 \times 10^{-5}$ s^{-1}. $t_{1/2} = 0.693/2.2 \times 10^{-5}$ s$^{-1} = 3.15 \times 10^4 = 3.2 \times 10^4$ s

14.45 *Analyze/Plan.* Follow the logic in Sample Exercise 14.7. In this reaction, pressure is a measure of concentration. In (a) we are given k, $[A]_0$, t and asked to find $[A]_t$, using Equation [14.13], the integrated form of the first-order rate law. In (b), $[A_t] = 0.1[A_0]$, find t. *Solve.*

(a) $\ln P_t = -kt + \ln P_0$; $P_0 = 450$ torr; $t = 60$ s

$\ln P_{60} = -4.5 \times 10^{-2}$ s$^{-1}(60) + \ln(450) = -2.70 + 6.109 = 3.409$

$P_{60} = 30.24 = 30$ torr

(b) $P_t = 0.10 \, P_0$; $\ln(P_t/P_0) = -kt$

$\ln(0.10 \, P_0/P_0) = -kt$, $\ln(0.10) = -kt$; $-\ln(0.10)/k = t$

$t = -(-2.303)/4.5 \times 10^{-2}$ s$^{-1} = 51.2 = 51$ s

Check. From part (a), the pressure at 60 s is 30 torr, $P_t \sim 0.07 \, P_0$. In part (b) we calculate the time where $P_t = 0.10 \, P_0$ to be 51 s. This time should be smaller than 60 s, and it is. Data and results in the two parts are consistent.

14.47 *Analyze/Plan.* Given reaction order, various values for t and P_t, find the rate constant for the reaction at this temperature. For a first-order reaction, a graph of lnP vs t is linear with as slope of –k.   *Solve.*

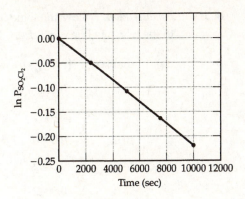

t(s)	$P_{SO_2Cl_2}$	$\ln P_{SO_2Cl_2}$
0	1.000	0
2500	0.947	–0.0545
5000	0.895	–0.111
7500	0.848	–0.165
10000	0.803	–0.219

Graph $\ln P_{SO_2Cl_2}$ vs. time. (Pressure is a satisfactory unit for a gas, since the concentration in moles/liter is proportional to P.) The graph is linear with slope $-2.19 \times 10^{-5}\ s^{-1}$ as shown on the figure. The rate constant k = –slope = $2.19 \times 10^{-5}\ s^{-1}$.

14.49 *Analyze/Plan.* Given: mol A, t. Change mol to *M* at various times. Make both first- and second-order plots to see which is linear.   *Solve.*

(a)

time(min)	mol A	[A] (*M*)	ln[A]	1/mol A
0	0.065	0.65	–0.43	1.5
10	0.051	0.51	–0.67	2.0
20	0.042	0.42	–0.87	2.4
30	0.036	0.36	–1.02	2.8
40	0.031	0.31	–1.17	3.2

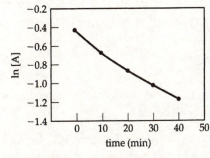 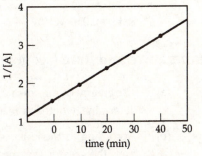

The plot of 1/[A] vs time is linear, so the reaction is second-order in [A].

(b) For a second-order reaction, a plot of 1/[A] vs. t is linear with slope k.

k = slope = $(3.2 - 2.0)\ M^{-1}$ / 30 min = $0.040\ M^{-1}\ min^{-1}$

(The best fit to the line yields slope = $0.042\ M^{-1}\ min^{-1}$.)

(c) $t_{1/2}$ = 1/k[A]$_0$ = 1/$(0.040\ M^{-1}\ min^{-1})(0.65\ M)$ = 38.46 = 38 min

(Using the "best-fit" slope, $t_{1/2}$ = 37 min.)

14.51   *Analyze/Plan.* Follow the logic in Solution 14.49. Make both first and second order plots to see which is linear.   *Solve.*

(a)

time(s)	[NO$_2$](M)	ln[NO$_2$]	1/[NO$_2$]
0.0	0.100	–2.303	10.0
5.0	0.017	–4.08	59
10.0	0.0090	–4.71	110
15.0	0.0062	–5.08	160
20.0	0.0047	–5.36	210

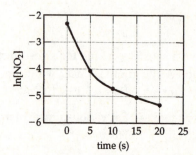
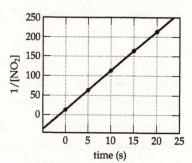

The plot of 1/[NO$_2$] vs time is linear, so the reaction is second order in NO$_2$.

(b)   The slope of the line is (210 – 59) M^{-1} / 15.0 s = 10.07 = 10 $M^{-1}s^{-1}$ = k. (The slope of the best-fit line is 10.02 = 10 $M^{-1}s^{-1}$.)

(c)   From the results above, the rate law is: rate = k[NO$_2$]^2 = 10 $M^{-1}s^{-1}$[NO$_2$]^2
Using the rate law, calculate the rate at each of the given initial concentrations.

Rate @ 0.200 M = 10 $M^{-1}s^{-1}$[NO$_2$]^2 = 10 $M^{-1}s^{-1}$[0.200 M]^2 = 0.400 M/s

Rate @ 0.100 M = 10 $M^{-1}s^{-1}$[NO$_2$]^2 = 10 $M^{-1}s^{-1}$[0.100 M]^2 = 0.100 M/s

Rate @ 0.050 M = 10 $M^{-1}s^{-1}$[NO$_2$]^2 = 10 $M^{-1}s^{-1}$[0.050 M]^2 = 0.025 M/s

Temperature and Rate (section 14.5)

14.53   (a)   The energy of the collision and the orientation of the molecules when they collide determine whether a reaction will occur.

(b)   According to the kinetic-molecular theory (Chapter 10), the higher the temperature, the greater the speed and kinetic energy of the molecules. Therefore, at a higher temperature, there are more total collisions and each collision is more energetic.

(c)   Assuming other conditions remain the same, the rate and therefore the rate constant usually increase with an increase in reaction temperature.

14.55   *Analyze/Plan.* Given the temperature and energy, use Equation [14.18] to calculate the fraction of Ar atoms that have at least this energy.   *Solve.*

$f = e^{-E_a/RT}$ $E_a = 10.0$ kJ/mol $= 1.00 \times 10^4$ J/mol; $T = 400$ K (127°C)

$$-E_a/RT = -\frac{1.00 \times 10^4 \text{ J/mol}}{400 \text{ K}} \times \frac{\text{mol - K}}{8.314 \text{ J}} = -3.0070 = -3.01$$

$f = e^{-3.0070} = 4.9 \times 10^{-2}$

At 400 K, approximately 1 out of 20 molecules has this kinetic energy.

14.57   *Analyze/Plan.* Use the definitions of activation energy ($E_{max} - E_{react}$) and ΔE ($E_{prod} - E_{react}$) to sketch the graph and calculate E_a for the reverse reaction.   *Solve.*

(a)   (b) E_a(reverse) = 73 kJ

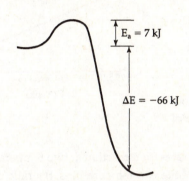

14.59   (a)   False. If you compare two reactions with similar collision factors, the one with the larger activation energy will be *slower*.

        (b)   False. A reaction that has a small rate constant will have either a small frequency factor (A), a large activation energy (E_a), or both.

        (c)   True.

14.61   Assuming all collision factors (A) to be the same, reaction rate depends only on E_a; it is independent of ΔE. Based on the magnitude of E_a, reaction (b) is fastest and reaction (c) is slowest.

14.63   *Analyze/Plan.* Given k_1, at T_1, calculate k_2 at T_2. Change T to Kelvins, then use the Equation [14.21] to calculate k_2.   *Solve.*

$T_1 = 20°C + 273 = 293$ K; $T_2 = 60°C + 273 = 333$ K; $k_1 = 2.75 \times 10^{-2} \text{s}^{-1}$

        (a) $\ln\left(\dfrac{k_1}{k_2}\right) = \dfrac{E_a}{R}\left(\dfrac{1}{333} - \dfrac{1}{293}\right) = \dfrac{75.5 \times 10^3 \text{ J/mol}}{8.314 \text{ J/mol}}(-4.1 \times 10^{-4})$

                $\ln(k_1/k_2) = -3.7229 = -3.7$; $k_1/k_2 = 0.0242 = 0.02$; $k_2 = \dfrac{0.0275 \text{ s}^{-1}}{0.0242} = 1.14 = 1 \text{ s}^{-1}$

        (b) $\ln\left(\dfrac{k_1}{k_2}\right) = \dfrac{125 \times 10^3 \text{ J/mol}}{8.314 \text{ J/mol}}\left(\dfrac{1}{333} - \dfrac{1}{293}\right) = -6.1638 = -6.2$

                $k_1/k_2 = 2.104 \times 10^{-3} = 2 \times 10^{-3}$; $k_2 = \dfrac{0.0275 \text{ s}^{-1}}{2.104 \times 10^{-3}} = 13.07 = 1 \times 10 \text{ s}^{-1}$

(c)   The method in parts (a) and (b) assumes that the collision model and thus the Arrhenious equation describe the kinetics of the reactions. That is, activation energy is constant over the temperature range under consideration. There is no assumption about temperature dependence of the frequency factor, because it drops out of the difference equation by subtraction.

14.65   *Analyze/Plan.* Follow the logic in Sample Exercise 14.11.   *Solve.*

k	ln k	T(K)	1/T(× 10³)
0.0521	–2.955	288	3.47
0.101	–2.293	298	3.36
0.184	–1.693	308	3.25
0.332	–1.103	318	3.14

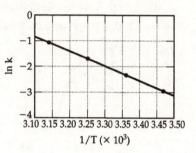

The slope, -5.71×10^3, equals $-E_a/R$. Thus,
$E_a = 5.71 \times 10^3 \times 8.314$ J/mol $= 47.5$ kJ/mol.

14.67   *Analyze/Plan.* Given E_a, find the ratio of rates for a reaction at two temperatures. Assuming initial concentrations are the same at the two temperatures, the ratio of rates will be the ratio of rate constants, k_1/k_2. Use Equation [14.21] to calculate this ratio.   *Solve.*

$T_1 = 50°C + 273 = 323$ K; $T_2 = 0°C + 273 = 273$ K

$$\ln\left(\frac{k_1}{k_2}\right) = \frac{E_a}{R}\left[\frac{1}{T_2} - \frac{1}{T_1}\right] = \frac{65.7 \text{ kJ/mol}}{8.314 \text{ J/mol}} \times \frac{1000 \text{ J}}{1 \text{ kJ}}\left[\frac{1}{273} - \frac{1}{323}\right]$$

$\ln(k_1/k_2) = 7.902 \times 10^3 (5.670 \times 10^{-4}) = 4.481 = 4.5$; $k_1/k_2 = 88.3 = 9 \times 10^1$

The reaction will occur 90 times faster at 50°C, assuming equal initial concentrations and Arrhenius kinetics.

Reaction Mechanisms (section 14.6)

14.69   (a)   An *elementary reaction* is a process that occurs in a single event; the order is given by the coefficients in the balanced equation for the reaction.

(b)   A *unimolecular* elementary reaction involves only one reactant molecule; the activated complex is derived from a single molecule. A *bimolecular* elementary reaction involves two reactant molecules in the activated complex and the overall process.

(c)   A *reaction mechanism* is a series of elementary reactions that describe how an overall reaction occurs and explain the experimentally determined rate law.

14.71   A *transition state* is a high energy complex formed when one or more reactants collide and distort in a way that can lead to formation of product(s). An *intermediate* is the product of an early elementary reaction in a multistep reaction mechanism. A transition state occurs at an an energy maximum or peak of a reaction profile such as Figure

14.20.   An intermediate exists at an energy minimum or trough of a reaction profile. Every reaction, single- or multi-step, has a transition state. Only multistep reactions have intermediates.

14.73   *Analyze/Plan.* Elementary reactions occur as a single step, so the molecularity is determined by the number of reactant molecules; the rate law reflects reactant stoichiometry. *Solve.*

   (a)   unimolecular, rate = $k[Cl_2]$

   (b)   bimolecular, rate = $k[OCl^-][H_2O]$

   (c)   bimolecular, rate = $k[NO][Cl_2]$

14.75   *Analyze/Plan.* Use the definitions of the terms 'intermediate' and 'exothermic', along with the characteristics of reaction profiles, to answer the questions.   *Solve.*

   This is a three-step mechanism, $A \rightarrow B$, $B \rightarrow C$, and $C \rightarrow D$.

   (a)   There are 2 intermediates, B and C.

   (b)   There are 3 energy maxima in the reaction profile, so there are 3 transition states.

   (c)   Step $C \rightarrow D$ has the lowest activation energy, so it is fastest.

   (d)   The energy of D is slightly greater than the energy of A, so the overall reaction is endothermic.

14.77   *Analyze/Plan.* Follow the logic in Sample Exercise 14.14.   *Solve.*

   (a) $H_2(g) + ICl(g) \rightarrow HI(g) + HCl(g)$

   $\underline{HI(g) + ICl(g) \rightarrow I_2(g) + HCl(g)}$

   $H_2(g) + 2ICl(g) \rightarrow I_2(g) + 2HCl(g)$

   (b)   Intermediates are produced and consumed during reaction. HI is the intermediate.

   (c)   The slow step determines the rate law for the overall reaction. If the first step is slow, the observed rate law is: rate = $k[H_2][ICl]$.

14.79   *Analyze.* Given data on concentration of a reactant vs time, determine whether the proposed reaction mechanism is consistent with the data. *Plan.* Based on the graph, decide the order of reaction with respect to [NO]. Write the two possible rate laws, depending on which step is rate-determining. Decide if one of the rate laws, and thus the mechanism, is consistent with the rate data. *Solve.*

   The graph of 1/[NO] vs time is linear with positive slope, indicating that the reaction is second order in [NO]. The rate law will include $[NO]^2$. We have no information about reaction order with respect to $[Cl_2]$.

   If the first step is slow, the observed rate law is the rate law for this step: rate = $k[NO][Cl_2]$. Since the observed rate law is second-order in [NO], the second step must be slow relative to the first step. Follow the logic in Sample Exercise 14.15 for determining the rate law of a mechanism with a fast initial step.

   From the rate-determining second step, rate = $k[NOCl_2][NO]$.

Assuming the first step is a fast equilibrium, $k_1[NO][Cl_2] = k_{-1}[NOCl_2]$.

Solving for $[NOCl_2]$ in terms of $[NO][Cl_2]$, $[NOCl_2] = \dfrac{k_1}{k_{-1}}[NO][Cl_2]$

$$\text{rate} = \frac{k_2 k_1}{k_{-1}}[NO][Cl_2][NO] = [NO]^2[Cl_2]$$

This rate law is second order in [NO]. It is consistent with the observed data.

[The graph of 1/[NO] vs time is linear with positive slope, indicating that the reaction is second order in [NO]. The rate law obtained by assuming the second step is rate determining is: rate = $[NO]^2[Cl_2]$. The two step mechanism is consistent with the data.]

Catalysis (section 14.7)

14.81   (a)   A catalyst is a substance that changes (usually increases) the speed of a chemical reaction without undergoing a permanent chemical change itself.

   (b)   A homogeneous catalyst is in the same phase as the reactants; a heterogeneous catalyst is in a different phase and is usually a solid.

   (c)   A catalyst has no effect on the overall enthalpy change for a reaction. A catalyst does affect activation energy, E_a, which is one way that it changes reaction rate. It can also affect the frequency factor, A.

14.83   *Analyze/Plan.* Use the structure and unit cell edge of Pt, along with the formulas for volume and surface area of a sphere, to calculate the number of Pt atoms in a 2-nm sphere and on the surface of a 2-nm sphere.

   (a)   For a Pt sphere with a 2.0 nm diameter, radius = 1.0 nm.

$$V = 4/3\pi r^3 = \frac{4\pi(1.0\,\text{nm})^3}{3} \times \frac{10^3\,\text{Å}^3}{1^3\,\text{nm}^3} = 4.188879 \times 10^3 = 4.2 \times 10^3\,\text{Å}^3$$

In a face-centered cubic metal structure, there are 4 metal atoms per unit cell. The volume of the unit cell is $(3.924\,\text{Å})^3 = 60.42\,\text{Å}^3$

$$\frac{4\,\text{Pt atoms}}{60.42\,\text{Å}^3} \times 4.1889 \times 10^3\,\text{Å}^3 = 277.3 = 2.8 \times 10^2\,\text{Pt atoms in a 2.0 - nm sphere}$$

   (b)   Assume that the "footprint" of an atom is its cross-sectional area, the area of a circle with the radius of the atom. The area of this circle is πr^2. The diameter, d, of a Pt atom is 2.8 Å, so r = d/2 = 1.4 Å. The footprint of the Pt atom is then

$\pi(1.4\,\text{Å})^2 = 6.1575 = 6.2\,\text{Å}^2$

The surface area of the 2.0-nm sphere is

$$4\pi r^2 = 4\pi(1.0\,\text{nm})^2 \times \frac{10^2\,\text{Å}^2}{1^2\,\text{nm}^2} = 12.56637 \times 10^2 = 1.3 \times 10^3\,\text{Å}^2$$

$$\frac{1\,\text{Pt atoms}}{6.1575\,\text{Å}^2} \times 1.2566 \times 10^3\,\text{Å}^2 = 204.1 = 2.0 \times 10^2\,\text{surface Pt atoms on a 2.0 - nm sphere.}$$

   (c) $\dfrac{204\,\text{surface Pt atoms}}{277\,\text{total Pt atoms}} \times 100 = 74\%\,\text{Pt atoms on the surface}$

(d)   For a 5.0-nm Pt sphere, radius = 2.5 nm

$$V = 4/3\pi r^3 = \frac{4\pi (2.50 \text{ nm})^3}{3} \times \frac{10^3 \text{ Å}^3}{1^3 \text{ nm}^3} = 65.4498 \times 10^3 = 6.5 \times 10^4 \text{ Å}^3$$

$$\frac{4 \text{ Pt atoms}}{60.42 \text{ Å}^3} \times 65.4498 \times 10^3 \text{ Å}^3 = 4,333 = 4.3 \times 10^3 \text{ Pt atoms in a } 5.0 \text{ - nm sphere}$$

The surface area of the 5.0-nm sphere is

$$4\pi r^2 = 4\pi (2.5 \text{ nm})^2 \times \frac{10^2 \text{ Å}^2}{1^2 \text{ nm}^2} = 7853.98 = 7.9 \times 10^3 \text{ Å}^2$$

$$\frac{1 \text{ Pt atoms}}{6.1575 \text{ Å}^2} \times 7.854 \times 10^3 \text{ Å}^2 = 1275.5 = 1.3 \times 10^3 \text{ surface Pt atoms on a } 5.0 \text{ - nm sphere}$$

$$\frac{1276 \text{ surface Pt atoms}}{4333 \text{ total Pt atoms}} \times 100 = 29\% \text{ Pt atoms on the surface}$$

The calculations in parts (b) and (d) overestimate the number of Pt atoms on the surface of the sphere, because they do not account for empty space between atoms. For the purpose of comparison, it is most important that we use the same method for both spheres.

[Alternatively, use one face of a face-entered cubic unit cell as a model for the surface area that Pt atoms will occupy. On a face, there is the cross-section of 1 Pt atoms in the center and ¼ Pt atom at each corner. This amounts to the cross-sections of two Pt atoms in $(3.924 \text{ Å})^2 = 15.398 = 15.4 \text{ Å}^2$

$$\frac{2 \text{ Pt atoms}}{15.40 \text{ Å}^3} \times 1.2566 \times 10^3 \text{ Å}^2 = 163.2 = 1.6 \times 10^2 \text{ surface Pt atoms on a } 2.0 \text{ - nm sphere.}$$

$$\frac{163 \text{ surface Pt atoms}}{277 \text{ total Pt atoms}} \times 100 = 59\% \text{ Pt atoms on the surface}$$

Similarly, in a 5.0-nm sphere, there are 1.0×10^3 Pt atoms on the surface, 24% of the total Pt atoms.]

(e)   Both surface models predict that the 2.0-nm sphere will be more catalytically active, because it has a much greater percentage of its atoms on the surface, where they can participate in the chemical reaction.

14.85   (a)

$$2[NO_2(g) + SO_2(g) \rightarrow NO(g) + SO_3(g)]$$
$$2NO(g) + O_2(g) \rightarrow 2NO_2(g)$$
$$\overline{}$$
$$2SO_2(g) + O_2(g) \rightarrow 2SO_3(g)$$

(b) $NO_2(g)$ is a catalyst because it is consumed and then reproduced in the reaction sequence. (NO(g) is an intermediate because it is produced and then consumed.)

(c)   Since NO_2 is in the same state as the other reactants, this is homogeneous catalysis.

14.87 (a) When using a powdered metal catalyst, only a small percentage of the metal atoms are at the surface of the bulk material and catalytically active. Use of chemically stable supports such as alumina and silica makes it possible to obtain very large surface areas per unit mass of the precious metal catalyst. This is so because the metal can be deposited in a very thin, even monomolecular, layer on the surface of the support.

(b) The greater the surface area of the catalyst, the more reaction sites, and the greater the rate of the catalyzed reaction.

14.89 As illustrated in Figure 14.24, the two C–H bonds that exist on each carbon of the ethylene molecule before adsorption are retained in the process in which a D atom is added to each C (assuming we use D_2 rather than H_2). To put two deuteriums on a single carbon, it is necessary that one of the already existing C–H bonds in ethylene be broken while the molecule is adsorbed, so the H atom moves off as an adsorbed atom, and is replaced by a D. This requires a larger activation energy than simply adsorbing C_2H_4 and adding one D atom to each carbon.

14.91 (a) Living organisms operate efficiently in a very narrow temperature range; heating to increase reaction rate is not an option. Therefore, the role of enzymes as homogeneous catalysts that speed up desirable reactions without heating and undesirable side-effects is crucial for biological systems.

(b) *catalase*: $2H_2O_2 \rightarrow 2H_2O + O_2$; *nitrogenase*: $N_2 \rightarrow 2NH_3$ (nitrogen fixation)

(c) This model for enzyme kinetics is similar to the mechanism detailed in Sample Exercise 14.15, where step one is a fast equilibrium and step two is slow and rate-determining. This model assumes that the rate of the bound substrate, ES, being chemically transformed into bound product is slow and rate-determining.

14.93 Let k and E_a equal the rate constant and activation energy without the enzyme (uncatalyzed). Let k_c and E_{ac} equal the rate constant and activation energy with the enzyme (catalyzed). A is the same for the uncatalyzed and catalyzed reactions. The difference in activation energies is $E_{ac} - E_a$, $k_c = 1.0 \times 10^6\,s^{-1}$, $k = 0.039\,s^{-1}$, T = 25°C = 298 K.

According to Equation [14.20], ln k = E_a/RT + ln A. Subtracting ln k from ln k_c

$$\ln k_c - \ln k = \left[\frac{-E_{ac}}{RT}\right] + \ln A - \left[\frac{-E_a}{RT}\right] - \ln A$$

$$\ln (k_c/k) = \frac{E_a - E_{ac}}{RT}; \quad E_a - E_{ac} = RT \ln (k_c/k)$$

$$E_a - E_{ac} = \frac{8.314\,J}{mol\text{-}K} \times 298\,K \times \ln\frac{1.0 \times 10^6}{0.039} = 42{,}267\,J = 42.267\,kJ = 42\,kJ$$

Carbonic anyhdrase lowers the activation energy of the reaction by 42 kJ.

14.95 *Analyze/Plan.* Let k = the rate constant for the uncatalyzed reaction,
k_c = the rate constant for the catalyzed reaction

According to Equation [14.20], ln k = $-E_a/RT$ + ln A

Subtracting ln k from ln k_c,

$$\ln k_c - \ln k = -\left[\frac{55 \text{ kJ/mol}}{RT} + \ln A\right] - \left[-\frac{95 \text{ kJ/mol}}{RT} + \ln A\right]. \quad \textit{Solve.}$$

(a) $RT = 8.314 \text{ J/mol-K} \times 298 \text{ K} \times 1 \text{ kJ/1000 J} = 2.478 \text{ kJ/mol}$; ln A is the same for both reactions.

$$\ln (k_c / k) = \frac{95 \text{ kJ/mol} - 55 \text{ kJ/mol}}{2.478 \text{ kJ/mol}}; \quad k_c / k = 1.024 \times 10^7 = 1 \times 10^7$$

The catalyzed reaction is approximately 10,000,000 (ten million) times faster at 25°C.

(b) $RT = 8.314 \text{ J/mol-K} \times 398 \text{ K} \times 1 \text{ kJ/1000 J} = 3.309 \text{ kJ/mol}$

$$\ln (k_c / k) = \frac{40 \text{ kJ/mol}}{3.309 \text{ kJ/mol}}; \quad k_c / k = 1.778 \times 10^5 = 2 \times 10^5$$

The catalyzed reaction is 200,000 times faster at 125°C.

Additional Exercises

14.99   (a) $\text{rate} = \dfrac{-\Delta[\text{NO}]}{2\Delta t} = \dfrac{-\Delta[\text{O}_2]}{\Delta t} = \dfrac{9.3 \times 10^{-5} \, M/s}{2} = 4.7 \times 10^{-5} \, M/s$

(b,c) $\text{rate} = k[\text{NO}]^2[\text{O}_2]; \quad k = \text{rate}/[\text{NO}]^2[\text{O}_2]$

$$k = \frac{4.7 \times 10^{-5} \, M/s}{(0.040 \, M)^2 (0.035 \, M)} = 0.8393 = 0.84 \, M^{-2} \, s^{-1}$$

(d)   Since the reaction is second order in NO, if the [NO] is increased by a factor of 1.8, the rate would increase by a factor of 1.8^2, or $(3.24) = 3.2$.

14.101   The units of rate are M/s. The reaction must be second order overall if the units of the rate constant are $M^{-1} \, s^{-1}$. If rate = $k[\text{NO}_2]^x$, then the cumulative units of $[\text{NO}_2]^x$ must be M^2, and x = 2.

If $[\text{NO}_2]_0 = 0.100 \, M$ and $[\text{NO}_2]_t = 0.025 \, M$, use the integrated form of the second order rate equation, $\dfrac{1}{[A]_t} = kt + \dfrac{1}{[A]_0}$, Equation [14.14], to solve for t.

$$\frac{1}{0.025 \, M} = 0.63 \, M^{-1}s^{-1} \, (t) + \frac{1}{0.100 \, M}; \quad \frac{(40 - 10) \, M^{-1}}{0.63 \, M^{-1}s^{-1}} = t = 47.62 = 48 \text{ s}.$$

14.105   *Analyze.* Given rate constants for the decay of two radioisotopes, determine half-lives, decay rates, and amount remaining after three half-lives. *Plan.* Determine reaction order.   Based on reaction-order, select the appropriate relationships for (a) rate constant and half-life and (c) rate-constant, time and concentration.   In this example, mass is a measure of concentration.

Solve. Decay of radioiosotopes is a first-order process, since only one species is involved and the decay is not initiated by collision.

(a)   For a first-order process, $t_{1/2} = 0.693/k$.

^{241}Am: $t_{1/2} = 0.693/1.6 \times 10^{-3} \text{ yr}^{-1} = 433.1 = 4.3 \times 10^2$ yr

^{125}I: $t_{1/2} = 0.693/0.011 \text{ day}^{-1} = 63.00 = 63$ days

(b)   For a given sample size, half of the ^{241}Am sample decays in 433 years, whereas half of the ^{125}I sample decays in 63 days. ^{125}I decays at a much faster rate.

(c)   For a first order process, $\ln[A]_t - \ln[A]_0 = -kt$. $\ln[A]_t = -kt + \ln[A]_0$.

$[A]_0 = 1.0$ mg; $t = 3\, t_{1/2}$.

^{241}Am: $t = 3\, t_{1/2} = 3(433.1 \text{ yr}) = 1.299 \times 10^3 = 1.3 \times 10^3$ yr

$\ln[\text{Am}]_t = -1.6 \times 10^{-3} \text{ yr}^{-1} (1.299 \times 10^3 \text{ yr}) - \ln(1.0) = -2.079 - 0 = -2.08$

$[\text{Am}]_t = 0.125 = 0.13$ mg

or, mass ^{241}Am remaining $= 1.0 \text{ mg}/2^3 = 0.125 = 0.13$ mg

^{125}I: For the same size starting sample and number of elapsed half-lives, the same mass, 0.13 mg ^{125}I, will remain. (The difference is that the elapsed time of 3 half-lives (for ^{125}I is 3(63) = 189 days = 0.52 yr, vs. 433 yr for ^{241}Am.)

(d)   Again, for a first order process, $\ln[A]_t - \ln[A]_0 = -kt$. $\ln[A]_t = -kt + \ln[A]_0$.

$[A]_0 = 1.0$ mg; $t = 4$ days.

$k_{AM} = 1.6 \times 10^{-3} \text{ yr}^{-1} (1 \text{ yr}/365 \text{ days}) = 4.3836 \times 10^{-6} = 4.4 \times 10^{-6} \text{ day}^{-1}$

^{241}Am: $\ln[\text{Am}]_t = -4.4 \times 10^{-6} \text{ day}^{-1} (4 \text{ days}) - \ln(1.0) = -1.7 \times 10^{-5} - 0 = -1.7 \times 10^{-5}$

The amount of ^{241}Am remaining after 4 days is 0.99998 mg, to three significant figures, 1.00 mg.

^{125}I: $\ln[\text{I}]_t = -0.011 \text{ day}^{-1} (4 \text{ days}) - \ln(1.0) = -0.044 - 0 = -0.044$

The amount of ^{125}I remaining after 4 days is 0.957 grams.

14.109

Time (s)	$[C_5H_6]$ (M)	$\ln[C_5H_6]$	$1/[C_5H_6]$
0	0.0400	−3.219	25.0
50	0.0300	−3.507	33.3
100	0.0240	−3.730	41.7
150	0.0200	−3.912	50.0
200	0.0174	−4.051	57.5

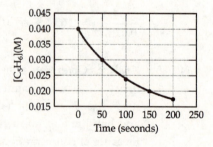

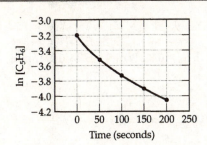 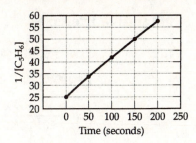

The plot of $1/[C_5H_6]$ vs time is linear and the reaction is second order.

The slope of this line is k. k = slope = $(50.0 - 25.0) M^{-1}/(150-0)s = 0.167 M^{-1} s^{-1}$

(The best-fit slope and k value is $0.163 M^{-1} s^{-1}$.)

14.112   (a) $NO(g) + NO(g) \rightarrow N_2O_2(g)$

                        $N_2O_2(g) + H_2(g) \rightarrow N_2O(g) + H_2O(g)$

             $\overline{2NO(g) + N_2O_2(g) + H_2(g) \rightarrow N_2O_2(g) + N_2O(g) + H_2O(g)}$

                     $2NO(g) + H_2(g) \rightarrow N_2O(g) + H_2O(g)$

   (b)   First reaction: $-\Delta[NO]/\Delta t = k[NO][NO] = k[NO]^2$

           Second reaction: $-\Delta[H_2]/\Delta t = k[H_2][N_2O_2]$

   (c) N_2O_2 is the intermediate: it is produced in the first step and consumed in the second.

   (d)   Since $[H_2]$ appears in the rate law, the second step must be slow relative to the first.

14.115   (a) $Cl_2(g) \;\rightleftharpoons\; 2Cl(g)$

                 $Cl(g) + CHCl_3(g) \rightarrow HCl(g) + CCl_3(g)$

                 $Cl(g) + CCl_3(g) \rightarrow CCl_4(g)$

       $\overline{Cl_2(g) + 2Cl(g) + CHCl_3(g) + CCl_3(g) \rightarrow 2Cl(g) + HCl(g) + CCl_3(g) + CCl_4(g)}$

               $Cl_2(g) + CHCl_3(g) \rightarrow HCl(g) + CCl_4(g)$

   (b) $Cl(g)$, $CCl_3(g)$

   (c)   Reaction 1 - unimolecular, Reaction 2 - bimolecular, Reaction 3 - bimolecular

   (d)   Reaction 2, the slow step, is rate determining.

   (e)   If Reaction 2 is rate determining, rate = $k_2[CHCl_3][Cl]$. Cl is an intermediate formed in reaction 1, an equilibrium. By definition, the rates of the forward and reverse processes are equal; $k_1[Cl_2] = k_{-1}[Cl]^2$. Solving for [Cl] in terms of $[Cl_2]$,

$$[Cl]^2 = \frac{k_1}{k_{-1}}[Cl_2]; \quad [Cl] = \left(\frac{k_1}{k_{-1}}[Cl_2]\right)^{1/2}$$

Substituting into the overall rate law

$$rate = k_2\left(\frac{k_1}{k_{-1}}\right)^{1/2}[CHCl_3][Cl_2]^{1/2} = k[CHCl_3][Cl_2]^{1/2} \text{ (The overall order is 3/2.)}$$

Integrative Exercises

14.122   (a) $\ln k = -E_a/RT + \ln A$, Equation [14.20]. $E_a = 6.3$ kJ/mol $= 6.3 \times 10^3$ J/mol

           $T = 100°C + 273 = 373$ K

$$\ln k = \frac{-6.3 \times 10^3 \text{ J/mol}}{8.314 \text{ J/K - mol} \times 373 \text{ K}} + \ln(6.0 \times 10^8 \ M^{-1}s^{-1})$$

           $\ln k = -2.032 + 20.212 = 18.181 = 18.2;\ k = 7.87 \times 10^7 = 8 \times 10^7 \ M^{-1}s^{-1}$

   (b)   NO, 11 valence e^-, 5.5 e^- pair   ONF, 18 valence e^-, 9 e^- pr
(Assume the less electronegative N atom will be electron deficient.)

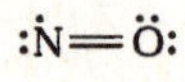

                    $:\ddot{O}=\ddot{N}-\ddot{F}: \longleftrightarrow \left(:\ddot{O}-\ddot{N}=\ddot{F}\right)$

            $:\dot{N}=\ddot{O}:$ The resonance form on the right is a very minor contributor to the true bonding picture, due to high formal charges and the unlikely double bond involving F.

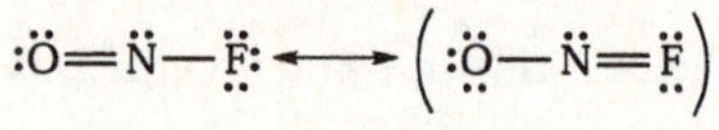

   (c)   ONF has trigonal planar electron domain geometry, which leads to a "bent" structure with a bond angle of approximately 120°.

   (d) $\left[\begin{array}{c} O=N \\ \ \ \ F \dashv F \end{array}\right]$

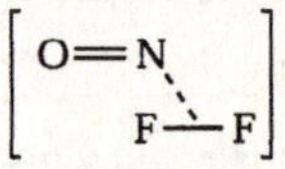

   (e)   The electron deficient NO molecule is attracted to electron-rich F_2, so the driving force for formation of the transition state is greater than simple random collisions.

15 Chemical Equilibrium

Visualizing Concepts

15.1 (a) $k_f > k_r$. According to the Arrhenius equation [14.19], $k = Ae^{-E_a/RT}$. As the magnitude of E_a increases, k decreases. On the energy profile, E_a is the difference in energy between the starting point and the energy at the top of the barrier. Clearly this difference is smaller for the forward reaction, so $k_f > k_r$.

(b) From the Equation [15.5], the equilibrium constant = k_f/k_r. Since $k_f > k_r$, the equilibrium constant for the process shown in the energy profile is greater than 1.

15.7 *Analyze/Plan.* The reaction with the largest equilibrium constant has the largest ratio of products to reactants. Count product and reactant molecules. Calculate ratios and compare. *Solve.*

$K = \dfrac{[C_2H_4X_2]}{[C_2H_4][X_2]}$. Use numbers of molecules as an adequate measure of concentration.

(While the volume terms don't cancel, they are the same for all parts. For the purpose of comparison, we can ignore volume.) *Solve.*

(a) 8 $C_2H_4Cl_2$, 2 Cl_2, 2 C_2H_4. $K = \dfrac{8}{(2)(2)} = 2$

(b) 6 $C_2H_4Br_2$, 4 Br_2, 4 C_2H_4. $K = \dfrac{6}{(4)(4)} = 0.375 = 0.4$

(c) 3 $C_2H_4I_2$, 7 I_2, 7 C_2H_4. $K = \dfrac{3}{(7)(7)} = 0.0612 = 0.06$

From the smallest to the largest equilibrium constant, (c) < (b) < (a).

Check. By inspection, there are the fewest product molecules and the most reactant molecules in (c); most product and least reactant in (a).

15.11 If temperature increases, K of an endothermic reaction increases and K of an exothermic reaction decreases. Calculate the value of K for the two temperatures and compare. For this reaction, $\Delta n = 0$ and $K_p = K_c$. We can ignore volume and use number of particles as a measure of moles and molarity. $K_c = [A][AB]/[A_2][B]$

(1) 300 K, 3A, 5AB, 1A_2, 1B; $K_c = (3)(5)/(1)(1) = 15$

(2) 500 K, 1A, 3AB, 3A_2, 3B; $K_c = (1)(3)/(3)(3) = 0.33$

K_c decreases as T increases, so the reaction is exothermic.

Equilibrium; The Equilibrium Constant (sections 15.1 – 15.4)

15.13   *Analyze/Plan.* Given the forward and reverse rate constants, calculate the equilibrium constant using Equation [15.5]. At equilibrium, the rates of the forward and reverse reactions are equal. Write the rate laws for the forward and reverse reactions and use their equality to answer part (b).   *Solve.*

(a) $K_c = \dfrac{k_f}{k_r}$, Equation [15.5]; $K_c = \dfrac{4.7 \times 10^{-3} \text{ s}^{-1}}{5.8 \times 10^{-1} \text{ s}^{-1}} = 8.1 \times 10^{-3}$

For this reaction, $K_p = K_c = 8.1 \times 10^{-3}$

(b) $\text{rate}_f = \text{rate}_r$; $k_f[A] = k_r[B]$

Since $k_f < k_r$, in order for the two rates to be equal, [A] must be greater than [B] and the partial pressure of A is greater than the partial pressure of B.

15.15   *Analyze/Plan.* Follow the logic in Sample Exercises 15.1 and 15.6.   *Solve.*

(a) $K_c = \dfrac{[N_2O][NO_2]}{[NO]^3}$ (b) $K_c = \dfrac{[CS_2][H_2]^4}{[CH_4][H_2S]^2}$

(c) $K_c = \dfrac{[CO]^4}{[Ni(CO)_4]}$ (d) $K_c = \dfrac{[H^+][F^-]}{[HF]}$

(e) $K_c = \dfrac{[Ag^+]^2}{[Zn^{2+}]}$ (f) $K_c = [H^+][OH^-]$

(g) $K_c = [H^+][OH^-]$

homogeneous: (a), (b), (d), (f), (g); heterogeneous: (c), (e)

15.17   *Analyze.* Given the value of K_c or K_p, predict the contents of the equilibrium mixture.

Plan. If K_c or $K_p \gg 1$, products dominate; if K_c or $K_p \ll 1$, reactants dominate. *Solve.*

(a)   mostly reactants ($K_c \ll 1$)

(b)   mostly products ($K_p \gg 1$)

15.19   No, the equilibrium constant can never be a negative number. The equilibrium constant is a ratio of concentrations of products to concentration of reactants, or of partial pressures of reactants to partial pressures of products. Concentrations and partial pressures are never negative, and neither is K. (Or, K is a ratio of rate constants, which also cannot have negative values.)

15.21   *Analyze/Plan.* Follow the logic in Sample Exercise 15.2.   *Solve.*

$PCl_3(g) + Cl_2(g) \rightleftharpoons PCl_5(g)$, $K_c = 0.042$. $\Delta n = 1 - 2 = -1$

$K_p = K_c(RT)^{\Delta n} = 0.042(RT)^{-1} = 0.042/RT$

$K_p = \dfrac{0.042}{(0.08206)(500)} = 0.001024 = 1.0 \times 10^{-3}$

15.23   *Analyze.* Given K_c for a chemical reaction, calculate K_c for the reverse reaction.

Plan. Evaluate which species are favored by examining the magnitude of K_c. The equilibrium expressions for the reaction and its reverse are the reciprocals of each other, and the values of K_c are also reciprocal.   *Solve.*

(a) For the reaction as written, $K_c < 1$, which means that reactants are favored. At this temperature, the equilibrium favors NO and Br_2.

(b) $K_c(\text{forward}) = \dfrac{[NOBr]^2}{[NO]^2[Br_2]} = 1.3 \times 10^{-2}$

$K_c(\text{reverse}) = \dfrac{[NO]^2[Br_2]}{[NOBr]^2} = \dfrac{1}{1.3 \times 10^{-2}} = 76.92 = 77$

(c) $K_{c2}(\text{reverse}) = \dfrac{[NO][Br_2]^{1/2}}{[NOBr]} = (K_c(\text{reverse}))^{1/2} = (76.92)^{1/2} = 8.8$

15.25   *Analyze.* Given K_p for a reaction, calculate K_p for a related reaction.

Plan. The algebraic relationship between the K_p values is the same as the algebraic relationship between equilibrium expressions.

Solve. $K_p = \dfrac{P_{SO_3}}{P_{SO_2} \times P_{O_2}^{1/2}} = 1.85$

(a) $K_p = \dfrac{P_{SO_2} \times P_{O_2}^{1/2}}{P_{SO_3}} = \dfrac{1}{1.85} = 0.541$

(b) $K_p = \dfrac{P_{SO_3}^2}{P_{SO_2}^2 \times P_{O_2}} = (1.85)^2 = 3.4225 = 3.42$

(c) $K_p = K_c(RT)^{\Delta n}$; $\Delta n = 2 - 3 = -1$; $T = 1000$ K

$K_p = K_c(RT)^{-1} = K_c/RT$; $K_c = K_p(RT)$

$K_c = 3.4225(0.08206)(1000) = 280.85 = 281$

15.27   *Analyze/Plan.* Follow the logic in Sample Exercise 15.5.   *Solve.*

$$CoO(s) + H_2(g) \rightleftharpoons Co(s) + H_2O(g) \qquad\qquad K_1 = 67$$

$$Co(s) + CO_2(g) \rightleftharpoons CoO(s) + CO(g) \qquad\qquad K_2 = 1/490$$

$$CoO(s) + H_2(g) + Co(s) + CO_2(g) \rightleftharpoons Co(s) + H_2O(g) + CoO(s) + CO(g)$$

$$H_2(g) + CO_2(g) \rightleftharpoons H_2O(g) + CO(g)$$

$$K_c = K_1 \times K_2 = 67 \times \dfrac{1}{490} = 0.1367 = 0.14$$

15.29   Pure solids and liquids are normally excluded from equilibrium-constant expressions because their concentrations are constant. Molar concentration is the ratio of moles of substance to volume occupied by substance. For a pure solid or liquid, if moles increase, volume occupied increases (and vice versa), so the concentration remains the same.

An alternate explanation involves the use of activities to express amounts of the various reactants and products in an equilibrium mixture. By definition, the activitiy of a pure solid or liquid is one, and it need not appear in the equilibrium-constant expression.

15.31   *Analyze/Plan.* Follow the logic in Sample Exercise 15.6.   *Solve.*

(a) $K_p = P_{O_2}$

(b) $K_c = [Hg(solv)]^4[O_2(solv)]$

Calculating Equilibrium Constants (section 15.5)

15.33   *Analyze/Plan.* Calculate molarity of reactants and products. Follow the logic in Sample Exercise 15.8 using concentrations rather than pressures.   *Solve.*

$$[CH_3OH] = \frac{0.0406 \text{ mol}}{2.00 \text{ L}} = 0.0203 \text{ M}$$

$$[CO] = \frac{0.170 \text{ mol CO}}{2.00 \text{ L}} = 0.0850 \text{ M}; \quad [H_2] = \frac{0.302 \text{ mol H}_2}{2.00 \text{ L}} = 0.151 \text{ M}$$

$$K_c = \frac{[CH_3OH]}{[CO][H_2]^2} = \frac{0.0203}{(0.0850)(0.151)^2} = 10.4743 = 10.5$$

15.35   *Analyze/Plan.* Follow the logic in Sample Exercise 15.8.   *Solve.*

(a) $2NO(g) + Cl_2(g) \rightleftharpoons 2NOCl(g)$

$$K_p = \frac{P_{NOCl}^2}{P_{NO}^2 \times P_{Cl_2}} = \frac{(0.28)^2}{(0.095)^2(0.171)} = 50.80 = 51$$

(b) $K_p = K_c(RT)^{\Delta n}$; $\Delta n = 2 - 3 = -1$; $K_p = K_c(RT)^{-1} = K_c/(RT)$

$K_c = K_p(RT) = 50.80(0.08206 \times 500) = 2.1 \times 10^3$

15.37   *Analyze/Plan.* Follow the logic in Sample Exercise 15.9. Since the container volume is 1.0 L, mol = *M*.   *Solve.*

(a)   First calculate the change in [NO], $0.062 - 0.10 = -0.038 = -0.04$ *M*. From the stoichiometry of the reaction, calculate the changes in the other pressures. Finally, calculate the equilibrium pressures.

	$2NO(g)$ +	$2H_2(g)$ $\rightleftharpoons$	$N_2(g)$ +	$2H_2O(g)$
initial	0.10 *M*	0.050 *M*	0 *M*	0.10 *M*
change	−0.038 *M*	−0.038 *M*	+0.019 *M*	+0.038 *M*
equil.	0.062 *M*	0.012 *M*	0.019 *M*	0.138 *M*

Strictly speaking, the change in [NO] has two decimal places and thus one sig fig. This limits equilibrium pressures to one sig fig for all but H₂O, and K_c to one sig fig. We compute the extra figures and then round.

(b) $K_c = \frac{[N_2][H_2O]^2}{[NO]^2[H_2]^2} = \frac{(0.019)(0.138)^2}{(0.062)^2(0.012)^2} = \frac{(0.02)(0.14)^2}{(0.06)^2(0.01)^2} = 653.7 = 7 \times 10^2$

15.39   *Analyze/Plan.* Follow the logic in Sample Exercise 15.9, using partial pressures, rather than concentrations.   *Solve.*

(a) $P = nRT/V; P_{CO_2} = 0.2000 \text{ mol} \times \dfrac{500 \text{ K}}{2.000 \text{ L}} \times \dfrac{0.08206 \text{ L} \cdot \text{atm}}{\text{mol} \cdot \text{K}} = 4.1030 = 4.10 \text{ atm}$

$P_{H_2} = 0.1000 \text{ mol} \times \dfrac{500 \text{ K}}{2.000 \text{ L}} \times \dfrac{0.08206 \text{ L} \cdot \text{atm}}{\text{mol} \cdot \text{K}} = 2.0515 = 2.05 \text{ atm}$

$P_{H_2O} = 0.1600 \times \dfrac{500 \text{ K}}{2.000 \text{ L}} \times \dfrac{0.08206 \text{ L} \cdot \text{atm}}{\text{mol} \cdot \text{K}} = 3.2824 = 3.28 \text{ atm}$

(b) The change in P_{H_2O} is $3.51 - 3.28 = 0.2276 = 0.23$ atm. From the reaction stoichiometry, calculate the change in the other pressures and the equilibrium pressures.

	$CO_2(g)$	+	$H_2(g)$	$\rightleftharpoons$	$CO(g)$	+	$H_2O(g)$
initial	4.10 atm		2.05 atm		0 atm		3.28 atm
change	−0.23 atm		−0.23 atm		+0.23		+0.23 atm
equil	3.87 atm		1.82 atm		0.23 atm		3.51 atm

(c) $K_p = \dfrac{P_{CO} \times P_{H_2O}}{P_{CO_2} \times P_{H_2}} = \dfrac{(0.23)(3.51)}{(3.87)(1.82)} = 0.1146 = 0.11$

Without intermediate rounding, equilibrium pressures are $P_{H_2O} = 3.51$, $P_{CO} = 0.2276$, $P_{H_2} = 1.8239$, $P_{CO_2} = 3.8754$ and $K_p = 0.1130 = 0.11$, in good agreement with the value above.

(d) $K_p = K_c(RT)^{\Delta n}$; $\Delta n = 2 - 2 = 0$; $K_p = K_c(RT)^0$; $K_c = K_p = 0.11$

15.41   *Analyze/Plan.* Follow the logic in Sample Exercise 15.9. $mM = 10^{-3} M$

	$X(aq)$	+	$Y(aq)$	$\rightleftharpoons$	$XY(aq)$
initial	1.0 mM		1.0 mM		0
change	−0.80 mM		−0.80 mM		+0.80 mM
equil.	0.20 mM		0.20 mM		0.80 mM

$K_c = \dfrac{[XY]}{[X][Y]} = \dfrac{(0.80 \times 10^{-3})}{(0.20 \times 10^{-3})(0.20 \times 10^{-3})} = 2.0 \times 10^4$

Applications of Equilibrium Constants (section 15.6)

15.43   (a)   A reaction quotient is the result of the law of mass action for a general set of concentrations, whereas the equilibrium constant requires equilibrium concentrations.

(b)   In the direction of more products, to the right.

(c)   If $Q_c = K_c$, the system is at equilibrium; the concentrations used to calculate Q must be equilibrium concentrations.

15.45   *Analyze/Plan.* Follow the logic in Sample Exercise 15.10. We are given molarities, so we calculate Q directly and decide on the direction to equilibrium.   *Solve.*

$$K_c = \frac{[CO][Cl_2]}{[COCl_2]} = 2.19 \times 10^{-10} \text{ at } 100°C$$

(a) $$Q = \frac{(3.3 \times 10^{-6})(6.62 \times 10^{-6})}{(2.00 \times 10^{-3})} = 1.1 \times 10^{-8}; Q > K$$

The reaction will proceed left to attain equilibrium.

(b) $$Q = \frac{(1.1 \times 10^{-7})(2.25 \times 10^{-6})}{(4.50 \times 10^{-2})} = 5.5 \times 10^{-12}; Q < K$$

The reaction will proceed right to attain equilibrium.

(c) $$Q = \frac{(1.48 \times 10^{-6})^2}{(0.0100)} = 2.19 \times 10^{-10}; Q = K$$

The reaction is at equilibrium.

15.47   *Analyze/Plan.* Follow the logic in Sample Exercise 15.11. We are given concentrations, so write the K_c expression and solve for $[Cl_2]$. Change molarity to partial pressure using the ideal gas equation and the definition of molarity.   *Solve.*

$$K_c = \frac{[SO_2][Cl_2]}{[SO_2Cl_2]}; \quad [Cl_2] = \frac{K_c[SO_2Cl_2]}{[SO_2]} = \frac{(0.078)(0.108)}{0.052} = 0.16200 = 0.16 \ M$$

$$PV = nRT, \quad P = \frac{n}{V}RT; \quad \frac{n}{V} = M; \quad P = M\,RT; \quad T = 100°C + 273 = 373 \ K$$

$$P_{Cl_2} = \frac{0.16200 \text{ mol}}{L} \times \frac{0.08206 \text{ L-atm}}{\text{mol-K}} \times 373 \ K = 4.959 = 5.0 \text{ atm}$$

Check. $K_c = \dfrac{(0.052)(0.162)}{(0.108)} = 0.078$. Our values are self-consistent.

15.49   *Analyze/Plan.* Follow the logic in Sample Exercise 15.11. In each case, change given masses to molarities solve for the equilibrium molarity of the desired component, and calculate mass of that substance present at equilibrium.   *Solve.*

(a) $$K_c = \frac{[Br]^2}{[Br_2]} = 1.04 \times 10^{-3}$$

$$[Br_2] = \frac{0.245 \text{ g Br}_2}{0.200 \text{ L}} \times \frac{1 \text{ mol Br}_2}{159.8 \text{ g Br}_2} = 0.007666 = 0.00767 \ M$$

$$[Br] = (K_c[Br_2])^{1/2} = [(1.04 \times 10^{-3})(0.007666)]^{1/2} = 0.002824 = 0.00282 \ M$$

$$\frac{0.002824 \text{ mol Br}}{L} \times 0.200 \text{ L} \times \frac{79.90 \text{ g Br}}{\text{mol}} = 0.0451 \text{ g Br(g)}$$

Check. $K_c = (0.002824)^2 / (0.007666) = 1.04 \times 10^{-3}$

(b) $K_c = \dfrac{[HI]^2}{[H_2][I_2]} = 55.3; \quad [HI] = (K_c[H_2][I_2])^{1/2}$

$[H_2] = \dfrac{0.056 \text{ g } H_2}{2.00 \text{ L}} \times \dfrac{1 \text{ mol } H_2}{2.016 \text{ g } H_2} = 0.01389 = 0.014 \ M$

$[I_2] = \dfrac{4.36 \text{ g } I_2}{2.00 \text{ L}} \times \dfrac{1 \text{ mol } I_2}{253.8 \text{ g } I_2} = 0.008589 = 0.00859 \ M$

$[HI] = [(55.3)(0.01389)(0.008589)]^{1/2} = 0.08122 = 0.081 \ M$

$0.08122 \ M \text{ HI} \times 2.00 \text{ L} \times \dfrac{127.9 \text{ g HI}}{\text{mol HI}} = 20.78 = 21 \text{ g HI}$

Check. $K_c = \dfrac{(0.08122)^2}{(0.01389)(0.008589)} = 55.3$

15.51   *Analyze/Plan.* Follow the logic in Sample Exercise 15.12. Since molarity of NO is given directly, we can construct the equilibrium table straight away.   *Solve.*

	$2NO(g)$	$\rightleftharpoons$	$N_2(g)$	$+$	$O_2(g)$	$K_c = \dfrac{[N_2][O_2]}{[NO]^2} = 2.4 \times 10^3$
initial	0.175 M		0		0	
change	$-2x$		$+x$		$+x$	
equil.	$0.175 - 2x$		$+x$		$+x$	

$2.4 \times 10^3 = \dfrac{x^2}{(0.175 - 2x)^2}; (2.4 \times 10^3)^{1/2} = \dfrac{x}{0.175 - 2x}$

$x = (2.4 \times 10^3)^{1/2}(0.175 - 2x); x = 8.573 - 97.98x; 98.98x = 8.573, x = 0.08662 = 0.087 \ M$

$[N_2] = [O_2] = 0.087 \ M; [NO] = 0.175 - 2(0.08662) = 0.00177 = 0.002 \ M$

Check. $K_c = (0.08662)^2 / (0.00177)^2 = 2.4 \times 10^3$

15.53   *Analyze/Plan.* Write the K_p expression, substitute the stated pressure relationship, and solve for P_{Br_2}.   *Solve.*

$K_p = \dfrac{P_{NO}^2 \times P_{Br_2}}{P_{NOBr}^2}$

When $P_{NOBr} = P_{NO}$, these terms cancel and $P_{Br_2} = K_p = 0.416$ atm. This is true for all cases where $P_{NOBr} = P_{NO}$.

15.55   (a) $CaSO_4(s) \rightleftharpoons Ca^{2+}(aq) + SO_4^{2-}(aq) \qquad K_c = [Ca^{2+}][SO_4^{2-}] = 2.4 \times 10^{-5}$

At equilibrium, $[Ca^{2+}] = [SO_4^{2-}] = x$

$K_c = 2.4 \times 10^{-5} = x^2; x = 4.9 \times 10^{-3} \ M \ Ca^{2+}$ and SO_4^{2-}

(b)   A saturated solution of $CaSO_4(aq)$ is 4.9×10^{-3} *M*.

1.4 L of this solution contain:

$$\frac{4.9 \times 10^{-3} \text{ mol}}{L} \times 1.4\,L \times \frac{136.14 \text{ g CaSO}_4}{\text{mol}} = 0.9337 = 0.94 \text{ g CaSO}_4$$

A bit more than 1.0 g $CaSO_4$ is needed in order to have some undissolved $CaSO_4(s)$ in equilibrium with 1.4 L of saturated solution.

15.57   *Analyze/Plan.* Follow the approach in Solution 15.51. Calculate [IBr] from mol IBr and construct the equilibrium table.

Solve. [IBr] = 0.500 mol/2.00 L = 0.250 *M*

Since no I_2 or Br_2 was present initially, the amounts present at equilibrium are produced by the reverse reaction and stoichiometrically equal. Let these amounts equal x. The amount of HBr that reacts is then 2x. Substitute the equilibrium molarities (in terms of x) into the equilibrium expression and solve for x.

	I_2 +	Br_2 ⇌	2IBr	$K_c = \dfrac{[IBr]^2}{[I_2][Br_2]} = 280$
initial	0 *M*	0 *M*	0.250 *M*	
change	+x *M*	+x *M*	–2x *M*	
equil.	x *M*	x *M*	(0.250 – 2x) *M*	

$K_c = 280 = \dfrac{(0.250 - 2x)^2}{x^2}$; taking the square root of both sides

$16.733 = \dfrac{0.250 - 2x}{x}$; $16.733x + 2x = 0.250$; $18.733x = 0.250$

$x = 0.013345 = 0.0133$ *M*; $[I_2] = [Br_2] = 0.0133$ *M*

[IBr] = 0.250 – 2(0.013345) = 0.2233 = 0.223 *M*

Check. $\dfrac{(0.2233)^2}{(0.013345)^2} = 280$. Our values are self-consistent.

15.59   *Analyze/Plan.* Follow the logic in sample Exercise 15.12, using torr in place of *M*. For this reaction, $\Delta n = 0$, and $K_p = K_c$, so we use the more convenient measure of concentration.

	$CH_4(g)$ +	$I_2(g)$ ⇌	$CH_3I(g)$ +	HI(g)
initial	105.1 torr	7.96 torr	0 torr	0 torr
change	–x torr	–x torr	+x torr	+x torr
equil	105.1–x torr	7.96–x torr	+x torr	+x torr

$K_p = 2.26 \times 10^{-4} = \dfrac{x^2}{(105.1-x)(7.96-x)}$; $x^2 = 2.26 \times 10^{-4}(836.6 - 113.1x + x^2)$

$$0.999774\,x^2 + 0.02555\,x - 0.18907 = 0; \quad x = \frac{-0.02555 \pm \sqrt{(0.02555)^2 - 4(0.999774)(-0.18907)}}{2(0.999774)}$$

(The negative solution is not chemically meaningful.)

x = 0.422 torr; at equilibirum : P_{CH_3I} = P_{HI} = 0.422 torr; P_{CH_4} = 104.7 torr; P_{I_2} = 7.54 torr

LeChâtelier's Principle (section 15.7)

15.61　*Analyze/Plan.* Follow the logic in Sample Exercise 15.13.　　*Solve.*

(a)　Shift equilibrium to the right; more $SO_3(g)$ is formed, the amount of $SO_2(g)$ decreases.

(b)　Heating an exothermic reaction decreases the value of K. More SO_2 and O_2 will form, the amount of SO_3 will decrease. This is fundamentally different than shifting the relative amounts of reactants and products to maintain K; here, the equilibrium position itself changes.

(c)　Since, Δn = –1, a change in volume will affect the equilibrium position and favor the side with more moles of gas. The amounts of SO_2 and O_2 increase and the amount of SO_3 decreases; equilibrium shifts to the left.

(d)　No effect. Speeds up the forward and reverse reactions equally.

(e)　No effect. Does not appear in the equilibrium expression.

(f)　Shift equilibrium to the right; amounts of SO_2 and O_2 decrease.

15.63　*Analyze/Plan.* Given certain changes to a reaction system, determine the effect on K_p, if any. Only changes in temperature cause changes to the value of K_p.　　*Solve.*

(a)　no effect　　(b)　no effect　　(c)　no effect

(d)　increase equilibrium constant　　(e)　no effect

15.65　*Analyze/Plan.* Use Hess's Law, $\Delta H° = \Sigma \Delta H_f°$ products $- \Sigma \Delta H_f°$ reactants, to calculate $\Delta H°$. According to the sign of $\Delta H°$, describe the effect of temperature on the value of K. According to the value of Δn, describe the effect of changes to container volume.

Solve.

(a)　$\Delta H° = \Delta H_f° NO_2(g) + \Delta H_f° N_2O(g) - 3\Delta H_f° NO(g)$

　　$\Delta H°$ = 33.84 kJ + 81.6 kJ – 3(90.37 kJ) = –155.7 kJ

(b)　The reaction is exothermic because it has a negative value of $\Delta H°$. The equilibrium constant will decrease with increasing temperature.

(c)　Δn does not equal zero, so a change in volume at constant temperature will affect the fraction of products in the equilibrium mixture. An increase in container volume would favor reactants, while a decrease in volume would favor products.

15.67　For this reaction, there are more moles of product gas than moles of reactant gas. An increase in total pressure increases the partial pressure of each gas, shifting the equilibrium towards reactants. An increase in pressure favors formation of ozone.

Additional Exercises

15.71 $CH_4(g) + H_2O(g) \rightarrow CO(g) + 3H_2(g)$

$$K_p = \frac{P_{CO} \times P_{H_2}^3}{P_{CH_4} \times P_{H_2O}} ; P = \frac{g\,RT}{MM\,V} ; T = 1000\,K$$

$$P_{CO} = \frac{8.62\,g}{28.01\,g/mol} \times \frac{0.08206\,L\text{-}atm}{mol\text{-}K} \times \frac{1000\,K}{5.00\,L} = 5.0507 = 5.05\,atm$$

$$P_{H_2} = \frac{2.60\,g}{2.016\,g/mol} \times \frac{0.08206\,L\text{-}atm}{mol\text{-}K} \times \frac{1000\,K}{5.00\,L} = 21.1663 = 21.2\,atm$$

$$P_{CH_4} = \frac{43.0\,g}{16.04\,g/mol} \times \frac{0.08206\,L\text{-}atm}{mol\text{-}K} \times \frac{1000\,K}{5.00\,L} = 43.9973 = 44.0\,atm$$

$$P_{H_2O} = \frac{48.4\,g}{18.02\,g/mol} \times \frac{0.08206\,L\text{-}atm}{mol\text{-}K} \times \frac{1000\,K}{5.00\,L} = 44.0811 = 44.1\,atm$$

$$K_p = \frac{(5.0507)(21.1663)^3}{(43.9973)(44.0811)} = 24.6949 = 24.7$$

$K_p = K_c(RT)^{\Delta n}, K_c = K_p/(RT)^{\Delta n} ; \Delta n = 4 - 2 = 2$

$K_c = (24.6949)/[(0.08206)(1000)]^2 = 3.6673 \times 10^{-3} = 3.67 \times 10^{-3}$

15.74   (a) $K_p = \dfrac{P_{Br_2} \times P_{NO}^2}{P_{NOBr}^2} ; P = \dfrac{gRT}{MM \times V} ; T = 100\degree C + 273 = 373\,K$

$$P_{Br_2} = \frac{4.19g}{159.8\,g/mol} \times \frac{0.08206\,L\text{-}atm}{mol\text{-}K} \times \frac{373}{5.00\,L} = 0.16051 = 0.161\,atm$$

$$P_{NO} = \frac{3.08g}{30.01\,g/mol} \times \frac{0.08206\,L\text{-}atm}{mol\text{-}K} \times \frac{373}{5.00\,L} = 0.62828 = 0.628\,atm$$

$$P_{NOBr} = \frac{3.22\,g\,NOBr}{109.9\,g/mol} \times \frac{0.08206\,L\text{-}atm}{mol\text{-}K} \times \frac{373}{5.00\,L} = 0.17936 = 0.179\,atm$$

$$K_p = \frac{(0.16051)(0.62828)^2}{(0.17936)^2} = 1.9695 = 1.97 \qquad K_p = K_c(RT)^{\Delta n}, \Delta n = 3 - 2 = 1$$

$K_c = K_p/RT = 1.9695/(0.08206)(373) = 0.064345 = 0.0643$

(b) $P_t = P_{Br_2} + P_{NO} + P_{NOBr} = 0.16051 + 0.62828 + 0.17936 = 0.96815 = 0.968\,atm$

(c)   All NO and Br_2 present at equilibrium came from the decomposition of the original NOBr. The mass of original NOBr is the sum of the masses of all compounds at equilibrium.

Original g NOBr = 4.19 gBr_2 + 3.08 g NO + 3.22 g NOBr = 10.49 g

15.77

	2IBr	$\rightleftharpoons$	I_2	+	Br_2
initial	0.025 atm		0		0
change	–2x		x		x
equil.	(0.025 – 2x) atm		x		x

$$K_p = 8.5 \times 10^{-3} = \frac{P_{I_2} \times P_{Br_2}}{P_{IBr}^2} = \frac{x^2}{(0.025 - 2x)^2}; \quad \text{Taking the square root of both sides}$$

$$\frac{x}{0.025 - 2x} = (8.5 \times 10^{-3})^{1/2} = 0.0922; \; x = 0.0922(0.025 - 2x)$$

$$x + 0.184x = 0.002305; \; 1.184x = 0.002305; \; x = 0.001947 = 1.9 \times 10^{-3}$$

At equilibrium, $P_{I_2} = P_{Br_2} = x = 1.9 \times 10^{-3}$ atm

P_{IBr} at equilibrium = $0.025 - 2(1.947 \times 10^{-3}) = 0.02111 = 0.021$ atm

Check. $K_p = (0.001947)^2/(0.02111)^2 = 8.5 \times 10^{-3}$; the calculated concentrations are self-consistent.

15.80   Initial $P_{SO_3} = \dfrac{gRT}{MM\,V} = \dfrac{0.831\,g}{80.07\,g/mol} \times \dfrac{0.08206\,L\text{-}atm}{mol\text{-}K} \times \dfrac{1100\,K}{1.00\,L} = 0.9368 = 0.937$ atm

	2SO$_3$	$\rightleftharpoons$	2SO$_2$	+	O$_2$
initial	0.9368 atm		0		0
change	–2x		+2x		+x
equil.	0.9368–2x		2x		x
[equil.]	0.2104 atm		0.7264 atm		0.3632 atm

$P_t = (0.9368 - 2x) + 2x + x; \; 0.9368 + x = 1.300$ atm; $x = 1.300 - 0.9368 = 0.3632 = 0.363$ atm

$$K_p = \frac{P_{SO_2}^2 \times P_{O_2}}{P_{SO_3}^2} = \frac{(0.7264)^2 (0.3632)}{(0.2104)^2} = 4.3292 = 4.33$$

$K_p = K_c(RT)^{\Delta n}; \; \Delta n = 3 - 2 = 1; \; K_p = K_c(RT)$

$K_c = K_p/RT = 4.3292/[(0.08206)(1100)] = 0.04796 = 0.0480$

15.83

	CO$_2$(g)	+	H$_2$(g)	$\rightleftharpoons$	CO(g)	+	H$_2$O(g)
initial	1.50 mol		1.50 mol		0		0
change	–x		–x		+x		+x
equil.	(1.50 – x)mol		(1.50 – x)mol		x		x

Since $\Delta n = 0$, the volume terms cancel and we can use moles in place of molarity in the K expression.

$$K_c = 0.802 = \frac{[CO][H_2O]}{[CO_2][H_2O]} = \frac{x^2}{(1.50 - x)^2}$$

Take the square root of both sides.

$(0.802)^{1/2} = x/(1.50 - x); \quad 0.8955(1.50 - x) = x$

$1.3433 = 1.8955x, \quad x = 0.7087 = 0.709 \text{ mol}$

$[CO] = [H_2O] = 0.7087 \text{ mol}/3.00 \text{ L} = 0.236 \ M$

$[CO_2] = [H_2] = (1.50 - 0.709)\text{mol}/3.00 \text{ L} = 0.264 \ M$

15.87   (a)

$$CCl_4(g) \rightleftharpoons C(s) + 2Cl_2(g)$$

initial	2.00 atm	0 atm
change	–x atm	+2x atm
equil.	(2.00–x) atm	2x atm

$$K_p = 0.76 = \frac{P_{Cl_2}^2}{P_{CCl_4}} = \frac{(2x)^2}{(2.00 - x)}$$

$1.52 - 0.76x = 4x^2; \quad 4x^2 + 0.76x - 1.52 = 0$

Using the quadratic formula, a = 4, b = 0.76, c = –1.52

$$x = \frac{-0.76 \pm \sqrt{(0.76)^2 - 4(4)(-1.52)}}{2(4)} = \frac{-0.76 + 4.99}{8} = 0.5287 = 0.53 \text{ atm}$$

$$\text{Fraction } CCl_4 \text{ reacted} = \frac{x \text{ atm}}{2.00 \text{ atm}} = \frac{0.5287}{2.00} = 0.264 = 26\%$$

(b) $P_{Cl_2} = 2x = 2(0.5287) = 1.06 \text{ atm}$

$P_{CCl_4} = 2.00 - x = 2.00 - 0.5287 = 1.47 \text{ atm}$

15.91 $K_p = \dfrac{P_{O_2} \times P_{CO}^2}{P_{CO_2}^2} \approx 1 \times 10^{-13}; P_{O_2} = (0.03)(1 \text{ atm}) = 0.03 \text{ atm}$

$P_{CO} = (0.002)(1 \text{ atm}) = 0.002 \text{ atm}; \quad P_{CO_2} = (0.12)(1 \text{ atm}) = 0.12 \text{ atm}$

$Q = \dfrac{(0.03)(0.002)^2}{(0.12)^2} = 8.3 \times 10^{-6} = 8 \times 10^{-6}$

Since Q > K_p, the system will shift to the left to attain equilibrium. Thus a catalyst that promoted the attainment of equilibrium would result in a lower CO content in the exhaust.

Integrative Exercises

15.93   Calculate the initial $[IO_4^-]$, and then construct an equilibrium table to determine $[H_4IO_6^-]$ at equilibrium.

$$M_c \times V_c = M_d \times L_d; \quad \frac{0.905\ M \times 25.0\ mL}{500.0\ mL} = M_d = 0.04525 = 0.0453\ M\ IO_4^-$$

$$IO_4^-(aq) + 2H_2O(l) \rightleftharpoons H_4IO_6^-(aq)$$

initial	$0.0453\ M$	0
change	$-x$	$+x$
equil.	$0.0453 - x$	$+x$

$$K_c = 3.5 \times 10^{-2} = \frac{[H_4IO_6^-]}{[IO_4^-]} = \frac{x}{(0.0453 - x)}$$

Because K_c is relatively large and $[IO_4^-]$ is relatively small, we cannot assume x is small relative to 0.0453.

$0.035(0.04525 - x) = x; \quad 0.001584 - 0.035x = x; \quad 0.001584 = 1.035x$

$x = 0.001584/1.035 = 0.001530 = 0.0015\ M\ H_4IO_6^-$ at equilibrium

15.97   Mole % = pressure %. Since the total pressure is 1 atm, mol %/100 = mol fraction = partial pressure. $K_p = P_{CO}^2 / P_{CO_2}$.

Temp (K)	P_{CO_2} (atm)	P_{CO} (atm)	K_P
1123	0.0623	0.9377	14.1
1223	0.0132	0.9868	73.8
1323	0.0037	0.9963	2.7×10^2
1473	0.0006	0.9994	1.7×10^3 (2×10^3)

Because K grows larger with increasing temperature, the reaction must be endothermic in the forward direction.

16 Acid-Base Equilibria

Visualizing Concepts

16.1   *Analyze.* From the structures decide which reactant fits the description of a Brønsted-Lowry (B-L) acid, a B-L base, a Lewis acid, and a Lewis base. *Plan.* A B-L acid is an H^+ donor, and a B-L base is an H^+ acceptor. A Lewis acid is an electron pair acceptor and a Lewis base is an electron pair donor.   *Solve.*

(a)   H–X is a B-L acid, because it loses H^+ during reaction. NH_3 is a B-L base, because it gains H^+ during reaction.

(b)   By virtue of its unshared electron pair, NH_3 is the electron pair donor and Lewis base. HX is the electron pair acceptor and Lewis acid.

16.3   *Plan.* Strong acids are completely ionized. The acid that is least ionized is weakest, and has the smallest K_a value. At equal concentrations, the weakest acid has the smallest $[H^+]$ and highest pH.   *Solve.*

(a)   HY is a strong acid. There are no neutral HY molecules in solution, only H^+ cations and Y^- anions.

(b)   HX has the smallest K_a value. It has most neutral acid molecules and fewest ions.

(c)   HX has the fewest H^+ and highest pH.

16.5   (a)   True. Solution A is the color of methyl orange in acidic solution.

(b)   False. Methyl orange turns yellow at a pH slightly greater than 4, so solution B could be at any pH greater than 4.

(c)   True. The basic color of any indicator occurs at higher pH than the acidic color does.

16.7   *Analyze/Plan.* Write the formula of each molecule and compare them to the entries in Tables 16.2 and 16.4. Select the molecule that is fits the definition of an acid, and the one that fits the definition of a base.   *Solve.*

(a)   Molecule A is hydroxyl amine, NH_2OH. It is an entry in Table 16.4. Molecule A is an H^+ acceptor because of the nonbonded electron pair on the N atom of the amine ($-NH_2$) group, not because it contains an $-OH$ group. The presence of an $-OH$ group in an organic molecule does not mean that the molecule is a base.

(b)   Molecule B is formic acid, $HCOOH$. It is similar to CH_3COOH, an entry in Table 16.2. The H atom bonded to O is ionizable and $HCOOH$ is an H^+ donor. In general, organic molecules that contain a carboxyl ($-COOH$) group are acids.

(c)   Molecule C is methanol, CH_3OH. In organic molecules, the $-OH$ functional group is an alcohol. The H atom bonded to O is not ionizable, and the $-OH$

group does not dissociate in aqueous solution. An alcohol is neither an acid nor a base.

16.9   *Plan.* Evaluate the molecular structures to determine if the acids are binary acids or oxyacids. Consider the trends in acid strength for both classes of acids.   *Solve.*

(a)   If X is the same atom on both molecules, the molecule (b) is more acidic. The carboxylate anion, the conjugate base of this carboxylic acid, is stabilized by resonance, while the conjugate base of (a) is not resonance-stabilized. Stabilization of the conjugate base causes the ionization equilibrium to favor products, and (b) is the stronger acid.

(b)   Increasing the electronegativity of X increases the strength of both acids. As X becomes more electronegative and attracts more electron density, the O–H bond becomes weaker and more polar. This increases the likelihood of ionization and increases acid strength. An electronegative X group also stabilizes the anionic conjugate bases by delocalizing the negative charge. This causes the ionization equilibrium to favor products, and the values of K_a to increase.

Arrhenius and Brønsted-Lowry Acids and Bases (sections 16.1 and 16.2)

16.11   Solutions of HCl and H_2SO_4 taste sour, turn litmus paper red (are acidic), neutralize solutions of bases, react with active metals to form $H_2(g)$ and conduct electricity. The two solutions have these properties in common because both solutes are strong acids. That is, they both ionize completely in H_2O to form $H^+(aq)$ and an anion. (The first ionization step for H_2SO_4 is complete, but the second is not.) The presence of ions enables the solutions to conduct electricity; the presence of $H^+(aq)$ in excess of $1 \times 10^{-7} M$ accounts for all the other listed properties.

16.13   (a)   According to the Arrhenius definition, an *acid* when dissolved in water increases $[H^+]$. According to the Brønsted-Lowry definition, an *acid* is capable of donating H^+, regardless of physical state. The Arrhenius definition of an acid is confined to an aqueous solution; the Brønsted-Lowry definition applies to any physical state.

(b) $HCl(g) + NH_3(g) \rightarrow NH_4^+Cl^-(s)$ HCl is the B-L (Brønsted-Lowry) acid; it donates an H^+ to NH_3 to form NH_4^+. NH_3 is the B-L base; it accepts the H^+ from HCl.

16.15   *Analyze/Plan.* Follow the logic in Sample Exercise 16.1. A conjugate base has one less H^+ than its conjugate acid. A conjugate acid has one more H^+ than its conjugate base. *Solve.*

(a)   (i) IO_3^- (ii) NH_3

(b)   (i) OH^- (ii) H_3PO_4

16.17   *Analyze/Plan.* Use the definitions of B-L acids and bases, and conjugate acids and bases to make the designations. Evaluate the changes going from reactant to product to inform your choices. *Solve.*

	B-L acid	+	**B-L base**	⇌	**Conjugate acid**	+	**Conjugate base**
(a)	$NH_4^+(aq)$		$CN^-(aq)$		$HCN(aq)$		$NH_3(aq)$
(b)	$H_2O(l)$		$(CH_3)_3N(aq)$		$(CH_3)_3NH^+(aq)$		$OH^-(aq)$
(c)	$HCOOH(aq)$		$PO_4^{3-}(aq)$		$HPO_4^{2-}(aq)$		$HCOO^-(aq)$

16.19   *Analyze/Plan.* Follow the logic in Sample Exercise 16.2.   *Solve.*

(a)   Acid: $HC_2O_4^-(aq) + H_2O(l) \rightleftharpoons C_2O_4^{2-}(aq) + H_3O^+(aq)$

            B-L acid   B-L base   conj. base   conj. acid

      Base: $HC_2O_4^-(aq) + H_2O(l) \rightleftharpoons H_2C_2O_4(aq) + OH^-(aq)$

            B-L base   B-L acid   conj. acid   conj. Base

(b) $H_2C_2O_4$ is the conjugate acid of $HC_2O_4^-$.

      $C_2O_4^{2-}$ is the conjugate base of $HC_2O_4^-$.

16.21   *Analyze/Plan.* Based on the chemical formula, decide whether the base is strong, weak or negligible. Is it the conjugate of a strong acid (negligible base), weak acid (weak base) or negligible acid (strong base)? Also check Figure 16.3. To write the formula of the conjugate acid, add a single H and increase the particle charge by one.

(a) CH_3COO^-, weak base; CH_3COOH, weak acid

(b) HCO_3^-, weak base; H_2CO_3, weak acid

(c) O^{2-}, strong base; OH^-, negligible acid

(d) Cl^-, negligible base; HCl, strong acid

(e) NH_3, weak base; NH_4^+, weak acid

16.23   *Analyze/Plan.* Given chemical formula, determine strength of acids and bases by checking the known strong acids (Section 16.5). Recall the paradigm "The stronger the acid, the weaker its conjugate base, and vice versa."   *Solve.*

(a)   HBr. It is one of the seven strong acids (Section 16.5).

(b) F^-. HCl is a stronger acid than HF, so F^- is the stronger conjugate base.

16.25   *Analyze/Plan.* Acid-base equilibria favor formation of the weaker acid and base. Compare the relative strengths of the substances acting as acids on opposite sides of the equation. (Bases can also be compared; the conclusion should be the same.)   *Solve.*

	Base	+	**Acid**	⇌	**Conjugate acid**	+	**Conjugate base**
(a)	$O^{2-}(aq)$		$H_2O(l)$		$OH^-(aq)$		$OH^-(aq)$

H_2O is a stronger acid than OH^-, so the equilibrium lies to the right.

	Base	+	**Acid**	⇌	**Conjugate acid**	+	**Conjugate base**
(b)	$HS^-(aq)$		$CH_3COOH(aq)$		$H_2S(aq)$		$CH_3COO^-(aq)$

CH_3COOH is a stronger acid than H_2S, so the equilibrium lies to the right.

(c)	$NO_2^-(aq)$	+	$H_2O(l)$	⇌	$HNO_2(aq)$	+	$OH^-(aq)$

HNO_2 is a stronger acid than H_2O, so the equilibrium lies to the left.

Autoionization of Water (section 16.3)

16.27 No. In pure water, the only source of H^+ is the autoionization reaction, which produces equal concentrations of H^+ and OH^-. As the temperature of water changes, the value of K_w changes, and the pH at which $[H^+] = [OH^-]$ changes. At 50 °C, if pH = 6.63,

$[H^+] = [OH^-] = 10^{-6.63} = 2.34 \times 10^{-7}$; $K_w = (2.34 \times 10^{-7})(2.34 \times 10^{-7}) = 5.5 \times 10^{-14}$

16.29 *Analyze/Plan.* Follow the logic in Sample Exercise 16.5. In pure water at 25°C, $[H^+] = [OH^-] = 1 \times 10^{-7}$ M. If $[H^+] > 1 \times 10^{-7}$ M, the solution is acidic; if $[H^+] < 1 \times 10^{-7}$ M, the solution is basic. *Solve.*

(a) $[H^+] = \dfrac{K_w}{[OH^-]} = \dfrac{1.0 \times 10^{-14}}{4.5 \times 10^{-4} \, M} = 2.2 \times 10^{-11} \, M < 1 \times 10^{-7} \, M$; basic

(b) $[H^+] = \dfrac{K_w}{[OH^-]} = \dfrac{1.0 \times 10^{-14}}{8.8 \times 10^{-9} \, M} = 1.1 \times 10^{-6} \, M > 1 \times 10^{-7} \, M$; acidic

(c) $[OH^-] = 100[H^+]$; $K_w = [H^+] \times 100[H^+] = 100[H^+]^2$;

$[H^+] = (K_w/100)^{1/2} = 1.0 \times 10^{-8} \, M < 1 \times 10^{-7} \, M$; basic

16.31 *Analyze/Plan.* Follow the logic in Sample Exercise 16.4. Note that the value of the equilibrium constant (in this case, K_w) changes with temperature. *Solve.*

At 0°C, $K_w = 1.2 \times 10^{-15} = [H^+][OH^-]$.

In pure water, $[H^+] = [OH^-]$; $1.2 \times 10^{-15} = [H^+]^2$; $[H^+] = (1.2 \times 10^{-15})^{1/2}$

$[H^+] = [OH^-] = 3.5 \times 10^{-8} \, M$

The pH Scale (section 16.4)

16.33 *Analyze/Plan.* A change of one pH unit (in either direction) is:

$\Delta pH = pH_2 - pH_1 = -(\log[H^+]_2 - \log[H^+]_1) = -\log\dfrac{[H^+]_2}{[H^+]_1} = \pm 1$. The antilog of +1 is 10;

the antilog of –1 is 1×10^{-1}. Thus, a ΔpH of one unit represents an increase or decrease in $[H^+]$ by a factor of 10. *Solve.*

(a) $\Delta pH = \pm 2.00$ is a change of $10^{2.00}$; $[H^+]$ changes by a factor of 100.

(b) $\Delta pH = \pm 0.5$ is a change of $10^{0.50}$; $[H^+]$ changes by a factor of 3.2.

16.35 (a) $K_w = [H^+][OH^-]$. If NaOH is added to water, it dissociates into $Na^+(aq)$ and $OH^-(aq)$. This increases $[OH^-]$ and necessarily decreases $[H^+]$. When $[H^+]$ decreases, pH increases.

(b) $0.0006 \, M = 6 \times 10^{-4} \, M$. On Figure 16.5, this is $[H^+] > 1 \times 10^{-4}$ but $< 1 \times 10^{-3}$. The pH is between 3 and 4, closer to 3. We estimate 3.3. If pH < 7, the solution is acidic.

By calculation: $pH = -\log[H^+] = -\log(6 \times 10^{-4} \, M) = 3.2$

(c) pH = 5.2 is between pH 5 and pH 6 on Figure 16.5, closer to pH = 5. At pH = 6, $[H^+] = 1 \times 10^{-6}$; at pH = 5, $[H^+] = 1 \times 10^{-5} = 10 \times 10^{-6}$. A good estimate is

$7 \times 10^{-6} M \, H^+$.

By calculation: $[H^+] = 10^{-pH} = 10^{-5.2} = 6 \times 10^{-6} M$

At pH = 5, $[OH^-] = 1 \times 10^{-9}$; at pH = 6, $[OH^-] = 1 \times 10^{-8} = 10 \times 10^{-9}$.

Since pH = 5.2 is closer to pH = 5, we estimate $3 \times 10^{-9} M \, OH^-$.

By calculation: pOH = 14.0 − 5.2 = 8.8

$[OH^-] = 10^{-pOH} = 10^{-8.8} = 2 \times 10^{-9} M \, OH^-$

16.37   *Analyze/Plan.* At 25°C, $[H^+][OH^-] = 1 \times 10^{-14}$; pH + pOH = 14. Use these relationships to complete the table. If pH < 7, the solution is acidic; if pH > 7, the solution is basic. *Solve.*

$[H^+]$	$[OH^-]$	pH	pOH	acidic or basic
$7.5 \times 10^{-3} M$	$1.3 \times 10^{-12} M$	2.12	11.88	acidic
$2.8 \times 10^{-5} M$	$3.6 \times 10^{-10} M$	4.56	9.44	acidic
$5.6 \times 10^{-9} M$	$1.8 \times 10^{-6} M$	8.25	5.75	basic
$5.0 \times 10^{-9} M$	$2.0 \times 10^{-6} M$	8.30	5.70	basic

Check. pH + pOH = 14; $[H^+][OH^-] = 1 \times 10^{-14}$.

16.39   *Analyze/Plan.* Given pH and a new value of the equilibrium constant K_w, calculate equilibrium concentrations of $H^+(aq)$ and $OH^-(aq)$. The definition of pH remains pH = −log$[H^+]$.   *Solve.*

pH = 7.40; $[H^+] = 10^{-pH} = 10^{-7.40} = 4.0 \times 10^{-8} M$

$K_w = 2.4 \times 10^{-14} = [H^+][OH^-]$; $[OH^-] = 2.4 \times 10^{-14} / [H^+]$

$[OH^-] = 2.4 \times 10^{-14} / 4.0 \times 10^{-8} = 6.0 \times 10^{-7} M$, pOH = −log$(6.0 \times 10^{-7}) = 6.22$

Alternately, pH + pOH = pK_w. At 37°C, pH + pOH = −log(2.4×10^{-14})

pH + pOH = 13.62; pOH = 13.62 − 7.40 = 6.22

$[OH^-] = 10^{-pOH} = 10^{-6.22} = 6.0 \times 10^{-7} M$

Strong Acids and Bases (section 16.5)

16.41   (a)   A strong acid is completely ionized in aqueous solution; a strong acid is a strong electrolyte.

        (b)   For a strong acid such as HCl, $[H^+]$ = initial acid concentration. $[H^+] = 0.500 M$

        (c)   HCl, HBr, HI

16.43   *Analyze/Plan.* Follow the logic in Sample Exercise 16.8. Strong acids are completely ionized, so $[H^+]$ = original acid concentration, and pH = −log$[H^+]$. For the solutions obtained by dilution, use the "dilution" formula, $M_1V_1 = M_2V_2$, to calculate molarity of the acid.   *Solve.*

        (a) $8.5 \times 10^{-3} M \, HBr = 8.5 \times 10^{-3} M \, H^+$; pH = −log $(8.5 \times 10^{-3}) = 2.07$

(b) $\dfrac{1.52 \text{ g HNO}_3}{0.575 \text{ L soln}} \times \dfrac{1 \text{ mol HNO}_3}{63.02 \text{ g HNO}_3} = 0.041947 = 0.0419 \ M \text{ HNO}_3$

$[\text{H}^+] = 0.0419 \ M$; pH $= -\log(0.041947) = 1.377$

(c) $M_c \times V_c = M_d \times V_d$; $0.250 \ M \times 0.00500 \text{ L} = ? \ M \times 0.0500 \text{ L}$

$M_d = \dfrac{0.250 \ M \ \times \ 0.00500 \text{ L}}{0.0500 \text{ L}} = 0.0250 \ M \text{ HCl}$

$[\text{H}^+] = 0.0250 \ M$; pH $= -\log(0.0250) = 1.602$

(d) $[\text{H}^+]_{\text{total}} = \dfrac{\text{mol H}^+ \text{ from HBr} + \text{mol H}^+ \text{ from HCl}}{\text{total L solution}}$

$[\text{H}^+]_{\text{total}} = \dfrac{(0.100 \ M \text{ HBr} \ \times \ 0.0100 \text{ L}) + (0.200 \ M \ \times \ 0.0200 \text{ L})}{0.0300 \text{ L}}$

$[\text{H}^+]_{\text{total}} = \dfrac{1.00 \ \times \ 10^{-3} \text{ mol H}^+ + 4.00 \ \times \ 10^{-3} \text{ mol H}^+}{0.0300 \text{ L}} = 0.1667 = 0.167 \ M$

pH $= -\log(0.1667 \ M) = 0.778$

16.45   *Analyze/Plan.* Follow the logic in Sample Exercise 16.9. Strong bases dissociate completely upon dissolving. pOH $= -\log[\text{OH}^-]$; pH $= 14 - \text{pOH}$.

(a)   Pay attention to the formula of the base to get $[\text{OH}^-]$.   *Solve.*

$[\text{OH}^-] = 2[\text{Sr(OH)}_2] = 2(1.5 \times 10^{-3} \ M) = 3.0 \times 10^{-3} \ M \text{ OH}^-$ (see Exercise 16.42(b))

pOH $= -\log(3.0 \times 10^{-3}) = 2.52$; pH $= 14 - \text{pOH} = 11.48$

(b)   mol/LiOH = g LiOH/molar mass LiOH. $[\text{OH}^-] = [\text{LiOH}]$.   *Solve.*

$\dfrac{2.250 \text{ g LiOH}}{0.2500 \text{ L soln}} \times \dfrac{1 \text{ mol LiOH}}{23.948 \text{ g LiOH}} = 0.37581 = 0.3758 \ M \text{ LiOH} = [\text{OH}^-]$

pOH $= -\log(0.37581) = 0.4250$; pH $= 14 - \text{pOH} = 13.5750$

(c)   Use the dilution formula to get the $[\text{NaOH}] = [\text{OH}^-]$.   *Solve.*

$M_c \times V_c = M_d \times V_d$; $0.175 \ M \times 0.00100 \text{ L} = ? \ M \times 2.00 \text{ L}$

$M_d = \dfrac{0.175 \ M \ \times \ 0.00100 \text{ L}}{2.00 \text{ L}} = 8.75 \times 10^{-5} \ M \text{ NaOH} = [\text{OH}^-]$

pOH $= -\log(8.75 \times 10^{-5}) = 4.058$; pH $= 14 - \text{pOH} = 9.942$

(d)   Consider total mol OH^- from KOH and Ca(OH)_2, as well as total solution volume. *Solve.*

$[\text{OH}^-]_{\text{total}} = \dfrac{\text{mol OH}^- \text{ from KOH} + \text{mol OH}^- \text{ from Ca(OH)}_2}{\text{total L soln}}$

$[\text{OH}^-]_{\text{total}} = \dfrac{(0.105 \ M \ \times \ 0.00500 \text{ L}) + 2(9.5 \ \times \ 10^{-2} \ M \times 0.0150 \text{ L})}{0.0200 \text{ L}}$

$$[OH^-]_{total} = \frac{0.525 \times 10^{-3} \text{ mol OH}^- + 2.85 \times 10^{-3} \text{ mol OH}^-}{0.0200 \text{ L}} = 0.16875 = 0.17 \text{ M}$$

pOH = –log (0.16875) = 0.77; pH = 14 – pOH = 13.23

(9.5×10^{-2} M has 2 sig figs, so the [OH⁻] has 2 sig figs and pH and pOH have 2 decimal places.)

16.47   *Analyze/Plan.* pH → pOH → [OH⁻] = [NaOH].   *Solve.*

pOH = 14 – pH = 14.00 – 11.50 = 2.50

pOH = 2.50 = –log[OH⁻]; [OH⁻] = $10^{-2.50}$ = 3.2×10^{-3} M

[OH⁻] = [NaOH] = 3.2×10^{-3} M

Weak Acids (section 16.6)

16.49   *Analyze/Plan.* Remember that K_a = [products]/[reactants]. If $H_2O(l)$ appears in the equilibrium reaction, it will **not** appear in the K_a expression, because it is a pure liquid. *Solve.*

(a) $HBrO_2(aq) \rightleftharpoons H^+(aq) + BrO_2^-(aq)$; $K_a = \dfrac{[H^+][BrO_2^-]}{[HBrO_2]}$

        $HBrO_2(aq) + H_2O(l) \rightleftharpoons H_3O^+(aq) + BrO_2^-(aq)$; $K_a = \dfrac{[H_3O^+][BrO_2^-]}{[HBrO_2]}$

(b) $C_2H_5COOH(aq) \rightleftharpoons H^+(aq) + C_2H_5COO^-(aq)$; $K_a = \dfrac{[H^+][C_2H_5COO^-]}{[C_2H_5COOH]}$

        $C_2H_5COOH(aq) + H_2O(l) \rightleftharpoons H_3O^+(aq) + C_2H_5COO^-(aq)$;

$$K_a = \frac{[H_3O^+][C_2H_5COO^-]}{[C_2H_5COOH]}$$

16.51   *Analyze/Plan.* Follow the logic in Sample Exercise 16.10.   *Solve.*

$CH_3CH(OH)COH \rightleftharpoons H^+(aq) + CH_3CH(OH)COO^-(aq)$; $K_a = \dfrac{[H^+][CH_3CH(OH)COO^-]}{[CH_3CH(OH)COOH]}$

$[H^+] = [CH_3CH(OH)COO^-] = 10^{-2.44} = 3.63 \times 10^{-3} = 3.6 \times 10^{-3}$ M

$[CH_3CH(OH)COOH] = 0.10 – 3.63 \times 10^{-3} = 0.0964 = 0.096$ M

$$K_a = \frac{(3.63 \times 10^{-3})^2}{(0.0964)} = 1.4 \times 10^{-4}$$

16.53   *Analyze/Plan.* Write the equilibrium reaction and the K_a expression. Use % ionization to get equilibrium concentration of [H⁺], and by stoichiometry, [X⁻] and [HX]. Calculate K_a *Solve.*

$[H^+] = 0.110 \times [CH_2ClCOOH]_{initial} = 0.0110$ M

	$CH_2ClCOOH(aq)$	$\rightleftharpoons$	$H^+(aq)$	+	$CH_2ClCOO^-(aq)$
initial	0.100 M		0		0
equil.	0.089 M		0.0110 M		0.0110 M

$$K_a = \frac{[H^+][CH_2ClCOO^-]}{[CH_2ClCOOH]} = \frac{(0.0110)^2}{0.089} = 1.4 \times 10^{-3}$$

16.55   *Analyze/Plan.* Write the equilibrium reaction and the K_a expression.

$[H^+] = 10^{-pH} = [CH_3COO^-]$; $[CH_3COOH] = x - [H^+]$.

Substitute into the K_a expression and solve for x.   *Solve.*

$[H^+] = 10^{-pH} = 10^{-2.90} = 1.26 \times 10^{-3} = 1.3 \times 10^{-3}\,M$

$$K_a = 1.8 \times 10^{-5} = \frac{[H^+][CH_3COO^-]}{[CH_3COOH]} = \frac{(1.26 \times 10^{-3})^2}{(x - 1.26 \times 10^{-3})}$$

$1.8 \times 10^{-5}(x - 1.26 \times 10^{-3}) = (1.26 \times 10^{-3})^2$;

$1.8 \times 10^{-5}\,x = 1.585 \times 10^{-6} + 2.266 \times 10^{-8} = 1.608 \times 10^{-6}$;

$x = 0.08931 = 0.089\,M\ CH_3COOH$

16.57   *Analyze/Plan.* Follow the logic in Sample Exercise 16.12. Write K_a, construct the equilibrium table, solve for $x = [H^+]$, then get equilibrium $[C_6H_5COO^-]$ and $[C_6H_5COOH]$ by substituting $[H^+]$ for x.   *Solve.*

	$C_6H_5COOH(aq)$	$\rightleftharpoons$	$H^+(aq)$	$+$	$C_6H_5COO^-(aq)$
initial	0.050 M		0		0
equil.	$(0.050 - x)\,M$		$x\,M$		$x\,M$

$$K_a = \frac{[H^+][C_6H_5COO^-]}{[C_6H_5COOH]} = \frac{x^2}{(0.050 - x)} \approx \frac{x^2}{0.050} = 6.3 \times 10^{-5}$$

$x^2 = 0.050\,(6.3 \times 10^{-5})$; $x = 1.8 \times 10^{-3}\,M = [H^+] = [H_3O^+] = [C_6H_5COO^-]$

$[C_6H_5COOH] = 0.050 - 0.0018 = 0.048\,M$

Check. $\dfrac{1.8 \times 10^{-3}\,M\,H^+}{0.050\,M\,C_6H_5COOH} \times 100 = 3.6\%$ ionization; the approximation is valid

16.59   *Analyze/Plan.* Follow the logic in Sample Exercise 16.12.

(a)   *Solve.*

	$C_2H_5COOH(aq)$	$\rightleftharpoons$	$H^+(aq)$	$+$	$C_2H_5COO^-(aq)$
initial	0.095 M		0		0
equil	$(0.095 - x)\,M$		$x\,M$		$x\,M$

$$K_a = \frac{[H^+][C_2H_5COO^-]}{[C_2H_5COOH]} = \frac{x^2}{(0.095 - x)} \approx \frac{x^2}{0.095} = 1.3 \times 10^{-5}$$

$x^2 = 0.095(1.3 \times 10^{-5})$; $x = 1.111 \times 10^{-3} = 1.1 \times 10^{-3}\,M\,H^+$; pH = 2.95

Check. $\dfrac{1.1 \times 10^{-3}\,M\,H^+}{0.095\,M\,C_2H_5COOH} \times 100 = 1.2\%$ ionization; the approximation is valid

(b)   *Solve.*

$$K_a = \frac{[H^+][CrO_4^{2-}]}{[HCrO_4^-]} = \frac{x^2}{(0.100-x)} \approx \frac{x^2}{0.100} = 3.0 \times 10^{-7}$$

$$x^2 = 0.100(3.0 \times 10^{-7}); \; x = 1.732 \times 10^{-4} = 1.7 \times 10^{-4} \, M \, H^+$$

$$pH = -\log(1.732 \times 10^{-4}) = 3.7614 = 3.76$$

Check. $\dfrac{1.7 \times 10^{-4} \, M \, H^+}{0.100 \, M \, HCrO_4^-} \times 100 = 0.17\%$ ionization; the approximation is valid

(c)   Follow the logic in Sample Exercise 16.15. $pOH = -\log[OH^-]$. $pH = 14 - pOH$

Solve.

	$C_5H_5N(aq) + H_2O(l)$	$\rightleftharpoons$	$C_5H_5NH^+(aq)$	+	$OH^-(aq)$
initial	0.120 M		0		0
equil	(0.120 − x) M		x M		x M

$$K_b = \frac{[C_5H_5NH^+][OH^-]}{[C_5H_5N]} = \frac{x^2}{(0.120-x)} \approx \frac{x^2}{0.120} = 1.7 \times 10^{-9}$$

$$x^2 = 0.120(1.7 \times 10^{-9}); \; x = 1.428 \times 10^{-5} = 1.4 \times 10^{-5} \, M \, OH^-; \; pH = 9.15$$

Check. $\dfrac{1.4 \times 10^{-5} \, M \, OH^-}{0.120 \, M \, C_5H_5N} \times 100 = 0.012\%$ ionization; the approximation is valid

16.61   *Analyze/Plan.* $K_a = 10^{-pK_a}$. Follow the logic in Sample Exercise 16.13.   *Solve.*

Let $[H^+] = [NC_7H_4SO_3^-] = z$. $K_a = $ antilog $(-2.32) = 4.79 \times 10^{-3} = 4.8 \times 10^{-3}$

$\dfrac{z^2}{0.10-z} = 4.79 \times 10^{-3}$.   Since K_a is relatively large, solve the quadratic.

$$z^2 + 4.79 \times 10^{-3} \, z - 4.79 \times 10^{-4} = 0$$

$$z = \frac{-4.79 \times 10^{-3} \pm \sqrt{(4.79 \times 10^{-3})^2 - 4(1)(-4.79 \times 10^{-4})}}{2(1)} = \frac{-4.79 \times 10^{-3} \pm \sqrt{1.937 \times 10^{-3}}}{2}$$

$$z = 1.96 \times 10^{-2} = 2.0 \times 10^{-2} \, M \, H^+; \; pH = -\log(1.96 \times 10^{-2}) = 1.71$$

16.63   *Analyze/Plan.* Follow the logic in Sample Exercise 16.13.   *Solve.*

(a)

	$HN_3(aq)$	$\rightleftharpoons$	$H^+(aq)$	+	$N_3^-(aq)$
initial	0.400 M		0		0
equil	(0.400 − x) M		x M		x M

$$K_a = \frac{[H^+][N_3^-]}{[HN_3]} = 1.9 \times 10^{-5}; \; \frac{x^2}{(0.400-x)} \approx \frac{x^2}{0.400} = 1.9 \times 10^{-5}$$

$$x = 0.00276 = 2.8 \times 10^{-3} \, M = [H^+]; \; \% \text{ ionization} = \frac{2.76 \times 10^{-3}}{0.400} \times 100 = 0.69\%$$

(b) $1.9 \times 10^{-5} \approx \dfrac{x^2}{0.100}$; $x = 0.00138 = 1.4 \times 10^{-3} \ M \ H^+$

$\text{\% ionization} = \dfrac{1.38 \times 10^{-3} \ M \ H^+}{0.100 \ M \ HN_3} \times 100 = 1.4\%$

(c) $1.9 \times 10^{-5} \approx \dfrac{x^2}{0.0400}$; $x = 8.72 \times 10^{-4} = 8.7 \times 10^{-4} \ M \ H^+$

$\text{\% ionization} = \dfrac{8.72 \times 10^{-4} \ M \ H^+}{0.0400 \ M \ HN_3} \times 100 = 2.2\%$

Check. Notice that a tenfold dilution [part (a) versus part (c)] leads to a slightly more than threefold increase in percent ionization.

16.65   *Analyze/Plan.* Let the weak acid be HX. $HX(aq) \rightleftharpoons H^+(aq) + X^-\ (aq)$. Solve the K_a expression symbolically for $[H^+]$ in terms of $[HX]$. Substitute into the formula for % ionization, $([H^+]/[HX]) \times 100$.   *Solve.*

$K_a = \dfrac{[H^+][X^-]}{[HX]}$; $[H^+] = [X^-] = y$; $K_a = \dfrac{y^2}{[HX] - y}$; assume that % ionization is small

$K_a = \dfrac{y^2}{[HX]}$; $y = K_a^{1/2}[HX]^{1/2}$

$\text{\% ionization} = \dfrac{y}{[HX]} \times 100 = \dfrac{K_a^{1/2}[HX]^{1/2}}{[HX]} \times 100 = \dfrac{K_a^{1/2}}{[HX]^{1/2}} \times 100$

That is, percent ionization varies inversely as the square root of concentration HX.

16.67   *Analyze/Plan.* Follow the logic in Sample Exercise 16.14. Citric acid is a triprotic acid with three K_a values that do not differ by more than 10^3. We must consider all three steps. Also, $C_6H_5O_7^{3-}$ is only produced in step 3.

Solve. <u>Assumptions</u> are explained as they are used in the solution.

$H_3C_6H_5O_7(aq) \rightleftharpoons H^+(aq) + H_2C_6H_5O_7^-(aq)$ $K_{a1} = 7.4 \times 10^{-4}$

$H_2C_6H_5O_7^-(aq) \rightleftharpoons H^+(aq) + HC_6H_5O_7^{2-}(aq)$ $K_{a2} = 1.7 \times 10^{-5}$

$HC_6H_5O_7^{2-}(aq) \rightleftharpoons H^+(aq) + C_6H_5O_7^{3-}(aq)$ $K_{a3} = 4.0 \times 10^{-7}$

To calculate the pH of a 0.040 *M* solution, <u>assume</u> initially that only the first ionization is important:

	$H_3C_6H_5O_7(aq)$	$\rightleftharpoons$	$H^+(aq)$	$+$	$H_2C_6H_5O_7^-(aq)$
initial	0.040 *M*		0		0
equil.	(0.040 − x) *M*		x *M*		x *M*

$K_{a_1} = \dfrac{[H^+][H_2C_6H_5O_7^-]}{[H_3C_6H_5O_7]} = \dfrac{x^2}{(0.040 - x)} = 7.4 \times 10^{-4}$

$x^2 = (0.040 - x)(7.4 \times 10^{-4})$; $\underline{x^2 \approx (0.040)(7.4 \times 10^{-4})}$; $x = 0.00544 = 5.4 \times 10^{-3} \, M$

Since this value for x is rather large in relation to 0.050, a better approximation for x can be obtained by substituting this first estimate into the expression for x^2, then solving again for x:

$$x^2 = (0.040 - x)(7.4 \times 10^{-4}) = (0.040 - 5.44 \times 10^{-3})(7.4 \times 10^{-4})$$

$$x^2 = 2.557 \times 10^{-5}; x = 5.057 \times 10^{-3} = 5.1 \times 10^{-3} \, M$$

(This is the same result obtained from the quadratic formula.)

The correction to the value of x, though not large, is significant. Does the second ionization produce a significant additional concentration of H^+?

	$H_2C_6H_5O_7^-(aq)$	$\rightleftharpoons$	$H^+(aq)$	+	$HC_6H_5O_7^{2-}(aq)$
initial	$5.1 \times 10^{-3} \, M$		$5.1 \times 10^{-3} \, M$		0
equil.	$(5.1 \times 10^{-3} - y)$		$(5.1 \times 10^{-3} + y)$		y

$$K_{a_2} = \frac{[H^+][HC_6H_5O_7^{2-}]}{[H_2C_6H_5O_7^-]} = 1.7 \times 10^{-5}; \frac{(5.1 \times 10^{-3} + y)(y)}{(5.1 \times 10^{-3} - y)} = 1.7 \times 10^{-5}$$

Assume that y is small relative to 5.1×10^{-3}; that is, that additional ionization of $H_2C_6H_5O_7^-$ is small, then

$$\frac{(5.1 \times 10^{-3})y}{(5.1 \times 10^{-3})} = 1.7 \times 10^{-5} \, M; y = 1.7 \times 10^{-5} \, M$$

This value is indeed small compared to $5.1 \times 10^{-3} \, M$; $[H^+]$ and pH are determined by the first ionization step. If we were only interested in pH, we could stop here. However, to calculate $[C_6H_5O_7^{3-}]$, we must consider the third ionization, with adjusted $[H^+] = 5.1 \times 10^{-3} + 1.7 \times 10^{-5} = 5.12 \times 10^{-3} \, M \, (= 5.1 \times 10^{-3})$

	$HC_6H_5O_7^{2-}$	$\rightleftharpoons$	$H^+(aq)$	+	$C_6H_5O_7^{3-}(aq)$
initial	$1.7 \times 10^{-5} \, M$		$5.12 \times 10^{-3} \, M$		0
equil.	$1.7 \times 10^{-5} - z$		$5.12 \times 10^{-3} + z$		z

$$K_{a_3} = \frac{[H^+][C_6H_5O_7^{3-}]}{[HC_6H_5O_7^{2-}]} = \frac{(5.12 \times 10^{-3} + z)(z)}{(1.7 \times 10^{-5} - z)} = 4.0 \times 10^{-7}$$

Assume z is small relative to 5.12×10^{-3}, but not relative to 1.7×10^{-5}.

$(4.0 \times 10^{-7})(1.7 \times 10^{-5} - z) = 5.12 \times 10^{-3} \, z; 6.8 \times 10^{-12} - 4.0 \times 10^{-7} z = 5.12 \times 10^{-3} z;$

$6.8 \times 10^{-12} = 5.12 \times 10^{-3} z + 4.0 \times 10^{-7} z = 5.12 \times 10^{-3} z; z = 1.33 \times 10^{-9} = 1.3 \times 10^{-9} \, M$

$[C_6H_5O_7^{3-}] = 1.3 \times 10^{-9} \, M; [H^+] = 5.12 \times 10^{-3} \, M + 1.3 \times 10^{-9} \, M = 5.1 \times 10^{-3} \, M$

$pH = -\log(5.12 \times 10^{-3}) = 2.29$

The concentration of citrate ion, $[C_6H_5O_7^{3-}]$, is much less than $[H^+]$. Note that neither the second nor third ionizations contributed significantly to $[H^+]$ and pH.

Weak Bases (section 16.7)

16.69   (a) $HONH_3^+$

       (b)   When hydroxylamine acts as a base, the nitrogen atom accepts a proton.

       (c)   14 e⁻, 7 e⁻ pairs

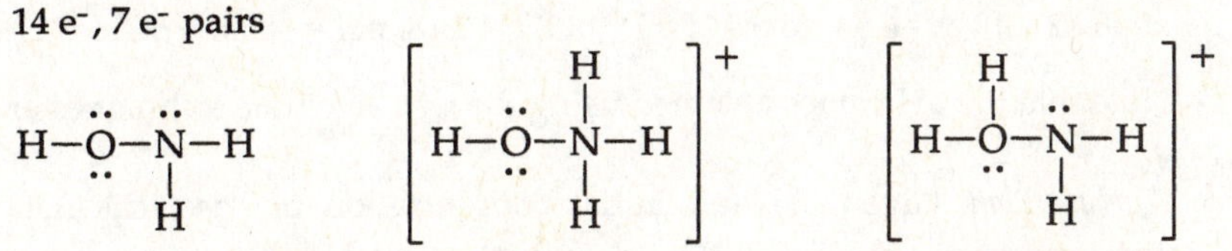

                                             FC on N is +1   FC on O is +1

            In neutral hydroxylamine, both O and N have zero formal charges. Nitrogen is less electronegative than oxygen, and more likely to share a lone pair of electrons with an incoming (and electron deficient) H^+. The resulting cation with the +1 formal charge on N is more stable than the one with the +1 formal charge on O.

16.71   *Analyze/Plan.* Remember that K_b = [products]/[reactants]. If $H_2O(l)$ appears in the equilibrium reaction, it will not appear in the K_b expression, because it is a pure liquid. *Solve.*

       (a) $(CH_3)_2NH(aq) + H_2O(l) \rightleftharpoons (CH_3)_2NH_2^+(aq) + OH^-(aq); K_b = \dfrac{[(CH_3)_2NH_2^+][OH^-]}{[(CH_3)_2NH]}$

       (b) $CO_3^{2-}(aq) + H_2O(l) \rightleftharpoons HCO_3^-(aq) + OH^-(aq); K_b = \dfrac{[HCO_3^-][OH^-]}{[CO_3^{2-}]}$

       (c) $HCOO^-(aq) + H_2O(l) \rightleftharpoons HCOOH(aq) + OH^-(aq); K_b = \dfrac{[HCOOH][OH^-]}{[HCOO^-]}$

16.73   *Analyze/Plan.* Follow the logic in Sample Exercise 16.15.   *Solve.*

$$C_2H_5NH_2(aq) + H_2O(l) \rightleftharpoons C_2H_5NH_3^+(aq) + OH^-(aq)$$

initial	0.075 M	0	0
equil.	(0.075 − x) M	x M	x M

$$K_b = \frac{[C_2H_5NH_3^+][OH^-]}{[C_2H_5NH_2]} = \frac{(x)(x)}{(0.075 - x)} = \frac{x^2}{0.075} = 6.4 \times 10^{-4}$$

$$x^2 = 0.075\,(6.4 \times 10^{-4}); \quad x = [OH^-] = 6.9 \times 10^{-3}\,M; \quad pH = 11.84$$

Check. $\dfrac{6.9 \times 10^{-3}\,M\,OH^-}{0.075\,M\,C_2H_5NH_2} \times 100 = 9.2\%$ ionization; the assumption is not valid

To obtain a more precise result, the K_b expression is rewritten in standard quadratic form and solved via the quadratic formula.

$$\frac{x^2}{0.075-x} = 6.4 \times 10^{-4}; \; x^2 + 6.4 \times 10^{-4}\,x - 4.8 \times 10^{-5} = 0$$

$$x = \frac{-b \pm \sqrt{b^2 - 4ac}}{2a} = \frac{-6.4 \times 10^{-4} \pm \sqrt{(6.4 \times 10^{-4})^2 - 4(1)(-4.8 \times 10^{-5})}}{2}$$

$x = 6.62 \times 10^{-3} = 6.6 \times 10^{-3}\,M\,OH^-$; pOH = 2.18, pH = 14.00 – pOH = 11.82

Note that the pH values obtained using the two algebraic techniques are very similar.

16.75   *Analyze/Plan.* Given pH and initial concentration of base, calculate all equilibrium concentrations. pH $\rightarrow$ pOH $\rightarrow$ [OH$^-$] at equilibrium. Construct the equilibrium table and calculate other equilibrium concentrations. Substitute into the K_b expression and calculate K_b. *Solve.*

(a)   [OH$^-$] = 10^{-pOH}; pOH = 14 – pH = 14.00 – 11.33 = 2.67

[OH$^-$] = $10^{-2.67}$ = 2.138×10^{-3} = $2.1 \times 10^{-3}\,M$

$$C_{10}H_{15}ON(aq) + H_2O(l) \;\rightleftharpoons\; C_{10}H_{15}ONH^+(aq) \;\; + \;\; OH^-(aq)$$

initial	0.035 M	0	0
equil.	0.033 M	$2.1 \times 10^{-3}\,M$	$2.1 \times 10^{-3}\,M$

(b) $K_b = \dfrac{[C_{10}H_{15}ONH^+][OH^-]}{[C_{10}H_{15}ON]} = \dfrac{(2.138 \times 10^{-3})^2}{(0.03286)} = 1.4 \times 10^{-4}$

The K_a – K_b Relationship; Acid-Base Properties of Salts
(sections 16.8 and 16.9)

16.77   (a)   For a conjugate acid/conjugate base pair such as $C_6H_5OH/C_6H_5O^-$, K_b for the conjugate base is always K_w/K_a for the conjugate acid. K_b for the conjugate base can always be calculated from K_a for the conjugate acid, so a separate list of K_b values is not necessary.

(b) $K_b = K_w/K_a = 1.0 \times 10^{-14} / 1.3 \times 10^{-10} = 7.7 \times 10^{-5}$

(c) K_b for phenolate (7.7×10^{-5}) > K_b for ammonia (1.8×10^{-5}).

Phenolate is a stronger base than NH_3.

16.79   *Analyze/Plan.* Given K_a, determine relative strengths of the acids and their conjugate bases. The greater the magnitude of K_a, the stronger the acid and the weaker the conjugate base. K_b (conjugate base) = K_w/K_a. *Solve.*

(a)   Acetic acid is stronger, because it has the larger K_a value.

(b)   Hypochlorite ion is the stronger base because the weaker acid, hypochlorous acid, has the stronger conjugate base.

(c) K_b for $CH_3COO^- = K_w/K_a$ for $CH_3COOH = 1.0 \times 10^{-14}/1.8 \times 10^{-5} = 5.6 \times 10^{-10}$

K_b for $ClO^- = K_w/K_a$ for $HClO = 1 \times 10^{-14}/3.0 \times 10^{-8} = 3.3 \times 10^{-7}$

Note that K_b for ClO^- is greater than K_b for CH_3COO^-.

16.81   *Analyze.* When the solute in an aqueous solution is a salt, evaluate the acid/base properties of the component ions.

(a)   *Plan.* NaBrO is a soluble salt and thus a strong electrolyte. When it is dissolved in H_2O, it dissociates completely into Na^+ and BrO^-. [NaBrO] = $[Na^+]$ = $[BrO^-]$ = 0.10 M. Na^+ is the conjugate acid of the strong base NaOH and thus does not influence the pH of the solution. BrO^-, on the other hand, is the conjugate base of the weak acid HBrO and **does** influence the pH of the solution. Like any other weak base, it hydrolyzes water to produce OH^-(aq). Solve the equilibrium problem to determine $[OH^-]$.   *Solve.*

$$BrO^-(aq) \ + \ H_2O(l) \ \rightleftharpoons \ HBrO(aq) \ + \ OH^-(aq)$$

initial	0.10 M	0	0
equil.	(0.10 – x) M	x M	x M

$$K_b \text{ for } BrO^- = \frac{[HBrO][OH^-]}{[BrO^-]} = \frac{K_w}{K_a \text{ for } HBrO} = \frac{1 \times 10^{-14}}{2.5 \times 10^{-9}} = 4.00 \times 10^{-6} = 4.0 \times 10^{-6}$$

$$4.00 \times 10^{-6} = \frac{(x)(x)}{(0.10-x)}; \text{ assume the percent of } BrO^- \text{ that hydrolyzes is small}$$

$$x^2 = 0.10 \, (4.00 \times 10^{-6}); \, x = [OH^-] = 6.32 \times 10^{-4} = 6.3 \times 10^{-4} \, M$$

pOH = 3.20; pH = 14 – 3.20 = 10.80

(b)   *Plan.* NaHS(aq) $\rightarrow$ Na^+(aq) + HS^-(aq)

HS^- is the conjugate base of H_2S and its hydrolysis reaction will determine the $[OH^-]$ and pH of the solution (see similar explanation for NaBrO in part (a)). We will assume the process HS^-(aq) $\rightleftharpoons$ H^+(aq) + S^-(aq) will not significantly affect the $[OH^-]$ in solution because K_{a2} for H_2S is so small. Solve the equilibrium problem for $[OH^-]$.   *Solve.*

$$HS^-(aq) \ + \ H_2O(l) \ \rightleftharpoons \ H_2S(aq) + OH^-(aq)$$

initial	0.080 M	0	0
equil.	(0.080 – x) M	x	x

$$K_b = \frac{[H_2S][OH^-]}{[HS^-]} = \frac{K_w}{K_a \text{ for } H_2S} = \frac{1.0 \times 10^{-14}}{9.5 \times 10^{-8}} = 1.053 \times 10^{-7} = 1.1 \times 10^{-7}$$

$$1.053 \times 10^{-7} = \frac{x^2}{(0.080-x)}; \, x^2 = 0.080 \, (1.053 \times 10^{-7}); \, x = 9.177 \times 10^{-5} = 9.2 \times 10^{-5} \, M \, OH^-$$

(Assume x is small compared to 0.080); pOH = 4.04; pH = 14 – 4.04 = 9.96

Check. $\frac{9.2 \times 10^{-5} \, M \, OH^-}{0.080 \, M \, HS^-} \times 100 = 0.12\%$ hydrolysis; the approximation is valid

(c)   *Plan.* For the two salts present, Na^+ and Ca^{2+} are negligible acids. NO_2^- is the conjugate base of HNO_2 and will determine the pH of the solution. *Solve.*

Calculate total $[NO_2^-]$ present initially.

$[NO_2^-]_{total} = [NO_2^-]$ from $NaNO_2 + [NO_2^-]$ from $Ca(NO_2)_2$

$[NO_2^-]_{total} = 0.10\ M + 2(0.20\ M) = 0.50\ M$

The hydrolysis equilibrium is:

$$NO_2^-(aq) + H_2O(l) \rightleftharpoons HNO_2 + OH^-(aq)$$

initial	$0.50\ M$	0	0
equil.	$(0.50 - x)\ M$	$x\ M$	$x\ M$

$$K_b = \frac{[HNO_2][OH^-]}{[NO_2^-]} = \frac{K_w}{K_a \text{ for } HNO_2} = \frac{1.0 \times 10^{-14}}{4.5 \times 10^{-4}} = 2.22 \times 10^{-11} = 2.2 \times 10^{-11}$$

$$2.2 \times 10^{-11} = \frac{x^2}{(0.50 - x)} \approx \frac{x^2}{0.50}; x^2 = 0.50\ (2.22 \times 10^{-11})$$

$$x = 3.33 \times 10^{-6} = 3.3 \times 10^{-6}\ M\ OH^-; pOH = 5.48; pH = 14 - 5.48 = 8.52$$

16.83   *Analyze/Plan.* Given the formula of a salt, predict whether an aqueous solution will be acidic, basic, or neutral. Evaluate the acid-base properties of both ions and determine the overall effect on solution pH. *Solve.*

(a)   acidic; NH_4^+ is a weak acid, Br^- is negligible.

(b)   acidic; Fe^{3+} is a highly charged metal cation and a Lewis acid; Cl^- is negligible.

(c)   basic; CO_3^{2-} is the conjugate base of HCO_3^-; Na^+ is negligible.

(d)   neutral; both K^+ and ClO_4^- are negligible.

(e)   acidic; $HC_2O_4^-$ is amphoteric, but K_a for the acid dissociation (6.4×10^{-5}) is much greater than K_b for the base hydrolysis $(1.0 \times 10^{-14} / 5.9 \times 10^{-2} = 1.7 \times 10^{-13})$.

16.85   *Plan.* Estimate pH using relative base strength and then calculate to confirm prediction. NaCl is a neutral salt, so it is not the unknown. The unknown is a relatively weak base, because a pH of 8.08 is not very basic. Since F^- is a weaker base than OCl^-, the unknown is probably NaF. Calculate K_b for the unknown from the data provided. *Solve.*

$[OH^-] = 10^{-pOH}; pOH = 14.00 - pH = 14.00 - 8.08 = 5.92$

$[OH^-] = 10^{-5.92} = 1.202 \times 10^{-6} = 1.2 \times 10^{-6}\ M = [HX]$

$[NaX] = [X^-] = 0.050\ \text{mol salt}/0.500\ L = 0.10\ M$

$$K_b = \frac{[OH^-][HX]}{[X^-]} = \frac{(1.202 \times 10^{-6})^2}{(0.10 - 1.2 \times 10^{-6})} = \frac{(1.202 \times 10^{-6})^2}{0.10} = 1.4 \times 10^{-11}$$

K_b for $F^- = K_w/K_a$ for $HF = 1.0 \times 10^{-14}/6.8 \times 10^{-4} = 1.5 \times 10^{-11}$

The unknown is NaF.

Acid-Base Character and Chemical Structure (section 16.10)

16.87   (a)   As the electronegativity of the central atom (X) increases, more electron density is withdrawn from the X–O and O–H bonds, respectively. In water, the O–H bond is ionized to a greater extent and the strength of the oxyacid increases.

       (b)   As the number of nonprotonated oxygen atoms in the molecule increases, they withdraw electron density from the other bonds in the molecule and the strength of the oxyacid increases.

16.89   (a) HNO_3 is a stronger acid than HNO_2 because it has one more nonprotonated oxygen atom, and thus a higher oxidation number on N.

       (b)   For binary hydrides, acid strength increases going down a family, so H_2S is a stronger acid than H_2O.

       (c) H_2SO_4 is a stronger acid because H^+ is much more tightly held by the anion HSO_4^-.

       (d)   For oxyacids, the greater the electronegativity of the central atom, the stronger the acid, so H_2SO_4 is a stronger acid than H_2SeO_4.

       (e) CCl_3COOH is stronger because the electronegative Cl atoms withdraw electron density from other parts of the molecule, which weakens the O–H bond and makes H^+ easier to remove. Also, the electronegative Cl delocalizes negative charge on the carboxylate anion. This stabilizes the conjugate base, favoring products in the ionization equilibrium and increasing K_a.

16.91   (a) BrO^- (HClO is the stronger acid due to a more electronegative central atom, so BrO^- is the stronger base.)

       (b) BrO^- ($HBrO_2$ has more nonprotonated O atoms and is the stronger acid, so BrO^- is the stronger base.)

       (c) HPO_4^{2-} (larger negative charge, greater attraction for H^+)

16.93   (a)   True.

       (b)   False. In a series of acids that have the same central atom, acid strength increases with the number of nonprotonated oxygen atoms bonded to the central atom.

       (c)   False. H_2Te is a stronger acid than H_2S because the H–Te bond is longer, weaker, and more easily dissociated than the H–S bond.

Lewis Acids and Bases (section 16.11)

16.95   Yes. If a substance is an Arrhenius base, it must also be a Brønsted-Lowry base and a Lewis base. The Arrhenius definition (hydroxide ion) is the most restrictive, the Brønsted-Lowry (H^+ acceptor) more general and the Lewis (electron pair donor) most general. Since a hydroxide ion is both an H^+ acceptor and an electron pair donor, any substance that fits the narrow Arrhenius definition will fit the broader Brønsted-Lowry and Lewis definitions.

16.97   *Analyze/Plan.* Identify each reactant as an electron pair donor (Lewis base) or electron pair acceptor (Lewis acid). Remember that a Brønsted-Lowry acid is necessarily a Lewis acid, and a Brønsted-Lowry base is necessarily a Lewis base (Solution 16.95).   *Solve.*

	Lewis Acid	**Lewis Base**
(a)	$Fe(ClO_4)_3$ or Fe^{3+}	H_2O
(b)	H_2O	CN^-
(c)	BF_3	$(CH_3)_3N$
(d)	HIO	NH_2^-

16.99   (a) Cu^{2+}, higher cation charge

(b) Fe^{3+}, higher cation charge

(c) Al^{3+}, smaller cation radius, same charge

Additional Exercises

16.101   To compare base strength, compare either K_b or pK_b values. The stronger the base, the larger the K_b and the smaller the pK_b. (pK_b is the negative log of K_b; pK_b increases as base strength decreases.)

pK_b for NH_3 = $-\log (1.8 \times 10^{-5})$ = 4.74. pK_b for $(C_2H_5)_3N$ = 2.99

$(C_2H_5)_3N$ is a stronger base than NH_3 by virtue of its smaller pK_b.

16.104 $H_3C_6H_5O_7$ + CH_3NH_2 → $CH_3NH_3^+$ + $H_2C_6H_5O_7^-$

citric acid   methylamine   odorless salt

$$H_3C_6H_5O_7 \rightleftharpoons H^+ + H_2C_6H_5O_7^- \qquad K_{a_1} = 7.4 \times 10^{-4}$$

$$CH_3NH_2 + H_2O \rightleftharpoons CH_3NH_3^+ + OH^- \qquad K_b = 4.4 \times 10^{-4}$$

$$H^+ + OH^- \rightleftharpoons H_2O \qquad 1/K_w = 1/1.0 \times 10^{-14}$$

$$\overline{H_3C_6H_5O_7 + CH_3NH_2 + H_2O + H^+ + OH^- \rightleftharpoons H_2C_6H_5O_7^- + CH_3NH_3^+ + H^+ + OH^- + H_2O}$$

$$H_3C_6H_5O_7 + CH_3NH_2 \rightleftharpoons H_2C_6H_5O_7^- + CH_3NH_3^+$$

$$K = \frac{K_{a_1} \times K_b}{K_w} = \frac{(7.4 \times 10^{-4})(4.4 \times 10^{-4})}{1.0 \times 10^{-14}} = 3.256 \times 10^7 = 3.3 \times 10^7$$

16.107   The solution with the higher pH has the lower $[H^+]$.

(a)   For solutions with equal concentrations, the weaker acid will have a lower $[H^+]$ and higher pH.

(b)   The acid with $K_a = 8 \times 10^{-6}$ is the weaker acid, so it has the higher pH.

(c)   The base with $pK_b = 4.5$ is the stronger base, has greater $[OH^-]$ and smaller $[H^+]$, so higher pH.

16.109 $K_a = \dfrac{[H^+][C_5H_{11}COO^-]}{[C_5H_{11}COOH]}$; $[H^+] = [C_5H_{11}COO^-] = 10^{-pH} = 10^{-2.94} = 0.001148$

$$= 1.1 \times 10^{-3} \, M$$

$[C_5H_{11}COOH] = \dfrac{11 \, g \, C_5H_{11}COOH}{L} \times \dfrac{1 \, mol \, C_5H_{11}COOH}{116.16 \, g \, C_5H_{11}COOH} = 0.09470 = 0.095 \, M$

$K_a = \dfrac{[H^+][C_5H_{11}COO^-]}{[C_5H_{11}COOH]} = \dfrac{(0.001148)^2}{(0.09470 - 0.001148)} = 1.4092 \times 10^{-5} = 1.4 \times 10^{-5}$

Integrative Exercises

16.115   At 25°C, $[H^+] = [OH^-] = 1.0 \times 10^{-7} \, M$

$$\dfrac{1.0 \times 10^{-7} \, mol \, H^+}{1 L \, H_2O} \times 0.0010 \, L \times \dfrac{6.022 \times 10^{23} \, H^+ \, ions}{mol \, H^+} = 6.0 \times 10^{13} \, H^+ \, ions$$

16.118   *Analyze.* If pH were directly related to CO_2 concentration, this exercise would be simple. Unfortunately, we must solve the equilibrium problem for the diprotic acid H_2CO_3 to calculate $[H^+]$ and pH. We are given ppm CO_2 in the atmosphere at two different times, and the pH that corresponds to one of these CO_2 levels. We are asked to find pH at the other atmospheric CO_2 level.

Plan. Assume all dissolved CO_2 is present as H_2CO_3 (aq) (Sample Exercise 16.14).

pH $\rightarrow$ $[H^+]$ $\rightarrow$ $[H_2CO_3]$. While H_2CO_3 is a diprotic acid, the two K_a values differ by more than 10^3, so we can ignore the second ionization when calculating $[H_2CO_3]$. Change 380 ppm CO_2 to pressure and calculate the Henry's law constant for CO_2. Calculate the dissolved $[CO_2] = [H_2CO_3]$ at 315 ppm, then solve the K_{a1} expression for $[H^+]$ and pH.

(a)   *Solve.* $H_2CO_3(aq) \rightleftharpoons H^+(aq) + HCO_3^-$ (aq)

$K_{a1} = 4.3 \times 10^{-7} = \dfrac{[H^+][HCO_3^-]}{[H_2CO_3]}$; $[H^+] = 10^{-5.4} = 3.98 \times 10^{-6} = 4 \times 10^{-6} \, M$

$[H^+] = [HCO_3^-]$; $[H_2CO_3] = x - 4 \times 10^{-6}$

$4.3 \times 10^{-7} = \dfrac{(3.98 \times 10^{-6})^2}{(x - 3.98 \times 10^{-6})}$; $4.3 \times 10^{-7} x = 1.585 \times 10^{-11} + 1.711 \times 10^{-12}$

$x = 1.756 \times 10^{-11}/4.3 \times 10^{-7} = 4.084 \times 10^{-5} = 4 \times 10^{-5} \, M \, H_2CO_3$

380 ppm = 380 mol CO_2/1 × 10^6 mol air = 0.000380 mol % CO_2

Because of the properties of gases, mol % = pressure %. $P_{CO_2} = 0.000380$ atm.

According to Equation [13.4], $S_{CO_2} = kP_{CO_2}$;

4.084×10^{-5} mol/L = k(3.80×10^{-4} atm)   k = 0.1075 = 0.1 mol/L-atm.

Forty years ago, $S_{CO_2} = 0.1075 \dfrac{mol}{L \cdot atm} \times 3.15 \times 10^{-4} \, atm = 3.385 \times 10^{-5}$

$$= 3 \times 10^{-5} \, M$$

Now solve K_{a1} for $[H^+]$ at this $[H_2CO_3]$. $[H^+] = x$

We cannot assume x is small, because $[H_2CO_3]$ is so low.

$4.3 \times 10^{-7} = x^2 / (3.385 \times 10^{-5} - x); \; x^2 + 4.3 \times 10^{-7} x - 1.456 \times 10^{-11} = 0$

$$x = \frac{-4.3 \times 10^{-7} \pm \sqrt{(4.3 \times 10^{-7})^2 - 4(-1.456 \times 10^{-11})}}{2} = \frac{-4.3 \times 10^{-7} + 7.644 \times 10^{-6}}{2}$$

$$= 3.607 \times 10^{-6} = 4 \times 10^{-6} \, M \, H^+$$

$[H^+] = 4 \times 10^{-6} \, M$, pH = 5.443 = 5.4

(Note that, to the precision that the pH data is reported, the change in atmospheric CO_2 leads to no change in pH.)

(b)   From part (a), $[H_2CO_3]$ today $= 4.084 \times 10^{-5} \, M$

$$\frac{4.084 \times 10^{-5} \, mol \, H_2CO_3}{1 \, L} \times 20.0 \, L = 8.168 \times 10^{-4} = 8 \times 10^{-4} \, mol \, CO_2$$

$$V = \frac{nRT}{P} = 8.168 \times 10^{-4} \, mol \times \frac{298 \, K}{1.0 \, atm} \times \frac{0.08206 \, L \cdot atm}{mol \cdot K} = 0.01997 = 0.02 \, L = 20 \, mL$$

16.119 (a)   24 valence e^-, 12 e^- pairs

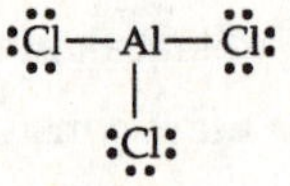

The formal charges on all atoms are zero. Structures with multiple bonds lead to nonzero formal charges. There are three electron domains about Al. The electron-domain geometry and molecular structure are trigonal planar.

(b)   The Al atom in $AlCl_3$ has an incomplete octet and is electron deficient. It "needs" to accept another electron pair, to act like a Lewis acid.

(c)

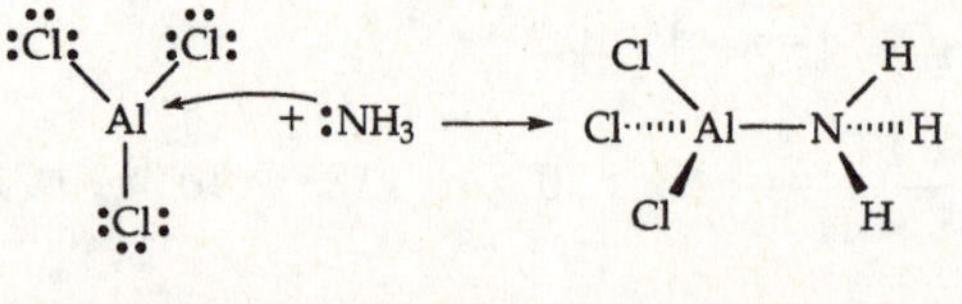

Both the Al and N atoms in the product have tetrahedral geometry.

(d)   The Lewis theory is most appropriate. H^+ and $AlCl_3$ are both electron pair acceptors, Lewis acids.

16.121   Rx 1: $\Delta H = D(H–F) + 2D(H–O) – 3D(H–O) = D(H–F) – D(H–O)$

$$\Delta H = 567 \text{ kJ} – 463 \text{ kJ} = 104 \text{ kJ}$$

Rx 2: $\Delta H = D(H–Cl) + 2D(H–O) – 3D(H–O) = D(H–Cl) – D(H–O)$

$$\Delta H = 431 \text{ kJ} – 463 \text{ kJ} = –32 \text{ kJ}$$

The reaction involving HCl is exothermic, while the reaction involving HF is endothermic, owing to the smaller bond dissociation enthalpy of H–Cl. HCl is a stronger acid than HF, and the enthalpy of ionization for HCl is exothermic, while that of HF is endothermic. This is consistent with the trend in acid strength for binary acids with heavy atoms (X) in the same family. That is, the longer and weaker the H–X bond, the stronger the acid (and the more exothermic the ionization reaction).

16.124   (a)   (i) $HCO_3^-(aq) \rightleftharpoons H^+(aq) + CO_3^{2-}(aq)$ $K_1 = K_{a2}$ for $H_2CO_3 = 5.6 \times 10^{-11}$

                        $H^+(aq) + OH^-(aq) \rightleftharpoons H_2O(l)$ $K_2 = 1/K_w = 1 \times 10^{14}$

                        $HCO_3^-(aq) + OH^-(aq) \rightleftharpoons CO_3^{2-}(aq) + H_2O(l)$ $K = K_1 \times K_2 = 5.6 \times 10^3$

            (ii) $NH_4^+(aq) \rightleftharpoons H^+(aq) + NH_3(aq)$ $K_1 = K_a$ for $NH_4^+ = 5.6 \times 10^{-10}$

                        $CO_3^{2-}(aq) + H^+(aq) \rightleftharpoons HCO_3^-(aq)$ $K_2 = 1/K_{a2}$ for $H_2CO_3 = 1.8 \times 10^{10}$

                        $NH_4^+(aq) + CO_3^{2-}(aq) \rightleftharpoons HCO_3^-(aq) + NH_3(aq)$ $K = K_1 \times K_2 = 10$

   (b)   Both (i) and (ii) have K > 1, although K = 10 is not **much** greater than 1. Both could be written with a single arrow. (This is true in general when a strong acid or strong base, $H^+(aq)$ or $OH^-(aq)$, is a reactant.)

17 Additional Aspects of Aqueous Equilibria

Visualizing Concepts

17.1   *Analyze.* Given diagrams showing equilibrium mixtures of HX and X⁻ with different compositions, decide which has the highest pH. HX is a weak acid and X⁻ is its conjugate base. *Plan.* Evaluate the contents of the boxes. Use acid-base equilibrium principles to relate [H⁺] to box composition.   *Solve.*

Use the following acid ionization equilibrium to describe the mixtures:
$HX(aq) \rightleftharpoons H^+(aq) + X^-(aq)$. Each box has 4 HX molecules, but differing amounts of X⁻ ions. The greater the amount of X⁻ (conjugate base), for the same amount of HX (weak acid), the lower the amount of H⁺ and the higher the pH. The middle box, with most X⁻, has least H⁺ and highest pH.

17.4   *Analyze/Plan.* When strong acid is added to a buffer, it reacts with conjugate base (CB) to produce conjugate acid (CA). [CA] increases and [CB] decreases. The opposite happens when strong base is added to a buffer, [CB] increases and [CA] decreases. Match these situations to the drawings.   *Solve.*

The buffer begins with equal concentrations of HX and X⁻.

   (a)   After addition of strong acid, [HX] will increase and [X⁻] will decrease. Drawing (3) fits this description.

   (b)   Adding of strong base causes [HX] to decrease and [X⁻] to increase. Drawing (1) matches the description.

   (c)   Drawing (2) shows both [HX] and [X⁻] to be smaller than the initial concentrations shown on the left. This situation cannot be achieved by adding strong acid or strong base to the original buffer.

17.7   *Analyze.* Given two titration curves where 0.10 *M* NaOH is the titrant, decide which represents the more concentrated acid, and which the stronger acid.

   Plan. For equal volumes of acid, concentration is related to volume of titrant (0.10 *M* NaOH) at the equivalence points. To determine K_a, pH = pK_a half-way to the equivalence point.   *Solve.*

   (a)   Both acids have one ionizable hydrogen, because there is one "jump" in each titration curve. For equal volumes of acid, and the same titrant, the more concentrated acid requires a greater volume of titrant to reach equivalence. The equivalence point of the blue curve is at 25 mL NaOH, the red curve at 35 mL NaOH. The red acid is more concentrated.

(b) According to the Henderson-Hasselbach equation, $pH = pK_a + \log \dfrac{[\text{conj. base}]}{[\text{conj. acid}]}$.

At half-way to the equivalence point, [conj. acid] = [conj. base] and $pH = pK_a$ of the conjugate acid. For the blue curve, half-way is 12.5 mL NaOH. The pH at this volume is approximately 7.0. For the red curve, half-way is 17.5 mL NaOH. The pH at this volume is approximately 4.2. A pK_a of 7 corresponds to K_a of 1×10^{-7}, while pK_a of 4.2 corresponds to K_a of 6×10^{-5}. The red acid has the larger K_a value.

Note that the stronger acid, the one with the larger K_a value, has a larger change in pH (jump) at the equivalence point. Also note that initial acid pH was not a definitive measure of acid strength, because the acids have different starting concentrations. Both K_a values and concentration contribute to solution pH.

17.10 *Analyze/Plan.* Calculate the molarity of the solution assuming all $Ca(OH)_2(s)$ dissolves. Use this concentration along with the K_{sp} expression for $Ca(OH)_2$ to answer the questions.

(a) $[Ca(OH)_2] = \dfrac{0.370 \text{ g } Ca(OH)_2}{0.500 \text{ L soln}} \times \dfrac{1 \text{ mol } Ca(OH)_2}{74.093 \text{ g } Ca(OH)_2} = 0.00998745 = 0.00999 \; M$

$[Ca^{2+}] = 0.00999 \; M$; $[OH^-] = 2(0.00998745) = 0.0199749 = 0.0200 \; M$;

$K_{sp} = [Ca^{2+}][OH^-]^2$. Calculate the reaction quotient using the calculated molarities. If it is equal to or greater than K_{sp}, the resulting solution is saturated.

$Q = (0.0098745)(0.0199749)^2 = 3.99 \times 10^{-6}$.

$Q < K_{sp}$ (6.5×10^{-6}) and the solution is not saturated.

(b) Consider the beakers individually.

(i) The 50 mL of 1.0 M HCl is more than enough to neutralize 50 mL of 0.0200 M OH^-(aq). No precipitate forms.

(ii) NaCl does not react with $Ca(OH)_2$ and the two compounds contain no common ions. No precipitate forms.

(iii) $CaCl_2$ does contain a common ion. Calculate Q for the resulting solution to see if $Ca(OH)_2$ precipitates. $[OH^-]$ in the new solution is 0.00999 M, because it is diluted by a factor of 2. $[Ca^{2+}] = (1.0 + 0.00999)/2 = 0.5050 \; M$.

$Q = (0.5050)(0.00999)^2 = 5.04 \times 10^{-5}$.

$Q > K_{sp}$ (6.5×10^{-6}) and $Ca(OH)_2$ precipitates.

(iv) A common ion with a different concentration; $[Ca^{2+}] = (0.10 + 0.00999)/2 = 0.0550 = 0.055 \; M$. $Q = (0.0550)(0.0999)^2 = 5.49 \times 10^{-6}$. $Q \approx K_{sp}$ (6.5×10^{-6}); the solution is very nearly saturated, but no precipitate forms.

Common-Ion Effect (section 17.1)

17.13 (a) The extent of ionization of a weak electrolyte is decreased when a strong electrolyte containing an ion in common with the weak electrolyte is added to it.

(b) $NaNO_2$

17.15   *Analyze/Plan.* Follow the logic in Sample Exercise 17.1.

(a)

	$C_2H_5COOH(aq)$	$\rightleftharpoons$	$H^+(aq)$	+	$C_2H_5COO^-(aq)$
i	0.085 M				0.060 M
c	$-x$		$+x$		$+x$
e	$(0.085 - x)\, M$		$+x\, M$		$(0.060 + x)\, M$

$$K_a = 1.3 \times 10^{-5} = \frac{[H^+][C_2H_5COO^-]}{[C_2H_5COOH]} = \frac{(x)(0.060 + x)}{(0.085 - x)}$$

Assume x is small compared to 0.060 and 0.085.

$$1.3 \times 10^{-5} = \frac{0.060\, x}{0.085}; x = 1.8 \times 10^{-5} = [H^+], \text{pH} = 4.73$$

Check. Since the extent of ionization of a weak acid or base is suppressed by the presence of a conjugate salt, the 5% rule usually holds true in buffer solutions.

(b)

	$(CH_3)_3N(aq) + H_2O(l)$	$\rightleftharpoons$	$(CH_3)_3NH^+(aq)$	+	$OH^-(aq)$
i	0.075 M		0.10 M		
c	$-x$		$+x$		$+x$
e	$(0.075 - x)\, M$		$(0.10 + x)\, M$		$+x\, M$

$$K_b = 6.4 \times 10^{-5} = \frac{[OH^-][(CH_3)_3NH^+]}{[(CH_3)_3N]} = \frac{(x)(0.10 + x)}{(0.075 - x)} \approx \frac{0.10\, x}{0.075}$$

$x = 4.8 \times 10^{-5} = [OH^-]$, pOH = 4.32, pH = 14.00 − 4.32 = 9.68

Check. In a buffer, if [conj. acid] > [conj. base], pH < pK_a of the conj. acid. If [conj. acid] < [conj. base], pH > pK_a of the conj. acid. In this buffer, pK_a of $(CH_3)_3NH^+$ is 9.81. $[(CH_3)_3NH^+] > [(CH_3)_3N]$ and pH = 9.68, less than 9.81.

(c) mol = $M \times$ L; mol CH_3COOH = 0.15 $M \times$ 0.0500 L = 7.5×10^{-3} mol

mol CH_3COO^- = 0.20 $M \times$ 0.0500 L = 0.010 mol

	$CH_3COOH(aq)$	$\rightleftharpoons$	$H^+(aq)$	+	$CH_3COO^-(aq)$
i	7.5×10^{-3} mol		0		0.010 mol
c	$-x$		$+x$		$+x$
e	$(7.5 \times 10^{-3} - x)$ mol		$+x$ mol		$(0.010 + x)$ mol

$[CH_3COOH(aq)] = (7.5 \times 10^{-3} - x)$ mol/0.1000 L;

$[CH_3COO^-(aq)] = (0.010 + x)$ mol/0.1000 L

$$K_a = 1.8 \times 10^{-5} = \frac{[H^+][CH_3COO^-]}{[CH_3COOH]} = \frac{(x)(0.010 + x)/0.1000\ \text{L}}{(0.0075 - x)/0.1000\ \text{L}} \approx \frac{x(0.010)}{0.0075}$$

$x = 1.35 \times 10^{-5}\, M = 1.4 \times 10^{-5}\, M\, H^+$; pH = 4.87

Check. pK_a for CH_3COOH = 4.74. $[CH_3COO^-] > [CH_3COOH]$, pH of buffer = 4.87, greater than 4.74.

17.17 *Analyze/Plan.* We are asked to calculate % ionization of (a) a weak acid and (b) a weak acid in a solution containing a common ion, its conjugate base. Calculate % ionization as in Sample Exercise 16.13. In part (b), the concentration of the common ion is 0.085 M, not x, as in part (a).   *Solve.*

$$buCOOH(aq) \rightleftharpoons H^+(aq) + buCOO^-(aq) \quad K_a = \frac{[H^+][buCOO^-]}{[buCOOH]} = 1.5 \times 10^{-5}$$

equil (a)   0.0075– x M x M x M

equil (b)   0.0075– x M x M 0.085 + x M

(a) $K_a = 1.5 \times 10^{-5} = \frac{x^2}{0.0075-x} \approx \frac{x^2}{0.0075}; x = [H^+] = 3.354 \times 10^{-4} = 3.4 \times 10^{-4} \, M \, H^+$

$\% \text{ ionization} = \frac{3.4 \times 10^{-4} \, M \, H^+}{0.0075 \, M \, buCOOH} \times 100 = 4.5\% \text{ ionization}$

(b) $K_a = 1.5 \times 10^{-5} = \frac{(x)(0.085+x)}{0.0075-x} \approx \frac{0.085 \, x}{0.0075}; x = 1.3 \times \quad H^+$

$\% \text{ ionization} = \frac{1.3 \times 10^{-6} \, M \, H^+}{0.0075 \, M \, buCOOH} \times 100 = 0.018\% \text{ i} \quad \text{n}$

Check. Percent ionization is much smaller when the "comm   is present.

Buffered Solutions (section 17.2)

17.19 CH$_3$COOH and CH$_3$COONa are a weak conjugate acid/con   ase pair which acts as a buffer because unionized CH$_3$COOH reacts with add   e, while CH$_3$COO$^-$ combines with added acid, leaving [H$^+$] relatively unchanged. Although HCl and NaCl are a conjugate acid/conjugate base pair, Cl$^-$ is a negligible base. That is, it has no tendency to combine with added acid to form molecular HCl. Any added acid simply increases [H$^+$] in an HCl/NaCl mixture. In general, the conjugate bases of strong acids are negligible and mixtures of strong acids and their conjugate salts do not act as buffers.

17.21 *Analyze/Plan.* Follow the logic in Sample Exercise 17.3. Assume that % ionization is small in these buffers (Solutions 17.17 and 17.18).   *Solve.*

(a) $K_a = \frac{[H^+][CH_3CH(OH)COO^-]}{[CH_3CH(OH)COOH]}; \quad [H^+] = \frac{[K_a][CH_3CH(OH)COOH]}{[CH_3CH(OH)COO^-]}$

$[H^+] = \frac{1.4 \times 10^{-4}(0.12)}{(0.11)}; [H^+] = 1.53 \times 10^{-4} = 1.5 \times 10^{-4} M; pH = 3.82$

(b)   mol = $M \times$ L; total volume = 85 mL + 95 mL = 180 mL

$[H^+] = \frac{K_a[CH_3CH(OH)COOH]}{[CH_3CH(OH)COO^-]} = \frac{1.4 \times 10^{-4}(0.13 \, M \times 0.085 \, L)/0.180 \, L}{(0.15 \, M \times 0.095 \, L)/0.180 \, L}$

$[H^+] = \frac{1.4 \times 10^{-4}(0.13 \times 0.085)}{(0.15 \times 0.095)}; [H^+] = 1.086 \times 10^{-4} = 1.1 \times 10^{-4} M; pH = 3.96$

17.23　(a)　*Analyze/Plan.* Follow the logic in Sample Exercises 17.1 and 17.3. As in Sample Exercise 17.1, start by calculating concentrations of the components.　*Solve.*

$$CH_3COOH(aq) \rightleftharpoons H^+(aq) + CH_3COO^-(aq); K_a = 1.8 \times 10^{-5} = \frac{[H^+][CH_3COO^-]}{[CH_3COOH]}$$

$[CH_3COOH] = 0.150 \, M$

$$[CH_3COO^-] = \frac{20.0 \text{ g } CH_3COONa}{0.500 \text{ L soln}} \times \frac{1 \text{ mol } CH_3COONa}{82.04 \text{ g } CH_3COONa} = 0.488 \, M$$

$$[H^+] = \frac{K_a[CH_3COOH]}{[CH_3COO^-]} = \frac{1.8 \times 10^{-5} (0.150 - x)}{(0.488 + x)} \approx \frac{1.8 \times 10^{-5} (0.150)}{(0.488)}$$

$[H^+] = 5.533 \times 10^{-6} = 5.5 \times 10^{-6} \, M$, pH = 5.26

(b)　*Plan.* On the left side of the equation, write all ions present in solution after HCl or NaOH is added to the buffer. Using acid-base properties and relative strengths, decide which ions will combine to form new products.　*Solve.*

$$Na^+(aq) + CH_3COO^-(aq) + H^+(aq) + Cl^-(aq) \rightarrow CH_3COOH(aq) + Na^+(aq) + Cl^-(aq)$$

(c)　$CH_3COOH(aq) + Na^+(aq) + OH^-(aq) \rightarrow CH_3COO^-(aq) + H_2O(l) + Na^+(aq)$

17.25　*Analyze/Plan.* Follow the logic in Sample Exercise 16.12 and 17.4.　*Solve.*

(a)　$K_a = 6.8 \times 10^{-4} = \dfrac{x^2}{1.00 - x} = \dfrac{x^2}{1.00}$; $x = [H^+] = 0.02608 = 0.026 \, M$; pH = 1.58

There is 2.6% ionization, so the approximation is valid.

(b)　In this problem, $[F^-]$ is the unknown.

pH = 3.00, $[H^+] = 10^{-3.00} = 1.0 \times 10^{-3}$; $[HF] = 1.00 - 0.0010 = 0.999 \, M$

$K_a = 6.8 \times 10^{-4} = \dfrac{1.0 \times 10^{-3} [F^-]}{0.999}$; $[F^-] = 0.6793 = 0.68 \, M$

$$\frac{0.6793 \text{ mol NaF}}{1 \text{ L}} \times \frac{41.990 \text{ g NaF}}{1 \text{ mol NaF}} \times 1.25 \text{ L} = 35.654 = 36 \text{ g NaF}$$

17.27　*Analyze/Plan.* Follow the logic in Sample Exercise 17.3 and 17.5.　*Solve.*

(a)　$K_a = \dfrac{[H^+][CH_3COO^-]}{[CH_3COOH]}$; $[H^+] = \dfrac{K_a[CH_3COOH]}{[CH_3COO^-]}$

$[H^+] \approx \dfrac{1.8 \times 10^{-5} (0.10)}{(0.13)} = 1.385 \times 10^{-5} = 1.4 \times 10^{-5}$ M; pH = 4.86

(b)

$CH_3COOH(aq)$	+	$KOH(aq)$	$\rightarrow$	$CH_3COO^-(aq) + H_2O(l) + K^+(aq)$
0.10 mol		0.02 mol		0.13 mol
–0.02 mol		–0.02 mol		+0.02 mol
0.08 mol		0　mol		0.15 mol

$$[H^+] = \frac{1.8 \times 10^{-5}\ (0.08\ \text{mol}/1.00\ \text{L})}{(0.15\ \text{mol}/1.00\ \text{L})} = 9.60 \times 10^{-6} = 1 \times 10^{-5}\ M;\ pH = 5.02 = 5.0$$

(c)

	CH$_3$COO$^-$(aq)	+	HNO$_3$(aq)	$\rightarrow$	CH$_3$COOH(aq)	+	NO$_3^-$(aq)
	0.13 mol		0.02 mol		0.10 mol		
	–0.02 mol		–0.02 mol		+0.02 mol		
	0.11 mol		0 mol		0.12 mol		

$$[H^+] = \frac{1.8 \times 10^{-5}\ (0.12\ \text{mol}/1.00\ \text{L})}{(0.11\ \text{mol}/1.00\ \text{L})} = 1.96 \times 10^{-5} = 2.0 \times 10^{-5}\ M;\ pH = 4.71$$

17.29 *Analyze/Plan.* Calculate the [conj. base]/[conj. acid] ratio in the H_2CO_3/HCO_3^- blood buffer. Write the acid dissociation equilibrium and K_a expression. Find K_a for H_2CO_3 in Appendix D. Calculate [H$^+$] from the pH and solve for the ratio. *Solve.*

$$H_2CO_3(aq) \rightleftharpoons H^+(aq) + HCO_3^-\ (aq) \quad K_a = \frac{[H^+][HCO_3^-]}{[H_2CO_3]};\ \frac{[HCO_3^-]}{[H_2CO_3]} = \frac{K_a}{[H^+]}$$

(a) at pH = 7.4, [H$^+$] = $10^{-7.4}$ = $4.0 \times 10^{-8}\ M$; $\dfrac{[HCO_3^-]}{[H_2CO_3]} = \dfrac{4.3 \times 10^{-7}}{4.0 \times 10^{-8}} = 11$

(b) at pH = 7.1, [H$^+$] = $7.9 \times 10^{-8}\ M$; $\dfrac{[HCO_3^-]}{[H_2CO_3]} = 5.4$

17.31 *Analyze.* Given six solutions, decide which two should be used to prepare a pH 3.50 buffer. Calculate the volumes of the two 0.10 M solutions needed to make approximately 1 L of buffer.

Plan. A buffer must contain a conjugate acid/conjugate base (CA/CB) pair. By examining the chemical formulas, decide which pairs of solutions could be used to make a buffer. If there is more than one possible pair, calculate pK_a for the acids. A buffer is most effective when its pH is within 1 pH unit of pK_a for the conjugate acid component. Select the pair with pK_a nearest to 3.50. Use Equation [17.9] to calculate the [CB]/[CA] ratio and the volumes of 0.10 *M* solutions needed to prepare 1 L of buffer. *Solve.*

There are three CA/CB pairs:

HCOOH/HCOONa, pK_a = 3.74

CH$_3$COOH/CH$_3$COONa, pK_a = 4.74

H$_3$PO$_4$/NaH$_2$PO$_4$, pK_a = 2.12

The most appropriate solutions are HCOOH/HCOONa, because pK_a for HCOOH is nearest to 3.50.

$$pH = pKa + \log\frac{[CB]}{[CA]};\quad 3.50 = 3.7447 + \log\frac{[HCOONa]}{[HCOOH]}$$

$$\log \frac{[\text{HCOONa}]}{[\text{HCOOH}]} = -0.2447; \quad \frac{[\text{HCOONa}]}{[\text{HCOOH}]} = 0.5692 = 0.57$$

Since we are making a total of 1 L of buffer,

let y = vol HCOONa and (1 − y) = vol HCOOH.

$$0.5692 = \frac{[\text{HCOONa}]}{[\text{HCOOH}]} = \frac{(0.10\,M \times y)/1L}{[0.10\,M \times (1-y)]/1\,L}; \quad 0.5692[0.10(1-y)] = 0.10\,y;$$

$0.05692 = 0.15692\,y; \quad y = 0.3627 = 0.36$ L

360 mL of 0.10 M HCOONa, 640 mL of 0.10 M HCOOH.

Check. The pH of the buffer is less than pK_a for the conjugate acid, indicating that the amount of CA in the buffer is greater than the amount of CB. This agrees with our result.

Acid-Base Titrations (section 17.3)

17.33 (a) Curve B. The initial pH is lower and the equivalence point region is steeper.

(b) pH at the approximate equivalence point of curve A = 8.0

pH at the approximate equivalence point of curve B = 7.0

(c) Volume of base required to reach the equivalence point depends only on moles of acid present; it is independent of acid strength. Since acid B requires 40 mL and acid A requires only 30 mL, more moles of acid B are being titrated. For equal volumes of A and B, the concentration of acid B is greater.

17.35 (a) False. The same volume of NaOH(aq) is required to reach the equivalence point of both titrations, because moles of acid to be titrated are the same in both flasks.

(b) True. CH_3COONa, the salt formed in the titration of CH_3COOH, produces a basic solution, while $NaNO_3$, formed in the titration of HNO_3 produces a neutral solution.

(c) True. Even though the pH values at the equivalence points of the two titrations are different, phenolphthalein changes color over a wide range of pH values and is appropriate for both titrations.

17.37 *Analyze.* Given reactants, predict whether pH at the equivalence point of a titration is less than, equal to or greater than 7.

Plan. At the equivalence point of a titration, only product is present in solution; there is no excess of either reactant. Determine the product of each reaction and whether a solution of it is acidic, basic or neutral. *Solve.*

(a) $NaHCO_3(aq) + NaOH(aq) \rightarrow Na_2CO_3(aq) + H_2O(l)$

At the equivalence point, the major species in solution are Na^+ and CO_3^{2-}. Na^+ is negligible and CO_3^{2-} is the CB of HCO_3^-. The solution is basic, above pH 7.

(b) $NH_3(aq) + HCl(aq) \rightarrow NH_4Cl(aq)$

At the equivalence point, the major species are NH_4^+ and Cl^-. Cl^- is negligible, NH_4^+ is the CA of NH_3. The solution is acidic, below pH 7.

(c) $KOH(aq) + HBr(aq) \rightarrow KBr(aq) + H_2O(l)$

At the equivalence point, the major species are K^+ and Br^-; both are negligible. The solution is at pH 7.

17.39   The second color change, from yellow to blue near pH = 8.5, is more suitable for the titration of a weak acid with a strong base. The salt present at the equivalence point of this type of titration produces a slightly basic solution. The second color change of Thymol blue is in the correct pH range to show (indicate) the equivalence point.

17.41   *Analyze/Plan.* We are asked to calculate the volume of 0.0850 *M* NaOH required to titrate various acid solutions to their equivalence point. At the equivalence point, moles base added equals moles acid initially present. Solve the stoichiometry problem, recalling that mol = M × L. In part (c) calculate molarity of HCl from g/L and proceed as outlined above. *Solve.*

(a) $40.0 \text{ mL } HNO_3 \times \dfrac{0.0900 \text{ mol } HNO_3}{1000 \text{ mL soln}} \times \dfrac{1 \text{ mol NaOH}}{1 \text{ mol } HNO_3} \times \dfrac{1000 \text{ mL soln}}{0.0850 \text{ mol NaOH}}$

$= 42.353 = 42.4 \text{ mL NaOH soln}$

(b) $35.0 \text{ mL } CH_3COOH \times \dfrac{0.0850 \text{ } M \text{ } CH_3COOH}{1000 \text{ mL soln}} \times \dfrac{1 \text{ mol NaOH}}{1 \text{ mol } CH_3COOH} \times \dfrac{1000 \text{ mL soln}}{0.0850 \text{ mol NaOH}}$

$= 35.0 \text{ mL NaOH soln}$

(c) $\dfrac{1.85 \text{ g } HCl}{1 \text{ L soln}} \times \dfrac{1 \text{ mol } HCl}{36.46 \text{ g } HCl} = 0.05074 = 0.0507 \text{ } M \text{ } HCl$

$50.0 \text{ mL } HCl \times \dfrac{0.05074 \text{ mol } HCl}{1000 \text{ mL}} \times \dfrac{1 \text{ mol NaOH}}{1 \text{ mol } HCl} \times \dfrac{1000 \text{ mL soln}}{0.0850 \text{ mol NaOH}}$

$= 29.847 = 29.8 \text{ mL NaOH soln}$

17.43   *Analyze/Plan.* Follow the logic in Sample Exercise 17.6 for the titration of a strong acid with a strong base. *Solve.*

moles $H^+ = M_{HBr} \times L_{HBr} = 0.200 \text{ } M \times 0.0200 \text{ L} = 4.00 \times 10^{-3} \text{ mol}$

moles $OH^- = M_{NaOH} \times L_{NaOH} = 0.200 \text{ } M \times L_{NaOH}$

	mL_{HBr}	mL_{NaOH}	Total Volume	Moles H^+	Moles OH^-	Molarity Excess Ion	pH
(a)	20.0	15.0	35.0	4.00×10^{-3}	3.00×10^{-3}	$0.0286(H^+)$	1.544
(b)	20.0	19.9	39.9	4.00×10^{-3}	3.98×10^{-3}	$5 \times 10^{-4}(H^+)$	3.3
(c)	20.0	20.0	40.0	4.00×10^{-3}	4.00×10^{-3}	$1 \times 10^{-7}(H^+)$	7.0
(d)	20.0	20.1	40.1	4.00×10^{-3}	4.02×10^{-3}	$5 \times 10^{-4}(OH^-)$	10.7
(e)	20.0	35.0	55.0	4.00×10^{-3}	7.00×10^{-3}	$0.0545(OH^-)$	12.737

molarity of excess ion = moles ion / total vol in L

(a) $\dfrac{4.00 \times 10^{-3} \text{ mol H}^+ - 3.00 \times 10^{-3} \text{ mol OH}^-}{0.0350 \text{ L}} = 0.0286 \; M \text{ H}^+$

(b) $\dfrac{4.00 \times 10^{-3} \text{ mol H}^+ - 3.98 \times 10^{-3} \text{ mol OH}^-}{0.0339 \text{ L}} = 5.01 \times 10^{-4} = 5 \times 10^{-4} \; M \text{ H}^+$

(c) equivalence point, mol H$^+$ = mol OH$^-$

NaBr does not hydrolyze, so $[H^+] = [OH^-] = 1 \times 10^{-7} \; M$

(d) $\dfrac{4.02 \times 10^{-3} \text{ mol OH}^- - 4.00 \times 10^{-3} \text{ mol H}^+}{0.0401 \text{ L}} = 4.99 \times 10^{-4} = 5 \times 10^{-4} \; M \text{ OH}^-$

(e) $\dfrac{7.00 \times 10^{-3} \text{ mol OH}^- - 4.00 \times 10^{-3} \text{ mol H}^+}{0.0550 \text{ L}} = 0.054545 = 0.0545 \; M \text{ OH}^-$

17.45   *Analyze/Plan.* Follow the logic in Sample Exercise 17.7 for the titration of a weak acid with a strong base.   *Solve.*

(a) At 0 mL, only weak acid, CH_3COOH, is present in solution. Using the acid ionization equilibrium

$$CH_3COOH(aq) \;\; \rightleftharpoons \;\; H^+(aq) \;\; + \;\; CH_3COO^-(aq)$$

Initial	0.150 M	0	0
equil	0.150 – x M	x M	x M

$K_a = \dfrac{[H^+][CH_3COO^-]}{[CH_3COOH]} = 1.8 \times 10^{-5}$ (Appendix D)

$1.8 \times 10^{-5} = \dfrac{x^2}{(0.150 - x)} \approx \dfrac{x^2}{0.150} \; ; \; x^2 = 2.7 \times 10^{-6} \; ; \; x = [H^+] = 0.001643$

$= 1.6 \times 10^{-3} \; M; \; \text{pH} = 2.78$

(b)–(f) Calculate the moles of each component after the acid-base reaction takes place.
Moles CH_3COOH originally present = $M \times L = 0.150 \; M \times 0.0350 \; L = 5.25 \times 10^{-3}$ mol.
Moles NaOH added = $M \times L = 0.150 \; M \times y$ mL.

$$\text{NaOH(aq)} \;\; + \;\; CH_3COOH \text{ (aq)} \rightarrow \;\;\;\;\;\; CH_3COONa(aq) + H_2O(l)$$

(0.150 $M \times$ 0.0175 L) =

(b) before rx	2.625 $\times 10^{-3}$ mol	5.25 $\times 10^{-3}$ mol	
after rx	0	2.625 $\times 10^{-3}$ mol	**2.63 $\times 10^{-3}$ mol**

(0.150 $M \times$ 0.0345 L) =

(c) before rx	5.175 $\times 10^{-3}$ mol	5.25 $\times 10^{-3}$ mol	
after rx	0	0.075 $\times 10^{-3}$ mol	**5.18 $\times 10^{-3}$ mol**

$$(0.150\ M \times 0.0350\ L) =$$

(d)	before rx	5.25×10^{-3} mol	5.25×10^{-3} mol	
	after rx	**0**	**0**	5.25×10^{-3} mol

$$(0.150\ M \times 0.0355\ L) =$$

(e)	before rx	5.325×10^{-3} mol	5.25×10^{-3} mol	
	after rx	0.075×10^{-3} mol	**0**	5.25×10^{-3} mol

$$(0.150\ M \times 0.0500\ L) =$$

(f)	before rx	7.50×10^{-3} mol	5.25×10^{-3} mol	
	after rx	2.25×10^{-3} mol	**0**	5.25×10^{-3} mol

Calculate the molarity of each species (M = mol/L) and solve the appropriate equilibrium problem in each part.

(b) $V_T = 35.0$ mL $CH_3COOH + 17.5$ mL NaOH $= 52.5$ mL $= 0.0525$ L

$$[CH_3COOH] = \frac{2.625 \times 10^{-3}\ mol}{0.0525} = 0.0500\ M$$

$$[CH_3COO^-] = \frac{2.625 \times 10^{-3}\ mol}{0.0525} = 0.0500\ M$$

$$CH_3COOH(aq) \rightleftharpoons H^+(aq) + CH_3COO^-(aq)$$

equil $0.0500 - x\ M$ $x\ M$ $0.0500 + x\ M$

$$K_a = \frac{[H^+][CH_3COO^-]}{[CH_3COOH]};\ [H^+] = \frac{K_a[CH_3COOH]}{[CH_3COO^-]}$$

$$[H^+] = \frac{1.8 \times 10^{-5}\ (0.0500 - x)}{(0.0500 + x)} = 1.8 \times 10^{-5}\ M\ H^+;\ pH = 4.74$$

(c) $$[CH_3COOH] = \frac{7.5 \times 10^{-5}\ mol}{0.0695\ L} = 0.001079 = 1.1 \times 10^{-3}\ M$$

$$[CH_3COO^-] = \frac{5.175 \times 10^{-3}\ mol}{0.0695\ L} = 0.07446 = 0.074\ M$$

$$[H^+] = \frac{1.8 \times 10^{-5}\ (1.079 \times 10^{-3} - x)}{(0.07446 + x)} \approx 2.6 \times 10^{-7}\ M\ H^+;\ pH = 6.58$$

(d)   At the equivalence point, only CH_3COO^- is present.

$$[CH_3COO^-] = \frac{5.25 \times 10^{-3}\ mol}{0.0700\ L} = 0.0750\ M$$

The pertinent equilibrium is the base hydrolysis of CH_3COO^-.

$$CH_3COO^-(aq) + H_2O(l) \rightleftharpoons CH_3COOH(aq) + OH^-(aq)$$

initial $0.0750\,M$ 0   0

equil $0.0750 - x\,M$ x   x

$$K_b = \frac{K_w}{K_a \text{ for } CH_3COOH} = \frac{1.0 \times 10^{-14}}{1.8 \times 10^{-5}} = 5.56 \times 10^{-10} = 5.6 \times 10^{-10} = \frac{[CH_3COOH][OH^-]}{[CH_3COO^-]}$$

$$5.56 \times 10^{-10} = \frac{x^2}{0.0750 - x}; x^2 \approx 5.56 \times 10^{-10}(0.0750); x = 6.458 \times 10^{-6}$$

$$= 6.5 \times 10^{-6}\,M\,OH^-$$

$pOH = -\log(6.458 \times 10^{-6}) = 5.19; pH = 14.00 - pOH = 8.81$

(e)   After the equivalence point, the excess strong base determines the pOH and pH. The $[OH^-]$ from the hydrolysis of CH_3COO^- is small and can be ignored.

$$[OH^-] = \frac{0.075 \times 10^{-3}\text{ mol}}{0.0705\text{ L}} = 1.064 \times 10^{-3} = 1.1 \times 10^{-3}\,M; pOH = 2.97$$

$$pH = 14.00 - 2.97 = 11.03$$

(f) $[OH^-] = \dfrac{2.25 \times 10^{-3}\text{ mol}}{0.0850\text{ L}} = 0.0265\,M\,OH^-; pOH = 1.577; pH = 14.00 - 1.577 = 12.423$

17.47   *Analyze/Plan.* Calculate the pH at the equivalence point for the titration of several bases with $0.200\,M$ HBr. The volume of $0.200\,M$ HBr required in all cases equals the volume of base and the final volume $= 2V_{base}$. The concentration of the salt produced at the equivalence point is $\dfrac{0.200\,M \times V_{base}}{2V_{base}} = 0.100\,M$.

In each case, identify the salt present at the equivalence point, determine its acid-base properties (Section 16.9), and solve the pH problem.   *Solve.*

(a)   NaOH is a strong base; the salt present at the equivalence point, NaBr, does not affect the pH of the solution. $0.100\,M$ NaBr, pH = 7.00

(b) $HONH_2$ is a weak base, so the salt present at the equivalence point is $HONH_3^+Br^-$. This is the salt of a strong acid and a weak base, so it produces an acidic solution.

$0.100\,M\,HONH_3^+Br^-$; $HONH_3^+(aq) \rightleftharpoons H^+(aq) + HONH_2$

                             [equil] $0.100 - x$ x   x

$$K_a = \frac{[H^+][HONH_2]}{[HONH_3^+]} = \frac{K_w}{K_b} = \frac{1.0 \times 10^{-14}}{1.1 \times 10^{-8}} = 9.09 \times 10^{-7} = 9.1 \times 10^{-7}$$

Assume x is small with respect to [salt].

$K_a = x^2 / 0.100; x = [H^+] = 3.02 \times 10^{-4} = 3.0 \times 10^{-4}\,M, pH = 3.52$

(c) $C_6H_5NH_2$ is a weak base and $C_6H_5NH_3{}^+Br^-$ is an acidic salt.

0.100 M $C_6H_5NH_3{}^+Br^-$.   Proceeding as in (b):

$$K_a = \frac{[H^+][C_6H_5NH_2]}{[C_6H_5NH_3{}^+]} = \frac{K_w}{K_b} = 2.33 \times 10^{-5} = 2.3 \times 10^{-5}$$

$[H^+]^2 = 0.100(2.33 \times 10^{-5}); [H^+] = 1.52 \times 10^{-3} = 1.5 \times 10^{-3}\,M$, pH = 2.82

Solubility Equilibria and Factors Affecting Solubility
(sections 17.4 and 17.5)

17.49   (a)   The concentration of undissolved solid does not appear in the solubility product expression because it is constant as long as there is solid present. Concentration is a ratio of moles solid to volume of the solid; solids occupy a specific volume not dependent on the solution volume. As the amount (moles) of solid changes, the volume changes proportionally, so that the ratio of moles solid to volume solid is constant.

(b)   *Analyze/Plan.* Follow the example in Sample Exercise 17.9.   *Solve.*

$K_{sp} = [Ag^+][I^-]$; $K_{sp} = [Sr^{2+}][SO_4{}^{2-}]$; $K_{sp} = [Fe^{2+}][OH^-]^2$; $K_{sp} = [Hg_2{}^{2+}][Br^-]^2$

17.51   *Analyze/Plan.* Follow the logic in Sample Exercise 17.10.   *Solve.*

(a) $CaF_2(s) \;\rightleftharpoons\; Ca^{2+}(aq) + 2F^-(aq)$; $K_{sp} = [Ca^{2+}][F^-]^2$

The molar solubility is the moles of CaF_2 that dissolve per liter of solution. Each mole of CaF_2 produces **1** mol Ca^{2+}(aq) and **2** mol F^-(aq).

$[Ca^{2+}] = 1.24 \times 10^{-3}\,M$; $[F^-] = 2 \times 1.24 \times 10^{-3}\,M = 2.48 \times 10^{-3}\,M$

$K_{sp} = (1.24 \times 10^{-3})(2.48 \times 10^{-3})^2 = 7.63 \times 10^{-9}$

(b) $SrF_2(s) \;\rightleftharpoons\; Sr^{2+}(aq) + 2F^-(aq)$; $K_{sp} = [Sr^{2+}][F^-]^2$

Transform the gram solubility to molar solubility.

$$\frac{1.1 \times 10^{-2}\text{ g SrF}_2}{0.100\text{ L}} \times \frac{1\text{ mol SrF}_2}{125.6\text{ g SrF}_2} = 8.76 \times 10^{-4} = 8.8 \times 10^{-4}\text{ mol SrF}_2/\text{L}$$

$[Sr^{2+}] = 8.76 \times 10^{-4}\,M$; $[F^-] = 2(8.76 \times 10^{-4}\,M)$

$K_{sp} = (8.76 \times 10^{-4})(2(8.76 \times 10^{-4}))^2 = 2.7 \times 10^{-9}$

(c) $Ba(IO_3)_2(s) \;\rightleftharpoons\; Ba^{2+}(aq) + 2IO_3{}^-(aq)$; $K_{sp} = [Ba^{2+}][IO_3{}^-]^2$

Since 1 mole of dissolved $Ba(IO_3)_2$ produces 1 mole of Ba^{2+}, the molar solubility of $Ba(IO_3)_2 = [Ba^{2+}]$. Let x = $[Ba^{2+}]$; $[IO_3{}^-] = 2x$

$K_{sp} = 6.0 \times 10^{-10} = (x)(2x)^2$; $4x^3 = 6.0 \times 10^{-10}$; $x^3 = 1.5 \times 10^{-10}$; $x = 5.3 \times 10^{-4}\,M$

The molar solubility of $Ba(IO_3)_2$ is 5.3×10^{-4} mol/L.

17.53   *Analyze/Plan.* Given gram solubility of a compound, calculate K_{sp}. Write the dissociation equilibrium and K_{sp} expression. Change gram solubility to molarity of the individual ions, taking the stoichiometry of the compound into account. Calculate K_{sp}. *Solve.*

$CaC_2O_4(s) \rightleftharpoons Ca^{2+}(aq) + C_2O_4^{2-}(aq)$; $K_{sp} = [Ca^{2+}][C_2O_4^{2-}]$

$[Ca^{2+}] = [C_2O_4^{2-}] = \dfrac{0.0061 \text{ g } CaC_2O_4}{1.00 \text{ L soln}} \times \dfrac{1 \text{ mol } CaC_2O_4}{128.1 \text{ g } CaC_2O_4} = 4.76 \times 10^{-5} = 4.8 \times 10^{-5} \, M$

$K_{sp} = (4.76 \times 10^{-5} \, M)(4.76 \times 10^{-5} \, M) = 2.3 \times 10^{-9}$

17.55   *Analyze/Plan.* Follow the logic in Sample Exercises 17.11 and 17.12.   *Solve.*

(a) $AgBr(s) \rightleftharpoons Ag^+(aq) + Br^-(aq)$; $K_{sp} = [Ag^+][Br^-] = 5.0 \times 10^{-13}$

molar solubility $= x = [Ag^+] = [Br^-]$; $K_{sp} = x^2$

$x = (5.0 \times 10^{-13})^{1/2}$; $x = 7.1 \times 10^{-7}$ mol AgBr/L

(b)   Molar solubility $= x = [Br^-]$; $[Ag^+] = 0.030 \, M + x$

$K_{sp} = (0.030 + x)(x) \approx 0.030(x)$

$5.0 \times 10^{-13} = 0.030(x)$; $x = 1.7 \times 10^{-11}$ mol AgBr/L

(c)   Molar solubility $= x = [Ag^+]$

There are two sources of Br^-: $NaBr(0.10 \, M)$ and $AgBr(x \, M)$

$K_{sp} = (x)(0.10 + x)$; Assuming x is small compared to 0.10 M

$5.0 \times 10^{-13} = 0.10 \, (x)$; $x \approx 5.0 \times 10^{-12}$ mol AgBr/L

17.57   *Analyze/Plan.* Given a saturated solution of CaF_2 in contact with undissolved $CaF_2(s)$, consider the effect of adding $CaCl_2(s)$. The two salts have the Ca^{2+} ion in common. *Solve.* As $CaCl_2$ is added, $[Ca^{2+}]$ increases, K_{sp} is exceeded, and additional CaF_2 precipitates until equilibrium is reestablished. At the new equilibrium position:

(a)   The amount of $CaF_2(s)$ on the bottom of the beaker increases because the added Ca^{2+} from $CaCl_2$ decreases the solubility of CaF_2.

(b)   The $[Ca^{2+}]$ in solution increases because of the added $CaCl_2$.

(c)   The $[F^-]$ in solution decreases because $CaF_2(s)$ precipitates upon addition of $CaCl_2$. The product of $[Ca^{2+}]$ and $[F^-]$, the K_{sp}, is the same.

17.59   *Analyze/Plan.* We are asked to calculate the solubility of a slightly-soluble hydroxide salt at various pH values. This is a common ion problem; pH tells us not only $[H^+]$ but also $[OH^-]$, which is an ion common to the salt. Use pH to calculate $[OH^-]$, then proceed as in Sample Exercise 17.12.   *Solve.*

$Mn(OH)_2(s) \rightleftharpoons Mn^{2+}(aq) + 2OH^-(aq)$; $K_{sp} = 1.6 \times 10^{-13}$

Since $[OH^-]$ is set by the pH of the solution, the solubility of $Mn(OH)_2$ is just $[Mn^{2+}]$.

(a) $pH = 7.0$, $pOH = 14 - pH = 7.0$, $[OH^-] = 10^{-pOH} = 1.0 \times 10^{-7}\ M$

$$K_{sp} = 1.6 \times 10^{-13} = [Mn^{2+}](1.0 \times 10^{-7})^2;\ [Mn^{2+}] = \frac{1.6 \times 10^{-13}}{1.0 \times 10^{-14}} = 16\ M$$

$$\frac{16\ mol\ Mn(OH)_2}{1\ L} \times \frac{88.95\ g\ Mn(OH)_2}{1\ mol\ Mn(OH)_2} = 1423 = 1.4 \times 10^3\ g\ Mn(OH)_2\ /\ L$$

Check. Note that the solubility of $Mn(OH)_2$ in pure water is $3.6 \times 10^{-5}\ M$, and the pH of the resulting solution is 9.0. The relatively low pH of a solution buffered to pH 7.0 actually increases the solubility of $Mn(OH)_2$.

(b) $pH = 9.5$, $pOH = 4.5$, $[OH^-] = 3.16 \times 10^{-5} = 3.2 \times 10^{-5}\ M$

$$K_{sp} = 1.6 \times 10^{-13} = [Mn^{2+}](3.16 \times 10^{-5})^2;\ [Mn^{2+}] = \frac{1.6 \times 10^{-13}}{1.0 \times 10^{-9}} = 1.6 \times 10^{-4}\ M$$

$$1.6 \times 10^{-4}\ M\ Mn(OH)_2 \times 88.95\ g/mol = 0.0142 = 0.014\ g/L$$

(c) $pH = 11.8$, $pOH = 2.2$, $[OH^-] = 6.31 \times 10^{-3} = 6.3 \times 10^{-3}\ M$

$$K_{sp} = 1.6 \times 10^{-13} = [Mn^{2+}](6.31 \times 10^{-3})^2;\ [Mn^{2+}] = \frac{1.6 \times 10^{-13}}{3.98 \times 10^{-5}} = 4.0 \times 10^{-9}\ M$$

$$4.02 \times 10^{-9}\ M\ Mn(OH)_2 \times 88.95\ g/mol = 3.575 \times 10^{-7} = 3.6 \times 10^{-7}\ g/L$$

17.61   *Analyze/Plan.* Follow the logic in Sample Exercise 17.13.   *Solve.*

If the anion of the salt is the conjugate base of a weak acid, it will combine with H^+, reducing the concentration of the free anion in solution, thereby causing more salt to dissolve. More soluble in acid: (a) $ZnCO_3$, (b) ZnS, (d) $AgCN$, (e) $Ba_3(PO_4)_2$

17.63   *Analyze/Plan.* Follow the logic in Sample Exercise 17.14.   *Solve.*

The formation equilibrium is

$$Ni^{2+}(aq) + 6NH_3(aq) \rightleftharpoons Ni(NH_3)_6^{2+}(aq) \qquad K_f = \frac{[Ni(NH_3)_6^{2+}]}{[Ni^{2+}][NH_3]^6} = 1.2 \times 10^9$$

Assuming that nearly all the Ni^{2+} is in the form $Ni(NH_3)_6^{2+}$,

$[Ni(NH_3)_6^{2+}] = 1 \times 10^{-3}\ M;\ [Ni^{2+}] = x;\ [NH_3] = 0.20\ M$

$$1.2 \times 10^9 = \frac{(1 \times 10^{-3})}{x(0.20)^6};\ x = 1.30 \times 10^{-8} = 1 \times 10^{-8}\ M = [Ni^{2+}]$$

17.65   *Analyze/Plan.* Calculate the solubility of AgI in pure water according to the method in Sample Exercise 17.11. Obtain K_{eq} for the complexaction reaction, making use of pertinent K_{sp} and K_f values from Appendix D and Table 17.1. Write the dissociation equilibrium for AgI and the formation reaction for $Ag(CN)_2^-$. Use algebra to manipulate these equations and their associated equilibrium constants to obtain the desired reaction and its equilibrium constant.   Finally, use this K_{eq} value to calculate the solubility of AgI in 0.100 M NaCN solution. *Solve.*

(a) $AgI(s) \rightleftharpoons Ag^+(aq) + I^-(aq)$; $K_{sp} = [Ag^+][I^-] = 8.3 \times 10^{-17}$

molar solubility $= x = [Ag^+] = [I^-]$; $K_{sp} = x^2$

$x = (8.3 \times 10^{-17})^{1/2}$; $x = 9.1 \times 10^{-9}$ mol AgI/L

(b) $AgI(s) \rightleftharpoons Ag^+(aq) + I^-(aq)$

$\underline{Ag^+(aq) + 2CN^-(aq) \rightleftharpoons Ag(CN)_2^-(aq)}$

$AgI(s) + 2CN^-(aq) \rightleftharpoons Ag(CN)_2^-(aq) + I^-(aq)$

$$K = K_{sp} \times K_f = [Ag^+][I^-] \times \frac{[Ag(CN)_2^-]}{[Ag^+][CN^-]^2} = (8.3 \times 10^{-17})(1 \times 10^{21}) = 8 \times 10^4$$

(c)   K is much greater than one for the reaction of AgI(s) with CN^-. This means that the reaction goes to completion. For a AgI(s) in 0.100 M NaCN solution, CN^- is the limiting reactant. Two moles of CN^- react with one mole of AgI, so the solublility of AgI in 0.100 M NaCN is $(0.100/2) = 0.0500$ M.

Precipitation and Separation of Ions (section 17.6)

17.67   *Analyze/Plan.* Follow the logic in Sample Exercise 17.15. Precipitation conditions: will Q (see Chapter 15) exceed K_{sp} for the compound?   *Solve.*

(a)   In base, Ca^{2+} can form $Ca(OH)_2(s)$.

$Ca(OH)_2(s) \rightleftharpoons Ca^{2+}(aq) + 2OH^-(aq)$; $K_{sp} = [Ca^{2+}][OH^-]^2$

$Q = [Ca^{2+}][OH^-]^2$; $[Ca^{2+}] = 0.050$ M; $pOH = 14 - 8.0 = 6.0$; $[OH^-] = 1.0 \times 10^{-6}$ M

$Q = (0.050)(1.0 \times 10^{-6})^2 = 5.0 \times 10^{-14}$; $K_{sp} = 6.5 \times 10^{-6}$ (Appendix D)

$Q < K_{sp}$, no $Ca(OH)_2$ precipitates.

(b) $Ag_2SO_4(s) \rightleftharpoons 2Ag^+(aq) + SO_4^{2-}(aq)$; $K_{sp} = [Ag+]^2[SO_4^{2-}]$

$$[Ag^+] = \frac{0.050\ M \times 100\ mL}{110\ mL} = 4.545 \times 10^{-2} = 4.5 \times 10^{-2}\ M$$

$$[SO_4^{2-}] = \frac{0.050\ M \times 10\ mL}{110\ mL} = 4.545 \times 10^{-3} = 4.5 \times 10^{-3}\ M$$

$Q = (4.545 \times 10^{-2})^2\ (4.545 \times 10^{-3}) = 9.4 \times 10^{-6}$; $K_{sp} = 1.5 \times 10^{-5}$

$Q < K_{sp}$, no Ag_2SO_4 precipitates.

17.69   *Analyze/Plan.* We are asked to calculate pH necessary to precipitate $Mn(OH)_2(s)$ if the resulting Mn^{2+} concentration is ≤ 1 µg/L.

$Mn(OH)_2(s) \rightleftharpoons Mn^{2+}(aq) + 2OH^-(aq)$; $K_{sp} = [Mn^{2+}][OH^-]^2 = 1.6 \times 10^{-13}$

At equilibrium, $[Mn^{2+}][OH^-]^2 = 1.6 \times 10^{-13}$. Change concentration $Mn^{2+}(aq)$ to mol/L and solve for $[OH^-]$.   *Solve.*

$$\frac{1\,\mu g\,Mn^{2+}}{1\,L} \times \frac{1 \times 10^{-6}\,g}{1\,\mu g} \times \frac{1\,mol\,Mn^{2+}}{54.94\,g\,Mn^{2+}} = 1.82 \times 10^{-8} = 2 \times 10^{-8}\,M\,Mn^{2+}$$

$$1.6 \times 10^{-13} = (1.82 \times 10^{-8})[OH^-]^2; [OH^-]^2 = 8.79 \times 10^{-6}; [OH^-] = 2.96 \times 10^{-3} = 3 \times 10^{-3}\,M$$

$$pOH = 2.53; pH = 14 - 2.53 = 11.47 = 11.5$$

$$\frac{2.08 \times 10^{-3}\,mol\,I^-}{1\,L} \times \frac{126.90\,g\,I^-}{1\,mol\,I^-} \times 0.0102\,L = 2.69 \times 10^{-3}\,g\,I^- = 3 \times 10^{-3}\,g\,I^-$$

17.71   *Analyze/Plan.* We are asked which ion will precipitate first from a solution containing $Pb^{2+}(aq)$ and $Ag^+(aq)$ when $I^-(aq)$ is added. Follow the logic in Sample Exercise 17.16. Calculate $[I^-]$ needed to initiate precipitation of each ion. The cation that requires lower $[I^-]$ will precipitate first.   *Solve.*

$$Ag^+: K_{sp} = [Ag^+][I^-]; 8.3 \times 10^{-17} = (2.0 \times 10^{-4})[I^-]; [I^-] = \frac{8.3 \times 10^{-17}}{2.0 \times 10^{-4}} = 4.2 \times 10^{-13}\,M\,I^-$$

$$Pb^{2+}: K_{sp} = [Pb^{2+}][I^-]^2; 7.9 \times 10^{-9} = (1.5 \times 10^{-3})[I^-]^2; [I^-] = \left(\frac{7.9 \times 10^{-9}}{1.5 \times 10^{-3}}\right)^{1/2} = 2.3 \times 10^{-3}\,M\,I^-$$

AgI will precipitate first, at $[I^-] = 4.2 \times 10^{-13}\,M$.

17.73   *Analyze/Plan.* We are asked which ion will precipitate first when dilute $Ag^+(aq)$ is added to a solution containing $0.20\,M\,CrO_4^{2-}$, $0.10\,M\,CO_3^{2-}$ and $0.10\,M\,Cl^-$. The anions are present at different concentrations and their silver compounds have different stoichiometry, so we cannot directly compare K_{sp} values. Follow the logic in Sample Exercise 17.16. Calculate $[Ag^+]$ needed to initiate precipitation of each ion. The anion that requires lowest $[Ag^+]$ will precipitate first, and so on.   *Solve.*

$$Ag_2CrO_4: K_{sp} = [Ag^+]^2[CrO_4^{2-}] = 1.2 \times 10^{-12}$$

$$1.2 \times 10^{-12} = [Ag^+]^2(0.20); [Ag^+]^2 = 6.0 \times 10^{-12}; [Ag^+] = 2.4 \times 10^{-6}\,M$$

$$Ag_2CO_3: K_{sp} = [Ag^+]^2[CO_3^{2-}] = 8.1 \times 10^{-12}$$

$$8.1 \times 10^{-12} = [Ag^+]^2(0.10); [Ag^+]^2 = 8.1 \times 10^{-11}; [Ag^+] = 9.0 \times 10^{-6}\,M$$

$$AgCl: K_{sp} = [Ag^+][Cl^-] = 1.8 \times 10^{-10}$$

$$1.8 \times 10^{-10} = [Ag^+](0.10); [Ag^+] = 1.8 \times 10^{-9}$$

AgCl requires the smallest $[Ag^+]$ for precipitation and it will precipitate first. The other two will precipitate almost simultaneously.

Qualitative Analysis for Metallic Elements (section 17.7)

17.75   *Analyze/Plan.* Use Figure 17.23 and the description of the five qualitative analysis "groups" in Section 17.7 to analyze the given data.   *Solve.*

The first two experiments eliminate Group 1 and 2 ions (Figure 17.23). The fact that no insoluble phosphates form in the filtrate from the third experiment rules out Group 4 ions. The ions which might be in the sample are those of Group 3, that is, Al^{3+}, Fe^{3+}, Cr^{3+}, Zn^{2+}, Ni^{2+}, Mn^{2+}, or Co^{2+}, and those of Group 5, NH_4^+, Na^+ or K^+.

17.77   *Analyze/Plan.* We are asked to devise a procedure to separate various pairs of ions in aqueous solutions. In each case, refer to Figure 17.23 to find a set of conditions where the solubility of the two ions differs. Construct a procedure to generate these conditions.   *Solve.*

    (a) Cd^{2+} is in Gp. 2, but Zn^{2+} is not. Make the solution acidic using 0.2 *M* HCl; saturate with H_2S. CdS will precipitate, ZnS will not.

    (b) $Cr(OH)_3$ is amphoteric but $Fe(OH)_3$ is not. Add excess base; $Fe(OH)_3(s)$ precipitates, but Cr^{3+} forms the soluble complex $Cr(OH)_4^-$.

    (c) Mg^{2+} is a member of Gp. 4, but K^+ is not. Add $(NH_4)_2HPO_4$ to a basic solution; Mg^{2+} precipitates as $MgNH_4PO_4$, K^+ remains in solution.

    (d) Ag^+ is a member of Gp. 1, but Mn^{2+} is not. Add 6 *M* HCl, precipitate Ag^+ as AgCl(s).

17.79   (a)   Because phosphoric acid is a weak acid, the concentration of free $PO_4^{3-}(aq)$ in an aqueous phosphate solution is low except in strongly basic media. In less basic media, the solubility product of the phosphates of interest is not exceeded.

    (b) K_{sp} for those cations in Group 3 is much larger. Thus, to exceed K_{sp}, a higher $[S^{2-}]$ is required. This is achieved by making the solution more basic.

    (c)   They should all redissolve in strongly acidic solution, e.g., in 12 *M* HCl (the chlorides of all Group 3 metals are soluble).

Additional Exercises

17.81   *Analyze/Plan.* Follow the approach for deriving the Henderson-Hasselbach (H–H) equation from the K_a expression shown in Section 17.2. Begin with a general K_b expression. *Solve.*

$$B(aq) + H_2O(l) \rightleftharpoons BH^+(aq) + OH^-(aq); \quad K_b = \frac{[BH^+][OH^-]}{[B]}$$

$pOH = -\log[OH^-]$; rearrange K_b to solve for $[OH^-]$.

$$[OH^-] = \frac{K_b[B]}{[BH^+]}; \text{ take the } -\log \text{ of both sides}$$

$$-\log[OH^-] = -\log K_b + (-\log[B] - (-\log[BH^+]))$$

$$pOH = pK_b + \log[BH^+] - \log[B]$$

$$pOH = pK_b + \log\frac{[BH^+]}{[B]}$$

17.83   The equilibrium of interest is

$$HC_5H_3O_3(aq) \rightleftharpoons H^+(aq) + C_5H_3O_3^-(aq); \quad K_a = 6.76 \times 10^{-4} = \frac{[H^+][C_5H_3O_3^-]}{[HC_5H_3O_3]}$$

Begin by calculating $[HC_5H_3O_3]$ and $[C_5H_3O_3^-]$ for each case.

(a) $$\frac{25.0 \text{ g } HC_5H_3O_3}{0.250 \text{ L soln}} \times \frac{1 \text{ mol } HC_5H_3O_3}{112.1 \text{ g } HC_5H_3O_3} = 0.8921 = 0.892 \; M \, HC_5H_3O_3$$

$$\frac{30.0 \text{ g } NaC_5H_3O_3}{0.250 \text{ L soln}} \times \frac{1 \text{ mol } NaC_5H_3O_3}{134.1 \text{ g } NaC_5H_3O_3} = 0.8949 = 0.895 \; M \, C_5H_3O_3^-$$

$$[H^+] = \frac{K_a[HC_5H_3O_3]}{[C_5H_3O_3^-]} = \frac{6.76 \times 10^{-4} \, (0.8921 - x)}{(0.8949 + x)} \approx \frac{6.76 \times 10^{-4} (0.8921)}{(0.8949)}$$

$[H^+] = 6.74 \times 10^{-4} \, M$, pH = 3.171

(b) For dilution, $M_1V_1 = M_2V_2$

$$[HC_5H_3O_3] = \frac{0.250 \; M \times 30.0 \text{ mL}}{125 \text{ mL}} = 0.0600 \; M$$

$$[C_5H_3O_3^-] = \frac{0.220 \; M \times 20.0 \text{ mL}}{125 \text{ mL}} = 0.0352 \; M$$

$$[H^+] \approx \frac{6.76 \times 10^{-4} \, (0.0600)}{0.0352} = 1.15 \times 10^{-3} \; M, \text{ pH} = 2.938$$

(yes, $[H^+]$ is < 5% of 0.0352 M)

(c) $0.0850 \; M \times 0.500 \text{ L} = 0.0425 \text{ mol } HC_5H_3O_3$

$1.65 \; M \times 0.0500 \text{ L} = 0.0825 \text{ mol NaOH}$

	$HC_5H_3O_3(aq)$	+	NaOH(aq)	$\rightarrow$	$NaC_5H_3O_3(aq) + H_2O(l)$
initial	0.0425 mol		0.0825 mol		
reaction	–0.0425 mol		–0.0425 mol		+0.0425 mol
after	0 mol		0.0400 mol		0.0425 mol

The strong base NaOH dominates the pH; the contribution of $C_5H_3O_3^-$ is negligible. This combination would be "after the equivalence point" of a titration. The total volume is 0.550 L.

$$[OH^-] = \frac{0.0400 \text{ mol}}{0.550 \text{ L}} = 0.0727 \; M; \text{ pOH} = 1.138, \text{ pH} = 12.862$$

17.86   (a) $$K_a = \frac{[H^+][HCOO^-]}{[HCOOH]}; [H^+] = \frac{K_a[HCOOH]}{[HCOO^-]}$$

Buffer A: $[HCOOH] = [HCOO^-] = \dfrac{1.00 \text{ mol}}{1.00 \text{ L}} = 1.00 \; M$

$$[H^+] = \frac{1.8 \times 10^{-4} \, (1.00 \; M)}{(1.00 \; M)} = 1.8 \times 10^{-4} \; M, \text{ pH} = 3.74$$

Buffer B: $[HCOOH] = [HCOO^-] = \dfrac{0.010 \text{ mol}}{1.00 \text{ L}} = 0.010 \; M$

$$[H^+] = \frac{1.8 \times 10^{-4}\ (0.010\ M)}{(0.010\ M)} = 1.8 \times 10^{-4}\ M, pH = 3.74$$

The pH of a buffer is determined by the identity of the conjugate acid/conjugate base pair (that is, the relevant K_a value) and the ratio of concentrations of the conjugate acid and conjugate base. The absolute concentrations of the components is not relevant. The pH values of the two buffers are equal because they both contain HCOOH and HCOONa and the [HCOOH] / [HCOO⁻] *ratio* is the same in both solutions.

(b)   Buffer capacity is determined by the absolute amount of conjugate acid and conjugate base available to absorb strong acid (H^+) or strong base (OH^-) that is added to the buffer. Buffer A has the greater capacity because it contains the greater absolute concentrations of HCOOH and HCOO⁻.

(c)   Buffer A: HCOO⁻   +   HCl   →   HCOOH + Cl⁻

                1.00 mol   0.001 mol   1.00 mol

                0.999 mol   0   1.001 mol

$$[H^+] = \frac{1.8 \times 10^{-4}\ (1.001)}{(0.999)} = 1.8 \times 10^{-4}\ M, pH = 3.74$$

(In a buffer calculation, volumes cancel and we can substitute moles directly into the K_a expression.)

Buffer B: HCOO⁻   +   HCl   →   HCOOH + Cl⁻

              0.010 mol   0.001 mol   0.010 mol

              0.009 mol   0   0.011 mol

$$[H^+] = \frac{1.8 \times 10^{-4}\ (0.011)}{(0.009)} = 2.2 \times 10^{-4}\ M, pH = 3.66$$

(d)   Buffer A: 1.00 M HCl × 0.010 L = 0.010 mol H^+ added

    mol HCOOH = 1.00 + 0.010 = 1.01 mol

    mol HCOO⁻ = 1.00 – 0.010 = 0.99 mol

$$[H^+] = \frac{1.8 \times 10^{-4}\ (1.01)}{(0.99)} = 1.8 \times 10^{-4}\ M, pH = 3.74$$

Buffer B: mol HCOOH = 0.010 + 0.010 = 0.020 mol = 0.020 M

  mol HCOO⁻ = 0.010 – 0.010 = 0.000 mol

The solution is no longer a buffer; the only source of HCOO⁻ is the dissociation of HCOOH.

$$K_a = \frac{[H^+][COO^-]}{[HCOOH]} = \frac{x^2}{(0.020 - x)\ M}$$

The extent of ionization is greater than 5%; from the quadratic formula, $x = [H^+] = 1.8 \times 10^{-3}$, pH = 2.74.

(e)   Adding 10 mL of 1.00 M HCl to buffer B exceeded its capacity, while the pH of buffer A was unaffected. This is quantitative confirmation that buffer A has a significantly greater capacity than buffer B. In fact, 1.0 L of 1.0 M HCl would be required to exceed the capacity of buffer A. Buffer A, with 100 times more HCOOH and HCOO⁻ has 100 times the capacity of buffer B.

17.88   (a)   For a monoprotic acid (one H^+ per mole of acid), at the equivalence point moles OH^- added = moles H^+ originally present

$M_B \times V_B$ = g acid/molar mass

$$MM = \frac{\text{g acid}}{M_B \times V_B} = \frac{0.2140\,\text{g}}{0.0950\,M \times 0.0274\,\text{L}} = 82.21 = 82.2\,\text{g/mol}$$

(b)   initial mol HA = $\dfrac{0.2140\,\text{g}}{82.21\,\text{g/mol}} = 2.603 \times 10^{-3} = 2.60 \times 10^{-3}$ mol HA

mol OH^- added to pH 6.50 = $0.0950\,M \times 0.0150\,\text{L} = 1.425 \times 10^{-3}$

$$= 1.43 \times 10^{-3}\,\text{mol OH}^-$$

	HA(aq)	+	NaOH(aq)	→	NaA(aq) + H₂O
before rx	2.603×10^{-3} mol		1.425×10^{-3} mol		0
change	-1.425×10^{-3} mol		-1.425×10^{-3} mol		$+1.425 \times 10^{-3}$ mol
after rx	1.178×10^{-3} mol		0		1.425×10^{-3} mol

$$[HA] = \frac{1.178 \times 10^{-3}\,\text{mol}}{0.0400\,\text{L}} = 0.02945 = 0.0295\,M$$

$$[A^-] = \frac{1.425 \times 10^{-3}\,\text{mol}}{0.0400\,\text{L}} = 0.03563 = 0.0356\,M; [H^+] = 10^{-6.50} = 3.162 \times 10^{-7}$$

$$= 3.2 \times 10^{-7}\,M$$

The mixture after reaction (a buffer) can be described by the acid dissociation equilibrium.

	HA(aq)	⇌	H⁺(aq)	+	A⁻(aq)
initial	$0.0295\,M$		0		$0.0356\,M$
equil	$(0.0295 - 3.2 \times 10^{-7}\,M)$		$3.2 \times 10^{-7}\,M$		$(0.0356 + 3.2 \times 10^{-7})\,M$

$$K_a = \frac{[H^+][A^-]}{[HA]} \approx \frac{(3.162 \times 10^{-7})(0.03563)}{(0.02945)} = 3.8 \times 10^{-7}$$

(Although we have carried 3 figures through the calculation to avoid rounding errors, the data dictate an answer with 2 significant figures.)

17.90   At the equivalence point of a titration, moles strong base added equals moles weak acid initially present. $M_B \times V_B$ = mol base added = mol acid initial

At the half-way point, the volume of base is one-half of the volume required to reach the equivalence point, and the moles base delivered equals one-half of the mol acid initially present. This means that one-half of the weak acid HA is converted to the conjugate base A$^-$. If exactly half of the acid reacts, mol HA = mol A$^-$ and [HA] = [A$^-$] at the half-way point.

From Equation [17.9], $pH = pK_a + \log\dfrac{[\text{conj. base}]}{[\text{conj. acid}]} = pK_a + \log\dfrac{[A^-]}{[HA^-]}$.

If [A$^-$]/[HA] = 1, log(1) = 0 and pH = pK$_a$ of the weak acid being titrated.

17.92   Assume that H$_3$PO$_4$ will react with NaOH in a stepwise fashion: (This is not unreasonable, since the three K$_a$ values for H$_3$PO$_4$ are significantly different.)

	H$_3$PO$_4$(aq)	+	NaOH(aq)	→	H$_2$PO$_4^-$(aq) + Na$^+$(aq) + H$_2$O(l)
before	0.20 mol		0.30 mol		0 mol
after	0 mol		0.10 mol		0.20 mol

	H$_2$PO$_4^-$(aq)	+	NaOH(aq)	→	HPO$_4^-$(aq) + Na$^+$(aq) + H$_2$O(l)
before	0.20 mol		0.10 mol		0.25 mol
after	0.10 mol		0		0.35 mol

Thus, after all NaOH has reacted, the resulting 1.00 L solution is a buffer containing 0.10 mol H$_2$PO$_4^-$ and 0.35 mol HPO$_4^{2-}$.   H$_2$PO$_4^-$(aq) $\rightleftharpoons$ H$^+$(aq) + HPO$_4^{2-}$(aq)

$$K_a = 6.2 \times 10^{-8} = \frac{[HPO_4^{2-}][H^+]}{[H_2PO_4^-]}; [H^+] = \frac{6.2 \times 10^{-8}\,(0.10\,M)}{0.35\,M} = 1.77 \times 10^{-8} = 1.8 \times 10^{-8}\,M;$$

$$pH = 7.75$$

17.93   The pH of a buffer system is centered around pK$_a$ for the conjugate acid component. For a diprotic acid, two conjugate acid/conjugate base pairs are possible.

H$_2$X(aq) $\rightleftharpoons$ H$^+$(aq) + HX$^-$(aq); K$_{a1}$ = 2 × 10^{-2}; pK$_{a1}$ = 1.70

HX$^-$(aq) $\rightleftharpoons$ H$^+$(aq) + X^{2-}(aq);   K$_{a2}$ = 5.0 × 10^{-7}; pK$_{a2}$ = 6.30

Clearly HX$^-$ / X^{2-} is the more appropriate combination for preparing a buffer with pH = 6.50. The [H$^+$] in this buffer = 10$^{-6.50}$ = 3.16 × 10^{-7} = 3.2 × 10^{-7} M. Using the K$_{a2}$ expression to calculate the [X^{2-}] / [HX$^-$] ratio:

$$K_{a2} = \frac{[H^+][X^{2-}]}{[HX^-]}; \frac{K_{a2}}{[H^+]} = \frac{[X^{2-}]}{[HX^-]} = \frac{5.0 \times 10^{-7}}{3.16 \times 10^{-7}} = 1.58 = 1.6$$

Since X^{2-} and HX$^-$ are present in the same solution, the ratio of concentrations is also a ratio of moles.

$$\frac{[X^{2-}]}{[HX^-]} = \left(\frac{\text{mol X}^{2-}/\text{L soln}}{\text{mol HX}^-/\text{L soln}}\right) = \frac{\text{mol X}^{2-}}{\text{mol HX}^-} = 1.58; \text{ mol X}^{2-} = (1.58)\text{ mol HX}^-$$

In the 1.0 L of 1.0 M H$_2$X, there is 1.0 mol of material containing X^{2-}.

Thus, mol HX$^-$ + 1.58 (mol HX$^-$) = 1.0 mol. 2.58 (mol HX$^-$) = 1.0;

mol HX$^-$ = 1.0 / 2.58 = 0.39 mol HX$^-$; mol X^{2-} = 1.0 − 0.39 = 0.61 mol X^{2-}.

Thus enough 1.0 M NaOH must be added to produce 0.39 mol HX^- and 0.61 mol X^{2-}.

Considering the neutralization in a step-wise fashion (see discussion of titrations of polyprotic acids in Section 17.3).

	$H_2X(aq)$	+	$NaOH(aq)$	$\rightarrow$	$HX^-(aq) + H_2O(l)$
before	1.0 mol		1 mol		0
after	0		0		1.0 mol

	$HX^-(aq)$	+	$NaOH(aq)$	$\rightarrow$	$X^{2-}(aq) + H_2O(l)$
before	1.0				0.61
change	−0.61		−0.61		+0.61
after	0.39		0		0.61

Starting with 1.0 mol of H_2X, 1.0 mol of NaOH is added to completely convert it to 1.0 mol of HX^-. Of that 1.0 mol of HX^-, 0.61 mol must be converted to 0.61 mol X^{2-}. The total moles of NaOH added is $(1.00 + 0.61) = 1.61$ mol NaOH.

$$L\ NaOH = \frac{mol\ NaOH}{M\ NaOH} = \frac{1.61\ mol}{1.0\ M} = 1.6\ L\ of\ 1.0\ M\ NaOH$$

17.96   (a)   CdS: 8.0×10^{-28}; CuS: 6×10^{-37} CdS has greater molar solubility.

(b) $PbCO_3$: 7.4×10^{-14}; $BaCrO_4$: 2.1×10^{-10} $BaCrO_4$ has greater molar solubility.

(c)   Since the stoichiometry of the two complexes is not the same, K_{sp} values can't be compared directly; molar solubilities must be calculated from K_{sp} values.

$Ni(OH)_2$: $K_{sp} = 6.0 \times 10^{-16} = [Ni^{2+}][OH^-]^2$; $[Ni^{2+}] = x$, $[OH^-] = 2x$

$6.0 \times 10^{-16} = (x)(2x)^2 = 4x^3$; $x = 5.3 \times 10^{-6}\ M\ Ni^{2+}$

Note that $[OH^-]$ from the autoionization of water is less than 1% of $[OH^-]$ from $Ni(OH)_2$ and can be neglected.

$NiCO_3$: $K_{sp} = 1.3 \times 10^{-7} = [Ni^{2+}][CO_3^{2-}]$; $[Ni^{2+}] = [CO_3^{2-}] = x$

$1.3 \times 10^{-7} = x^2$; $x = 3.6 \times 10^{-4}\ M\ Ni^{2+}$

$NiCO_3$ has greater molar solubility than $Ni(OH)_2$, but the values are much closer than expected from inspection of K_{sp} values alone.

(d)   Again, molar solubilities must be calculated for comparison.

Ag_2SO_4: $K_{sp} = 1.5 \times 10^{-5} = [Ag^+]^2[SO_4^{2-}]$; $[SO_4^{2-}] = x$, $[Ag^+] = 2x$

$$1.5 \times 10^{-5} = (2x)^2(x) = 4x^3; x = 1.6 \times 10^{-2}\ M\ SO_4^{2-}$$

AgI: $K_{sp} = 8.3 \times 10^{-17} = [Ag^+][I^-]$; $[Ag^+] = [I^-] = x$

$$8.3 \times 10^{-17} = x^2; x = 9.1 \times 10^{-9}\ M\ Ag^+$$

Ag_2SO_4 has greater molar solubility than AgI.

17.100 *Analyze/Plan.* Calculate the solubility of $Mg(OH)_2$ in 0.50 M NH_4Cl. Find K_{sp} for $Mg(OH)_2$ in Appendix D. NH_4^+ is a weak acid, which will increase the solubility of $Mg(OH)_2$. Combine the various interacting equilibibria to obtain an overall reaction. Calculate K for this reaction and use it to calculate solubility (s) for $Mg(OH)_2$ in 0.50 M NH_4Cl. *Solve.*

$$Mg(OH)_2(s) \rightleftharpoons Mg^{2+}(aq) + 2\,OH^-(aq) \qquad K_{sp}$$

$$2NH_4^+(aq) \rightleftharpoons 2NH_3(aq) + 2H^+(aq) \qquad K_a$$

$$2H^+(aq) + 2OH^-(aq) \rightleftharpoons 2H_2O(l) \qquad 1/K_w$$

$$Mg(OH)_2(s) + 2NH_4^+(aq) + 2H^+(aq) + 2OH^-(aq) \rightleftharpoons Mg^{2+}(aq) + 2NH_3(aq)$$
$$+ 2OH^-(aq) + 2H^+(aq) + 2H_2O(l)$$

$$Mg(OH)_2(s) + 2NH_4^+(aq) \rightleftharpoons Mg^{2+}(aq) + 2NH_3(aq) + 2H_2O(l)$$

$$K = \frac{[Mg^{2+}][NH_3]^2}{[NH_4^+]^2} = \frac{K_{sp} \times K_a^2}{K_w^2}$$

$$K_a \text{ for } NH_4^+ = \frac{K_w}{K_b \text{ for } NH_3}; \frac{K_a}{K_w} = \frac{1}{K_b}$$

$$K = \frac{K_{sp} \times K_a^2}{K_w^2} = \frac{K_{sp}}{K_b^2} = \frac{1.8 \times 10^{-11}}{(1.8 \times 10^{-5})^2} = 5.556 \times 10^{-2} = 5.6 \times 10^{-2}$$

Let $[Mg^{2+}] = s$, $[NH_3] = 2s$, $[NH_4^+] = 0.50 - 2s$

$$K = 5.6 \times 10^{-2} = \frac{[Mg^{2+}][NH_3]^2}{[NH_4^+]^2} = \frac{s(2s)^2}{(0.5 - 2s)^2} = \frac{4s^3}{0.25 - 2s + 4s^2}$$

$5.6 \times 10^{-2}(0.25 - 2s + 4s^2) = 4s^3$; $4s^3 - 0.222s^2 + 0.111s - 1.39 \times 10^{-2} = 0$

Clearly, 2s is not small relative to 0.50. Solving the third-order equation, s = 0.1054 = 0.11 M. The solubility of $Mg(OH)_2$ in 0.50 M NH_4Cl is 0.11 mol/L.

Check. Substitute s = 0.1054 into the K expression.

$$K = \frac{4(0.1054)^3}{[0.50 - 2(0.1054)]^2} = 5.6 \times 10^{-2}.$$

The solubility and K value are consistent, to the precision of the K_{sp} and K_b values.

17.101 $K_{sp} = [Ba^{2+}][MnO_4^-]^2 = 2.5 \times 10^{-10}$

$[MnO_4^-]^2 = 2.5 \times 10^{-10} / 2.0 \times 10^{-8} = 0.0125$; $[MnO_4^-] = \sqrt{0.0125} = 0.11\,M$

17.104 $MgC_2O_4(s) \rightleftharpoons Mg^{2+}(aq) + C_2O_4^{2-}(aq)$

$K_{sp} = [Mg^{2+}][C_2O_4^{2-}] = 8.6 \times 10^{-5}$

If $[Mg^{2+}]$ is to be $3.0 \times 10^{-2}\,M$, $[C_2O_4^{2-}] = 8.6 \times 10^{-5} / 3.0 \times 10^{-2} = 2.87 \times 10^{-3} = 2.9 \times 10^{-3}\,M$

The oxalate ion undergoes hydrolysis:

$$C_2O_4^{2-}(aq) + H_2O(l) \;\rightleftharpoons\; HC_2O_4^-(aq) + OH^-(aq)$$

$$K_b = \frac{[HC_2O_4^-][OH^-]}{[C_2O_4^{2-}]} = 1.0 \times 10^{-14} / 6.4 \times 10^{-5} = 1.56 \times 10^{-10} = 1.6 \times 10^{-10}$$

$$[Mg^{2+}] = 3.0 \times 10^{-2}\,M,\; [C_2O_4^{2-}] = 2.87 \times 10^{-3} = 2.9 \times 10^{-3}\,M$$

$$[HC_2O_4^-] = (3.0 \times 10^{-2} - 2.87 \times 10^{-3})\,M = 2.71 \times 10^{-2} = 2.7 \times 10^{-2}\,M$$

$$[OH^-] = 1.56 \times 10^{-10} \times \frac{[C_2O_4^{2-}]}{[HC_2O_4^-]} = 1.56 \times 10^{-10} \times \frac{(2.87 \times 10^{-3})}{(2.71 \times 10^{-2})} = 1.652 \times 10^{-11}$$

$$[OH^-] = 1.7 \times 10^{-11}\,M;\; pOH = 10.78,\, pH = 3.22$$

17.107 **(a)** $Cd(OH)_2(s) \rightleftharpoons Cd^{2+}(aq) + 2OH^-(aq);\; K_{sp} = 2.5 \times 10^{-14} = [Cd^{2+}][OH^-]^2.$

$[Cd^{2+}] = s;\; [OH^-] = 2s;\; K_{sp} = 2.5 \times 10^{-14} = 4s^3.\quad s = 1.8 \times 10^{-5}\,M.$

(b)

$$Cd(OH)_2(s) \rightleftharpoons Cd^{2+}(aq) + 2OH^-(aq) \qquad K_{sp} = 2.5 \times 10^{-14}$$

$$Cd^{2+}(aq) + 4Br^-(aq) \rightleftharpoons CdBr_4^{2-}(aq) \qquad K_f = 5 \times 10^3$$

$$Cd(OH)_2(s) + 4Br^-(aq) \rightleftharpoons CdBr_4^{2-}(aq) + 2OH^-(aq) \qquad K$$

$$K = K_{sp} \times K_f = (2.5 \times 10^{-14})(5 \times 10^3) = 1.25 \times 10^{-10} = 1 \times 10^{-10}$$

The desired molar solubility of $Cd(OH)_2$ is 1.0×10^{-3}. Assume all soluble Cd^{2+} is present as $CdBr_4^{2-}$. $[CdBr_4^{2-}] = 1.0 \times 10^{-3};\; [OH^-] = 2(1.0 \times 10^{-3}) = 2.0 \times 10^{-3}.$

Let c = initial [NaBr] = initial [Br$^-$]; [Br$^-$] at equilibrium =
$c - 4(1.0 \times 10^{-3}) = (c - 4.0 \times 10^{-3}).$

$$K = 1.25 \times 10^{-10} = \frac{[CdBr_4^{2-}][OH^-]^2}{[Br^-]^4} = \frac{(1.0 \times 10^{-3})(2.0 \times 10^{-3})^2}{(c - 4.0 \times 10^{-3})^4}$$

Assume c is large relative to 4.0×10^{-3}.

$(1.25 \times 10^{-10})c^4 = 4.0 \times 10^{-9};\; c = (32)^{1/4} = 2.378 = 2\,M.$ The approximation is valid. 4.0×10^{-3} is about 0.2% of $2\,M$. Check this result in the equilibrium expression.

$$K = \frac{(1.0 \times 10^{-3})(2.0 \times 10^{-3})^2}{(2.378 - 4.0 \times 10^{-3})^4} = 1.26 \times 10^{-10}.$$ Our calculations are consistent.

Integrative Exercises

17.108 **(a)** Complete ionic $(CHO_2^- = HCOO^-)$

$$H^+(aq) + Cl^-(aq) + Na^+(aq) + HCOO^-(aq) \rightarrow HCOOH(aq) + Na^+(aq) + Cl^-(aq)$$

Na^+ and Cl^- are spectator ions.

Net ionic: $H^+(aq) + HCOO^-(aq) \rightleftharpoons HCOOH(aq)$

(b)   The net ionic equation in part (a) is the reverse of the dissociation of HCOOH.

$$K = \frac{1}{K_a} = \frac{1}{1.8 \times 10^{-4}} = 5.55 \times 10^3 = 5.6 \times 10^3$$

(c)   For Na^+ and Cl^-, this is just a dilution problem.

$M_1 V_1 = M_2 V_2$; V_2 is 50.0 mL + 50.0 mL = 100.0 mL

Cl^-: $\dfrac{0.15\ M \times 50.0\ mL}{100.0\ mL} = 0.075\ M$; Na^+: $\dfrac{0.15\ M \times 50.0\ mL}{100.0\ mL} = 0.075\ M$

H^+ and $HCOO^-$ react to form HCOOH. Since K >> 1, the reaction essentially goes to completion.

$0.15\ M \times 0.0500\ mL = 7.5 \times 10^{-3}\ mol\ H^+$

$\underline{0.15\ M \times 0.0500\ mL = 7.5 \times 10^{-3}\ mol\ HCOO^-}$

$\qquad\qquad\qquad = 7.5 \times 10^{-3}\ mol\ HCOOH$

Solve the weak acid problem to determine $[H^+]$, $[HCOO^-]$ and [HCOOH] at equilibrium.

$$K_a = \frac{[H^+][HCOO^-]}{[HCOOH]};\ [H^+] = [HCOO^-] = x\ M;\ [HCOOH] = \frac{(7.5 \times 10^{-3} - x)\ mol}{0.100\ L}$$

$$= (0.075 - x)\ M$$

$$1.8 \times 10^{-4} = \frac{x^2}{(0.075 - x)} \approx \frac{x^2}{0.075};\ x = 3.7 \times 10^{-3}\ M\ H^+\ and\ HCOO^-$$

[HCOOH] = (0.075 − 0.0037) = 0.071 M

$$\frac{[H^+]}{[HCOOH]} \times 100 = \frac{3.7 \times 10^{-3}}{0.075} \times 100 = 4.9\%\ dissociation$$

In summary:

$[Na^+] = [Cl^-] = 0.075\ M$, [HCOOH] = 0.071 M, $[H^+] = [HCOO^-] = 0.0037\ M$

17.114 $\Pi = MRT, M = \dfrac{\Pi}{RT} = \dfrac{21\,torr}{298\,K} \times \dfrac{1\,atm}{760\,torr} \times \dfrac{K - mol}{0.08206\ L - atm} = 1.13 \times 10^{-3} = 1.1 \times 10^{-3}\ M$

$SrSO_4(s) \ \rightleftharpoons\ Sr^{2+}(aq) + SO_4^{2-}(aq);\ K_{sp} = [Sr^{2+}][SO_4^{2-}]$

The total particle concentration is $1.13 \times 10^{-3}\ M$. Each mole of $SrSO_4$ that dissolves produces 2 mol of ions, so $[Sr^{2+}] = [SO_4^{2-}] = 1.13 \times 10^{-3}\ M/2 = 5.65 \times 10^{-4} = 5.7 \times 10^{-4}\ M$.

$K_{sp} = (5.65 \times 10^{-4})^2 = 3.2 \times 10^{-7}$

18 Chemistry of the Environment

Visualizing Concepts

18.1   *Analyze*. Given that one mole of an ideal gas at 1 atm and 298 K occupies 22.4 L, is the volume of one mole of ideal gas in the middle of the stratosphere greater than, equal to, or less than 22.4 L?

Plan. Consider the relationship between pressure, temperature, and volume of an ideal gas. Use Figure 18.1 to estimate the pressure and temperature in the middle of the stratosphere, and compare the two sets of temperature and pressure.

Solve. According to the ideal gas law, $PV = nRT$, so $V = nRT/P$. Since n and R are constant for this exercise, V is proportional to T/P.

(a)   The stratosphere ranges from 10 to 50 km, so the middle is at approximately 30 km. At this altitude, $T \approx 230$ K, $P \approx 40$ torr (from Figure 18.1). Since we are comparing T/P ratios, either atm or torr can be used as pressure units; we will use torr.

At sea level: T/P = 298 K/760 torr = 0.39

At 30 km: T/P = 230 K/40 torr = 5.75

The proportionality constant (T/P) is much greater at 30 km than sea level, so the volume of 1 mol of an ideal gas is greater at this altitude. The decrease in temperature at 30 km is more than offset by the substantial decrease in pressure.

(b)   Volume is proportional to T/P, not simply T. The relative volumes of one mole of an ideal gas at 50 km and 85 km depend on the temperature and pressure at the two altitudes. From Figure 18.1,

50 km: $T \approx 270$ K, $P \approx 20$ torr, T/P = 270 K/20 torr = 13.5

85 km: $T \approx 190$ K, $P < 0.01$ torr, T/P = 190 K/0.01 torr = 19,000

Again, the slightly lower temperature at 85 km is more than offset by a much lower pressure. One mole of an ideal gas will occupy a much larger volume at 85 km than 50 km.

(c)   Gases behave most ideally at high temperature and low pressure. Pressure is minimum and temperature is high in the thermosphere. The stratopause (the boundary between the stratosphere and mesosphere) and the troposphere at low altitude are other regions with temperature maxima and relatively low pressures.

18.6   *Saltwater*, *freshwater* and *groundwater* differ in salt content, location and percentage of Earth's total water.

Saltwater contains high concentrations of dissolved salts and solids and includes the world ocean (97.2% of all water) and brackish or salty water (0.1% of all water) in places such as the Great Salt Lake and the Chesapeake Bay. The world ocean averages about 35 g of dissolved salts per kg of water, or 35,000 ppm.

Freshwater (0.6% of all water on earth) refers to natural waters that have low concentrations (less than 500 ppm) of dissolved salts and solids. Freshwater includes the waters of lakes, rivers, ponds, and streams.

Groundwater is freshwater that is under the soil. It resides in aquifers, porous rock that holds water, and composes 20% of the world's freshwater.

18.9   The guiding principle of green chemistry is that "an ounce of prevention is worth a pound of cure." Processes should be designed to minimize or eliminate solvents and waste, generate nontoxic waste, be energy efficient, employ renewable starting materials, and take advantage of catalysts that enable the use of safe and common reagents.

Earth's Atmosphere (section 18.1)

18.11   (a)   The temperature profile of the atmosphere (Figure 18.1) is the basis of its division into regions. The center of each peak or trough in the temperature profile corresponds to a new region.

      (b)   Troposphere, 0–12 km; stratosphere, 12–50 km; mesosphere, 50–85 km; thermosphere, 85–110 km.

18.13   *Analyze/Plan.* Given O_3 concentration in ppm, calculate partial pressure. Use the definition of ppm to get mol fraction O_3. For gases mole fraction = pressure fraction. Use the ideal-gas law to find mol O_3/L air and Avogadro's number to get molecules/L.

$$P_{O_3} = \chi_{O_3} \times P_{atm}; \quad 0.441 \text{ ppm } O_3 = \frac{0.441 \text{ mol } O_3}{1 \times 10^6 \text{ mol air}} = 4.41 \times 10^{-7} = \chi_{O_3} \quad Solve.$$

(a) $P_{O_3} = \chi_{O_3} \times P_{atm} = 4.41 \times 10^{-7} (0.67 \text{ atm}) = 2.955 \times 10^{-7} = 3.0 \times 10^{-7} \text{ atm}$

(b) $n = \dfrac{PV}{RT} = \dfrac{2.955 \times 10^{-7} \text{ atm} \times 1.0 \text{ L}}{298 \text{ K}} \times \dfrac{\text{mol-K}}{0.08206 \text{ L-atm}} = 1.208 \times 10^{-8} = 1.2 \times 10^{-8} \text{ mol } O_3$

$$1.208 \times 10^{-8} \text{ mol } O_3 \times \frac{6.022 \times 10^{23} \text{ molecules}}{\text{mol}} = 7.277 \times 10^{15} = 7.3 \times 10^{15} \text{ } O_3 \text{ molecules}$$

18.15   *Analyze/Plan.* Given CO concentration in ppm, calculate number of CO molecules in 1.0 L air at given conditions. ppm CO → χ_{O_3} → atm CO → mol CO → molecules CO.
Use the ideal gas law to change atm CO to mol CO, then Avogadro's number to get molecules. *Solve.*

$$3.5 \text{ ppm } CO = \frac{3.5 \text{ mol } CO}{1 \times 10^6 \text{ mol air}} = 3.5 \times 10^{-6} = \chi_{CO}$$

$$P_{CO} = \chi_{CO} \times P_{atm} = 3.5 \times 10^{-6} \times 759 \text{ torr} \times \frac{1 \text{ atm}}{760 \text{ torr}} = 3.495 \times 10^{-6} = 3.5 \times 10^{-6} \text{ atm}$$

$$n_{CO} = \frac{P_{CO}V}{RT} = \frac{3.495 \times 10^{-6} \text{ atm} \times 1.0 \text{ L}}{295 \text{ K}} \times \frac{\text{mol} \cdot \text{K}}{0.08206 \text{ L} \cdot \text{atm}} = 1.444 \times 10^{-7} = 1.4 \times 10^{-7} \text{ mol CO}$$

$$1.444 \times 10^{-7} \text{ mol CO} \times \frac{6.022 \times 10^{23} \text{ molecules}}{\text{mol}} = 8.695 \times 10^{16} = 8.7 \times 10^{16} \text{ molecules CO}$$

18.17 *Analyze/Plan.* Given bond dissociation energy in kJ/mol, calculate the wavelength of a single photon that will rupture a C–Br bond. kJ/mol → J/molecule. $\lambda = hc/E$. ($\lambda = hc/E$ describes the energy/wavelength relationship of a single photon.)   *Solve.*

(a) $\dfrac{210 \times 10^3 \text{ J}}{1 \text{ mol}} \times \dfrac{1 \text{ mol}}{6.022 \times 10^{23} \text{ molecules}} = 3.487 \times 10^{-19} = 3.49 \times 10^{-19}$ J/molecule

$\lambda = c/\nu$ We also have that $E = h\nu$, so $\nu = E/h$. Thus,

$$\lambda = \frac{hc}{E} = \frac{(6.626 \times 10^{-34} \text{ J} \cdot \text{sec})(3.00 \times 10^8 \text{ m/sec})}{3.487 \times 10^{-19} \text{ J}} = 5.70 \times 10^{-7} \text{ m} = 570 \text{ nm}$$

(b) This 570 nm wavelength is visible electromagnetic radiation.

18.19 (a) *Photodissociation* is cleavage of the $O{=}O$ bond such that two neutral O atoms are produced: $O_2(g) \rightarrow 2O(g)$

Photoionization is absorption of a photon with sufficient energy to eject an electron from an O_2 molecule: $O_2(g) + h\nu \rightarrow O_2^+ + e^-$

(b) Photoionization of O_2 requires 1205 kJ/mol. Photodissociation requires only 495 kJ/mol. At lower elevations, solar radiation with wavelengths corresponding to 1205 kJ/mol or shorter has already been absorbed, while the longer wavelength radiation has passed through relatively well. Below 90 km, the increased concentration of O_2 and the availability of longer wavelength radiation cause the photodissociation process to dominate.

Human Activities and Earth's Atmosphere (section 18.2)

18.21 The oxidation state of oxygen in O_3, O_2 and O is zero (0). Reactions in which oxygen changes only from one of these species to another do not involve changes in oxidation state. Examples: $O(g) + O(g) \rightarrow O_2(g)$; $2\,O_3(g) \rightarrow 3\,O_2(g)$.

Ozone depletion reactions which involve a halogen oxide such as ClO do involve a change in oxidation state for oxygen. In ClO, the oxidation state of oxygen is either +1 or +2, but it is not zero. A reaction involving ClO and one of the oxygen species with a zero oxidation state does involve a change in the oxidation state of oxygen atoms.

18.23 (a) A *chlorofluorocarbon* is a compound tha contains chlorine, fluorine and carbon. A *hydrofluorocarbon* contains hydrogen, fluorine, and carbon; it contains hydrogen in place of chlorine.

(b) CFCs are harmful because they undergo photodissociation to produce Cl atoms that catalyze the destructions of ozone. HFCs are potentially less harmful because they contain no C—Cl bonds. Their relatively stronger C—F bonds require more energy to undergo photodissociation, energy that is unlikely to be available in the stratosphere.

18.25   (a)   In order to catalyze ozone depletion, the halogen must be present as single halogen atoms. These halogen atoms are produced in the stratosphere by photo-dissociation of a carbon-halogen bond. According to Table 8.4, the C–F average bond dissociation energy is 485 kJ/mol, while that of C–Cl is 328 kJ/mol. The C–F bond requires more energy for dissociation and is not readily cleaved by the available wavelengths of UV light.

       (b)   Chlorine is present as chlorine atoms and chlorine oxide molecules, Cl and ClO.

18.27   (a)   Methane, CH_4, arises from decomposition of organic matter by certain microorganisms; it also escapes from underground gas deposits.

       (b) SO_2 is released in volcanic gases, and also is produced by bacterial action on decomposing vegetable and animal matter.

       (c)   Nitric oxide, NO, results from oxidation of decomposing organic matter, and is formed in lightning flashes.

18.29   (a)   Acid rain is primarily $H_2SO_4(aq)$.

$$H_2SO_4(aq) + CaCO_3(s) \rightarrow CaSO_4(s) + H_2O(l) + CO_2(g)$$

       (b)   The $CaSO_4(s)$ would be much less reactive with acidic solution, since it would require a strongly acidic solution to shift the relevant equilibrium to the right.

$$CaSO_4(s) + 2H^+(aq) \rightleftharpoons Ca^{2+}(aq) + 2HSO_4^-(aq)$$

            Note, however, that $CaSO_4(s)$ is brittle and easily dislodged; it provides none of the structural strength of limestone.

18.31   *Analyze/Plan.* Given wavelength of a photon, place it in the electromagnetic spectrum, calculate its energy in kJ/mol, and compare it to an average bond dissociation energy. Use Figure 6.4; $E(J/photon) = hc/\lambda$. J/photon $\rightarrow$ kJ/mol.   *Solve.*

       (a)   Ultraviolet (Figure 6.4)

       (b) $E_{photon} = hc/\lambda = \dfrac{6.626 \times 10^{-34} \text{ J-s} \times 3.00 \times 10^8 \text{ m/s}}{335 \times 10^{-9} \text{ m}} = 5.934 \times 10^{-19}$

$$= 5.93 \times 10^{-19} \text{ J/photon}$$

$$\frac{5.934 \times 10^{-19} \text{ J}}{1 \text{ photon}} \times \frac{6.022 \times 10^{23} \text{ photons}}{1 \text{ mol}} \times \frac{1 \text{kJ}}{1000 \text{ J}} = 357 \text{ kJ/mol}$$

       (c)   The average C–H bond energy from Table 8.4 is 413 kJ/mol. The energy calculated in part (b), 357 kJ/mol, is the energy required to break 1 mol of C–H bonds in formaldehyde, CH_2O. The C–H bond energy in CH_2O must be less than the "average" C–H bond energy.

       (d)

```
     :O:                  :O:
      ||                   ||
 H — C — H  +  hν ———▶  H — C·  + H
```

18.33   Most of the energy entering the atmosphere from the sun is in the form of visible radiation, while most of the energy leaving the earth is in the form of infrared radiation. CO_2 is transparent to the incoming visible radiation, but absorbs the outgoing infrared radiation.

Earth's Water (section 18.3)

18.35   *Analyze/Plan.* Given salinity and density, calculate molarity. A salinity of 5.6 denotes that there are 5.6 g of dry salt per kg of water.   *Solve.*

$$\frac{5.6 \text{ g NaCl}}{1 \text{ kg soln}} \times \frac{1.03 \text{ kg soln}}{1 \text{ L soln}} \times \frac{1 \text{ mol NaCl}}{58.44 \text{ g NaCl}} \times \frac{1 \text{ mol Na}^+}{1 \text{ mol NaCl}} = 0.0987 = 0.099 \; M \text{ Na}^+$$

18.37   *Analyze/Plan.* Given the power of sunlight per square meter striking Earth's surface, the enthalpy of evaporation of water and specific heat capacity of water, calculate the amount of energy delivered by the Sun over a 12-hour day. Use this amount of energy to calculate: (a) how many grams of water can be evaporated and (b) the temperature of a 10.0 cm by 1 square meter volume of water after 12 hours in the sunlight, assuming no evaporation. Calculate the mass of this volume of water using density at 25 °C.   *Solve.*

(a) $$\frac{168 \text{ W}}{\text{m}^2} \times \frac{1 \text{ J/s}}{1 \text{ W}} = \frac{168 \text{ J}}{\text{m}^2 \text{-s}}$$

$$\frac{168 \text{ J}}{\text{m}^2 \text{-s}} \times 1.00 \text{ m}^2 \times 12 \text{ hr} \times \frac{60 \text{ min}}{1 \text{ hr}} \times \frac{60 \text{ s}}{1 \text{ min}} \times \frac{1 \text{ kJ}}{1000 \text{ J}} = 7257.6 = 7.26 \times 10^3 \text{ kJ}$$

$$7257.6 \text{ kJ} \times \frac{1 \text{ mol H}_2\text{O}}{40.67 \text{ kJ}} \times \frac{18.02 \text{ g H}_2\text{O}}{1 \text{ mol H}_2\text{O}} = 3215.7 = 3.22 \times 10^3 \text{ g H}_2\text{O}$$

(b) $$1.00 \text{ m}^2 \times 10.0 \text{ cm} \times \frac{(100)^2 \text{ cm}^2}{1 \text{ m}^2} \times \frac{0.99707 \text{ g}}{1 \text{ cm}^3} = 99,707 = 9.97 \times 10^4 \text{ g H}_2\text{O}$$

$$7257.6 \text{ kJ} \times \frac{1000 \text{ J}}{1 \text{ kJ}} \times \frac{1 \text{ g-}^\circ\text{C}}{4.184 \text{ J}} \times \frac{1}{99,707 \text{ g H}_2\text{O}} = 17.397 = 17.4 \;^\circ\text{C}$$

The final temperature is 26 °C + 17.4 °C = 43.4 °C.

18.39   *Analyze/Plan.* g $Mg(OH)_2 \rightarrow$ mol $Mg(OH)_2 \rightarrow$ mol ratio $\rightarrow$ mol $CaO \rightarrow$ g CaO.   *Solve.*

$$1000 \text{ lb Mg(OH)}_2 \times \frac{453.6 \text{ g}}{\text{lb}} \times \frac{1 \text{ mol Mg(OH)}_2}{58.33 \text{ g Mg(OH)}_2} \times \frac{1 \text{ mol CaO}}{1 \text{ mol Mg(OH)}_2} \times \frac{56.08 \text{ g CaO}}{1 \text{ mol CaO}}$$
$$= 4.361 \times 10^5 \text{ g CaO}$$

18.41   (a)   *Groundwater* is freshwater (less than 500 ppm total salt content) that is under the soil; it composes 20% of the world's freshwater.

(b)   An *aquifer* is a layer of porous rock that holds groundwater.

Human Activities and Earth's Water (section 18.4)

18.43   *Analyze/Plan.* Given temperature and the concentration difference between the two solutions, ($\Delta M = 0.22 - 0.01 = 0.21 \; M$), calculate the minimum pressure for reverse osmosis. Use the relationship $\Pi = MRT$ from Section 13.5. This is the pressure required to halt osmosis from the more dilute (0.01 M) to the more concentrated (0.22 M) solution. Slightly more pressure will initiate reverse osmosis.   *Solve.*

$$\Pi = \Delta MRT = \frac{0.21 \text{ mol}}{\text{L}} \times \frac{0.08206 \text{ L-atm}}{\text{mol-K}} \times 298 \text{ K} = 5.135 = 5.1 \text{ atm}$$

The minimum pressure required to initiate reverse osmosis is greater than 5.1 atm.

18.45   *Analyze/Plan.* Under aerobic conditions, excess oxygen is present and decomposition leads to oxidized products, the element in its maximum oxidation state combined with oxygen. Under anaerobic conditions, little or no oxygen is present so decomposition leads to reduced products, the element in its minimum oxidation state combined with hydrogen.   *Solve.*

(a) CO_2, HCO_3^-, H_2O, SO_4^{2-}, NO_3^-

(b) $CH_4(g)$, $H_2S(g)$, $NH_3(g)$

18.47   *Analyze/Plan.* Given the balanced equation, calculate the amount of one reactant required to react exactly with a certain amount of the other reactants. Solve the stoichiometry problem. $g\ C_{18}H_{29}SO_3^- \rightarrow mol \rightarrow mol\ ratio \rightarrow mol\ O_2 \rightarrow g\ O_2$.   *Solve.*

$$10.0\ g\ C_{18}H_{29}SO_3^- \times \frac{1\ mol\ C_{18}H_{29}SO_3^-}{325\ g\ C_{18}H_{29}SO_3^-} \times \frac{51\ mol\ O_2}{2\ mol\ C_{18}H_{29}SO_3^-} \times \frac{32.0\ g\ O_2}{1\ mol\ O_2} = 25.1\ g\ O_2$$

Notice that the mass of O_2 required is 2.5 times greater than the mass of biodegradable material.

18.49   *Analyze/Plan.* Slaked lime is $Ca(OH)_2(s)$. The reaction is metathesis.   *Solve.*

$$Mg^{2+}(aq) + Ca(OH)_2(s) \rightarrow Mg(OH)_2(s) + Ca^{2+}(aq)$$

The excess $Ca^{2+}(aq)$ is removed as $CaCO_3$ by naturally occurring bicarbonate or added Na_2CO_3.

18.51   *Analyze/Plan.* Given $[Ca^{2+}]$ and $[HCO_3^-]$ calculate mole $Ca(OH)_2$ and Na_2CO_3 needed to remove the Ca^{2+} and HCO_3^-. Consider the chemical equations and reaction stoichiometry in the stepwise process.   *Solve.*

$Ca(OH)_2$ is added to remove Ca^{2+} as $CaCO_3(s)$, and Na_2CO_3 removes the remaining Ca^{2+}.

$$Ca^{2+}(aq) + 2HCO_3^-(aq) + [Ca^{2+}(aq) + 2OH^-(aq)] \rightarrow 2CaCO_3(s) + 2H_2O(l)$$

One mole $Ca(OH)_2$ is needed for each 2 moles of $HCO_3^-(aq)$ present.

$$\frac{7.0 \times 10^{-4}\ mol\ HCO_3^-}{L} \times \frac{1\ mol\ Ca(OH)_2}{2\ mol\ HCO_3^-} \times 1.200 \times 10^3\ L\ H_2O = 0.42\ mol\ Ca(OH)_2$$

$$1.200 \times 10^3\ L\ H_2O \times \frac{5.0 \times 10^{-4}\ mol\ Ca^{2+}}{L} = 0.60\ mol\ Ca^{2+}(aq)\ total$$

0.42 mol $Ca(OH)_2$ removes 0.42 mol of the 0.60 mol $Ca^{2+}(aq)$ in the sample. This leaves 0.18 mol $Ca^{2+}(aq)$ to be removed by Na_2CO_3.

$$Ca^{2+}(aq) + Na_2CO_3(aq) \rightarrow CaCO_3(s) + 2Na^+(aq)$$

0.18 mol of Na_2CO_3 is needed to remove the remaining $Ca^{2+}(aq)$.

18.53 $4FeSO_4(aq) + O_2(aq) + 2H_2O(l) \rightarrow 4Fe^{3+}(aq) + 4OH^-(aq) + 4SO_4^{2-}(aq)$

SO_4^{2-} is a spectator, so the net ionic equation is

$4Fe^{2+}(aq) + O_2(aq) + 2H_2O(l) \rightarrow 4Fe^{3+}(aq) + 4OH^-(aq)$.

$Fe^{3+}(aq) + 3HCO_3^-(aq) \rightarrow Fe(OH)_3(s) + 3CO_2(g)$

In this reaction, Fe^{3+} acts as a Lewis acid, and HCO_3^- acts as a Lewis base.

18.55   (a)   *Trihalomethanes* are a class of molecules with one central carbon atom bound to one hydrogen and three halogen atoms. They are produced by the reaction of dissolved chlorine with organic matter naturally present in water, and are byproducts of water disinfection via chlorination.

(b)

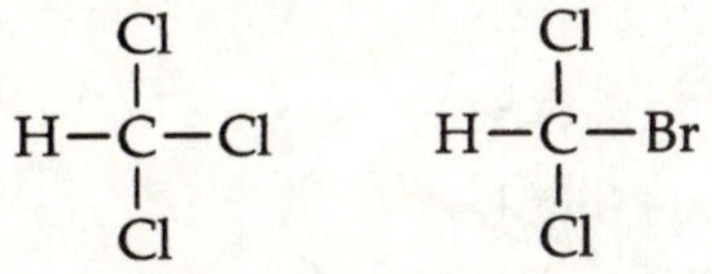

Green Chemistry (section 18.5)

18.57   The fewer steps in a process, the less waste (solvents as well as unusable by-products) is generated. It is probably true that a process with fewer steps requires less energy at the site of the process, and it is certainly true that the less waste the process generates, the less energy is required to clean or dispose of the waste.

18.59   (a)

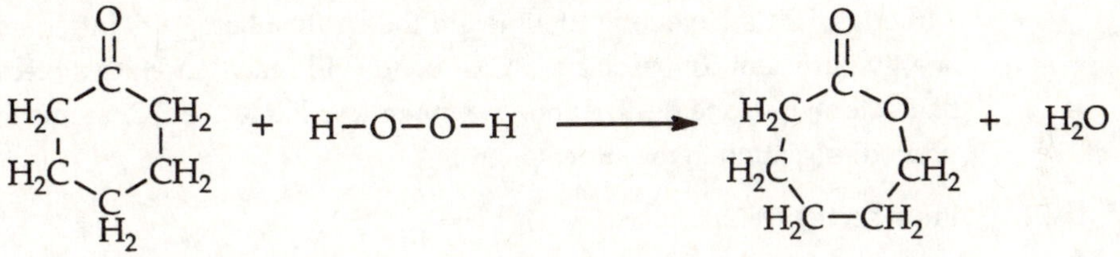

(b)   • **Prevention (1).** The alternative process eliminates production of 3-chlorobenzoic acid by-product, chlorine-containing waste that must be treated.

• **Atom Economy (2).** Most of the starting atoms are in the final product.

• **Less hazardous chemical synthesis (3)** and **Inherently safer for accident prevention (12).** The starting material of the alternative process is not shock-sensitive, and the by-product is nontoxic water. The low molar mass of water means that a small amount of "waste" is generated.

• **Catalysis (9)** and **Design for energy efficiency(6).** The alternative process is catalyzed, which could mean that the process will be more energy efficient than the Baeyer-Villiger reaction (see Solution 18.58).

• **Raw materials should be renewable (7).** The catalyst can be recovered from the reaction mixture and reused. We don't have information about solvents or other auxiliary substances.

18.61   (a)   Water as a solvent is much "greener" than benzene, which is a known carcinogen. Water fits criteria: (5) safer solvent, (7) renewable feedstock and (12) inherently safer for accident prevention.

(b)   Reaction temperature of 500 K rather than 1000 K is "greener", according to criteria (6) design for energy efficiency and (12) inherently safer chemistry for accident prevention. Also, low temperature is less likely to produce undesireable byproducts that have to be separated and treated as waste, which fits criterium (1).

(c)   Sodium chloride as a byproduct rather than chloroform ($CHCl_3$) is "greener", according to criteria: (1) prevention, (3) less hazardous chemical systems, and (12) inherently safer ($CHCl_3$ is flammable, while NaCl is not).

Additional Exercises

18.66

$$2[Cl(g) + O_3(g) \rightarrow ClO(g) + O_2(g)] \qquad [18.7]$$
$$2[ClO(g) + hv \rightarrow O(g) + Cl(g)] \qquad [18.9]$$
$$\underline{O(g) + O(g) \rightarrow O_2(g)}$$
$$2Cl(g) + 2O_3(g) + 2ClO(g) + 2O(g) \rightarrow 2ClO(g) + 3O_2(g) + 2Cl(g)$$
$$2O_3(g) \xrightarrow{Cl} 3O_2(g) \qquad [18.10]$$

Note that Cl(g) fits the definition of a catalyst in this reaction.

18.69   In an HFC, C–Cl bonds are replaced by C–F bonds. The bond dissociation enthalpy of a C–F bond is 485 kJ/mol, much more than for a C–Cl bond, 328 kJ/mol (Table 8.4). Although HFCs have long lifetimes in the stratosphere, it is infrequent that light with energy sufficient to dissociate a C–F bond will reach an HFC molecule. F atoms, the bad actors in ozone destruction, are much less likely than Cl atoms to be produced by photodissociation in the stratosphere.

18.71   From section 18.2:

$$N_2(g) + O_2(g) \rightleftharpoons 2NO(g) \quad \Delta H = +180.8 \text{ kJ} \quad [18.11]$$
$$2\,NO(g) + O_2(g) \rightleftharpoons 2NO_2(g) \quad \Delta H = -113.1 \text{ kJ} \quad [18.12]$$

In an endothermic reaction, heat is a reactant. As the temperature of the reaction increases, the addition of heat favors formation of products and the value of K increases. The reverse is true for exothermic reactions; as temperature increases, the value of K decreases. Thus, K for reaction [18.11], which is endothermic, increases with increasing temperature and K for reaction [18.12], which is exothermic, decreases with increasing temperature.

18.75   Given 168 watts/m^2 at 10% efficiency, find the land area needed to produce 12,000 megawatts. 12,000 megawatts = $12,000 \times 10^6 = 1.2 \times 10^{10}$ watts.

168 watts/m^2 (0.10) = 16.8 watts/m^2 solar energy possible with current technology.

$$1.2 \times 10^{10} \text{ watts} \times \frac{1\,m^2}{16.8\,\text{watts}} = 7.143 \times 10^8 = 7.1 \times 10^8 \text{ m}^2$$

The land area of New York City is 830 km^2, which is 830×10^6 m^2. The area needed for solar energy harvesting to provide peak power would then be $\dfrac{7.143 \times 10^8 \text{ m}^2}{830 \times 10^6 \text{ m}^2} = 0.86$

times the land area of New York City.

18.77   (a) CO_3^{2-} is a relatively strong Brønsted-Lowry base and produces OH^- in aqueous solution according to the hydrolysis reaction:

$$CO_3^{2-}(aq) + H_2O(l) \rightleftharpoons HCO_3^-(aq) + OH^-(aq), \quad K_b = 1.8 \times 10^{-4}$$

If [$OH^-(aq)$] is sufficient for the reaction quotient, Q, to exceed K_{sp} for $Mg(OH)_2$, the solid will precipitate.

(b) $\dfrac{125 \text{ mg Mg}^{2+}}{1 \text{ kg soln}} \times \dfrac{1 \text{ g Mg}^{2+}}{1000 \text{ mg Mg}^{2+}} \times \dfrac{1.00 \text{ kg soln}}{1.00 \text{ L soln}} \times \dfrac{1 \text{ mol Mg}^{2+}}{24.305 \text{ g Mg}^{2+}} = 5.143 \times 10^{-3}$

$= 5.14 \times 10^{-3} \; M \text{ Mg}^{2+}$

$\dfrac{4.0 \text{ g Na}_2\text{CO}_3}{1.0 \text{ L soln}} \times \dfrac{1 \text{ mol CO}_3^{2-}}{106.0 \text{ g Na}_2\text{CO}_3} = 0.03774 = 0.038 \; M \text{ CO}_3^{2-}$

$K_b = 1.8 \times 10^{-4} = \dfrac{[\text{HCO}_3^-][\text{OH}^-]}{[\text{CO}_3^{2-}]} \approx \dfrac{x^2}{0.03774}; \; x = [\text{OH}^-] = 2.606 \times 10^{-3}$

$= 2.6 \times 10^{-3} \; M$

(This represents 6.9% hydrolysis, but the result will not be significantly different using the quadratic formula.)

$Q = [\text{Mg}^{2+}][\text{OH}^-]^2 = (5.143 \times 10^{-3})(2.606 \times 10^{-3})^2 = 3.5 \times 10^{-8}$

K_{sp} for $\text{Mg(OH)}_2 = 1.6 \times 10^{-12}$; $Q > K_{sp}$, so Mg(OH)_2 will precipitate.

Integrative Exercises

18.81　(a)　$8{,}376{,}726 \text{ tons coal} \times \dfrac{83 \text{ ton C}}{100 \text{ ton coal}} \times \dfrac{44.01 \text{ ton CO}_2}{12.01 \text{ ton C}} = 2.5 \times 10^7 \text{ ton CO}_2$

$8{,}376{,}726 \text{ tons coal} \times \dfrac{2.5 \text{ ton S}}{100 \text{ ton coal}} \times \dfrac{64.07 \text{ ton SO}_2}{32.07 \text{ ton S}} = 4.2 \times 10^5 \text{ ton SO}_2$

(b)　$\text{CaO(s)} + \text{SO}_2(\text{g}) \rightarrow \text{CaSO}_3(\text{s})$

$4.18 \times 10^5 \text{ ton SO}_2 \times \dfrac{55 \text{ ton SO}_2 \text{ removed}}{100 \text{ ton SO}_2 \text{ produced}} \times \dfrac{120.15 \text{ ton CaSO}_3}{64.07 \text{ ton SO}_2}$

$= 4.3 \times 10^5 \text{ ton CaSO}_3$

18.84　(a)　$\text{H} - \ddot{\text{O}} - \text{H} \longrightarrow \text{H}\cdot + \cdot\ddot{\text{O}} - \text{H}$

(b)　$\Delta H = 2D(\text{O}-\text{H}) - D(\text{O}-\text{H}) = D(\text{O}-\text{H}) = 463 \text{ kJ/mol}$

$\dfrac{463 \text{ kJ}}{\text{mol H}_2\text{O}} \times \dfrac{1 \text{ mol H}_2\text{O}}{6.022 \times 10^{23} \text{ molecules}} \times \dfrac{1000 \text{ J}}{\text{kJ}} = 7.688 \times 10^{-19}$

$= 7.69 \times 10^{-19} \text{ J/H}_2\text{O molecule}$

$\lambda = \dfrac{hc}{\Delta E} = \dfrac{6.626 \times 10^{-34} \text{ J-sec} \times 2.998 \times 10^8 \text{ m/s}}{7.688 \times 10^{-19} \text{ J}} = 2.58 \times 10^{-7} \text{ m} = 258 \text{ nm}$

This wavelength is in the UV region of the spectrum, close to the visible.

(c)　$\text{OH(g)} + \text{O}_3(\text{g}) \rightarrow \text{HO}_2(\text{g}) + \text{O}_2(\text{g})$

$\underline{\text{HO}_2(\text{g}) + \text{O(g)} \rightarrow \text{OH(g)} + \text{O}_2(\text{g})}$

$\text{OH(g)} + \text{O}_3(\text{g}) + \text{HO}_2(\text{g}) + \text{O(g)} \rightarrow \text{HO}_2(\text{g}) + 2\text{O}_2(\text{g}) + \text{OH(g)}$

$\text{O}_3(\text{g}) + \text{O(g)} \rightarrow 2\text{O}_2(\text{g})$

OH(g) is the catalyst is this overall reaction, another pathway for the destruction of ozone.

18.86   (i) $ClO(g) + O_3(g) \rightarrow ClO_2(g) + O_2(g)$

$\Delta H_i = \Delta H_f^\circ ClO_2(g) + \Delta H_f^\circ O_2(g) - \Delta H_f^\circ ClO(g) - \Delta H_f^\circ O_3(g)$

$\Delta H_i = 102 + 0 - 101 - (142.3) = -141$ kJ

(ii) $ClO_2(g) + O(g) \rightarrow ClO(g) + O_2(g)$

$\Delta H_{ii} = \Delta H_f^\circ ClO(g) + \Delta H_f^\circ O_2(g) - \Delta H_f^\circ ClO_2(g) - \Delta H_f^\circ O(g)$

$\Delta H_{ii} = 101 + 0 - 102 - (247.5) = -249$ kJ

(overall) $ClO(g) + O_3(g) + ClO_2(g) + O(g) \rightarrow ClO_2(g) + O_2(g) + ClO(g) + O_2(g)$

$$O_3(g) + O(g) \rightarrow 2O_2(g)$$

$\Delta H = \Delta H_i + \Delta H_{ii} = -141$ kJ $+ (-249)$ kJ $= -390$ kJ

Because the enthalpies of both (i) and (ii) are distinctly exothermic, it is possible that the $ClO - ClO_2$ pair could be a catalyst for the destruction of ozone.

18.90   (a)   Holding one reactant concentration constant and changing the other, evaluate the effect this has on the initial rate. Use these observations to write the rate law.

Compare Experiments 1 and 3. $[O_3]$ is constant, $[H]$ doubles, initial rate doubles. The reaction is first order in $[H]$.

Compare Experiments 2 and 1. $[H]$ is constant, $[O_3]$ doubles, initial rate doubles. The reaction is first order in $[O_3]$.

rate $= k[O_3][H]$

(b)   Calculate a value for the rate constant for each experiment, then average them to obtain a single representative value.

rate $= k[O_3][H]$; $k = $ rate$/[O_3][H]$

$$k_1 = \frac{1.88 \times 10^{-14} \ M/s}{(5.17 \times 10^{-33} \ M)(3.22 \times 10^{-26} \ M)} = 1.1293 \times 10^{44} = 1.13 \times 10^{44}$$

$$k_2 = \frac{9.44 \times 10^{-15} \ M/s}{(2.59 \times 10^{-33} \ M)(3.25 \times 10^{-26} \ M)} = 1.1215 \times 10^{44} = 1.12 \times 10^{44}$$

$$k_3 = \frac{3.77 \times 10^{-14} \ M/s}{(5.19 \times 10^{-33} \ M)(6.46 \times 10^{-26} \ M)} = 1.1245 \times 10^{44} = 1.12 \times 10^{44}$$

$k_{avg} = (1.1293 \times 10^{44} + 1.1215 \times 10^{44} + 1.1245 \times 10^{44})/3 = 1.1251 \times 10^{44} =$

$1.13 \times 10^{44} \ M^{-1} \ s^{-1}$

18.95   (a)   Process (i) is greener, because it does not involve the toxic reactant phosgene $(COCl_2)$ and the by-product is water, not HCl.

(b)   Reaction (i): C in CO_2 is linear with *sp* hybridization; C in R—N=C=O is linear with *sp* hybridization; C in the urethane monomer is trigonal planar with sp^2 hybridization. Reaction (ii): C in $COCl_2$ is trigonal planar with sp^2 hybridization; C in R—N=C=O is linear with *sp* hybridization; C in the urethane monomer is trigonal planar with sp^2 hybridization.

(c)   Traditionally, industrial processes are conducted at higher temperatures to speed up reactions and encourage formation of product. However, this is not a green solution, because it requires additional energy. Using Le Chatelier's principle, we could either "push" or "pull" the reaction toward products. The "push" requires that we increase the amount of reactants, again not a green approach. The greenest way to promote formation of the isocyanate is to "pull" the reaction forward by removing by-product from the reaction mixture. In reaction (i), remove water; in reaction (ii), remove HCl.

19 Chemical Thermodynamics

Visualizing Concepts

19.1 (a)

(b) ΔS is positive, because the disorder of the system increases. Each gas has greater motional freedom as it expands into the second bulb, and there are many more possible arrangements for the mixed gases.

By definition, ideal gases experience no attractive or repulsive intermolecular interactions, so ΔH for the mixing of ideal gases is zero, assuming heat exchange only between the two bulbs.

(c) The process is spontaneous and therefore irreversible. It is inconceivable that the gases would reseparate.

(d) The entropy change of the surroundings is related to ΔH for the system. Since we are mixing ideal gases and $\Delta H = 0$, ΔH_{surr} is also zero, assuming heat exchange only between the two bulbs.

19.4 In the depicted reaction, both reactants and products are in the gas phase (they are far apart and randomly placed). There are twice as many molecules (or moles) of gas in the products, so ΔS is positive for this reaction.

19.7 (a) At 300 K, $\Delta H = T\Delta S$. Since $\Delta G = \Delta H - T\Delta S$, $\Delta G = 0$ at this point. When $\Delta G = 0$, the system is at equilibrium.

(b) The reaction is spontaneous when ΔG is negative. This condition is met when $T\Delta S > \Delta H$. From the diagram, $T\Delta S > \Delta H$ when $T > 300$ K. The reaction is spontaneous at temperatures above 300 K.

19.10 (a) The minimum in the plot is the equilibrium position of the reaction, where $\Delta G = 0$.

(b) The quantity x is the difference in free energy between reactant and products in their standard states, $\Delta G°$.

Spontaneous Processes (section 19.1)

19.11 *Analyze/Plan.* Follow the logic in Sample Exercise 19.1.   *Solve.*

(a) Spontaneous; at ambient temperature, ripening happens without intervention.

(b) Spontaneous; sugar is soluble in water, and even more soluble in hot coffee.

(c)   Spontaneous; N_2 molecules are stable relative to isolated N atoms.

(d)   Spontaneous; under certain atmospheric conditions, lightening occurs.

(e)   Nonspontaneous; CO_2 and H_2O are in contact continuously at atmospheric conditions in nature and do not form CH_4 and O_2.

19.13   (a) $NH_4NO_3(s)$ dissolves in water, as in a chemical cold pack. Naphthalene (mothballs) sublimes at room temperature.

(b)   Melting of a solid is spontaneous above its melting point but nonspontaneous below its melting point.

19.15   *Analyze/Plan.* Define the system and surroundings. Use the appropriate definition to answer the specific questions.   *Solve.*

(a)   Water is the system. Heat must be added to the system to evaporate the water. The process is endothermic.

(b)   At 1 atm, the reaction is spontaneous at temperatures above 100°C.

(c)   At 1 atm, the reaction is nonspontaneous at temperatures below 100°C.

(d)   The two phases are in equilibrium at 100°C.

19.17   *Analyze/Plan.* Define the system and surroundings. Use the appropriate definition to answer the specific questions.   *Solve.*

(a)   For a *reversible* process, the forward and reverse changes occur by the same path. In a reversible process, both the system and the surroundings are restored to their original condition by exactly reversing the change. A reversible change produces the maximum amount of work.

(b)   If a system is returned to its original state via a reversible path, the surroundings are also returned to their original state. That is, there is no net change in the surroundings.

(c)   The vaporization of water to steam is reversible if it occurs at the boiling temperature of water for a specified external (atmospheric) pressure, and if the required heat is added infinitely slowly.

(d)   No. Natural processes, such as a banana ripening or a lightning strike are spontaneous in the direction they occur and nonspontaneous in the opposite direction. By definition they are irreversible; they do not occur by reversible pathways. Neither the system nor the surroundings can be returned to their original condition by the same pathway that the change occurred. It is impossible to imagine a banana unripening.

19.19   *Analyze/Plan.* The related properties of a gas are pressure, volume, temperature and amount.

(a)   If T decreases while V is unchanged, either P or amount must change. For a closed system (Section 5.1) at constant volume, a decrease in external temperature leads to a decrease in the temperature of the system, in this case an ideal gas, as well as a decrease in pressure of the gas. An example is the decrease in air pressure in a tire or the first cold autumn day.

(b)   If T decreases while P stays constant, either amount or volume must change. For a closed system at constant pressure, if the temperature of the gas decreases, the volume also decreases.

(c)   No. ΔE is a state function. $\Delta E = q + w$; q and w are not state functions. Their values do depend on path, but their sum, ΔE, does not.

19.21   *Analyze/Plan.* Define the system and surroundings. Use the appropriate definition to answer the specific questions.   *Solve.*

(a)   An ice cube can melt reversibly at the conditions of temperature and pressure where the solid and liquid are in equilibrium. At 1 atm external pressure, the normal melting point of water is 0°C.

(b)   We know that melting is a process that increases the energy of the system, even though there is no change in temperature. ΔE is not zero for the process.

Entropy and the Second Law of Thermodynamics (section 19.2)

19.23   (a)   For a process that occurs at constant temperature, an isothermal process, $\Delta S = q_{rev}/T$. Here q_{rev} is the heat that would be transferred if the process were reversible. Since ΔS is a state function, it is independent of path, so ΔS for the reversible path must equal ΔS for any path.

(b)   No. ΔS is a state function, so it is independent of path.

19.25   (a) $Br_2(l) \rightarrow Br_2(g)$, entropy increases, more mol gas in products, greater motional freedom.

(b) $$\Delta S = \frac{\Delta H}{T} = \frac{29.6\,kJ}{mol\,Br_2(l)} \times 1.00\,mol\,Br_2(l) \times \frac{1}{(273.15 + 58.8)K} \times \frac{1000\,J}{1\,kJ} = 89.2\,J/K$$

19.27   (a)   For a spontaneous process, the entropy of the universe increases; for a reversible process, the entropy of the universe does not change.

(b)   In a reversible process, $\Delta S_{sys} + \Delta S_{surr} = 0$. If ΔS_{sys} is positive, ΔS_{surr} must be negative.

(c)   Since ΔS_{univ} must be positive for a spontaneous process, ΔS_{surr} must be greater than $-42\,J/K$.

19.29   *Analyze.* Consider ΔS for the isothermal expansion of 0.200 mol of an ideal gas at 27°C and an initial volume of 10.0 L.

(a)   Whenever an ideal gas expands isothermally, we expect an increase in entropy, or positive ΔS, owing to the greater volume available for motion of the particles.

(b)   *Plan.* Use the relationship $\Delta S_{sys} = nR\ln(V_2/V_1)$, Equation [19.3].

Solve. $\Delta S_{sys} = 0.200\,(8.314\,J/mol\text{-}K)(\ln\,[18.5\,L/10.0\,L]) = 1.02\,J/K$.

Check. We expect ΔS to be positive when the motional freedom of a gas increases, and our calculation agrees with this prediction.

(c)   The temperature at which the expansion occurs is not needed to calculate the entropy change, as long as the process is isothermal.

The Molecular Interpretation of Entropy (section 19.3)

19.31   (a)   Yes, the expansion is spontaneous.

(b)   The ideal gas is the system, and everything else, including the vessel containing the vacuum, is the surroundings. There is literally nothing inside the vessel containing the vacuum, no gas molecules and no physical barriers. As the ideal gas expands into the vacuum, there is nothing for it to "push back", so no work is done. Mathematically, $w = -P_{ext}\Delta V$. Since the gas expands into a vacuum, $P_{ext} = 0$ and $w = 0$.

(c)   The "driving force" for the expansion of the gas is the increase in entropy associated with greater volume, more motional freedom and more possible positions for the gas particles.

19.33   (a)   The higher the temperature, the broader the distribution of molecular speeds and kinetic energies available to the particles. At higher temperature, the wider range of accessible kinetic energies leads to more microstates for the system.

(b)   A decrease in volume reduces the number of possible positions for the particles and leads to fewer microstates for the system.

(c)   Going from liquid to gas, particles have greater translational motion, which increases the number of positions available to the particles and the number of microstates for the system.

19.35   *Analyze/Plan.* Consider the conditions that lead to an increase in entropy: more mol gas in products than reactants, increase in volume of sample and, therefore, number of possible arrangements, more motional freedom of molecules, etc.   *Solve.*

(a)   More gaseous particles means more possible arrangements and greater disorder; ΔS is positive.

(b) S_{sys} may increase slightly in 19.11 (a), where the sample softens; this is not definitive.

S_{sys} clearly increases in Exercise 19.11 (b), where there is an increase in volume and possible arrangements for the sample.

In 19.11 (c), the system goes from two moles of gaseous reactants to one mole of gaseous products, and S_{sys} decreases.

In 19.11 (d), the entropy of the universe clearly increases, but the definition of the system in a lightning strike is more problematic.

In 19.11 (e), the state is specified as room temperature and 1 atm pressure. This means that H_2O is present as a liquid; there is then one mol of gaseous reactants (CO_2) and three mol of gaseous products (CH_4 and 2 O_2), so S_{sys} increases. (The reaction is not spontaneous because of the very large positive ΔH_{sys} for the reaction as written.)

19.37   *Analyze/Plan.* Consider the conditions that lead to an increase in entropy: more mol gas in products than reactants, increase in volume of sample and, therefore, number of possible arrangements, more motional freedom of molecules, etc.   *Solve.*

(a)   S increases; translational motion is greater in the liquid than the solid.

(b)   S decreases; volume and translational motion decrease going from the gas to the liquid.

(c)   S increases; volume and translational motion are greater in the gas than the solid.

19.39   (a)   The entropy of a pure crystalline substance at absolute zero is zero.

(b)   In *translational* motion, the entire molecule moves in a single direction; in *rotational* motion, the molecule rotates or spins around a fixed axis. *Vibrational* motion is reciprocating motion. The bonds within a molecule stretch and bend, but the average position of the atoms does not change.

(c)

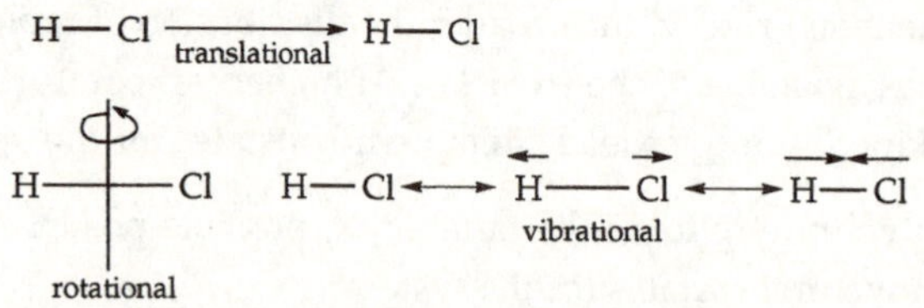

19.41   *Analyze/Plan.* Consider the factors that lead to higher entropy: more mol gas in products than reactants, increase in volume of sample and, therefore, number of possible arrangements, more motional freedom of molecules, etc.   *Solve.*

(a)   Ar(g) (gases have higher entropy due primarily to much larger volume)

(b)   He(g) at 1.5 atm (larger volume and more motional freedom)

(c)   1 mol of Ne(g) in 15.0 L (larger volume provides more motional freedom)

(d) CO_2(g) (more motional freedom)

19.43   *Analyze/Plan.* Consider the markers of an increase in entropy for a chemical reaction: liquids or solutions formed from solids, gases formed from either solids or liquids, increase in moles gas during reaction.   *Solve.*

(a)   ΔS negative (moles of gas decrease)

(b)   ΔS positive (gas produced, increased disorder)

(c)   ΔS negative (moles of gas decrease)

(d)   ΔS is small and probably positive (moles of gas same in reactants and products, H_2O(g) is more structurally complex than H_2(g)]

Entropy Changes in Chemical Reactions (section 19.4)

19.45   (a)

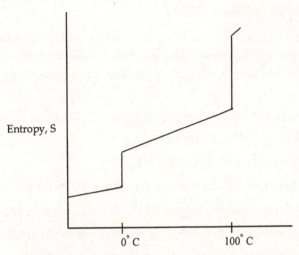

(b)   Boiling water, at 100 °C, has a much larger entropy change than melting ice at 0 °C. Before and after melting, H_2O molecules are touching. And there is actually a small decrease in volume going from solid to liquid water. Boiling drastically increases the distance between molecules and the volume of the sample. The increase in available molecular positions is much greater for boiling than melting, so the entropy change is also greater.

19.47   *Analyze/Plan.* Given two molecules in the same state, predict which will have the higher molar entropy. In general, for molecules in the same state, the more atoms in the molecule, the more degrees of freedom, the greater the number of microstates and the higher the standard entropy, S°.

(a) $C_2H_6(g)$ has more degrees of freedom and larger S°.

(b) $CO_2(g)$ has more degrees of freedom and larger S°.

19.49   *Analyze/Plan.* Consider the conditions that lead to an increase in entropy: more mol gas in products than reactants, increase in volume of sample and, therefore, number of possible arrangements, more motional freedom of molecules, etc. *Solve.*

(a)   Sc(s), 34.6 J/mol-K; Sc(g), 174.7 J/mol-K. In general, the gas phase of a substance has a larger S° than the solid phase because of the greater volume and motional freedom of the molecules.

(b) $NH_3(g)$, 192.5 J/mol-K; $NH_3(aq)$, 111.3 J/mol-K. Molecules in the gas phase have more motional freedom than molecules in solution.

(c)   1 mol of $P_4(g)$, 280 J/K; 2 mol of $P_2(g)$, 2(218.1) = 436.2 J/K. More particles have greater motional energy (more available microstates).

(d)   C(diamond), 2.43 J/mol-K; C(graphite) 5.69 J/mol-K. Diamond is a network covalent solid with each C atom tetrahedrally bound to four other C atoms. Graphite consists of sheets of fused planar 6-membered rings with each C atom bound in a trigonal planar arrangement to three other C atoms. The internal entropy in graphite is greater because there is translational freedom among the planar sheets of C atoms while there is very little vibrational freedom within the network covalent diamond lattice.

19.51   For elements with similar structures, the heavier the atoms, the lower the vibrational frequencies at a given temperature. This means that more vibrations can be accessed at a particular temperature resulting in a greater absolute entropy for the heavier elements.

19.53   *Analyze/Plan.* Follow the logic in Sample Exercise 19.5.   *Solve.*

(a) $\Delta S° = S° \, C_2H_6(g) - S° \, C_2H_4(g) - S° \, H_2(g)$

   $= 229.5 - 219.4 - 130.58 = -120.5 \, J/K$

$\Delta S°$ is negative because there are fewer moles of gas in the products.

(b) $\Delta S° = 2S° \, NO_2(g) - \Delta S° \, N_2O_4(g) = 2(240.45) - 304.3 = +176.6 \, J/K$

$\Delta S°$ is positive because there are more moles of gas in the products.

(c) $\Delta S° = \Delta S° \, BeO(s) + \Delta S° \, H_2O(g) - \Delta S° \, Be(OH)_2(s)$

   $= 13.77 + 188.83 - 50.21 = +152.39 \, J/K$

$\Delta S°$ is positive because the product contains more total particles and more moles of gas.

(d) $\Delta S° = 2S° \, CO_2(g) + 4S° \, H_2O(g) - 2S° \, CH_3OH(g) - 3S° \, O_2(g)$

   $= 2(213.6) + 4(188.83) - 2(237.6) - 3(205.0) = +92.3 \, J/K$

$\Delta S°$ is positive because the product contains more total particles and more moles of gas.

Gibbs Free Energy (sections 19.5 and 19.6)

19.55   (a) $\Delta G = \Delta H - T\Delta S$

(b)   If ΔG is positive, the process is nonspontaneous, but the reverse process is spontaneous.

(c)   There is no relationship between ΔG and rate of reaction. A spontaneous reaction, one with a $-\Delta G$, may occur at a very slow rate. For example: $2H_2(g) + O_2(g) \rightarrow 2H_2O(g)$, $\Delta G = -457 \, kJ$ is very slow if not initiated by a spark.

19.57   *Analyze/Plan.* Consider the definitions of $\Delta H°$, $\Delta S°$ and $\Delta G°$, along with sign conventions. $\Delta G° = \Delta H° - T\Delta S°$.   *Solve.*

(a) $\Delta H°$ is negative; the reaction is exothermic.

(b) $\Delta S°$ is negative; the reaction leads to decrease in disorder (increase in order) of the system.

(c) $\Delta G° = \Delta H° - T\Delta S° = -35.4 \, kJ - 298 \, K \, (-0.0855 \, kJ/K) = -9.921 = -9.9 \, kJ$

(d)   At 298 K, $\Delta G°$ is negative. If all reactants and products are present in their standard states, the reaction is spontaneous (in the forward direction) at this temperature.

19.59   *Analyze/Plan.* Follow the logic in Sample Exercise 19.7. Calculate $\Delta H°$ according to Equation [5.31], $\Delta S°$ by Equation [19.8] and $\Delta G°$ by Equation [19.14]. Then use $\Delta H°$ and $\Delta S°$ to calculate $\Delta G°$ using Equation [19.12], $\Delta G° = \Delta H° - T\Delta S°$.   *Solve.*

(a) $\Delta H° = 2(-268.61) - [0 + 0] = -537.22$ kJ

        $\Delta S° = 2(173.51) - [130.58 + 202.7] = 13.74 = 13.7$ J/K

        $\Delta G° = 2(-270.70) - [0 + 0] = -541.40$ kJ

        $\Delta G° = -537.22$ kJ $- 298(0.01374)$ kJ $= -541.31$ kJ

(b) $\Delta H° = -106.7 - [0 + 2(0)] = -106.7$ kJ

        $\Delta S° = 309.4 - [5.69 + 2(222.96)] = -142.21 = -142.2$ J/K

        $\Delta G° = -64.0 - [0 + 2(0)] = -64.0$ kJ

        $\Delta G° = -106.7$ kJ $- 298(-0.14221)$ kJ $= -64.3$ kJ

(c) $\Delta H° = 2(-542.2) - [2(-288.07) + 0] = -508.26 = -508.3$ kJ

        $\Delta S° = 2(325) - [2(311.7) + 205.0] = -178.4 = -178$ J/K

        $\Delta G° = 2(-502.5) - [2(-269.6) + 0] = -465.8$ kJ

        $\Delta G° = -508.26$ kJ $- 298(-0.1784)$ kJ $= -455.097 = -455.1$ kJ

        (The discrepancy in $\Delta G°$ values is due to experimental uncertainties in the tabulated thermodynamic data.)

(d) $\Delta H° = -84.68 + 2(-241.82) - [2(-201.2) + 0] = -165.92 = -165.9$ kJ

        $\Delta S° = 229.5 + 2(188.83) - [2(237.6) + 130.58] = 1.38 = 1.4$ J/K

        $\Delta G° = -32.89 + 2(-228.57) - [2(-161.9) + 0] = -166.23 = -166.2$ kJ

        $\Delta G° = -165.92$ kJ $- 298(0.00138)$ kJ $= -166.33 = -166.3$ kJ

19.61   *Analyze/Plan.* Follow the logic in Sample Exercise 19.7.   *Solve.*

(a) $\Delta G° = 2\Delta G° \, SO_3(g) - [2\Delta G° \, SO_2(g) + \Delta G° \, O_2(g)]$

        $= 2(-370.4) - [2(-300.4) + 0] = -140.0$ kJ, spontaneous

(b) $\Delta G° = 3\Delta G° \, NO(g) - [\Delta G° \, NO_2(g) + \Delta G° \, N_2O(g)]$

        $= 3(86.71) - [51.84 + 103.59] = +104.70$ kJ, nonspontaneous

(c) $\Delta G° = 4\Delta G° \, FeCl_3(s) + 3\Delta G° \, O_2(g) - [6\Delta G° \, Cl_2(g) + 2\Delta G° \, Fe_2O_3(s)]$

        $= 4(-334) + 3(0) - [6(0) + 2(-740.98)] = +146$ kJ, nonspontaneous

(d) $\Delta G° = \Delta G° \, S(s) + 2\Delta G° \, H_2O(g) - [\Delta G° \, SO_2(g) + 2\Delta G° \, H_2(g)]$

        $= 0 + 2(-228.57) - [(-300.4) + 2(0)] = -156.7$ kJ, spontaneous

19.63   *Analyze/Plan.* Follow the logic in Sample Exercise 19.8(a).   *Solve.*

(a) $2C_8H_{18}(l) + 25O_2(g) \rightarrow 16CO_2(g) + 18H_2O(l)$

(b)   Because there are more moles of gas in the reactants, $\Delta S°$ is negative, which makes $-T\Delta S$ positive. $\Delta G°$ is less negative than $\Delta H°$. (This argument is true for the reaction as written. If the products are all in the gas phase, there are more moles of gas in the products and $\Delta G°$ is more negative than $\Delta H°$.)

19.65   *Analyze/Plan.* Based on the signs of ΔH and ΔS for a particular reaction, assign a category from Table 19.3 to each reaction.   *Solve.*

(a) ΔG is negative at low temperatures, positive at high temperatures. That is, the reaction proceeds in the forward direction spontaneously at lower temperatures but spontaneously reverses at higher temperatures.

(b) ΔG is positive at all temperatures. The reaction is nonspontaneous in the forward direction at all temperatures.

(c) ΔG is positive at low temperatures, negative at high temperatures. That is, the reaction will proceed spontaneously in the forward direction at high temperature.

19.67   *Analyze/Plan.* We are told that the reaction is spontaneous and endothermic, and asked to estimate the sign and magnitude of ΔS. If a reaction is spontaneous, $\Delta G < 0$. Use this information with Equation [19.11] to solve the problem.   *Solve.*

At 390 K, $\Delta G < 0$; $\Delta G = \Delta H - T\Delta S < 0$

23.7 kJ – 390 K $(\Delta S) < 0$;   23.7 kJ < 390 K (ΔS); $\Delta S > 23.7$ kJ/390 K

$\Delta S > 0.06077$ kJ/K or $\Delta S > 60.8$ J/K

19.69   *Analyze/Plan.* Use Equation [19.11] to calculate T when $\Delta G = 0$. This is similar to calculating the temperature of a phase trasition in Sample Exercise 19.10. Use Table 19.3 to determine whether the reaction is spontaneous or non-spontaneous above this temperature. *Solve.*

(a) $\Delta G = \Delta H - T\Delta S$; $0 = -32$ kJ $- T(-98$ J/K$)$; 32×10^3 J $= T(98$ J/K$)$

     $T = 32 \times 10^3$ J/(98 J/K) $= 326.5 = 330$ K

(b)   Nonspontaneous. The sign of ΔS is negative, so as T increases, ΔG becomes more positive.

19.71   *Analyze/Plan.* Given a chemical equation and thermodynamic data (values of ΔH_f°, ΔG_f° and S°) for reactants and products, predict the variation of ΔG° with temperature and calculate ΔG° at 800 K and 1000 K. Use Equations [5.31] and [19.8] to calculate ΔH° and ΔS°, respectively; use these values to calculate ΔG° at various temperatures, using Equation [19.12]. The signs of ΔH° and ΔS° determine the variation of ΔG° with temperature.   *Solve.*

(a)   Calculate ΔH° and ΔS° to determine the sign of $T\Delta S^{\circ}$.

     $\Delta H^{\circ} = 3\Delta H^{\circ}$ NO(g) $- \Delta H^{\circ}$ NO$_2$(g) $- \Delta H^{\circ}$ N$_2$O(g)

         $= 3(90.37) - 33.84 - 81.6 = 155.7$ kJ

     $\Delta S^{\circ} = 3S^{\circ}$ NO(g) $- S^{\circ}$ NO$_2$(g) $- S^{\circ}$ N$_2$O(g)

         $= 3(210.62) - 240.45 - 220.0 = 171.4$ J/K

     $\Delta G^{\circ} = \Delta H^{\circ} - T\Delta S^{\circ}$. Since ΔS° is positive, $-T\Delta S^{\circ}$ becomes more negative as T increases and ΔG° becomes more negative.

(b) $\Delta G^{\circ} = \Delta H^{\circ} - T\Delta S^{\circ} = 155.7$ kJ $- (800$ K$)(0.1714$ kJ/K$)$

     $\Delta G^{\circ} = 155.7$ kJ $- 137$ kJ $= 19$ kJ

     Since ΔG° is positive at 800 K, the reaction is not spontaneous at this temperature.

(c) $\Delta G° = 155.7$ kJ $- (1000$ K$)(0.1714$ kJ/K$) = 155.7$ kJ $- 171.4$ kJ $= -15.7$ kJ

$\Delta G°$ is negative at 1000 K and the reaction is spontaneous at this temperature.

19.73   *Analyze/Plan.* Follow the logic in Sample Exercise 19.10.   *Solve.*

(a) $\Delta S°_{vap} = \Delta H°_{vap}/T_b$; $T_b = \Delta H°_{vap}/\Delta S°_{vap}$

$\Delta H°_{vap} = \Delta H° \; C_6H_6(g) - \Delta H° \; C_6H_6(l) = 82.9 - 49.0 = 33.9$ kJ

$\Delta S°_{vap} = S° \; C_6H_6(g) - S° \; C_6H_6(l) = 269.2 - 172.8 = 96.4$ J/K

$T_b = 33.9 \times 10^3$ J/96.4 J/K $= 351.66 = 352$ K $= 79°C$

(b)   From the *Handbook of Chemistry and Physics*, 74th Edition, $T_b = 80.1°C$. The values are remarkably close; the small difference is due to deviation from ideal behavior by $C_6H_6(g)$ and experimental uncertainty in the boiling point measurement and the thermodynamic data.

19.75   *Analyze/Plan.* We are asked to write a balanced equation for the combustion of acetylene, calculate $\Delta H°$ for this reaction and calculate maximum useful work possible by the system. Combustion is combination with O_2 to produce CO_2 and H_2O. Calculate $\Delta H°$ using data from Appendix C and Equation [5.31]. The maximum obtainable work is ΔG (Equation [19.18]), which can be calculated from data in Appendix C and Equation [19.14].   *Solve.*

(a) $C_2H_2(g) + 5/2 \, O_2(g) \rightarrow 2CO_2(g) + H_2O(l)$

(b) $\Delta H° = 2\Delta H° \; CO_2(g) + \Delta H° \; H_2O(l) - \Delta H° \; C_2H_2(g) - 5/2\Delta H° \; O_2(g)$

$= 2(-393.5) - 285.83 - 226.77 - 5/2(0)$

$= -1299.6$ kJ produced/mol C_2H_2 burned

(c) $w_{max} = \Delta G° = 2\Delta G° \; CO_2(g) + \Delta G° \; H_2O(l) - \Delta G° \; C_2H_2(g) - 5/2 \, \Delta G° \; O_2(g)$

$= 2(-394.4) - 237.13 - 209.2 - 5/2(0) = -1235.1$ kJ

The negative sign indicates that the system does work on the surroundings; the system can accomplish a maximum of 1235.1 kJ of work on its surroundings.

Free Energy and Equilibrium (section 19.6)

19.77   *Analyze/Plan.* We are given a chemical reaction and asked to predict the effect of the partial pressure of $O_2(g)$ on the value of ΔG for the system. Consider the relationship $\Delta G = \Delta G° + RT \ln Q$ where Q is the reaction quotient.   *Solve.*

(a) $O_2(g)$ appears in the denominator of Q for this reaction. An increase in pressure of O_2 decreases Q and ΔG becomes smaller or more negative. Increasing the concentration or partial pressure of a reactant increases the tendency for a reaction to occur.

(b) $O_2(g)$ appears in the numerator of Q for this reaction. Increasing the pressure of O_2 increases Q and ΔG becomes more positive. Increasing the concentration or partial pressure of a product decreases the tendency for the reaction to occur.

(c) $O_2(g)$ appears in the numerator of Q for this reaction. An increase in pressure of O_2 increases Q and ΔG becomes more positive. Since pressure of O_2 is raised to the third power in Q, an increase in pressure of O_2 will have the largest effect on ΔG for this reaction. Increasing the concentration or partial pressure of a product decreases the tendency for the reaction to occur.

19.79   *Analyze/Plan.* Given a chemical reaction, we are asked to calculate $\Delta G°$ from Appendix C data, and ΔG for a given set of initial conditions. Use Equation [19.14] to calculate $\Delta G°$, and Equation [19.19] to calculate ΔG. Follow the logic in Sample Exercise 19.11 when calculating ΔG.   *Solve.*

(a) $\Delta G° = \Delta G° \, N_2O_4(g) - 2\Delta G° \, NO_2(g) = 98.28 - 2(51.84) = -5.40 \text{ kJ}$

(b) $\Delta G = \Delta G° + RT \ln P_{N_2O_4} / P_{NO_2}^2$

$$= -5.40 \text{ kJ} + \frac{8.314 \times 10^{-3} \text{ kJ}}{\text{mol-K}} \times 298 \text{ K} \times \ln[1.60/(0.40)^2] = 0.3048 = 0.30 \text{ kJ}$$

19.81   *Analyze/Plan.* Given a chemical reaction, we are asked to calculate K using $\Delta G_f°$ data from Appendix C. Calculate $\Delta G°$ using Equation [19.14]. Then $\Delta G° = -RT \ln K$, Equation [19.20]; $\ln K = -\Delta G°/RT$ *Solve.*

(a) $\Delta G° = 2\Delta G° \, HI(g) - \Delta G° \, H_2(g) - \Delta G° \, I_2(g)$

$\qquad = 2(1.30) - 0 - 19.37 = -16.77 \text{ kJ}$

$$\ln K = \frac{-(-16.77 \text{ kJ}) \times 10^3 \text{ J/kJ}}{8.314 \text{ J/K} \times 298 \text{ K}} = 6.76876 = 6.769; \quad K = 870$$

(b) $\Delta G° = \Delta G° \, C_2H_4(g) + \Delta G° \, H_2O(g) - \Delta G° \, C_2H_5OH(g)$

$\qquad = 68.11 - 228.57 - (-168.5) = 8.04 = 8.0 \text{ kJ}$

$$\ln K = \frac{-(8.04 \text{ kJ}) \times 10^3 \text{ J/kJ}}{8.314 \text{ J/K} \times 298 \text{ K}} = -3.24511 = -3.25; K = 0.039$$

(c) $\Delta G° = \Delta G° \, C_6H_6(g) - 3\Delta G° \, C_2H_2(g) = 129.7 - 3(209.2) = -497.9 \text{ kJ}$

$$\ln K = \frac{-\Delta G°}{RT} = \frac{-(-497.9 \text{ kJ}) \times 10^3 \text{ J/KJ}}{8.314 \text{ J/K} \times 298 \text{ K}} = 200.963 = 201.0; K = 2 \times 10^{87}$$

19.83   *Analyze/Plan.* Given a chemical reaction and thermodynamic data in Appendix C, calculate the equilibrium pressure of $CO_2(g)$ at two temperatures. $K = P_{CO_2}$. Calculate $\Delta G°$ at the two temperatures using $\Delta G° = \Delta H° - T\Delta S°$ and then calculate K and P_{CO_2}. *Solve.*

$\Delta H° = \Delta H° \, BaO(s) + \Delta H° \, CO_2(g) - \Delta H° \, BaCO_3(s)$

$\qquad = -553.5 + -393.5 - (-1216.3) = +269.3 \text{ kJ}$

$\Delta S° = S° \, BaO(s) + S° \, CO_2(g) - S° \, BaCO_3(s)$

$\qquad = 70.42 + 213.6 - 112.1 = 171.92 \text{ J/K} = 0.1719 \text{ kJ/K}$

(a) ΔG at 298 K = 269.3 kJ – 298 K (0.17192 kJ/K) = 218.07 = 218.1 kJ

$$\ln K = \frac{-\Delta G^\circ}{RT} = \frac{-218.07 \times 10^3 \text{ J}}{8.314 \text{ J/K} \times 298 \text{ K}} = -88.017 = -88.02$$

$K = 6.0 \times 10^{-39}; \quad P_{CO_2} = 6.0 \times 10^{-39}$ atm

(b) ΔG at 1100 K = 269.3 kJ – 1100 K (0.17192 kJ) = 80.19 = +80.2 kJ

$$\ln K = \frac{-\Delta G^\circ}{RT} = \frac{-80.19 \times 10^3 \text{ J}}{8.314 \text{ J/K} \times 1100 \text{ K}} = -8.768 = -8.77$$

$K = 1.6 \times 10^{-4}; \quad P_{CO_2} = 1.6 \times 10^{-4}$ atm

19.85   *Analyze/Plan.* Given an acid dissociation equilibrium and the corresponding K_a value, calculate ΔG° and ΔG for a given set of concentrations. Use Equation [19.20] to calculate ΔG° and Equation [19.19] to calculate ΔG.   *Solve.*

(a) $HNO_2(aq) \;\rightleftharpoons\; H^+(aq) + NO_2^-(aq)$

(b) $\Delta G^\circ = -RT \ln K_a = -(8.314 \times 10^{-3})(298) \ln (4.5 \times 10^{-4}) = 19.0928 = 19.1$ kJ

(c) $\Delta G = 0$ at equilibrium

(d) $\Delta G = \Delta G^\circ + RT \ln Q$

$$= 19.09 \text{ kJ} + (8.314 \times 10^{-3})(298) \ln \frac{(5.0 \times 10^{-2})(6.0 \times 10^{-4})}{0.20} = -2.725 = -2.7 \text{ kJ}$$

Additional Exercises

19.87   (a)   The thermodynamic quantities T, E, and S are state functions. T is directly related to the distribution of molecular speeds, which does not depend on the path from one state to another.

(b)   The quantities q and w do depend on the path taken from one state to another.

(c)   There is only one *reversible* path between states.

(d)   Isothermal processes occur at constant T. Since the process is reversible, q is q_{rev} and w is w_{max}.

$$\Delta E = q_{rev} + w_{max}. \quad \Delta S = \frac{q_{rev}}{T}.$$

19.91   (a)   Each of the 4 molecules can be in either the left or the right bulb. Thus, there are $(2)^4 = 16$ arrangements.

(b)   Only one arrangement has all 4 molecules in the left flask.

(c)   The gas will spontaneously adopt the state with maximum disorder, the state with the most possible arrangements for the molecules.

19.96   (a)   Formation reactions are the synthesis of 1 mole of compound from elements in their standard states.

$1/2 \, N_2(g) + 3/2 \, H_2(g) \rightarrow NH_3(g)$

$C(s) + 2Cl_2(g) \rightarrow CCl_4(l)$

$K(s) + 1/2 \, N_2(g) + 3/2 \, O_2(g) \rightarrow KNO_3(s)$

In each of these formation reactions, there are fewer moles of gas in the products than the reactants, so we expect $\Delta S°$ to be negative. If $\Delta G_f° = \Delta H_f° - T\Delta S°$ and $\Delta S°$ is negative, $-T\Delta S°$ is positive and $\Delta G_f°$ is more positive than $\Delta H_f°$.

(b) $C(s) + 1/2\,O_2(g) \rightarrow CO(g)$

In this reaction, there are more moles of gas in products, $\Delta S_f°$ is positive, $-T\Delta S_f°$ is negative and $\Delta G_f°$ is more negative than $\Delta H_f°$.

19.100   (a) $K = \dfrac{\chi_{CH_3COOH}}{\chi_{CH_3OH}P_{CO}}$

$\Delta G° = -RT \ln K;\ \ln K = -\Delta G/RT$

$\Delta G° = \Delta G°\ CH_3COOH(l) - \Delta G°\ CH_3OH(l) - \Delta G°\ CO(g)$

$ = -392.4 - (-166.23) - (-137.2) = -89.0\ kJ$

$\ln K = \dfrac{-(-89.0\ kJ)}{(8.314 \times 10^{-3}\ kJ/K)(298\ K)} = 35.922 = 35.9;\ \ K = 4 \times 10^{15}$

(b) $\Delta H° = \Delta H°\ CH_3COOH(l) - \Delta H°\ CH_3OH(l) - \Delta H°\ CO(g)$

$ = -487.0 - (-238.6) - (-110.5) = -137.9\ kJ$

The reaction is exothermic, so the value of K will decrease with increasing temperature, and the mole fraction of CH_3COOH will also decrease. Elevated temperatures must be used to increase the speed of the reaction. Thermodynamics cannot predict the rate at which a reaction reaches equilibrium.

(c) $\Delta G° = -RT \ln K;\ K = 1,\ \ln K = 0,\ \Delta G° = 0$

$\Delta G° = \Delta H° - T\Delta S°;$ when $\Delta G° = 0,\ \Delta H° = T\Delta S°$

$\Delta S° = S°\ CH_3COOH(l) - S°\ CH_3OH(l) - S°\ CO(g)$

$ = 159.8 - 126.8 - 197.9 = -164.9\ J/K = -0.1649\ kJ/K$

$-137.9\ kJ = T(-0.1649\ kJ/K),\ T = 836.3\ K$

The equilibrium favors products up to 836 K or 563°C, so the elevated temperatures to increase the rate of reaction can be safely employed.

19.104   (a)   The equilibrium of interest here can be written as:

$K^+\ (plasma) \ \rightleftharpoons \ K^+\ (muscle)$

Since an aqueous solution is involved in both cases, assume that the equilibrium constant for the above process is exactly 1, that is, $\Delta G° = 0$. However, ΔG is not zero because the concentrations are not the same on both sides of the membrane. Use Equation [19.16] to calculate ΔG:

$\Delta G = \Delta G° + RT \ln \dfrac{[K^+\ (muscle)]}{[K^+\ (plasma)]}$

$ = 0 + (8.314)(310) \ln \dfrac{(0.15)}{(5.0 \times 10^{-3})} = 8766\ J = 8.8\ kJ$

(b)   Note that ΔG is positive. This means that work must be done on the system (blood plasma plus muscle cells) to move the K^+ ions "uphill," as it were. The minimum amount of work possible is given by the value for ΔG. This value represents the minimum amount of work required to transfer one mole of K^+ ions from the blood plasma at $5 \times 10^{-3} \, M$ to muscle cell fluids at $0.15 \, M$, assuming constancy of concentrations. In practice, a larger than minimum amount of work is required.

Integrative Exercises

19.108   (a)   At the boiling point, vaporization is a reversible process, so $\Delta S^o_{vap} = \Delta H^o_{vap}/T$.

acetone: $\Delta S^o_{vap} = \Delta H^o_{vap}/T = (29.1 \, \text{kJ/mol}) / 329.25 \, \text{K} = 88.4 \, \text{J/mol-K}$

dimethyl ether: $\Delta S^o_{vap} = (21.5 \, \text{kJ/mol}) / 248.35 \, \text{K} = 86.6 \, \text{J/mol-K}$

ethanol: $\Delta S^o_{vap} = (38.6 \, \text{kJ/mol}) / 351.6 \, \text{K} = 110 \, \text{J/mol-K}$

octane: $\Delta S^o_{vap} = (34.4 \, \text{kJ/mol}) / 398.75 \, \text{K} = 86.3 \, \text{J/mol-K}$

pyridine: $\Delta S^o_{vap} = (35.1 \, \text{kJ/mol}) / 388.45 \, \text{K} = 90.4 \, \text{J/mol-K}$

(b)   Ethanol is the only liquid listed that doesn't follow *Trouton's rule* and it is also the only substance that exhibits hydrogen bonding in the pure liquid. Hydrogen bonding leads to more ordering in the liquid state and a greater than usual increase in entropy upon vaporization. The rule appears to hold for liquids with London dispersion forces (octane) and ordinary dipole-dipole forces (acetone, dimethyl ether, pyridine), but not for those with hydrogen bonding.

(c)   Owing to strong hydrogen bonding interactions, water probably does not obey Trouton's rule.

From Appendix B, ΔH^o_{vap} at 100°C = 40.67 kJ/mol.

$\Delta S^o_{vap} = (40.67 \, \text{kJ/mol}) / 373.15 \, \text{K} = 109.0 \, \text{J/mol-K}$

(d)   Use $\Delta S^o_{vap} = 88 \, \text{J/mol-K}$, the middle of the range for Trouton's rule, to estimate ΔH^o_{vap} for chlorobenzene.

$\Delta H^o_{vap} = \Delta S^o_{vap} \times T = 88 \, \text{J/mol-K} \times 404.95 \, \text{K} = 36 \, \text{kJ/mol}$

19.113   (a) $K = P^2_{NO_2} / P_{N_2O_4}$

Assume equal amounts means equal number of moles. For gases, $P = n(RT/V)$. In an equilibrium mixture, RT/V is a constant, so moles of gas are directly proportional to partial pressure. Gases with equal partial pressures will have equal moles of gas present. The condition $P_{NO_2} = P_{N_2O_4}$ leads to the expression $K = P_{NO_2}$. The value of K then depends on P_t for the mixture. For any particular value of P_t, the condition of equal moles of the two gases can be achieved at some temperature. For example, $P_{NO_2} = P_{N_2O_4} = 1.0 \, \text{atm}, P_t = 2.0 \, \text{atm}$.

$$K = \frac{(1.0)^2}{1.0} = 1.0; \quad \ln K = 0; \quad \Delta G^\circ = 0 = \Delta H^\circ - T\Delta S^\circ; \quad T = \Delta H^\circ / \Delta S^\circ$$

$$\Delta H^\circ = 2\Delta H^\circ\, NO_2(g) - \Delta H^\circ\, N_2O_4(g) = 2(33.84) - 9.66 = +58.02 \text{ kJ}$$

$$\Delta S^\circ = 2S^\circ\, NO_2(g) - S^\circ\, N_2O_4(g) = 2(240.45) - 304.3 = 0.1766 \text{ kJ/K}$$

$$T = \frac{58.02 \text{ kJ}}{0.1766 \text{ kJ/K}} = 328.5 \text{ K or } 55.5^\circ C$$

(b) $P_t = 1.00$ atm; $P_{N_2O_4} = x, P_{NO_2} = 2x; \quad x + 2x = 1.00$ atm

$$x = P_{N_2O_4} = 0.3333 = 0.333 \text{ atm}; \quad P_{NO_2} = 0.6667 = 0.667 \text{ atm}$$

$$K = \frac{(0.6667)^2}{0.3333} = 1.334 = 1.33; \quad \Delta G^\circ = -RT \ln K = \Delta H^\circ - T\Delta S^\circ$$

$$-(8.314 \times 10^{-3} \text{ kJ/K})(\ln 1.334)\, T = 58.02 \text{ kJ} - (0.1766 \text{ kJ/K})\, T$$

$$(-0.00239 \text{ kJ/K})\, T + (0.1766 \text{ kJ/K})\, T = 58.02 \text{ kJ}$$

$$(0.1742 \text{ kJ/K})\, T = 58.02 \text{ kJ}; \quad T = 333.0 \text{ K}$$

(c) $P_t = 10.00$ atm; $x + 2x = 10.00$ atm

$$x = P_{N_2O_4} = 3.3333 = 3.333 \text{ atm}; \quad P_{NO_2} = 6.6667 = 6.667 \text{ atm}$$

$$K = \frac{(6.6667)^2}{3.3333} = 13.334 = 13.33; \quad -RT \ln K = \Delta H^\circ - T\Delta S^\circ$$

$$-(8.314 \times 10^{-3} \text{ kJ/K})(\ln 13.334)\, T = 58.02 \text{ kJ} - (0.1766 \text{ kJ/K})\, T$$

$$(-0.02154 \text{ kJ/K})\, T + (0.1766 \text{ kJ/K})\, T = 58.02 \text{ kJ}$$

$$(0.15506 \text{ kJ/K})\, T = 58.02 \text{ kJ}; \quad T = 374.2 \text{ K}$$

(d)   The reaction is endothermic, so an increase in the value of K as calculated in parts (b) and (c) should be accompanied by an increase in T.

20 Electrochemistry

Visualizing Concepts

20.1 Consider the Brønsted-Lowry acid-base reaction and the redox reaction below.

$$HA + B \rightarrow BH^+ + A^-$$
$$HA \rightleftharpoons H^+ + A^-$$
$$B + H^+ \rightleftharpoons BH^+$$
$$X(red) + Y^+(ox) \rightarrow X^+(ox) + Y(red)$$
$$X(red) \rightleftharpoons X^+(ox) + e^-$$
$$Y+(ox) + e^- \rightleftharpoons Y(red)$$

Just as acid-base reactions can be viewed as proton-transfer reactions, redox reactions can be viewed as electron-transfer reactions. In the Brønsted-Lowry acid-base reaction, H^+ is transferred from HA to B. In the redox reaction, X(red) loses electrons and Y^+(ox) gains electrons; the number of electrons gained and lost must be equal. The concept of electron transfer from one reactant to the other is clearly applicable to redox reactions. (The path of the transfer may or may not be direct, but ultimately electrons are transferred during redox reactions.)

Furthermore, the species in redox reactions each have an oxidized and reduced form, like a conjugate acid-conjugate base pair. The reduced form has more electrons; the oxidized form has less. The greater the tendency of HA to donate H^+ (the stronger the acid), the lesser the tendency of A^- to gain them (the weaker the conjugate base). Similarly, the greater the tendency of X(red) to donate electrons (the stronger the reducing agent) the lesser the tendency of X^+(ox) to gain electrons (the weaker the oxidizing agent). Similar arguments exist for B, BH^+ and Y^+(ox), Y(red).

20.4 *Analyze/Plan.* Consider the voltaic cell pictured in Figure 20.5 as a model. The reaction in a voltaic cell is spontaneous. To generate a <u>standard</u> emf, substances must be present in their standard states.

(a) A concentration of 1 M is the standard state for ions in solution. Ions, but not solution, must be able to flow between compartments in order to complete the circuit so that the cell can develop an emf. Add 1 M A^{2+}(aq) to the beaker with the A(s) electrode. Add 1 M B^{2+}(aq) to the beaker with the B(s) electrode. Add a salt bridge to enable the flow of ions from one compartment to the other.

(b) Reduction occurs at the cathode. In order for the reaction to occur spontaneously (and thus generate an emf), the half-reaction with the greater E°_{red} will be the reduction half-reaction. In this cell, it is the half-reaction involving A(s) and A^{2+}(aq). The A electrode functions as the cathode.

(c)   According to Figure 20.5, electrons flow through the external circuit from the anode to the cathode. In this example, B is the anode and A is the cathode, so electrons flow from B to A through the external circuit.

(d) $E_{cell.}^{o} = E_{red}^{o}(\text{cathode}) - E_{red}^{o}(\text{anode})$

         $E_{cell}^{o} = -0.10\,\text{V} - (-1.10\,\text{V}) = 1.00\,\text{V}$

20.7   *Analyze.* Given a redox reaction with a negative E^{o}, answer questions regarding ΔG^{o}, the equilibrium constant (K), and work (w). *Plan.* $\Delta G^\circ = -nFE^\circ$; $\Delta G^\circ = -RT\ln K$; $w_{max} = -nFE$. *Solve.*

(a)   The signs of ΔG° and E° are opposite. If E° is negative, ΔG° is positive. (The reaction is not spontaneous in the forward direction.)

(b)   If ΔG° is positive, $\ln K$ is negative and $K < 1$. Also, K is less than one for a nonspontaneous reaction.

(c)   No. If E° is negative, the sign of w is positive. A positive value for w means that work is done on the system by the surroundings. An electrochemical cell based on this reaction cannot accomplish work on its surroundings.

20.9   *Analyze/Plan.* Consider the Nernst equation, which describes the variation of potential (emf) with respect to changes in concentration.

Solve. $E = E^\circ - \dfrac{0.0592}{n}\log Q.$

(a)   For this half-reaction, $E_{red}^{o} = 0.799\,\text{V}$; $Q = 1/[Ag^+]$

         $E = 0.80 - \dfrac{0.0592}{1}\log\dfrac{1}{[Ag^+]}$; $E = 0.80 + \dfrac{0.0592}{1}\log[Ag^+]$

         The y-intercept of the graph is E°. The slope of the line is $+0.0592$. So, as $[Ag^+]$ and $\log[Ag^+]$ increase, E increases. The line that describes this behavior is line 1.

(b)   When $\log[Ag^+] = 0$, $[Ag^+] = 1\,M$; this is the standard state for $Ag^+(aq)$, so $E_{red} = E_{red}^{o} = 0.799\text{V}$.

Oxidation - Reduction Reactions (section 20.1)

20.13   (a)   *Oxidation* is the loss of electrons.

       (b)   The electrons appear on the products side (right side) of an oxidation half-reaction.

       (c)   The *oxidant* is the reactant that is reduced; it gains the electrons that are lost by the substance being oxidized.

       (d)   An *oxidizing agent* is the substance that promotes oxidation. That is, it gains electrons that are lost by the substance being oxidized. It is the same as the oxidant.

20.15   (a)   True.

       (b)   False. Fe^{3+} is reduced to Fe^{2+}, so it is the oxidizing agent, and Co^{2+} is the reducing agent.

       (c)   True.

20.17 *Analyze/Plan.* Given a chemical equation, we are asked to indicate which elements undergo a change in oxidation number and the magnitude of the change. Assign oxidation numbers according to the rules given in Section 4.4. Note the changes and report the magnitudes.   *Solve.*

   (a)   I is reduced from +5 to 0; C is oxidized from +2 to +4.

   (b)   Hg is reduced from +2 to 0; N is oxidized from –2 to 0.

   (c)   N is reduced from +5 to +2; S is oxidized from –2 to 0.

   (d)   Cl is reduced from +4 to +3; O is oxidized from –1 to 0.

Balancing Oxidation-Reduction Reactions (section 20.2)

20.19 *Analyze/Plan.* Write the balanced chemical equation and assign oxidation numbers. The substance oxidized is the reductant and the substance reduced is the oxidant.   *Solve.*

   (a) $TiCl_4(g) + 2Mg(l) \rightarrow Ti(s) + 2MgCl_2(l)$

   (b) $Mg(l)$ is oxidized; $TiCl_4(g)$ is reduced.

   (c) $Mg(l)$ is the reductant; $TiCl_4(g)$ is the oxidant.

20.21 *Analyze/Plan.* Follow the logic in Sample Exercises 20.2 and 20.3. If the half-reaction occurs in basic solution, balance as in acid, then add OH^- to each side.   *Solve.*

   (a) $Sn^{2+}(aq) \rightarrow Sn^{4+}(aq) + 2e^-$, oxidation

   (b) $TiO_2(s) + 4H^+(aq) + 2e^- \rightarrow Ti^{2+}(aq) + 2H_2O(l)$, reduction

   (c) $ClO_3^-(aq) + 6H^+(aq) + 6e^- \rightarrow Cl^-(aq) + 3H_2O(l)$, reduction

   (d) $N_2(g) + 8H^+(aq) + 6e^- \rightarrow 2NH_4^+(aq)$, reduction

   (e) $4OH^-(aq) \rightarrow O_2(g) + 2H_2O(l) + 4e^-$, oxidation

   (f) $SO_3^{2-}(aq) + 2OH^-(aq) \rightarrow SO_4^{2-}(aq) + H_2O(l) + 2e^-$, oxidation

   (g) $N_2(g) + 6H_2O + 6e^- \rightarrow 2NH_3(g) + 6OH^-(aq)$, reduction

20.23 *Analyze/Plan.* Follow the logic in Sample Exercises 20.2 and 20.3 to balance the given equations. Use the method in Sample Exercise 20.1 to identify oxidizing and reducing agents.   *Solve.*

   (a) $Cr_2O_7^{2-}(aq) + I^-(aq) + 8H^+ \rightarrow 2Cr^{3+}(aq) + IO_3^-(aq) + 4H_2O(l)$

        oxidizing agent, $Cr_2O_7^{2-}$; reducing agent, I^-

   (b)   The half-reactions are:

$$4[MnO_4^-(aq) + 8H^+(aq) + 5e^- \rightarrow Mn^{2+}(aq) + 4H_2O(l)]$$
$$5[CH_3OH(aq) + H_2O(l) \rightarrow HCO_2H(aq) + 4H^+(aq) + 4e^-]$$

$$\overline{4MnO_4^-(aq) + 5CH_3OH(aq) + 12H^+(aq) \rightarrow 4Mn^{2+}(aq) + 5HCO_2H(aq) + 11H_2O(l)}$$

        oxidizing agent, MnO_4^-; reducing agent, CH_3OH

(c)
$$I_2(s) + 6H_2O(l) \rightarrow 2IO_3^-(aq) + 12H^+(aq) + 10e^-$$
$$\frac{5[OCl^-(aq) + 2H^+(aq) + 2e^- \rightarrow Cl^-(aq) + H_2O(l)}{I_2(s) + 5OCl^-(aq) + H_2O(l) \rightarrow 2IO_3^-(aq) + 5Cl^-(aq) + 2H^+(aq)]}$$

oxidizing agent, OCl^-; reducing agent, I_2

(d)
$$As_2O_3(s) + 5H_2O(l) \rightarrow 2H_3AsO_4(aq) + 4H^+(aq) + 4e^-$$
$$\frac{2NO_3^-(aq) + 6H^+(aq) + 4e^- \rightarrow N_2O_3(aq) + 3H_2O(l)}{As_2O_3(s) + 2NO_3^-(aq) + 2H_2O(l) + 2H^+(aq) \rightarrow 2H_3AsO_4(aq) + N_2O_3(aq)}$$

oxidizing agent, NO_3^-; reducing agent, As_2O_3

(e)
$$2[MnO_4^-(aq) + 2H_2O(l) + 3e^- \rightarrow MnO_2(s) + 4OH^-]$$
$$\frac{Br^-(aq) + 6OH^-(aq) \rightarrow BrO_3^-(aq) + 3H_2O(l) + 6e^-}{2MnO_4^-(aq) + Br^-(aq) + H_2O(l) \rightarrow 2MnO_2(s) + BrO_3^-(aq) + 2OH^-(aq)}$$

oxidizing agent, MnO_4^-; reducing agent, Br^-

(f) $Pb(OH)_4^{2-}(aq) + ClO^-(aq) \rightarrow PbO_2(s) + Cl^-(aq) + 2OH^-(aq) + H_2O(l)$

oxidizing agent, ClO^-; reducing agent, $Pb(OH)_4^{2-}$

Voltaic Cells (section 20.3)

20.25   (a)   The reaction $Cu^{2+}(aq) + Zn(s) \rightarrow Cu(s) + Zn^{2+}(aq)$ is occurring in both figures. In Figure 20.3, the reactants are in contact, and the concentrations of the ions in solution aren't specified. In Figure 20.4 the oxidation half-reaction and reduction half-reaction are occurring in separate compartments, joined by a porous connector. The concentrations of the two solutions are initially 1.0 M. In Figure 20.4, electrical current is isolated and flows through the voltmeter. In Figure 20.3, the flow of electrons cannot be isolated or utilized.

     (b)   In the cathode compartment of the voltaic cell in Figure 20.5, Cu^{2+} cations are reduced to Cu atoms, decreasing the number of positively charged particles in the compartment. Na^+ cations are drawn into the compartment to maintain charge balance as Cu^{2+} ions are removed.

20.27   *Analyze/Plan.* Follow the logic in Sample Exercise 20.4.   *Solve.*

     (a)   Fe(s) is oxidized, Ag^+(aq) is reduced.

     (b) $Ag^+(aq) + 1e^- \rightarrow Ag(s)$; $Fe(s) \rightarrow Fe^{2+}(aq) + 2e^-$

     (c)   Fe(s) is the anode, Ag(s) is the cathode.

     (d)   Fe(s) is negative; Ag(s) is positive.

     (e)   Electrons flow from the Fe(–) electrode toward the Ag(+) electrode.

     (f)   Cations migrate toward the Ag(s) cathode; anions migrate toward the Fe(s) anode.

Cell Potentials under Standard Conditions (section 20.4)

20.29 (a) *Electromotive force*, emf, is the driving force that causes electrons to flow through the external circuit of a voltaic cell. It is the potential energy difference between an electron at the anode and an electron at the cathode.

(b) One *volt* is the potential energy difference required to impart 1 J of energy to a charge of 1 coulomb. 1 V = 1 J/C.

(c) *Cell potential*, E_{cell}, is the emf of an electrochemical cell.

20.31 (a) $2H^+(aq) + 2e^- \rightarrow H_2(g)$

(b) A *standard* hydrogen electrode is a hydrogen electrode where the components are at standard conditions, $1\ M\ H^+(aq)$ and $H_2(g)$ at 1 atm.

(c) The platinum foil in an SHE serves as an inert electron carrier and a solid reaction surface.

20.33 (a) A *standard reduction potential* is the relative potential of a reduction half-reaction measured at standard conditions, $1\ M$ aqueous solution and 1 atm gas pressure.

(b) $E_{red}^o = 0\ V$ for a standard hydrogen electrode.

20.35 *Analyze/Plan.* Follow the logic in Sample Exercise 20.5.   *Solve.*

(a) The two half-reactions are:

$Tl^{3+}(aq) + 2e^- \rightarrow Tl^+(aq)$ cathode $E_{red}^o = ?$

$2[Cr^{2+}(aq) \rightarrow Cr^{3+}(aq) + e^-]$ anode $E_{red}^o = -0.41\ V$

(b) $E_{cell}^o = E_{red}^o\ (\text{cathode}) - E_{red}^o\ (\text{anode}); 1.19\ V = E_{red}^o - (-0.41\ V);$

$E_{red}^o = 1.19\ V - 0.41\ V = 0.78\ V$

(c)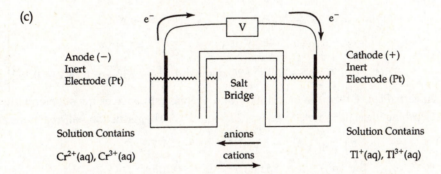

Note that because $Cr^{2+}(aq)$ is readily oxidized, it would be necessary to keep oxygen out of the left-hand cell compartment.

20.37 *Analyze/Plan.* Follow the logic in Sample Exercise 20.6.   *Solve.*

(a) $Cl_2(g) \rightarrow 2Cl^-(aq) + 2e^-$ $E_{red}^o = 1.359\ V$

$I_2(s) + 2e^- \rightarrow 2I^-(aq)$ $E_{red}^o = 0.536\ V$

$E^o = 1.359\ V - 0.536\ V = 0.823\ V$

(b)
$$Ni(s) \rightarrow Ni^{2+}(aq) + 2e^- \qquad E^o_{red} = -0.28 \text{ V}$$
$$2[Ce^{4+}(aq) + 1e^- \rightarrow Ce^{3+}(aq)] \qquad E^o_{red} = 1.61 \text{ V}$$
$$E^\circ = 1.61 \text{ V} - (-0.28 \text{ V}) = 1.89 \text{ V}$$

(c)
$$Fe(s) \rightarrow Fe^{2+}(aq) + 2e^- \qquad E^o_{red} = -0.440 \text{ V}$$
$$2[Fe^{3+}(aq) + 1e^- \rightarrow Fe^{2+}(aq)] \qquad E^o_{red} = 0.771 \text{ V}$$
$$E^\circ = 0.771 \text{ V} - (-0.440 \text{ V}) = 1.211 \text{ V}$$

(d)
$$2[NO_3^-(aq) + 4H^+ + 3e^- \rightarrow NO(g) + 2H_2O(l)] \qquad E^o_{red} = 0.96 \text{ V}$$
$$3[Cu(s) \rightarrow Cu^{2+}(aq) + 2e^-] \qquad E^o_{red} = 0.337 \text{ V}$$
$$E^\circ = 0.96 \text{ V} - (0.337 \text{ V}) = 0.623 = 0.62 \text{ V}$$

20.39   *Analyze/Plan.* Given four half-reactions, find E^o_{red} from Appendix E and combine them to obtain a desired E_{cell}. (a) The largest E_{cell} will combine the half-reaction with the most positive E^o_{red} as the cathode reaction and the one with the most negative E^o_{red} as the anode reaction. (b) The smallest positive E^o_{cell} will combine two half-reactions whose E^o_{red} values are closest in magnitude **and sign.**   *Solve.*

(a)
$$3[Ag^+(aq) + 1e^- \rightarrow Ag(s)] \qquad E^o_{red} = 0.799$$
$$\underline{\quad Cr(s) \rightarrow Cr^{3+}(aq) + 3e^- \qquad E^o_{red} = -0.74 \quad}$$
$$3Ag^+(aq) + Cr(s) \rightarrow 3Ag(s) + Cr^{3+}(aq) \qquad E^\circ = 0.799 - (-0.74) = 1.54 \text{ V}$$

(b)   Two of the combinations have essentially equal E° values.

$$2[Ag^+(aq) + 1e^- \rightarrow Ag(s)] \qquad E^o_{red} = 0.799 \text{ V}$$
$$\underline{\quad Cu(s) \rightarrow Cu^{2+}(aq) + 2e^- \qquad E^o_{red} = 0.337 \text{ V} \quad}$$
$$2Ag^+(aq) + Cu(s) \rightarrow 2Ag(s) + Cu^{2+}(aq) \qquad E^\circ = 0.799 \text{ V} - 0.337 \text{ V} = 0.462 \text{ V}$$

$$3[Ni^{2+}(aq) + 2e^- \rightarrow Ni(s)] \qquad E^o_{red} = -0.28 \text{ V}$$
$$\underline{\quad 2[Cr(s) \rightarrow Cr^{3+}(aq) + 3e^-] \qquad E^o_{red} = -0.74 \text{ V} \quad}$$
$$3Ni^{2+}(aq) + 2Cr(s) \rightarrow 3Ni(s) + 2Cr^{3+}(aq) \qquad E^\circ = -0.28 \text{ V} - (-0.74 \text{ V}) = 0.46 \text{ V}$$

20.41   *Analyze/Plan.* Given the description of a voltaic cell, answer questions about this cell. Combine ideas in Sample Exercises 20.4 and 20.7. The reduction half-reactions are:

$$Cu^{2+}(aq) + 2e^- \rightarrow Cu(s) \qquad E^\circ = 0.337 \text{ V}$$
$$Sn^{2+}(aq) + 2e^- \rightarrow Sn(s) \qquad E^\circ = -0.136 \text{ V}$$

Solve.

(a)   It is evident that Cu^{2+} is more readily reduced. Therefore, Cu serves as the cathode, Sn as the anode.

(b)   The copper electrode gains mass as Cu is plated out, the Sn electrode loses mass as Sn is oxidized.

(c)   The overall cell reaction is $Cu^{2+}(aq) + Sn(s) \rightarrow Cu(s) + Sn^{2+}(aq)$

(d) $E^\circ = 0.337 \text{ V} - (-0.136 \text{ V}) = 0.473 \text{ V}$

Strengths of Oxidizing and Reducing Agents (section 20.4)

20.43 *Analyze/Plan.* The more readily a substance is oxidized, the stronger it is as a reducing agent. In each case choose the half-reaction with the more negative reduction potential and the given substance on the right. *Solve.*

    (a)   Mg(s) (–2.37 V vs. –0.440 V)

    (b)   Ca(s) (–2.87 V vs. –1.66 V)

    (c) H_2(g, acidic) (0.000 V vs. 0.141 V)

    (d) IO_3^-(aq) [Both IO_3^-(aq) and BrO_3^-(aq) are good oxidizing agents, but IO_3^-(aq) has the smaller positive reduction potential. (1.195 V vs. 1.52 V)]

20.45 *Analyze/Plan.* If the substance is on the left of a reduction half-reaction, it will be an oxidant; if it is on the right, it will be a reductant. The sign and magnitude of the E_{red}^o determines whether it is strong or weak. *Solve.*

    (a) Cl_2(aq): strong oxidant (on the left, $E_{red}^o = 1.359\ V$)

    (b) MnO_4^-(aq, acidic): strong oxidant (on the left, $E_{red}^o = 1.51\ V$)

    (c)   Ba(s): strong reductant (on the right, $E_{red}^o = -2.90\ V$)

    (d)   Zn(s): reductant (on the right, $E_{red}^o = -0.763\ V$)

20.47 *Analyze/Plan.* Follow the logic in Sample Exercise 20.8. *Solve.*

    (a)   Arranged in order of increasing strength as oxidizing agents (and increasing reduction potential):

$$Cu^{2+}(aq) < O_2(g) < Cr_2O_7^{2-}(aq) < Cl_2(g) < H_2O_2(aq)$$

    (b)   Arranged in order of increasing strength as reducing agents (and decreasing reduction potential):

$$H_2O_2(aq) < I^-(aq) < Sn^{2+}(aq) < Zn(s) < Al(s)$$

20.49 *Analyze/Plan.* In order to reduce Eu^{3+} to Eu^{2+}, we need an oxidizing agent, one of the reduced species from Appendix E. It must have a greater tendency to be oxidized than Eu^{3+} has to be reduced. That is, E_{red}^o must be more negative than –0.43 V. *Solve.*

Any of the **reduced** species in Appendix E from a half-reaction with a reduction potential more negative than –0.43 V will reduce Eu^{3+} to Eu^{2+}. From the list of possible reductants in the exercise, Al and $H_2C_2O_4$ will reduce Eu^{3+} to Eu^{2+}.

Free Energy and Redox Reactions (section 20.5)

20.51 *Analyze/Plan.* In each reaction, $Fe^{2+} \rightarrow Fe^{3+}$ will be the oxidation half-reaction and one of the other given half-reactions will be the reduction half-reaction. Follow the logic in Sample Exercise 20.10 to calculate E° and ΔG° for each reaction. *Solve.*

    (a) $2Fe^{2+}(aq) + S_2O_6^{2-}(aq) + 4H^+(aq) \rightarrow 2Fe^{3+}(aq) + 2H_2SO_3(aq)$

            E° = 0.60 V – 0.77 V = –0.17 V

            $2Fe^{2+}(aq) + N_2O(aq) + 2H^+(aq) \rightarrow 2Fe^{3+}(aq) + N_2(g) + H_2O(l)$

            E° = –1.77 V – 0.77 V = –2.54 V

$$Fe^{2+}(aq) + VO_2^+(aq) + 2H^+(aq) \rightarrow Fe^{3+}(aq) + VO^{2+}(aq) + H_2O(l)$$

$$E° = 1.00\ V - 0.77\ V = +0.23\ V$$

(b) $\Delta G° = -nFE°$ For the first reaction,

$$\Delta G° = -2\ mol \times \frac{96,485\ J}{1\ V\text{-}mol} \times (-0.17\ V) = 3.280 \times 10^4 = 3.3 \times 10^4\ J\ or\ 33\ kJ$$

For the second reaction, $\Delta G° = -2(96,485)(-2.54) = 4.901 \times 10^5 = 4.90 \times 10^2\ kJ$

For the third reaction, $\Delta G° = -1(96,485)(0.23) = -2.22 \times 10^4\ J = -22\ kJ$

(c) $\Delta G° = -RT\ \ln K;\ \ln K = -\Delta G°/RT;\ K = e^{-\Delta G°/RT}$

For the first reaction,

$$\ln K = \frac{-3.281 \times 10^4\ J}{(8.314\ J/mol\text{-}K)(298\ K)} = -13.243 = -13;\ K = e^{-13.2428} = 1.78 \times 10^{-6} = 2 \times 10^{-6}$$

[Convert ln to log; the number of decimal places in the log is the number of sig figs in the result.]

For the second reaction,

$$\ln K = \frac{-4.902 \times 10^5\ J}{8.314\ J/mol\text{-}K \times 298\ K} = -197.86 = -198;\ K = e^{-198} = 1.23 \times 10^{-86} = 10^{-86}$$

For the third reaction,

$$\ln K = \frac{-(-2.22 \times 10^4\ J)}{8.314\ J/mol\text{-}K \times 298\ K} = 8.958 = 9.0;\ K = e^{9.0} = 7.77 \times 10^3 = 8 \times 10^3$$

Check. The equilibrium constants calculated here are indicators of equilibrium position, but are not particularly precise numerical values.

20.53   *Analyze/Plan.* Given K, calculate $\Delta G°$ and $E°$. Reverse the logic in Sample Exercise 20.10. According to Equation [19.20], $\Delta G° = -RT\ \ln K$. According to Equation [20.12], $\Delta G° = -nFE°$, $E° = -\Delta G°/nF$. *Solve.*

$K = 1.5 \times 10^{-4}$

$\Delta G° = -RT\ \ln K = -(8.314\ J/mol\text{-}K)(298)\ \ln\ (1.5 \times 10^{-4}) = 2.181 \times 10^4\ J = 21.8\ kJ$

$E° = -\Delta G°/nF;\ n = 2;\ F = 96.5\ kJ/mol\ e^-$

$$E° = \frac{-21.81\ kJ}{2\ mol\ e^- \times 96.5\ kJ/V\text{-}mol\ e^-} = -0.113\ V$$

Check. The unit of $\Delta G°$ is actually kJ/mol, which means kJ per 'mole of reaction', or for the reaction as written. Since we don't have a specific reaction, we interpret the unit as referring to the overall reaction.

20.55   *Analyze.* Given $E°_{red}$ values for half reactions, calculate the value of K for a given redox reaction.

Plan. Combine the relationships involving $E°$, $\Delta G°$ and K to get a direct relationship between $E°$ and K. For each reaction, calculate $E°$ from $E°_{red}$, then apply the relationship to calculate K.

Solve. $\Delta G^\circ = -nFE^\circ$, $\Delta G^\circ = -RT\ln K$; $\ln K = 2.303\log K$

$$-nFE^\circ = -RT\ln K, \quad E^\circ = \frac{RT}{nF}\ln K = \frac{2.303\,RT}{nF}\log K$$

From Equation [20.17] and [20.18], $2.303\,RT/F = 0.0592$.

$$E^\circ = \frac{0.0592}{n}\log K; \quad \log K = \frac{nE^\circ}{0.0592}; \quad K = 10^{\log K}$$

(a) $E^\circ = -0.28 - (-0.440) = 0.16$ V, $n = 2$ ($Ni^{2+} + 2e^- \rightarrow Ni$)

$$\log K = \frac{2(0.16)}{0.0592} = 5.4054 = 5.4; \quad K = 2.54 \times 10^5 = 3 \times 10^5$$

(b) $E^\circ = 0 - (-0.277) = 0.277$ V; $n = 2$ ($2H^+ + 2e^- \rightarrow H_2$)

$$\log K = \frac{2(0.277)}{0.0592} = 9.358 = 9.36; \quad K = 2.3 \times 10^9$$

(c) $E^\circ = 1.51 - 1.065 = 0.445 = 0.45$ V; $n = 10$ ($2MnO_4^- + 10e^- \rightarrow 2Mn^{+2}$)

$$\log K = \frac{10(0.445)}{0.0592} = 75.169 \approx 75; \quad K = 1.5 \times 10^{75} = 10^{75}$$

Check. Note that small differences in E° values lead to large changes in the magnitude of K. Sig fig rules limit precision of K values; using log instead of ln leads to more sig figs in the K value. This result is strictly numerical and does not indicate any greater precision in the data.

20.57   *Analyze/Plan.* $E^\circ = \dfrac{0.0592\ \text{V}}{n}\log K$.   See Solution 20.55 for a more complete development. $\log K = \dfrac{nE^\circ}{0.0592\ \text{V}}$.   *Solve.*

(a) $\log K = \dfrac{1(0.177\ \text{V})}{0.0592\ \text{V}} = 2.9899 = 2.99$; $K = 9.8 \times 10^2$

(b) $\log K = \dfrac{2(0.177\ \text{V})}{0.0592\ \text{V}} = 5.9797 = 5.98$; $K = 9.5 \times 10^5$

(c) $\log K = \dfrac{3(0.177\ \text{V})}{0.0592\ \text{V}} = 8.9696 = 8.97$; $K = 9.32 \times 10^8 = 9.3 \times 10^8$

20.59   *Analyze/Plan.* Given a spontaneous chemical reaction, calculate the maximum possible work for a given amount of reactant at standard conditions. Separate the equation into half-reactions and calculate cell emf. Use Equation [20.14], $w_{max} = -nFE$, to calculate maximum work. At standard conditions, $E = E^\circ$.   *Solve.*

$$I_2(s) + 2e^- \rightarrow 2I^-(aq) \qquad\qquad E^\circ_{red} = \ \ 0.536\ \text{V}$$
$$\underline{\quad Sn(s) \rightarrow Sn^{2+}(aq) + 2e^- \qquad\qquad E^\circ_{red} = -0.136\ \text{V}\quad}$$
$$I_2(s) + Sn(s) \rightarrow 2I^-(aq) + Sn^{2+}(aq) \qquad E^\circ = 0.536 - (-0.136) = 0.672\ \text{V}$$

$w_{max} = -2(96.5)(0.672) = -129.7 = -130$ kJ/mol Sn

$$\frac{-129.7\ \text{kJ}}{\text{mol Sn(s)}} \times \frac{1\ \text{mol Sn}}{118.71\ \text{g Sn}} \times 75.0\ \text{g Sn} \times \frac{1000\ \text{J}}{\text{kJ}} = -8.19 \times 10^4\ \text{J}$$

Check. The (–) sign indicates that work is done by the cell.

Cell EMF under Nonstandard Conditions (section 20.6)

20.61   (a)   The *Nernst equation* is applicable when the components of an electrochemical cell are at nonstandard conditions.

   (b) $Q = 1$ if all reactants and products are at standard conditions.

   (c)   If concentration of reactants increases, Q decreases, and E increases.

20.63   *Analyze/Plan.* Given a circumstance, determine its effect on cell emf. Each circumstance changes the value of Q. An increase in Q reduces emf; a decrease in Q increases emf. *Solve.*

$$Zn(s) + 2H^+(aq) \rightarrow Zn^{2+}(aq) + H_2(g); \quad E = E^\circ - \frac{0.0592}{n} \log Q; \quad Q = \frac{[Zn^{2+}]P_{H_2}}{[H^+]^2}$$

   (a) P_{H_2} increases, Q increases, E decreases

   (b) $[Zn^{2+}]$ increases, Q increases, E decreases

   (c) $[H^+]$ decreases, Q increases, E decreases

   (d)   No effect; does not appear in the Nernst equation

20.65   *Analyze/Plan.* Follow the logic in Sample Exercise 20.11.   *Solve.*

   (a)
$$\begin{array}{ll} Ni^{2+}(aq) + 2e^- \rightarrow Ni(s) & E^\circ_{red} = -0.28\ V \\ Zn(s) \rightarrow Zn^{2+}(aq) + 2e^- & E^\circ_{red} = -0.763\ V \\ \hline Ni^{2+}(aq) + Zn(s) \rightarrow Ni(s) + Zn^{2+}(aq) & E^\circ = -0.28 - (-0.763) = 0.483 = 0.48\ V \end{array}$$

   (b)
$$E = E^\circ - \frac{0.0592}{n} \log \frac{[Zn^{2+}]}{[Ni^{2+}]}; \quad n = 2$$

$$E = 0.483 - \frac{0.0592}{2} \log \frac{(0.100)}{(3.00)} = 0.483 - \frac{0.0592}{2} \log (0.0333)$$

$$E = 0.483 - \frac{0.0592(-1.477)}{2} = 0.483 + 0.0437 = 0.527 = 0.53\ V$$

   (c)
$$E = 0.483 - \frac{0.0592}{2} \log \frac{(0.900)}{(0.200)} = 0.483 - 0.0193 = 0.464 = 0.46\ V$$

20.67   *Analyze/Plan.* Follow the logic in Sample Exercise 20.11.   *Solve.*

   (a)
$$\begin{array}{ll} 4[Fe^{2+}(aq) \rightarrow Fe^{3+}(aq) + 1e^-] & E^\circ_{red} = 0.771\ V \\ O_2(g) + 4H^+(aq) + 4e^- \rightarrow 2H_2O(l) & E^\circ_{red} = 1.23\ V \\ \hline 4Fe^{2+}(aq) + O_2(g) + 4H^+(aq) \rightarrow 4Fe^{3+}(aq) + 2H_2O(l) & E^\circ = 1.23 - 0.771 = 0.459 = 0.46 \end{array}$$

   (b)
$$E = E^\circ - \frac{0.0592}{n} \log \frac{[Fe^{3+}]^4}{[Fe^{2+}]^4[H^+]^4 P_{O_2}}; \quad n = 4,\ [H^+] = 10^{-3.50} = 3.2 \times 10^{-4}\ M$$

$$E = 0.459\ V - \frac{0.0592}{4} \log \frac{(0.010)^4}{(1.3)^4(3.2 \times 10^{-4})^4(0.50)} = 0.459 - \frac{0.0592}{4} \log (7.0 \times 10^5)$$

$$E = 0.459 - \frac{0.0592}{4}(5.845) = 0.459 - 0.0865 = 0.3725 = 0.37\ V$$

20.69   *Analyze/Plan.* We are given a concentration cell with Zn electrodes. Use the definition of a concentration cell in Section 20.6 to answer the stated questions. Use Equation [20.18] to calculate the cell emf. For a concentration cell, Q = [dilute]/[concentrated].   *Solve.*

(a)   The compartment with the more dilute solution will be the anode. That is, the compartment with $[Zn^{2+}] = 1.00 \times 10^{-2}\ M$ is the anode.

(b)   Since the oxidation half-reaction is the opposite of the reduction half-reaction, E° is zero.

(c) $E = E° - \dfrac{0.0592}{n} \log Q; Q = [Zn^{2+}, \text{dilute}]/[Zn^{2+}, \text{conc.}]$

$E = 0 - \dfrac{0.0592}{2} \log \dfrac{(1.00 \times 10^{-2})}{(1.8)} = 0.0668\ V$

(d)   In the anode compartment, $Zn(s) \rightarrow Zn^{2+}(aq)$, so $[Zn^{2+}]$ increases from $1.00 \times 10^{-2}\ M$. In the cathode compartment, $Zn^{2+}(aq) \rightarrow Zn(s)$, so $[Zn^{2+}]$ decreases from 1.8 M.

20.71   *Analyze/Plan.* Follow the logic in Sample Exercise 20.12.   *Solve.*

$E = E° - \dfrac{0.0592}{2} \log \dfrac{[P_{H_2}][Zn^{2+}]}{[H^+]^2}; E° = 0.0\ V - (-0.763\ V) = 0.763\ V$

$0.684 = 0.763 - \dfrac{0.0592}{2} \times (\log [P_{H_2}][Zn^{2+}] - 2\log [H^+])$

$= 0.763 - \dfrac{0.0592}{2} \times (-0.5686 - 2\log [H^+])$

$0.684 = 0.763 + 0.0168 + 0.0592 \log [H^+]; \log [H^+] = \dfrac{0.684 - 0.0168 - 0.763}{0.0592}$

$\log [H^+] = -1.6188 = -1.6; [H^+] = 0.0241 = 0.02\ M; pH = 1.6$

Batteries and Fuel Cells (section 20.7)

20.73   (a)   The emf of a battery decreases as it is used. This happens because the concentrations of products increase and the concentrations of reactants decrease. According to the Nernst equation, these changes increase Q and decrease E_{cell}.

(b)   The major difference between AA- and D-size batteries is the amount of reactants present. The additional reactants in a D-size battery enable it to provide power for a longer time.

20.75   *Analyze/Plan.* Given mass of a reactant (Pb), calculate mass of product (PbO_2), and coulombs of charge transferred. This is a stoichiometry problem; we need the balanced equation for the chemical reaction that occurs in the lead-acid battery.

The overall cell reaction is:

$Pb(s) + PbO_2(s) + 2H^+ (aq) + 2HSO_4^- (aq) \rightarrow 2PbSO_4(s) + 2H_2O(l)$

Solve.

(a) $g\ Pb \rightarrow mol\ Pb \rightarrow mol\ PbO_2 \rightarrow g\ PbO_2$

$$402\ g\ Pb \times \frac{1\ mol\ Pb}{207.2\ g\ Pb} \times \frac{1\ mol\ PbO_2}{1\ mol\ Pb} \times \frac{239.2\ g\ PbO_2}{1\ mol\ PbO_2} = 464\ g\ PbO_2$$

(b)   From the half-reactions for the lead-acid battery, 2 mol electrons are transferred for each mol of Pb reacted. From section 20.5, 96,485 C/mol e$^-$.

$$402\ g\ Pb \times \frac{1\ mol\ Pb}{207.2\ g\ Pb} \times \frac{2\ mol\ e^-}{1\ mol\ Pb} \times \frac{96,485\ C}{1\ mol\ e^-} = 374,392 = 3.74 \times 10^5\ C$$

20.77   *Analyze/Plan.* We are given a redox reaction and asked to write half-reactions, calculate E°, and indicate whether Li(s) is the anode or cathode. Determine which reactant is oxidized and which is reduced. Separate into half-reactions, find E_{red}^{o} for the half-reactions from Appendix E and calculate E°.   *Solve.*

(a)   Li(s) is oxidized at the anode.

(b)
$Ag_2CrO_4(s) + 2e^- \rightarrow 2Ag(s) + CrO_4^{2-}(aq)$	$E_{red}^{o} = 0.446\ V$
$2[Li(s) \rightarrow Li^+(aq) + 1e^-]$	$E_{red}^{o} = -3.05\ V$

$$\overline{Ag_2CrO_4(s) + 2Li(s) \rightarrow 2Ag(s) + CrO_4^-(aq) + 2Li^+(aq)}$$

E° = 0.446 V – (–3.05 V) = 3.496 = 3.50 V

(c)   The emf of the battery, 3.5 V, is exactly the standard cell potential calculated in part (b).

(d)   For this battery at ambient conditions, E ≈ E°, so log Q ≈ 0. This makes sense because all reactants and products in the battery are solids and thus present in their standard states. Assuming that E° is relatively constant with temperature, the value of the second term in the Nernst equation is ≈ 0 at 37°C, and E ≈ 3.5 V.

20.79   *Analyze/Plan.* (a) Consider the function of Zn in an alkaline battery. What effect would it have on the redox reaction and cell emf if Cd replaces Zn? (b) Both batteries contain Ni. What is the difference in environmental impact between Cd and the metal hydride? *Solve.*

(a) E_{red}^{o} for Cd (–0.40 V) is less negative than E_{red}^{o} for Zn (–0.76 V), so E_{cell} will have a smaller (less positive) value.

(b)   NiMH batteries use an alloy such as $ZrNi_2$ as the anode material. This eliminates the use and concomitant disposal problems associated with Cd, a toxic heavy metal.

20.81   The main advantage of a H_2-O_2 fuel cell over an alkaline battery is that the fuel cell is not a closed system. Fuel, H_2, and oxidant, O_2 are continuously supplied to the fuel cell, so that it can produce electrical current for a time limited only by the amount of available fuel. An alkaline battery contains a finite amount of reactant and produces current only until the reactants are spent, or reach equilibrium.

Alkaline batteries are much more convenient, because they are self-contained. Fuel cells require a means to acquire and store volatile and explosive $H_2(g)$. Disposal of spent alkaline batteries, which contain zinc and manganese solids, is much more problematic. H_2-O_2 fuel cells produce only $H_2O(l)$, which is not a disposal problem.

Corrosion (section 20.8)

20.83 *Analyze/Plan.* (a) Decide which reactant is oxidized and which is reduced. Write the balanced half-reactions and assign the appropriate one as anode and cathode. (b) Write the balanced half-reaction for $Fe^{2+}(aq) \rightarrow Fe_2O_3 \cdot 3H_2O$. Use the reduction half-reaction from part (a) to obtain the overall reaction. *Solve.*

(a)   anode: $Fe(s) \rightarrow Fe^{2+}(aq) + 2e^-$

cathode: $O_2(g) + 4H^+(aq) + 4e^- \rightarrow 2H_2O(l)$

(b) $2Fe^{2+}(aq) + 6H_2O(l) \rightarrow Fe_2O_3 \cdot 3H_2O(s) + 6H^+(aq) + 2e^-$

$O_2(g) + 4H^+(aq) + 4e^- \rightarrow 2H_2O(l)$

(Multiply the oxidation half-reaction by two to balance electrons and obtain the overall balanced reaction.)

20.85 (a)   A "sacrificial anode" is a metal that is oxidized in preference to another when the two metals are coupled in an electrochemical cell; the sacrificial anode has a more negative E^o_{red} than the other metal. In this case, Mg acts as a sacrificial anode because it is oxidized in preference to the pipe metal; it is sacrificed to preserve the pipe.

(b) E^o_{red} for Mg^{2+} is –2.37 V, more negative than most metals present in pipes, including Fe ($E^o_{red} = -0.44$ V) and Zn ($E^o_{red} = -0.763$ V).

20.87 *Analyze/Plan.* Given the materials brass, composed of Zn and Cu, and galvanized steel, determine the possible spontaneous redox reactions that could occur when the materials come in contact. Calculate E° values for these reactions.

Solve. The main metallic component of steel is Fe. Galvanized steel is steel plated with Zn. The three metals in question are Fe, Zn, and Cu; their E^o_{red} values are shown below.

E^o_{red} $Fe^{2+}(aq) = -0.440$ V

E^o_{red} $Zn^{2+}(aq) = -0.763$ V

E^o_{red} $Cu^{2+}(aq) = 0.337$ V

Zn, with the most negative E^o_{red} value, can act as a sacrificial anode for either Fe or Cu. That is, Zn(s) will be preferentially oxidized when in contact with Fe(s) or Cu(s). For environmental corrosion, the oxidizing agent is usually $O_2(g)$ in acidic solution, $E^o_{red} = 1.23$ V. The pertinent reactions and their E° values are:

$2Zn(s) + O_2(g) + 4H^+(aq) \rightarrow 2Zn^{2+}(aq) + 2H_2O(l)$

$E° = 1.23$ V – (–0.763 V) = 1.99 V

$2Fe(s) + O_2(g) + 4H^+(aq) \rightarrow 2Fe^{2+}(aq) + 2H_2O(l)$

$E° = 1.23$ V – (–0.440 V) = 1.67 V

$2Cu(s) + O_2(g) + 4H^+(aq) \rightarrow 2Cu^{2+}(aq) + 2H_2O(l)$

$E° = 1.23$ V – (0.337 V) = 0.893 V

Note, however, that Fe has a more negative E_{red}^o than Cu so when the two are in contact Fe acts as the sacrificial anode, and corrosion (of Fe) occurs preferentially. This is verified by the larger E° value for the corrosion of Fe, 1.67 V, relative to the corrosion of Cu, 0.893 V. When the three metals Zn, Fe, and Cu are in contact, oxidation of Zn will happen first, followed by oxidation of Fe, and finally Cu.

Electrolysis; Electrical Work (section 20.9)

20.89　(a)　*Electrolysis* is an electrochemical process driven by an outside energy source.

　　　(b)　Electrolysis reactions are, by definition, nonspontaneous.

　　　(c)　$2Cl^-(l) \rightarrow Cl_2(g) + 2e^-$

　　　(d)　When an aqueous solution of NaCl undergoes electrolysis, H_2O is preferentially reduced to form $H_2(g)$.

20.91　*Analyze/Plan.* Follow the logic in Sample Exercise 20.14, paying close attention to units. Coulombs = **amps-s**; since this is a $3e^-$ reduction, each mole of Cr(s) requires 3 Faradays. *Solve.*

　　　(a)　$7.60 \text{ A} \times 2.00 \text{ d} \times \dfrac{24 \text{ hr}}{1 \text{ d}} \times \dfrac{60 \text{ min}}{1 \text{ hr}} \times \dfrac{60 \text{ s}}{1 \text{ min}} \times \dfrac{1 \text{ C}}{1 \text{ amp-s}} \times \dfrac{1 \text{ F}}{96,485 \text{ C}}$

$$\times \dfrac{1 \text{ mol Cr}}{3 \text{ F}} \times \dfrac{52.00 \text{ g Cr}}{1 \text{ mol Cr}} = 236 \text{ g Cr(s)}$$

　　　(b)　$0.250 \text{ mol Cr} \times \dfrac{3 \text{ F}}{1 \text{ mol Cr}} \times \dfrac{96,485 \text{ C}}{\text{F}} \times \dfrac{1 \text{ amp-s}}{1 \text{ C}} \times \dfrac{1}{8.00 \text{ hr}} \times \dfrac{1 \text{ hr}}{60 \text{ min}} \times \dfrac{1 \text{ min}}{60 \text{ s}}$

$$= 2.51 \text{ A}$$

20.93　*Analyze/Plan.* Combine the ideas in Sample Exercises 20.14 and 20.15, paying close attention to units. Li^+ is reduced at the anode; Cl^- is oxidized at the anode. *Solve.*

　　　(a)　If the cell is 85% efficient, $\dfrac{96,485 \text{ C}}{\text{F}} \times \dfrac{1 \text{ F}}{0.85 \text{ mol}} = 1.13512 \times 10^5$

$$= 1.1 \times 10^5 \text{ C/mol Li required}$$

$$7.5 \times 10^4 \text{ A} \times 24 \text{ h} \times \dfrac{3600 \text{ s}}{1 \text{ h}} \times \dfrac{1 \text{ C}}{1 \text{ amp-s}} \times \dfrac{1 \text{ mol Li}}{1.13512 \times 10^5 \text{ C}} \times \dfrac{6.94 \text{ g Li}}{1 \text{ mol Li}}$$

$$= 3.962 \times 10^5 = 4.0 \times 10^5 \text{ g Li}$$

　　　(b)　$E_{cell}^o = E_{red}^o(\text{cathode}) - E_{red}^o(\text{anode}) = -3.05 \text{ V} - (1.359 \text{ V}) = -4.409 = -4.41 \text{ V}$

　　　The minimum voltage required to drive the reaction is the magnitude of E_{cell}^o, 4.41 V.

20.95　Table 4.5 is "The Activity Series of the Metals". Gold is the least active metal on this table, less active than copper. This means that gold is more difficult to oxidize than copper (and that E_{red}^o for Au^{3+} is more positive than E_{red}^o for Cu^{2+}). When crude copper is refined by electrolysis, Cu is oxidized from the crude anode, but any metallic gold present in the crude copper is not oxidized, so it accumulates near the anode, along with other impurities less active than copper.

Additional Exercises

20.97   (a) $Ni^+(aq) + 1e^- \rightarrow Ni(s)$

$$\frac{Ni^+(aq) \quad\quad\quad\quad \rightarrow Ni^{2+}(aq) + 1e^-}{2Ni^+(aq) \quad\quad\quad\quad \rightarrow Ni(s) + Ni^{2+}(aq)}$$

(b) $MnO_4{}^{2-}(aq) + 4H^+(aq) + 2e^- \rightarrow MnO_2(s) + 2H_2O(l)$

$$\frac{2[MnO_4{}^{2-}(aq) \rightarrow MnO_4{}^-(aq) + 1e^-]}{3MnO_4{}^{2-}(aq) + 4H^+(aq) \rightarrow 2MnO_4{}^-(aq) + MnO_2(s) + 2H_2O(l)}$$

(c) $H_2SO_3(aq) + 4H^+(aq) + 4e^- \rightarrow S(s) + 3H_2O(l)$

$$\frac{2[H_2SO_3(aq) + H_2O(l) \rightarrow HSO_4{}^-(aq) + 3H^+(aq) + 2e^-]}{3H_2SO_3(aq) \rightarrow S(s) + 2HSO_4{}^-(aq) + 2H^+(aq) + H_2O(l)}$$

(d) $Cl_2(aq) + 2H_2O(l) \rightarrow 2ClO^-(aq) + 4H^+(aq) + 2e^-$

$$\frac{4OH^-(aq) \quad\quad\quad + 4OH^-(aq)}{Cl_2(aq) + 4OH^-(aq) \rightarrow 2ClO^-(aq) + 2H_2O(l) + 2e^-}$$

$$\frac{Cl_2(aq) + 2e^- \rightarrow 2Cl^-(aq)}{1/2[2Cl_2(aq) + 4OH^-(aq) \rightarrow 2Cl^-(aq) + 2ClO^-(aq) + 2H_2O(l)]}$$

$$Cl_2(aq) + 2OH^-(aq) \rightarrow Cl^-(aq) + ClO^-(aq) + H_2O(l)$$

20.100   We need in each case to determine whether E° is positive (spontaneous) or negative (nonspontaneous).

(a) $I_2(s) + 2e^- \rightarrow 2I^-(aq)$ $E_{red}^{\circ} = 0.536$ V

$$\frac{Sn(s) \rightarrow Sn^{2+}(aq) + 2e^- \quad\quad E_{red}^{\circ} = -0.136}{Sn(s) + I_2(s) \rightarrow Sn^{2+}(aq) + 2I^-(aq) \quad E° = 0.536 - (-0.136) = 0.672 \text{ V, spontaneous}}$$

(b) $Ni^{2+}(aq) + 2e^- \rightarrow Ni(s)$ $E_{red}^{\circ} = -0.28$ V

$$\frac{2I^-(aq) \rightarrow I_2(s) + 2e^- \quad\quad E_{red}^{\circ} = 0.536 \text{ V}}{Ni^{2+}(aq) + 2I^-(aq) \rightarrow Ni(s) + I_2(s) \quad E° = -0.28 - 0.536 = -0.82 \text{ V, nonspontaneous}}$$

(c) $2[Ce^{4+}(aq) + 1e^- \rightarrow Ce^{3+}(aq)]$ $E_{red}^{\circ} = 1.61$ V

$$\frac{H_2O_2(aq) \rightarrow O_2(g) + 2H^+(aq) + 2e^- \quad\quad E_{red}^{\circ} = 0.68 \text{ V}}{2Ce^{4+}(aq) + H_2O_2(aq) \rightarrow 2Ce^{3+}(aq) + O_2(g) + 2H^+(aq) \quad E° = 1.61 - 0.68 = 0.93 \text{ V, spontaneous}}$$

(d) $Cu^{2+}(aq) + 2e^- \rightarrow Cu(s)$ $E_{red}^{\circ} = 0.337$ V

$$\frac{Sn^{2+}(aq) \rightarrow Sn^{4+}(aq) + 2e^- \quad\quad E_{red}^{\circ} = 0.154 \text{ V}}{Cu^{2+}(aq) + Sn^{2+}(aq) \rightarrow Cu(s) + Sn^{4+}(aq) \quad E° = 0.337 - 1.54 = 0.183 \text{ V, spontaneous}}$$

20.104   Use the relationship developed in Solution 20.101 to calculate K from E°. Use data from Appendix E to calculate E° for the disproportionation.

$$Cu^+(aq) + 1e^- \rightarrow Cu(s) \qquad E^{\circ}_{red} = 0.521 \text{ V}$$

$$\underline{Cu^+(aq) \rightarrow Cu^{2+}(aq) + 1e^- \qquad E^{\circ}_{red} = 0.153 \text{ V}}$$

$$2Cu^+(aq) \rightarrow Cu(s) + Cu^{2+}(aq) \qquad E^{\circ} = 0.521 \text{ V} - 0.153 \text{ V} = 0.368 \text{ V}$$

$$E^{\circ} = \frac{0.0592}{n} \log K, \log K = \frac{nE^{\circ}}{0.0592} = \frac{1 \times 0.368}{0.0592} = 6.216 = 6.22$$

$$K = 10^{6.216} = 1.6 \times 10^6$$

20.107   The ship's hull should be made negative. By keeping an excess of electrons in the metal of the ship, the tendency for iron to undergo oxidation, with release of electrons, is diminished. The ship, as a negatively charged "electrode," becomes the site of reduction, rather than oxidation, in an electrolytic process.

20.110   *Analyze.* Given mass of aluminum desired, applied voltage and electrolysis efficiency, calculate kWh of electricity required. *Plan.* Beginning with mass Al and paying attention to units, calculate coulombs required if the process is 100% efficient. Then, take efficiency into account and then use V, C and the relationship between J and kWh to calculate kWh required. *Solve.*

$$1.0 \times 10^3 \text{ kg Al} \times \frac{1000 \text{ g}}{1 \text{ kg}} \times \frac{1 \text{ mol Al}}{26.98 \text{ g Al}} \times \frac{3 \text{ F}}{1 \text{ mol Al}} \times \frac{96,485 \text{ C}}{\text{F}} = 1.073 \times 10^{10} = 1.1 \times 10^{10} \text{ C}$$

If the cell is 45 efficient, $(1.073 \times 10^{10}/0.45) = 2.384 \times 10^{10} = 2.4 \times 10^{10}$ C are required to plate the bumper.

$$4.50 \text{ V} \times 2.384 \times 10^{10} \text{ C} \times \frac{1 \text{ J}}{1 \text{ C-V}} \times \frac{1 \text{ kWh}}{3.6 \times 10^6 \text{ J}} = 29,801 = 3.0 \times 10^4 \text{ kWh}$$

Integrative Exercises

20.112 $N_2(g) + 3H_2(g) \rightarrow 2NH_3(g)$

(a)   The oxidation number of $H_2(g)$ and $N_2(g)$ is 0. The oxidation number of N in NH_3 is –3, H in NH_3 is +1. H_2 is being oxidized and N_2 is being reduced.

(b)   Calculate ΔG° from ΔG°_f values in Appendix C. Use $\Delta G^{\circ} = -RT \ln K$ to calculate K.

$$\Delta G^{\circ} = 2\Delta G^{\circ}_f \text{ } NH_3(g) - \Delta G^{\circ}_f \text{ } N_2(g) - 3\Delta G^{\circ}_f \text{ } H_2(g)$$

$$\Delta G^{\circ} = 2(-16.66 \text{ kJ}) - 0 - 3(0) = -33.32 \text{ kJ}$$

$$\Delta G^{\circ} = -RT \ln K, \ln K = \frac{-\Delta G^{\circ}}{RT} = \frac{-(-33.32 \times 10^3 \text{ J})}{(8.314 \text{ J/mol - K})(298 \text{ K})} = 13.4487 = 13.45$$

$$K = e^{13.4487} = 6.9 \times 10^5$$

(c) $\Delta G^\circ = -nfE^\circ$; $E^\circ = \dfrac{-\Delta G^\circ}{nF}$; $n = ?$

2 N atoms change from 0 to –3, or 6 H atoms change from 0 to +1.

Either way, n = 6.

$E^\circ = \dfrac{-(-33.32 \text{ kJ})}{6 \times 96.5 \text{ kJ/V}} = 0.05755 \text{ V}$

20.115   (a) $Ag^+(aq) + e^- \rightarrow Ag(s)$ $E^\circ_{red} = 0.799 \text{ V}$

$\dfrac{Fe^{2+}(aq) \rightarrow Fe^{3+}(aq) + 1e^- \qquad E^\circ_{red} = 0.771 \text{ V}}{Ag^+(aq) + Fe^{2+}(aq) \rightarrow Ag(s) + Fe^{3+}(aq) \quad E^\circ = 0.799 \text{ V} - 0.771 \text{ V} = 0.028 \text{ V}}$

(b) $Ag^+(aq)$ is reduced at the cathode and $Fe^{2+}(aq)$ is oxidized at the anode.

(c) $\Delta G^\circ = -nFE^\circ = -(1)(96.5)(0.028) = -2.7 \text{ kJ}$

$\Delta S^\circ = S^\circ \, Ag(s) + S^\circ \, Fe^{3+}(aq) - S^\circ Ag^+(aq) - S^\circ \, Fe^{2+}(aq)$

$= 42.55 \text{ J} + 293.3 \text{ J} - 73.93 \text{ J} - 113.4 \text{ J} = 148.5 \text{ J}$

$\Delta G^\circ = \Delta H^\circ - T\Delta S^\circ$ Since ΔS° is positive, ΔG° will become more negative and E° will become more positive as temperature is increased.

20.118 $AgSCN(s) + e^- \rightarrow Ag(s) + SCN^-(aq)$ $E^\circ_{red} = 0.0895 \text{ V}$

$\dfrac{Ag(s) \rightarrow Ag^+(aq) + e^- \qquad\qquad E^\circ_{red} = 0.799 \text{ V}}{AgSCN(s) \rightarrow Ag^+(aq) + SCN^-(aq) \quad E^\circ = 0.0895 - 0.799 = -0.710 \text{ V}}$

$E^\circ = \dfrac{0.0592}{n} \log K_{sp}$; $\log K_{sp} = \dfrac{(-0.710)\,(1)}{0.0592} = -11.993 = -12.0$

$K_{sp} = 10^{-11.993} = 1.02 \times 10^{-12} = 1 \times 10^{-12}$

21 Nuclear Chemistry

Visualizing Concepts

21.1   *Analyze.* Given the name and mass number of a nuclide, decide if it lies within the belt of stability. If not, suggest a process that moves it toward the belt.

Plan. Calculate the number of protons and neutrons in each nuclide. Locate this point on Figure 21.2. If the point is above the belt, β-decay increases protons and decreases neutrons, decreasing the neutron-to-proton ratio. If the point is below the belt, either positron emission or neutron capture decreases protons and increases neutrons, increasing the neutron-to-proton ratio. *Solve.*

(a) ^{24}Ne: 10 p, 14 n, just above the belt of stability. Reduce the neutron-to proton ratio via β-decay.

(b) ^{32}Cl: 17 p, 15 n, just below the belt of stability. Increase the neutron-to-proton ratio via positron emission or orbital electron capture.

(c) ^{108}Sn: 50 p, 58 n, just below the belt of stability. Increase the neutron-to-proton ratio via positron emission or orbital electron capture.

(d) ^{216}Po: 84 p, 132 n, just beyond the belt of stability. Nuclei with atomic numbers ≥ 84 tend to decay via alpha emission, which decreases both protons and neutrons.

21.4   *Analyze/Plan.* Write the balanced equation for the decay. Nuclear decay is a first-order process; use appropriate relationships for first-order processes to determine $t_{1/2}$, k and remaining ^{88}Mo after 12 minutes.   *Solve.*

(a) $t_{1/2}$ is the time required for half of the original nuclide to decay. Relative to the graph, this is the time when the amount of ^{88}Mo is reduced from 1.0 to 0.5. This time is 7 minutes.

(b)   For a first-order process, $t_{1/2} = 0.693/k$ or $k = 0.693/t_{1/2}$.

$k = 0.693/7$ min $= 0.0990 = 0.1$ min^{-1}

(c)   From the graph, the fraction of ^{88}Mo remaining after 12 min is 0.3/1.0 = 0.3

Check. $\ln(N_t/N_o) = -kt = -(0.099)(12) = -1.188$; $N_t/N_o = e^{-1.188} = 0.30$.

(d) $^{88}_{42}$Mo $\rightarrow$ $^{88}_{41}$Nb $+$ $^{0}_{1}$e

21.5   *Analyze/Plan.*   Atomic number is number of protons, mass number is (protons + neutrons). Chemical symbol is determined by atomic number. Beta decay increases the number of protons while mass number stays constant. Positron emission decreases the

298

number of protons while mass number stays constant. Half-life, $t_{1/2}$, is the time required to reduce the amount of radioactive material by half. *Solve.*

(a) $^{10}_{5}B$, $^{11}_{5}B$; $^{12}_{6}C$, $^{13}_{6}C$; $^{14}_{7}N$, $^{15}_{7}N$; $^{16}_{8}O$, $^{17}_{8}O$, $^{18}_{8}O$; $^{19}_{9}F$

(b)   On the diagram, $^{14}_{6}C$ is the only radioactive nuclide above and left of the band of stable red nuclides. $^{14}_{6}C$ will reduce its neutron-to-proton ratio by β-decay.

$$^{14}_{6}C \rightarrow {}^{14}_{7}N + {}^{0}_{-1}e$$

(c)   On the diagram, radioactive nuclides right and below the band of stable red nuclides are likely to increase their neutron-to-proton ratio via positron emission. In order to be useful for positron emission tomography, the nuclides must have a half-life on the order of minutes. (Nuclides with very fast decay disappear before they can be imaged. Those with longer half-lives linger in the patient.) Four radioactive nuclides fit these criteria: $^{11}_{6}C$, $^{13}_{7}N$, $^{15}_{8}O$, $^{18}_{9}F$

(d)   A reduction of 12.5% in amount or concentration of a radionuclide amounts to three half-lives. If the total decay time is 1 hour, 1/3 of this time is 20 min. The radionuclide with a half-life of 20 min is $^{11}_{6}C$.

Radioactivity (section 21.1)

21.7   *Analyze/Plan.* Given various nuclide descriptions, determine the number of protons and neutrons in each nuclide. The left superscript is the mass number, protons plus neutrons. If there is a left subscript, it is the atomic number, the number of protons. Protons can always be determined from chemical symbol; all isotopes of the same element have the same number of protons. A number following the element name, as in part (c) is the mass number.   *Solve.*

p = protons, n = neutrons, e = electrons; number of protons = atomic number;
number of neutrons = mass number – atomic number

(a) $^{55}_{25}Mn$: 25p, 30n   (b) ^{201}Hg: 80p, 121n   (c) ^{39}K: 19p, 20n

21.9   *Analyze/Plan.* See definitions in Section 21.1. In each case, the left superscript is mass number, the left subscript is related to atomic number.   *Solve.*

(a) $^{1}_{0}n$ (b) $^{4}_{2}He$ or α   (c) $^{0}_{0}\gamma$ or γ

21.11   *Analyze/Plan.* Follow the logic in Sample Exercises 21.1 and 21.2. Pay attention to definitions of decay particles and conservation of mass and charge.   *Solve.*

(a) $^{90}_{37}Rb \rightarrow {}^{90}_{38}Sr + {}^{0}_{-1}e$ (b) $^{72}_{34}Se + {}^{0}_{-1}e$ (orbital electron) $\rightarrow {}^{72}_{33}As$

(c) $^{76}_{36}Kr \rightarrow {}^{76}_{35}Br + {}^{0}_{1}e$ (d) $^{226}_{88}Ra \rightarrow {}^{222}_{86}Rn + {}^{4}_{2}He$

21.13   *Analyze/Plan.* Using definitions of the decay processes and conservation of mass number and atomic number, work backwards to the reactants in the nuclear reactions. *Solve.*

(a) $^{211}_{82}Pb \rightarrow {}^{211}_{83}Bi + {}^{0}_{-1}\beta$ (b) $^{50}_{25}Mn \rightarrow {}^{50}_{24}Cr + {}^{0}_{1}e$

(c) $^{179}_{74}W + {}^{0}_{-1}e \rightarrow {}^{179}_{73}Ta$ (d) $^{230}_{90}Th \rightarrow {}^{226}_{88}Ra + {}^{4}_{2}He$

21.15 *Analyze/Plan.* Given the starting and ending nuclides in a nuclear decay sequence, we are asked to determine the number of alpha and beta emissions. Use the total change in A and Z, along with definitions of alpha and beta decay, to answer the question. *Solve.*

The total mass number change is (235–207) = 28. Since each α particle emission decreases the mass number by four, whereas emission of a β particle does not correspond to a mass change, there are 7 α particle emissions. The change in atomic number in the series is 10. Each α particle results in an atomic number lower by two. The 7 α particle emissions alone would cause a decrease of 14 in atomic number. Each β particle emission raises the atomic number by one. To obtain the observed lowering of 10 in the series, there must be 4 β emissions.

Nuclear Stability (section 21.2)

21.17 *Analyze/Plan.* Follow the logic in Sample Exercise 21.3, paying attention to the guidelines for neutron-to-proton ratio. *Solve.*

(a) $^{8}_{5}B$ - low neutron/proton ratio, positron emission (for low atomic numbers, positron emission is more common than orbital electron capture)

(b) $^{68}_{29}Cu$ - high neutron/proton ratio, beta emission

(c) $^{32}_{15}P$ - slightly high neutron/proton ratio, beta emission

(d) $^{39}_{17}Cl$ - high neutron/proton ratio, beta emission

21.19 *Analyze/Plan.* Use the criteria listed in Table 21.4. *Solve.*

(a) Stable: $^{39}_{19}K$ odd proton, even neutron more abundant than odd proton, odd neutron; 20 neutrons is a magic number.

(b) Stable: $^{209}_{83}Bi$ odd proton, even neutron more abundant than odd proton, odd neutron; 126 neutrons is a magic number.

(c) Stable: $^{58}_{28}Ni$ even proton, even neutron more likely to be stable than even proton, odd neutron; $^{65}_{28}Ni$ has high neutron/proton ratio

21.21 *Analyze/Plan.* For each nuclide, determine the number of protons and neutrons and decide if they are magic numbers. *Solve.*

(a) $^{4}_{2}He$, both   (b) $^{18}_{8}O$ has a magic number of protons, but not neutrons

(c) $^{40}_{20}Ca$, both   (d) $^{66}_{30}Zn$ has neither   (e) $^{208}_{82}Pb$, both

21.23 Consider the stability of the two emitted particles. The alpha particle, $^{4}_{2}He$, has a magic number of both protons and neutrons. The proton bas no magic numbers, and is an odd proton – even neutron particle. The alpha is a very stable emitted particle, Twhich makes alpha emission a favorable process. The proton is not a stable emitted particle and its formation does not encourage proton emission as a process.

Nuclear Transmutations (section 21.3)

21.25 Protons and alpha particles are positively charged and must be moving very fast to overcome electrostatic forces which would repel them from the target nucleus. Neutrons are electrically neutral and not repelled by the nucleus.

21.27 *Analyze/Plan.* Determine A and Z for the missing particle by conservation principles. Find the appropriate symbol for the particle.   *Solve.*

(a) $^{252}_{98}Cf + ^{10}_{5}B \rightarrow 3\,^{1}_{0}n + ^{259}_{103}Lr$ (b) $^{2}_{1}H + ^{3}_{2}He \rightarrow ^{4}_{2}He + ^{1}_{1}H$

(c) $^{1}_{1}H + ^{11}_{5}B \rightarrow 3\,^{4}_{2}He$ (d) $^{122}_{53}I \rightarrow ^{122}_{54}Xe + ^{0}_{-1}e$

(e) $^{59}_{26}Fe \rightarrow ^{0}_{-1}e + ^{59}_{27}Co$

21.29 *Analyze/Plan.* Follow the logic in Sample Exercise 21.5, paying attention to conservation of A and Z.   *Solve.*

(a) $^{238}_{92}U + ^{4}_{2}He \rightarrow ^{241}_{94}Pu + ^{1}_{0}n$ (b) $^{14}_{7}N + ^{4}_{2}He \rightarrow ^{17}_{8}O + ^{1}_{1}H$

(c) $^{56}_{26}Fe + ^{4}_{2}He \rightarrow ^{60}_{29}Cu + ^{0}_{-1}e$

Rates of Radioactive Decay (section 21.4)

21.31 (a) True. $k = 0.693/t_{1/2}$. The decay rate constant, k, and half-life, $t_{1/2}$ are inversely related.

(b) False. If X is not radioactive, it does not spontaneously decay and its half-life is essentially infinity.

(c) True. Changes in the amount of A would be measurable over the 40-year time frame, while changes in the amount of X would be very small and difficult to detect.

21.33 *Analyze/Plan.* The half-life is 12.3 yr. Use $t_{1/2}$ to calculate k and the mass fraction (N_t / N_0) remaining after 50 yr.

Solve. $k = 0.693/\,t_{1/2} = 0.693/12.3\ yr = 0.056341 = 0.0563\ yr^{-1}$

$\ln\dfrac{N_t}{N_0} = -kt;\ \ \ln\dfrac{N_t}{N_0} = -0.056341\ yr^{-1}(50yr) = -2.81707 = -2.8;\ \ \dfrac{N_t}{N_0} = 0.05978 = 0.06$

When the watch is 50 years old, only 6% (or 6.0%) of the tritium remains. The dial will be dimmed by 94%.

21.35 *Analyze/Plan.* We are given half-life of cobalt-60, and replacement time when the activity of the sample is 75% of the initial value. Consider the rate law for (first-order) nuclear decay: $\ln(N_t/N_o) = -kt$. *Solve.*

$k = 0.693\,/\,t_{1/2} = 0.693/5.26\ yr = 0.1317 = 0.132\ yr^{-1};\ N_t/N_o = 0.75$

$t = \dfrac{-1}{k}\ln\dfrac{N_t}{N_o} = -(1/0.1317\ yr^{-1})\ln(0.75) = 2.18\ yr\ = 26.2\ mo = 797\ d.$

The source would have been replaced sometime in the summer of 2012, probably in August.

21.37   (a)   *Analyze/Plan.* $^{226}_{88}Ra \rightarrow {}^{222}_{86}Rn + {}^{4}_{2}He$

1 α particle is produced for each ^{226}Ra that decays. Rate = kN. Calculate the number of ^{226}Ra particles in the 10.0 mg sample. Calculate $t_{1/2}$ and k in min, then rate in dis/min, then the number of disintegrations in 5.0 min. *Solve.*

$$10.0 \text{ mg Ra} \times \frac{1 \text{ g}}{1000 \text{ mg}} \times \frac{1 \text{ mol Ra}}{226 \text{ g Ra}} \times \frac{6.022 \times 10^{23} \text{ Ra atoms}}{1 \text{ mol Ra}} = 2.6646 \times 10^{19} = 2.66 \times 10^{19} \text{ atoms}$$

Calculate k in min^{-1}. $1600 \text{ yr} \times \frac{365 \text{ d}}{1 \text{ yr}} \times \frac{24 \text{ hr}}{1 \text{ d}} \times \frac{60 \text{ min}}{1 \text{ hr}} = 8.410 \times 10^8 \text{ min}^{-1}$

$$k = \frac{0.693}{t_{1/2}} = \frac{0.693}{8.410 \times 10^8 \text{ min}} = 8.241 \times 10^{-10} \text{ min}^{-1}$$

Rate = kN = $(8.241 \times 10^{-10} \text{ min}^{-1})(2.6646 \times 10^{19} \text{ atoms}) = 2.20 \times 10^{10} \text{ atoms/min}$

$(2.20 \times 10^{10} \text{ atoms/min })(5.0 \text{ min}) = 1.1 \times 10^{11} {}^{226}Ra$ atoms decay in 5.0 min

1.1×10^{11} α particles emitted in 5.0 min

    (b)   *Plan.* From part (a), the rate is 2.20×10^{10} disintegrations/min. Change this to dis/s and apply the definition 1 Ci = 3.7×10^{10} dis/s.

$$\frac{2.20 \times 10^{10} \text{ dis}}{1 \text{ min}} \times \frac{1 \text{ min}}{60 \text{ s}} \times \frac{1 \text{ Ci}}{3.7 \times 10^{10} \text{ dis/s}} \times \frac{1000 \text{ mCi}}{\text{Ci}} = 9.891 = 9.9 \text{ mCi}$$

21.39   *Analyze/Plan.* Calculate k in yr^{-1} and solve Equation [21.19] for t. N_o = 16.3/min/g, N_t = 9.7/min/g. *Solve.*

$k = 0.693/t_{1/2} = 0.693/5715 \text{ yr} = 1.213 \times 10^{-4} = 1.21 \times 10^{-4} \text{ yr}^{-1}$

$$t = \frac{-1}{k} \ln \frac{N_t}{N_o} = \frac{-1}{1.213 \times 10^{-4} \text{ yr}^{-1}} \ln \frac{9.7}{16.3} = 4.280 \times 10^3 = 4.3 \times 10^3 \text{ yr}$$

21.41   *Analyze/Plan.* Follow the procedure outlined in Sample Exercise 21.7. If the mass of ^{40}Ar is 4.2 times that of ^{40}K, the original mass of ^{40}K must have been 4.2 + 1 = 5.2 times the amount of ^{40}K present now. *Solve.*

$k = 0.693/1.27 \times 10^9 \text{ yr} = 5.457 \times 10^{-10} = 5.46 \times 10^{-10} \text{ yr}^{-1}$

$$t = \frac{-1}{5.457 \times 10^{-10} \text{ yr}^{-1}} \times \ln \frac{1}{(5.2)} = 3.0 \times 10^9 \text{ yr}$$

Energy Changes (section 21.6)

21.43   *Analyze/Plan.* Given a particle reaction, calculate Δm and ΔE. An α particle is $^{4}_{2}He$. Use Equation [21.22], E = mc². Compare the calculated energy to the energy given off by the thermite reaction. *Solve.*

$2 \, {}^{1}_{1}p + 2 \, {}^{1}_{0}n \rightarrow {}^{4}_{2}He$

Δm = mass of individual protons and neutrons – mass of $^{4}_{2}He$

Δm = 2(1.0072765 amu) + 2(1.0086649 amu) – 4.0015 amu = 0.030383 = 0.0304 amu

The mass change for the formation of 1 mol of $^4_2$He can be expressed as 0.0304 g.

$\Delta E = c^2 \Delta m; \quad 1 J = kg \text{-} m^2/s^2$

$\Delta E = 0.030383 \text{ g} \times \dfrac{1 \text{ kg}}{1000 \text{ g}} \times (2.9979 \times 10^8 \text{ m/s})^2 = \dfrac{2.73 \times 10^{12} \text{ kg-m}^2}{s^2} = 2.73 \times 10^{12} \text{ J}$

The energy released when one mole of Fe_2O_3 reacts is or 8.515×10^5 J. The energy released when one mole of $^4_2$He is formed from protons and neutrons is 2.73×10^{12} J. This is 3×10^6 or 3 million times as much energy as the thermite reaction.

21.45   *Analyze/Plan.* Given the mass of an ^{27}Al atom, subtract the mass of 13 electrons to get the mass of an ^{27}Al nucleus. Calculate the mass difference between the ^{27}Al nucleus and the separate nucleons, convert this to energy using Equation [21.22]. Use the molar mass of ^{27}Al to calculate the energy required for 100 g of ^{27}Al. *Solve.*

The mass of an electron is 5.485799×10^{-4} amu (inside back cover of the text). The mass of a ^{27}Al nucleus is then 26.9815386 amu $-13(5.485799 \times 10^{-4}$ amu $) = 26.9744071$ amu. $\Delta m = 13(1.0072765 \text{ amu}) + 14(1.0086649 \text{ amu}) - 26.9744071 \text{ amu} = 0.2414960 \text{ amu}.$

$\Delta E = (2.9979246 \times 10^8 \text{ m/s})^2 \times 0.2414960 \text{ amu} \times \dfrac{1 \text{ g}}{6.0221421 \times 10^{23} \text{ amu}} \times \dfrac{1 \text{ kg}}{1 \times 10^{23} \text{ g}}$

$= 3.604129 \times 10^{-11} = 3.604129 \times 10^{-11} \text{ J}/^{27}\text{Al nucleus required}$

If the mass change for a single ^{27}Al nucleus is 0.2414960 amu, the mass change for 1 mole of ^{27}Al is 0.2414960 g.

$\Delta E = 100 \text{ g } ^{27}\text{Al} \times \dfrac{1 \text{ mol } ^{27}\text{Al}}{26.9815386 \text{ g } ^{27}\text{Al}} \times \dfrac{0.241960}{\text{mol } ^{27}\text{Al}} \times \dfrac{1 \text{ kg}}{1000 \text{ g}} \times (2.9979246 \times 10^8 \text{ m/s})^2$

$= 8.044234 \times 10^{13} \text{ J} = 8.044234 \times 10^{10} \text{ kJ}/100 \text{ g } ^{27}\text{Al}$

21.47   *Analyze/Plan.* Given atomic mass, subtract mass of the electrons to get nuclear mass. Calculate the nuclear binding energy by finding the mass difference between the nucleus and the separate nucleons and converting this to energy using Equation [21.22]. Divide by the total number of nucleons to find binding energy per nucleon. *Solve.*

(a)   Nuclear mass

     ^{2}H: 2.014102 amu $- 1(5.485799 \times 10^{-4}$ amu) = 2.013553 amu

     ^{4}He: 4.002602 amu $- 2(5.485799 \times 10^{-4}$ amu) = 4.001505 amu

     ^{6}Li: 6.0151228 amu $- 3(5.4857991 \times 10^{-4}$ amu) = 6.0134771 amu

(b)   Nuclear binding energy

     ^{2}H: $\Delta m = 1(1.0072765) + 1(1.0086649) - 2.013553 = 0.002388$ amu

$\Delta E = 0.002388 \text{ amu} \times \dfrac{1 \text{ g}}{6.022142 \times 10^{23} \text{ amu}} \times \dfrac{1 \text{ kg}}{1000 \text{ g}} \times \dfrac{8.987551 \times 10^{16} \text{ m}^2}{s^2}$

$= 3.564490 \times 10^{-13} = 3.564 \times 10^{-13} \text{ J}$

^{4}He: $\Delta m = 2(1.0072765) + 2(1.0086649) - 4.001505 = 0.030378$ amu

$$\Delta E = 0.030378 \text{ amu} \times \frac{1\,g}{6.022142 \times 10^{23} \text{ amu}} \times \frac{1\,kg}{1000\,g} \times \frac{8.987551 \times 10^{16} \text{ m}^2}{\text{s}^2}$$

$$= 4.533636 \times 10^{-12} = 4.5336 \times 10^{-12} \text{ J}$$

^{6}Li: $\Delta m = 3(1.0072765) + 3(1.0086649) - 6.0134771 = 0.0343471$ amu

$$\Delta E = 0.0343471 \text{ amu} \times \frac{1\,g}{6.022142 \times 10^{23} \text{ amu}} \times \frac{1\,kg}{1000\,g} \times \frac{8.987551 \times 10^{16} \text{ m}^2}{\text{s}^2}$$

$$= 5.126021 \times 10^{-12} = 5.12602 \times 10^{-12} \text{ J}$$

(c)   Binding energy per nucleon

^{2}H: 3.564490×10^{-13} J/2 nucleons $= 1.782 \times 10^{-13}$ J/nucleon

^{4}He: 4.533636×10^{-12} J/4 nucleons $= 1.1334 \times 10^{-12}$ J/nucleon

^{6}Li: 5.126021×10^{-12} J/6 nucleons $= 8.54337 \times 10^{-13}$ J/nucleon

This trend in binding energy/nucleon agrees with the curve in Figure 21.12, which shows an irregular increase in binding energy/nucleon up to atomic number 56. The anomalously high value for ^{4}He calculated above is also apparent on the figure.

21.49   *Analyze/Plan.* Use Equation [21.22] to calculate the mass equivalence of the solar radiation.   *Solve.*

(a) $$\frac{1.07 \times 10^{16} \text{ kJ}}{1\,\text{min}} \times \frac{60\,\text{min}}{1\,h} \times \frac{24\,h}{1\,d} = 1.541 \times 10^{19} \frac{\text{kJ}}{d} = 1.54 \times 10^{22} \text{ J/d}$$

$$\Delta m = \frac{1.541 \times 10^{22} \text{ kg-m}^2/\text{s}^2/d}{(2.998 \times 10^8 \text{ m/s})^2} = 1.714 \times 10^5 = 1.71 \times 10^5 \text{ kg/d}$$

(b)   *Analyze/Plan.* Calculate the mass change in the given nuclear reaction, then a conversion factor for g ^{235}U to mass equivalent.   *Solve.*

$\Delta m = 140.8833 + 91.9021 + 2(1.0086649) - 234.9935 = -0.19077 = -0.1908$ amu

Converting from atoms to moles and amu to grams, it requires 1.000 mol or 235.0 g ^{235}U to produce energy equivalent to a change in mass of 0.1908 g.

0.10% of 1.714×10^5 kg is 1.714×10^2 kg $= 1.714 \times 10^5$ g

$$1.714 \times 10^5 \text{ g} \times \frac{235.0\,\text{g }^{235}\text{U}}{0.1908\,\text{g}} = 2.111 \times 10^8 = 2.1 \times 10^8 \text{ g }^{235}\text{U}$$

(This is about 230 tons of ^{235}U **per day**.)

21.51   We can use Figure 21.12 to see that the binding energy per nucleon (which gives rise to the mass defect) is greatest for nuclei of mass numbers around 50. Thus (a) $^{59}_{27}$Co should possess the greatest mass defect per nucleon.

Effects and Uses of Radioisotopes (sections 21.7-21.9)

21.53　(a)　NaI is a good source of iodine, because it is a strong electrolyte and completely dissociated into ions in aqueous solution. The $I^-(aq)$ are mobile and immediately available for bio-uptake. They do not need to be digested or processed in the body before uptake can occur. Also, iodine is a large percentage of the total mass of NaI.

(b)　After ingestion, $I^-(aq)$ must enter the bloodstream, travel to the thyroid and then be absorbed. This requires a finite amount of time. A Geiger counter placed near the thyroid immediately after ingestion will register background, then gradually increase in signal until the concentration of $I^-(aq)$ in the thyroid reaches a maximum. Then, over time, iodine-131 decays, and the signal decreases.

(c)　*Analyze/Plan.* The half-life of iodine-131 is 8.02 days. Use $t_{1/2}$ to calculate the decay rate constant, k. Then solve Equation [21.19] for t. $N_o = 0.12$ (12% of ingested iodine absorbed); $N_t = 0.0001$ (0.01% of the original ingested amount).

Solve.　$k = 0.693 \, t_{1/2} = 0.693/8.02 \, d = 0.086409 = 0.0864 \, d^{-1}$

$\ln(N_t/N_o) = -kt; \; t = -\ln(N_t/N_o)/k$

$t = \dfrac{-\ln(0.0001/0.12)}{0.086409 \, d^{-1}} = 82.05 = 82 \, d$

Check. N_t is given to 1 sig fig, so 8×10^1 (70 – 90) days is a more correct representation of the time frame for decay.

21.55　^{235}U

21.57　The *control rods* in a nuclear reactor regulate the flux of neutrons to keep the reaction chain self-sustaining and also prevent the reactor core from overheating. They are composed of materials such as boron or cadmium that absorb neutrons.

21.59　(a)　$^2_1H + ^2_1H \rightarrow ^3_2He + ^1_0n$

(b)　$^{239}_{92}U + ^1_0n \rightarrow ^{133}_{51}Sb + ^{98}_{41}Nb + 9\,^1_0n$

21.61　(a)　*Analyze/Plan.* At these temperatures, assume the reaction occurs between nuclei rather than atoms. From Table 21.7, the nuclear mass of ^4_2He is 4.00150. The nuclear mass of ^1_1H is simply the mass of a proton, 1.007276467 amu. Note that ^0_1e is a positron, which has the same mass as an electron, 5.4857991×10^{-4} amu. Calculate the difference in mass between product and reactant nuclei, and the energy released by this mass change. Do the calculation in terms of moles and grams, rather than nuclei and amu. 1 mol amu = 1 g. *Solve.*

$\Delta m = 4.00150 + 2(5.4857991 \times 10^{-4}) - 4(1.007276467) = -0.0265087 = -0.02651 \, g$

If the reaction is run with 1 mol of ^1_1H, the mass change is $(0.0265087/4) = 0.0066272 = 0.006627 \, g$

$$\Delta E = c^2 \Delta m = 0.0066272 \text{ g} \times \frac{1 \text{ kg}}{1000 \text{ g}} \times \frac{8.987551 \times 10^{16} \text{ m}^2}{\text{s}^2} = 5.95621 \times 10^{11}$$

$$= 5.956 \times 10^{11} \text{ J} = 5.956 \times 10^8 \text{ kJ}$$

(b) The extremely high temperature is required to overcome the electrostatic charge repulsions between the nuclei so that they come together to react.

21.63 (a) A *boiling water reactor* does not use a secondary coolant.

(b) A *fast breeder reactor* creates more fissionable material than it consumes.

(c) A *gas-cooled reactor* uses a gas as a primary coolant.

21.65 *Analyze/Plan.* Hydroxyl radical is electrically neutral but has an unpaired electron, •OH. Hydroxide is an anion, OH⁻. *Solve.*

Hydrogen abstraction: $RCOOH + \bullet OH \rightarrow RCOO\bullet + H_2O$

Deprotonation: $RCOOH + OH^- \rightarrow RCOO^- + H_2O$

Hydroxyl radical is more toxic to living systems, because it produces other radicals when it reacts with molecules in the organism. This often starts a disruptive chain of reactions, each producing a different free radical.

Hydroxide ion, OH⁻, on the other hand, will be readily neutralized in the buffered cell environment. Its most common reaction is ubiquitous and innocuous: $H^+ + OH^- \rightarrow H_2O$. The acid-base reactions of OH⁻ are usually much less disruptive to the organism than the chain of redox reactions initiated by •OH radical.

21.67 *Analyze/Plan.* Use definitions of the various radiation units and conversion factors to calculate the specified quantities. Pay particular attention to units.   *Solve.*

(a) $1 \text{ Ci} = 3.7 \times 10^{10}$ disintegrations(dis)/s; 1 Bq = 1 dis/s

$$14.3 \text{ mCi} \times \frac{1 \text{ Ci}}{1000 \text{ m Ci}} \times \frac{3.7 \times 10^{10} \text{ dis/s}}{\text{Ci}} = 5.29 \times 10^8 = 5.3 \times 10^8 \text{ dis/s} = 5.3 \times 10^8 \text{ Bq}$$

(b) $1 \text{ rad} = 1 \times 10^{-2}$ J/kg; 1 Gy = 1 J/kg = 100 rad. From part (a), the activity of the source is 5.3×10^8 dis/s.

$$5.29 \times 10^8 \text{ dis/s} \times 14.0 \text{ s} \times 0.35 \times \frac{9.12 \times 10^{-13} \text{ J}}{\text{dis}} \times \frac{1}{0.385 \text{ kg}} = 6.14 \times 10^{-3} = 6.1 \times 10^{-3} \text{ J/kg}$$

$$6.1 \times 10^{-3} \text{ J/kg} \times \frac{1 \text{ rad}}{1 \times 10^{-2} \text{ J/kg}} \times \frac{1000 \text{ mrad}}{\text{rad}} = 6.1 \times 10^2 \text{ mrad}$$

$$6.1 \times 10^{-3} \text{ J/kg} \times \frac{1 \text{ Gy}}{1 \text{ J/kg}} = 6.1 \times 10^{-3} \text{ Gy}$$

(c) rem = rad (RBE); Sv = Gy (RBE) , where 1 Sv = 100 rem

mrem = 6.14×10^2 mrad (9.5) = $5.83 \times 10^3 = 5.8 \times 10^3$ mrem (or 5.8 rem)

Sv = 6.14×10^{-3} Gy (9.5) = $5.83 \times 10^{-2} = 5.8 \times 10^{-2}$ Sv

Additional Exercises

21.69 $^{222}_{86}Rn \rightarrow X + 3\ ^{4}_{2}He + 2\ ^{0}_{-1}\beta$

This corresponds to a reduction in mass number of $(3 \times 4 =)$ 12 and a reduction in atomic number of $(3 \times 2 - 2) = 4$. The stable nucleus is $^{210}_{82}Pb$. (This is part of the sequence in Figure 21.5.)

21.71 (a) $^{36}_{17}Cl \rightarrow\ ^{36}_{18}Ar +\ ^{0}_{-1}e$

(b) According to Table 21.3, nuclei with even numbers of both protons and neutrons, or an even number of one kind of nucleon, are more stable. ^{35}Cl and ^{37}Cl both have an odd number of protons **but** an even number of neutrons. ^{36}Cl has an odd number of protons and neutrons (17 p, 19 n), so it is less stable than the other two isotopes. Also, ^{37}Cl has 20 neutrons, a nuclear closed shell.

21.73 (a) $^{6}_{3}Li +\ ^{56}_{28}Ni \rightarrow\ ^{62}_{31}Ga$

(b) $^{40}_{20}Ca +\ ^{248}_{96}Cm \rightarrow\ ^{147}_{62}Sm +\ ^{141}_{54}Xe$

(c) $^{88}_{38}Sr +\ ^{84}_{36}Kr \rightarrow\ ^{116}_{46}Pd +\ ^{56}_{28}Ni$

(d) $^{40}_{20}Ca +\ ^{238}_{92}U \rightarrow\ ^{70}_{30}Zn + 4\ ^{1}_{0}n + 2\ ^{102}_{41}Nb$

21.77 The C—OH bond of the acid and the O—H bond of the alcohol break in this reaction. Initially, ^{18}O is present in the C—^{18}OH group of the alcohol. In order for ^{18}O to end up in the ester, the ^{18}O—H bond of the alcohol must break. This requires that the C—OH bond in the acid also breaks. The unlabeled O from the acid ends up in the H_2O product.

21.79 Because of the relationship $\Delta E = \Delta mc^2$, the mass defect (Δm) is directly related to the binding energy (ΔE) of the nucleus.

^7Be: 4p, 3n; $4(1.0072765) + 3(1.0086649) = 7.05510$ amu

Total mass defect = $7.0551 - 7.0147 = 0.0404$ amu

0.0404 amu/7 nucleons = 5.77×10^{-3} amu/nucleon

$$\Delta E = \Delta m \times c^2 = \frac{5.77 \times 10^{-3}\ \text{amu}}{\text{nucleon}} \times \frac{1g}{6.022 \times 10^{23}\ \text{amu}} \times \frac{1\ kg}{1 \times 10^3\ g} \times \frac{8.988 \times 10^{16}\ m^2}{\sec^2}$$

$$= \frac{5.77 \times 10^{-3}\ \text{amu}}{\text{nucleon}} \times \frac{1.4925 \times 10^{-10}\ J}{1\ \text{amu}} = 8.612 \times 10^{-13} = 8.61 \times 10^{-13}\ \text{J/nucleon}$$

^9Be: 4p, 5n; $4(1.0072765) + 5(1.0086649) = 9.07243$ amu

Total mass defect = $9.0724 - 9.0100 = 0.06243 = 0.0624$ amu

0.0624 amu/9 nucleons = $6.937 \times 10^{-3} = 6.94 \times 10^{-3}$ amu/nucleon

6.937×10^{-3} amu/nucleon $\times 1.4925 \times 10^{-10}$ J/amu $= 1.035 \times 10^{-12} = 1.04 \times 10^{-12}$ J/nucleon

^{10}Be: 4p, 6n; 4(1.0072765) + 6(1.0086649) = 10.0811 amu

Total mass defect = 10.0811 – 10.0113 = 0.0698 amu

0.0698 amu/10 nucleons = 6.98×10^{-3} amu/nucleon

6.98×10^{-3} amu/nucleon $\times$ 1.4925×10^{-10} J/amu = 1.042×10^{-12} = 1.04×10^{-12} J/nucleon

The binding energies/nucleon for ^{9}Be and ^{10}Be are very similar; that for ^{10}Be is slightly higher.

Integrative Exercises

21.85   Calculate the amount of energy produced by the nuclear fusion reaction, the enthalpy of combustion, $\Delta H°$, of C_8H_{18}, and then the mass of C_8H_{18} required.

Δm for the reaction $4\,^1_1H \rightarrow\ ^4_2He + 2\,^0_1e$ is :

$4(1.00782) - 4.00260$ amu $- 2(5.4858 \times 10^{-4}$ amu$)$ = 0.027583 = 0.02758 amu

$\Delta E = \Delta mc^2 = 0.027583$ amu $\times \dfrac{1\,g}{6.02214 \times 10^{23}\ amu} \times \dfrac{1\,kg}{1000\,g} \times (2.9979246 \times 10^8\ m/s)^2$

$$= 4.11654 \times 10^{-12} = 4.117 \times 10^{-12}\ J/4\ ^1H\ nuclei$$

$1.0\,g\,^1H \times \dfrac{1\,^1H\ nucleus}{1.00782\ amu} \times \dfrac{6.02214 \times 10^{23}\ amu}{g} \times \dfrac{4.11654 \times 10^{-12}\ J}{4\ ^1H\ nuclei}$

$$= 6.1495 \times 10^{11}\ J = 6.1 \times 10^8\ kJ\ \text{produced by the fusion of 1.0 g }^1H.$$

$C_8H_{18}(l) + 25/2\,O_2(g) \rightarrow 8CO_2(g) + 9H_2O(g)$

$\Delta H° =\ 8(-393.5\ kJ) + 9(-241.82\ kJ) - (-250.1\ kJ) = -5074.3\ kJ$

$6.1495 \times 10^8\ kJ \times \dfrac{1\ mol\ C_8H_{18}(l)}{5074.3\ kJ} \times \dfrac{114.231\ g\ C_8H_{18}}{mol\ C_8H_{18}} = 1.384 \times 10^7\ g = 1.4 \times 10^4\ kg\ C_8H_{18}$

14,000 kg C_8H_{18}(l) would have to be burned to produce the same amount of energy as fusion of 1.0 g ^1H.

$$0.18\ Ci \times \dfrac{3.7 \times 10^{10}\ dis/s}{Ci} \times \dfrac{3600\ s}{hr} \times \dfrac{24\ hr}{d} \times 245\ d = 1.41 \times 10^{17} = 1.4 \times 10^{17}\ \alpha\ \text{particles}$$

22 Chemistry of the Nonmetals

Visualizing Concepts

22.1 (a) C_2H_4, the structure on the left, is the stable compound. Carbon, with a relatively small covalent radius owing to its location in the second row of the periodic chart, is able to closely approach other atoms. This close approach enables significant π overlap, so carbon can form strong multiple bonds to satisfy the octet rule. Silicon, in the third row of the periodic table, has a covalent radius too large for significant π overlap. Si does not form stable multiple bonds and Si_2H_4 is unstable.

(b) There are 3 electron domains about each central C atoms in C_2H_4. The geometry about these atoms is trigonal planar.

22.3 *Analyze.* The structure is a trigonal bipyramid where one of the five positions about the central atom is occupied by a lone pair, often called a see-saw.

Plan A: Count the valence electrons in each molecule, draw a correct Lewis structure, and count the electron domains about the central atom.

Plan B: Molecules (a)–(d) each contain four F atoms bound to a central atom through a single bond (F is unlikely to form multiple bonds because of its high electronegativity). This represents 16 electron pairs; the fifth position is occupied by a lone pair, for a total of 17 e⁻ pairs. A valence e⁻ count for (a)–(d) will tell us which molecules are likely to have the designated structure. Molecule (e), $HClO_4$, is not exactly of the type AX_4, so a Lewis structure will be required. *Solve.*

(a) XeF_4 36 e⁻, 16 e⁻ pairs. Plan B predicts that this molecule **will not** adopt the see-saw structure.

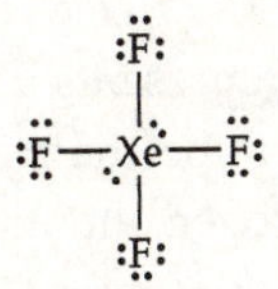

6 e⁻ domains about the Xe
octahedral domain geometry
square planar structure

(b) BrF_4^+ 34 e⁻, 17 e⁻ pairs; structure **will** be see-saw.

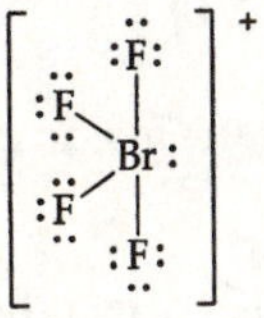

5 e⁻ domains about Br trigonal
bipyramidal domain geometry
see-saw structure

(c)   SiF$_4$ 32 e$^-$, 16 e$^-$ pairs; structure **will not** be see-saw

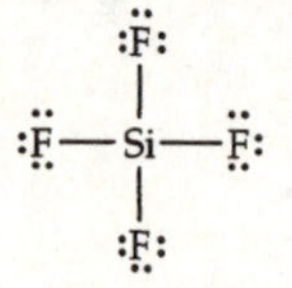

4 e$^-$ domains about Si
tetrahedral domain geometry and structure

(d)   TeCl$_4$ 34 e$^-$, 17 e$^-$ pairs; structure **will be** see-saw

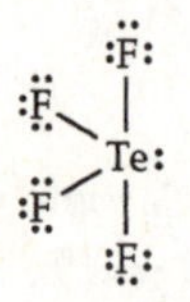

5 e$^-$ domains about Te
trigonal bipyramidal domain geometry
see-saw structure

(e)   HClO$_4$ 32 e$^-$, 16 e$^-$ pairs; structure **will not** be see-saw

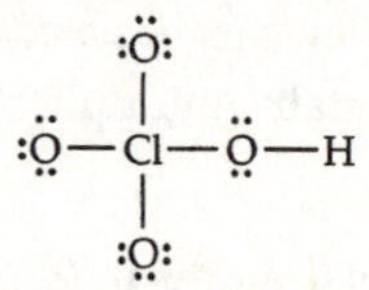

(HClO$_4$ is an oxyacid, so H is bound to O, not Cl.
Other Lewis structures that optimize formal charges
are possible; structure predictions are the same.)

22.6   The graph is applicable only to (c) density. Density depends on both atomic mass and volume (radius). Both increase going down a family, but atomic mass increases to a greater extent. Density, the ratio of mass to volume, increases going down the family; this trend is consistent with the data in the figure.

According to periodic trends, (a) electronegativity and (b) first ionization energy both decrease rather than increase going down the family. According to Table 22.5 both (d) X—X single bond enthalpy and (e) electron affinity are somewhat erratic, with the trends decreasing from S to Po, and anomalous values for the properties of O, probably owing to its small covalent radius.

22.9   The compound on the left, with the strained three-membered ring, will be the most generally reactive. For central atoms with four electron domains*, idealized bond angles are 109°. From left to right, the bond angles in the three molecules pictured are 60°, 90° and 108°. The larger the deviation from ideal bond angles, the more strain in the molecule and the more generally reactive it is.

*For the stick structures shown in the exercise, each line represents a C—C single bond and the intersection of two lines is a C atom. To determine the number of electron domains about each atom, visualize or draw the hydrogen atoms and nonbonded electron pairs in each molecule. Alternatively, note that both C and O atoms form only single bonds, so hybridization must be sp^3 and idealized bond angles are 109°.

Periodic Trends and Chemical Reactions (section 22.1)

22.11   *Analyze/Plan.* Use the color-coded periodic chart on the front-inside cover of the text to classify the given elements.   *Solve.*

Metals: (b) Sr, (c) Mn, (e) Na;   nonmetals: (a) P, (d) Se, (f) Kr;   metalloids: none

22.13   *Analyze/Plan.* Follow the logic in Sample Exercise 22.1.   *Solve.*

    (a)   O   (b)   Br   (c)   Ba

    (d)   O   (e)   Co   (f)   Br

22.15   *Analyze/Plan.* Use the position of the specified elements on the periodic chart, periodic trends, and the arguments in Sample Exercise 22.1 to explain the observations.   *Solve.*

    (a)   Nitrogen is too small to accommodate five fluorine atoms about it. The P and As atoms are larger. Furthermore, P and As have available 3d and 4d orbitals, respectively, to form hybrid orbitals that can accommodate more than an octet of electrons about the central atom.

    (b)   Si does not readily form π bonds, which would be necessary to satisfy the octet rule for both atoms in SiO.

    (c)   A reducing agent is a substance that readily loses electrons. As has a lower electronegativity than N; that is, it more readily gives up electrons to an acceptor and is more easily oxidized.

22.17   *Analyze/Plan.* Follow the logic in Sample Exercise 22.2.   *Solve.*

    (a) $NaOCH_3(s) + H_2O(l) \rightarrow NaOH(aq) + CH_3OH(aq)$

    (b) $CuO(s) + 2HNO_3(aq) \rightarrow Cu(NO_3)_2(aq) + H_2O(l)$

    (c) $WO_3(s) + 3H_2(g) \rightarrow W(s) + 3H_2O(g)$

    (d) $4NH_2OH(l) + O_2(g) \rightarrow 6H_2O(l) + 2N_2(g)$

    (e) $Al_4C_3(s) + 12H_2O(l) \rightarrow 4Al(OH)_3(s) + 3CH_4(g)$

Hydrogen, the Noble Gases, and the Halogens (sections 22.2-22.4)

22.19   *Analyze/Plan.* Use information on the isotopes of hydrogen in Section 22.2 to list their symbols, names, and relative abundances.   *Solve.*

    (a) $^{1}_{1}H$ - protium; $^{2}_{1}H$ - deuterium; $^{3}_{1}H$ - tritium

    (b)   The order of abundance is proteum > deuterium > tritium.

    (c) $^{3}_{1}H$ - tritium is radioactive.

    (d) $^{3}_{1}H \rightarrow {}^{3}_{2}He + {}^{0}_{-1}e$

22.21   *Analyze/Plan.* Consider the electron configuration of hydrogen and the Group 1A elements. *Solve.*

    Like other elements in group 1A, hydrogen has only one valence electron and its most common oxidation number is +1.

22.23   *Analyze/Plan.* Use information on the descriptive chemistry of hydrogen in Section 22.2 to formulate the required equations. Steam is $H_2O(g)$.   *Solve.*

    (a) $Mg(s) + 2H^+(aq) \rightarrow Mg^{2+}(aq) + H_2(g)$

(b) $C(s) + H_2O(g) \xrightarrow{1000°C} CO(g) + H_2(g)$

(c) $CH_4(g) + H_2O(g) \xrightarrow{1100°C} CO(g) + 3H_2(g)$

22.25   *Analyze/Plan.* Use information on the descriptive chemistry of hydrogen given in Section 22.2 to complete and balance the equations.   *Solve.*

(a) $NaH(s) + H_2O(l) \rightarrow NaOH(aq) + H_2(g)$

(b) $Fe(s) + H_2SO_4(aq) \rightarrow Fe^{2+}(aq) + H_2(g) + SO_4{}^{2-}(aq)$

(c) $H_2(g) + Br_2(g) \rightarrow 2HBr(g)$

(d) $2Na(l) + H_2(g) \rightarrow 2NaH(s)$

(e) $PbO(s) + H_2(g) \xrightarrow{\Delta} Pb(s) + H_2O(g)$

22.27   *Analyze/Plan.* If the element bound to H is a nonmetal, the hydride is molecular. If H is bound to a metal with integer stoichiometry, the hydride is ionic; with noninteger stoichiometry, the hydride is metallic.   *Solve.*

(a)   ionic (metal hydride)

(b)   molecular (nonmetal hydride)

(c)   metallic (nonstoichiometric transition metal hydride)

22.29   Vehicle fuels produce energy via combustion reactions. The reaction $H_2(g) + 1/2\,O_2(g) \rightarrow H_2O(g)$ is very exothermic, producing 242 kJ per mole of H_2 burned. The only product of combustion is H_2O, a nonpollutant (but like CO_2, a greenhouse gas).

22.31   *Analyze/Plan.* Consider the periodic properties of Xe and Ar.   *Solve.*

Xenon is larger, and can more readily accommodate an expanded octet. More important is the lower ionization energy of xenon; because the valence electrons are a greater average distance from the nucleus, they are more readily promoted to a state in which the Xe atom can form bonds with fluorine.

22.33   *Analyze/Plan.* Follow the rules for assigning oxidation numbers in Section 4.4 and the logic in Sample Exercise 4.8.   *Solve.*

(a) $Ca(OBr)_2$, Br, +1   (b) $HBrO_3$, Br, +5   (c)   **Xe**O_3, Xe, +6

(d) $ClO_4{}^-$, Cl, +7   (e) HIO_2, I, +3   (f)   **IF**$_5$; I, +5; F, –1

22.35   *Analyze/Plan.* Review the nomenclature rules and ion names in Section 2.8, as well as the rules for assignming oxidation numbers in Section 4.4.   *Solve.*

(a)   iron(III) chlorate, Cl, +5   (b)   chlorous acid, Cl, +3

(c)   xenon hexafluoride, F, –1   (d)   bromine pentafluoride; Br, +5; F, –1

(e)   xenon oxide tetrafluoride, F, –1   (f)   iodic acid, I, +5

22.37   *Analyze/Plan.* Consider intermolecular forces and periodic properties, including oxidizing power, of the listed substances.   *Solve.*

(a)   Van der Waals intermolecular attractive forces increase with increasing numbers of electrons in the atoms.

(b) F_2 reacts with water: $F_2(g) + H_2O(l) \rightarrow 2HF(aq) + 1/2\ O_2(g)$. That is, fluorine is too strong an oxidizing agent to exist in water.

(c)   HF has extensive hydrogen bonding.

(d)   Oxidizing power is related to electronegativity. Electronegativity decreases in the order given.

Oxygen and the Group 6A Elements (sections 22.5 and 22.6)

22.39   *Analyze/Plan.* Use information on the descriptive chemistry of oxygen given in Section 22.5 to complete and balance the equations.   *Solve.*

(a) $2HgO(s) \overset{\Delta}{\rightarrow} 2Hg(l) + O_2(g)$

(b) $2Cu(NO_3)_2(s) \overset{\Delta}{\rightarrow} 2CuO(s) + 4NO_2(g) + O_2(g)$

(c) $PbS(s) + 4O_3(g) \rightarrow PbSO_4(s) + 4O_2(g)$

(d) $2ZnS(s) + 3O_2(g) \overset{\Delta}{\rightarrow} 2ZnO(s) + 2SO_2(g)$

(e) $2K_2O_2(s) + 2CO_2(g) \rightarrow 2K_2CO_3(s) + O_2(g)$

(f) $3O_2(g) \overset{h\nu}{\rightarrow} 2O_3(g)$

22.41   *Analyze/Plan.* Oxides of metals are bases, oxides of nonmetals are acids, oxides that act as both acids and bases are amphoteric and oxides that act as neither acids nor bases are neutral.   *Solve.*

(a)   acidic (oxide of a nonmetal)

(b)   acidic (oxide of a nonmetal)

(c)   amphoteric

(d)   basic (oxide of a metal)

22.43   *Analyze/Plan.* Follow the rules for assigning oxidation numbers in Section 4.4 and the logic in Sample Exercise 4.8.   *Solve.*

(a) H_2SeO_3, +4   (b) $KHSO_3$, +4   (c) H_2Te, –2   (d) CS_2, –2

(e) $CaSO_4$, +6   (f) CdS, –2   (g) $ZnTe$, –2

Oxygen (a group 6A element) is in the –2 oxidation state in compounds (a), (b), and (e).

22.45   *Analyze/Plan.* The half-reaction for oxidation in all these cases is:

$H_2S(aq) \rightarrow S(s) + 2H^+ + 2e^-$ (The product could be written as $S_8(s)$, but this is not necessary. In fact it is not necessarily the case that S_8 would be formed, rather than some other allotropic form of the element.) Combine this half-reaction with the given reductions to write complete equations. The reduction in (c) happens only in acid solution. The reactants in (d) are acids, so the medium is acidic.   *Solve.*

(a) $2Fe^{3+}(aq) + H_2S(aq) \rightarrow 2Fe^{2+}(aq) + S(s) + 2H^+(aq)$

(b) $Br_2(l) + H_2S(aq) \rightarrow 2Br^-(aq) + S(s) + 2H^+(aq)$

(c) $2MnO_4^-(aq) + 6H^+(aq) + 5H_2S(aq) \rightarrow 2Mn^{2+}(aq) + 5S(s) + 8H_2O(l)$

(d) $2NO_3^-(aq) + H_2S(aq) + 2H^+(aq) \rightarrow 2NO_2(aq) + S(s) + 2H_2O(l)$

22.47   *Analyze/Plan.* For each substance, count valence electrons, draw the correct Lewis structure, and apply the rules of VSEPR to decide electron domain geometry and geometric structure.   *Solve.*

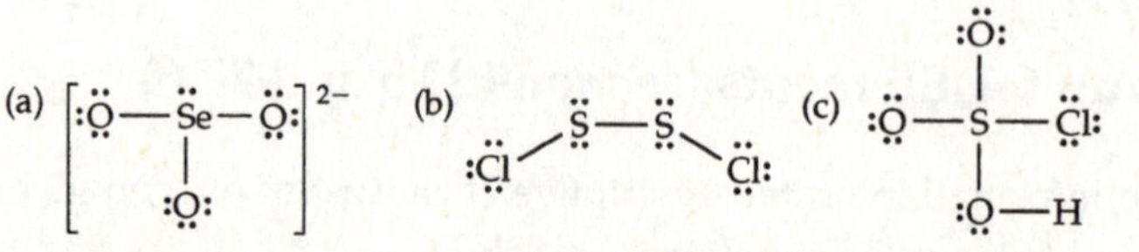

(a) trigonal pyramidal   (b) bent (free rotation around S-S bond)   (c) tetrahedral

22.49   *Analyze/Plan.* Use information on the descriptive chemistry of sulfur given in Section 22.6 to complete and balance the equations.   *Solve.*

(a) $SO_2(s) + H_2O(l) \rightarrow H_2SO_3(aq) \rightleftharpoons H^+(aq) + HSO_3^-(aq)$

(b) $ZnS(s) + 2HCl(aq) \rightarrow ZnCl_2(aq) + H_2S(g)$

(c) $8SO_3^{2-}(aq) + S_8(s) \rightarrow 8S_2O_3^{2-}(aq)$

(d) $SO_3(aq) + H_2SO_4(l) \rightarrow H_2S_2O_7(l)$

Nitrogen and the Group 5A Elements (sections 22.7 and 22.8)

22.51   *Analyze/Plan.* Follow the rules for assigning oxidation numbers in Section 4.4 and the logic in Sample Exercise 4.8.   *Solve.*

(a) $NaNO_2$, +3   (b) NH_3, -3   (c) N_2O, +1   (d) $NaCN$, -3

(e) HNO_3, +5   (f) NO_2, +4   (g) N_2, 0   (h) BN, -3

22.53   *Analyze/Plan.* For each substance, count valence electrons, draw the correct Lewis structure, and apply the rules of VSEPR to decide electron domain geometry and geometric structure.   *Solve.*

(a)

The molecule is bent around the central oxygen and nitrogen atoms; the four atoms need not lie in a plane. The right-most form does not minimize formal charges and is less important in the actual bonding model. The oxidation state of N is +3.

(b)

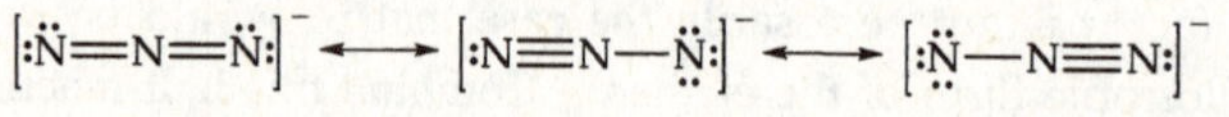

The molecule is linear. The oxidation state of N is −1/3.

(c)

$$\left[\begin{array}{c} H \quad H \\ | \quad | \\ H—N—N: \\ | \quad | \\ H \quad H \end{array} \right]^{+}$$

(d)

$$\left[\begin{array}{c} :\overset{..}{O}: \\ | \\ :\overset{..}{O}—N=\overset{..}{O} \end{array} \right]^{-}$$

The geometry is tetrahedral around the left nitrogen, trigonal pyramidal around the right. The oxidation state of N is −2.

(three equivalent resonance forms) The ion is trigonal planar. The oxidation state of N is +5.

22.55   *Analyze/Plan.* Use information on the descriptive chemistry of nitrogen given in Section 22.7 to complete and balance the equations.   *Solve.*

(a) $Mg_3N_2(s) + 6H_2O(l) \rightarrow 2NH_3(g) + 3Mg(OH)_2(s)$

Because $H_2O(l)$ is a reactant, the state of NH_3 in the products could be expressed as $NH_3(aq)$.

(b) $2NO(g) + O_2(g) \rightarrow 2NO_2(g)$, redox reaction

(c) $N_2O_5(g) + H_2O(l) \rightarrow 2H^+(aq) + 2NO_3^-(aq)$

(d) $NH_3(aq) + H^+(aq) \rightarrow NH_4^+(aq)$

(e) $N_2H_4(l) + O_2(g) \rightarrow N_2(g) + 2H_2O(g)$, redox reaction

22.57   *Analyze/Plan.* Follow the method for writing balanced half-reactions given in Section 20.2 and Sample Exercises 20.2 and 20.3   *Solve.*

(a) $HNO_2(aq) + H_2O(l) \rightarrow NO_3^-(aq) + 3H^+(aq) + 2e^-,\ E_{red}^\circ = 0.96\ V$

(b) $N_2(g) + H_2O(l) \rightarrow N_2O(g) + 2H^+(aq) + 2e^-,\ E_{red}^\circ = 1.77\ V$

22.59   *Analyze/Plan.* Follow the rules for assigning oxidation numbers in Section 4.4 and the logic in Sample Exercise 4.8.   *Solve.*

(a) H_3PO_3, +3   (b) $H_4P_2O_7$, +5   (c)   **SbCl$_3$**, +3

(d) Mg_3**As**$_2$, −3   (e) P_2O_5, +5   (f) Na_3PO_4, +5

22.61   *Analyze/Plan.* Consider the structures of the compounds of interest when explaining the observations.   *Solve.*

(a)   Phosphorus is a larger atom and can more easily accommodate five surrounding atoms and an expanded octet of electrons than nitrogen can. Also, P has energetically "available" 3d orbitals which participate in the bonding, but nitrogen does not.

(b)   Only one of the three hydrogens in H_3PO_2 is bonded to oxygen. The other two are bonded directly to phosphorus and are not easily ionized because the P—H bond is not very polar.

(c) PH_3 is a weaker base than H_2O (PH_4^+ is a stronger acid than H_3O^+). Any attempt to add H^+ to PH_3 in the presence of H_2O merely causes protonation of H_2O.

(d)   Refer to the structures of white and red phosphorus in Figure 22.27. White phosphorus consists of P_4 molecules, with P—P—P bond angles of 60°. Each P atom has four VSEPR pairs of electrons, so the predicted electron pair geometry is tetrahedral and the preferred bond angle is 109°. Because of the severely strained bond angles in P_4 molecules, white phosphorus is highly reactive. Red phosphorus is a chain of groups of four P atoms. It has fewer severely strained P—P—P bond angles and is less reactive than white phosphorus.

22.63   *Analyze/Plan.* Use information on the descriptive chemistry of phosphorus given in Section 22.8 to complete and balance the equations.   *Solve.*

(a) $2Ca_3(PO_4)_2(s) + 6SiO_2(s) + 10C(s) \overset{\Delta}{\rightarrow} P_4(g) + 6CaSiO_3(l) + 10CO(g)$

(b) $PBr_3(l) + 3H_2O(l) \rightarrow H_3PO_3(aq) + 3HBr(aq)$

(c) $4PBr_3(g) + 6H_2(g) \rightarrow P_4(g) + 12HBr(g)$

Carbon, the Other Group 4A Elements, and Boron (sections 22.9-22.11)

22.65   *Analyze/Plan.* Review the nomenclature rules and ion names in Section 2.8.   *Solve.*
(a)   HCN   (b) $Ni(CO)_4$ (c) $Ba(HCO_3)_2$ (d) CaC_2 (e) K_2CO_3

22.67   *Analyze/Plan.* Use information on the descriptive chemistry of carbon given in Section 22.9 to complete and balance the equations.   *Solve.*

(a) $ZnCO_3(s) \overset{\Delta}{\rightarrow} ZnO(s) + CO_2(g)$

(b) $BaC_2(s) + 2H_2O(l) \rightarrow Ba^{2+}(aq) + 2OH^-(aq) + C_2H_2(g)$

(c) $2C_2H_2(g) + 5O_2(g) \rightarrow 4CO_2(g) + 2H_2O(g)$

(d) $CS_2(g) + 3O_2(g) \rightarrow CO_2(g) + 2SO_2(g)$

(e) $Ca(CN)_2(s) + 2HBr(aq) \rightarrow CaBr_2(aq) + 2HCN(aq)$

22.69   *Analyze/Plan.* Use information on the descriptive chemistry of carbon given in Section 22.9 to complete and balance the equations.   *Solve.*

(a) $2CH_4(g) + 2NH_3(g) + 3O_2(g) \xrightarrow[\text{cat}]{800°C} 2HCN(g) + 6H_2O(g)$

(b) $NaHCO_3(s) + H^+(aq) \rightarrow CO_2(g) + H_2O(l) + Na^+(aq)$

(c) $2BaCO_3(s) + O_2(g) + 2SO_2(g) \rightarrow 2BaSO_4(s) + 2CO_2(g)$

22.71   *Analyze/Plan.* Follow the rules for assigning oxidation numbers in Section 4.4 and the logic in Sample Exercise 4.8.   *Solve.*

(a) H_3BO_3, +3   (b) $SiBr_4$, +4   (c) $PbCl_2$, +2

(d) $Na_2B_4O_7 \cdot 10H_2O$, +3   (e) B_2O_3, +3   (f) GeO_2, +4

22.73   *Analyze/Plan.* Consider periodic trends within a family, particularly metallic character, as well as descriptive chemistry in Sections 22.9 and 22.10.   *Solve.*

(a)   Tin; see Table 22.8. The filling of the 4f subshell at the beginning of the sixth row of the periodic table increases Z and Z_{eff} for later elements. This causes the ionization energy of Pb to be greater than that of Sn.

(b)   Carbon, silicon, and germanium; these are the nonmetal and metalloids in group 4A. They form compounds ranging from XH_4 (–4) to XO_2 (+4). The metals tin and lead are not found in negative oxidation states.

(c)   Silicon; silicates are the main component of sand.

22.75   *Analyze/Plan.* Consider the structural chemistry of silicates discussed in Section 22.10 and shown in Figure 22.34.   *Solve.*

(a)   Tetrahedral

(b)   Metasilicic acid will probably adopt the single-strand silicate chain structure shown in Figure 22.34(b). The empirical formula shows 3 O and 2 H atoms per Si atom. The chain has the same Si to O ratio as metasilicic acid. Furthermore, in the chain structure, there are two terminal (not bridging) O atoms on each Si. These can accommodate the 2 H atoms associated with each Si atom of the acid. The sheet structure does not fulfill these requirements.

22.77   (a)   Diborane (Figure 22.36 and below) has bridging H atoms linking the two B atoms. The structure of ethane shown below has the C atoms bound directly, with no bridging atoms.

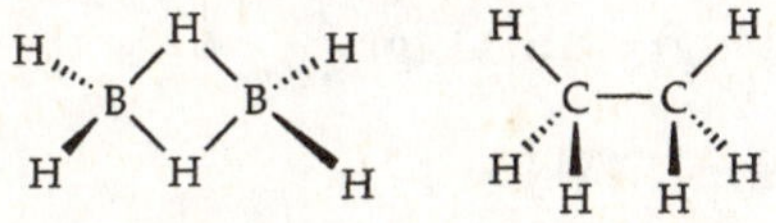

(b) B_2H_6 is an electron deficient molecule. It has 12 valence electrons, while C_2H_6 has 14 valence electrons. The 6 valence electron pairs in B_2H_6 are all involved in B—H sigma bonding, so the only way to satisfy the octet rule at B is to have the bridging H atoms shown in Figure 22.36.

(c)   A hydride ion, H^-, has two electrons while an H atom has one. The term *hydridic* indicates that the H atoms in B_2H_6 have more than the usual amount of electron density for a covalently bound H atom.

Additional Exercises

22.81   (a) $SO_2(g) + H_2O(l) \rightleftharpoons H_2SO_3$ (aq)

(b) $Cl_2O_7(g) + H_2O(l) \rightleftharpoons 2HClO_4(aq)$

(c) $Na_2O_2(s) + 2H_2O(l) \rightarrow H_2O_2(aq) + 2NaOH(aq)$

(d) $BaC_2(s) + 2H_2O(l) \rightarrow Ba^{2+}(aq) + 2OH^-(aq) + C_2H_2(g)$

(e) $2RbO_2(s) + 2H_2O(l) \rightarrow 2Rb^+(aq) + 2OH^-(aq) + O_2(g) + H_2O_2(aq)$

(f) $Mg_3N_2(s) + 6H_2O(l) \rightarrow 3Mg(OH)_2(s) + 2NH_3(g)$

(g) $NaH(s) + H_2O(l) \rightarrow NaOH(aq) + H_2(g)$

22.85   (a) PO_4^{3-}, + 5; NO_3^-, + 5

(b)   The Lewis structure for NO_4^{3-} would be:

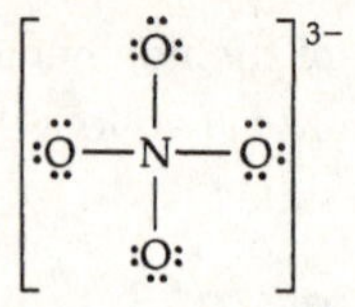

$$\left[\begin{array}{c} :\ddot{O}: \\ | \\ :\ddot{O}-N-\ddot{O}: \\ | \\ :\ddot{O}: \end{array}\right]^{3-}$$

The formal charge on N is +1 and on each O atom is –1. The four electronegative oxygen atoms withdraw electron density, leaving the nitrogen deficient. Since N can form a maximum of four bonds, it cannot form a π bond with one or more of the O atoms to regain electron density, as the P atom in PO_4^{3-} does. Also, the short $N{-}O$ distance would lead to a tight tetrahedron of O atoms subject to steric repulsion.

Integrative Exercises

22.89 (a) $100.0 \times 10^3 \text{ g FeTi} \times \dfrac{1 \text{ mol FeTi}}{103.7 \text{ g FeTi}} \times \dfrac{1 \text{ mol H}_2}{1 \text{ mol FeTi}} \times \dfrac{2.016 \text{ g H}_2}{1 \text{ mol H}_2} = 1944.1 = 1.94 \times 10^3 \text{ g H}_2$

(b) $V = \dfrac{1944.1 \text{ g H}_2}{2.016 \text{ g/mol H}_2} \times \dfrac{0.08206 \text{ L-atm}}{\text{mol-K}} \times \dfrac{273 \text{ K}}{1 \text{ atm}} = 21,603 = 2.16 \times 10^4 \text{ L H}_2$

(c) $2H_2(g) + O_2(g) \rightarrow 2H_2O(l)$

$\Delta H° = 2\,\Delta H_f° \ H_2O(l) - 2\,\Delta H_f° \ H_2(g) - \Delta H_f° \ O_2(g)$

$\Delta H° = 2(-285.83) - 2(0) - (0) = -571.66 \text{ kJ}$

$1944.1 \text{ g H}_2 \times \dfrac{1 \text{ mol H}_2}{2.016 \text{ g H}_2} \times \dfrac{-571.66 \text{ kJ}}{2 \text{ mol H}_2} = -275,636 = -2.76 \times 10^5 \text{ kJ}$

The minus sign indicates that energy is produced.

22.91 (a) $H_2(g) + 1/2\,O_2(g) \rightarrow H_2O(l);\ \Delta H = -285.83 \text{ kJ/mol H}_2$

$CH_4(g) + 2O_2(g) \rightarrow CO_2(g) + 2H_2O(l)$

$\Delta H = 2(-285.83) - 393.5 - (-74.8) = -890.4 \text{ kJ/ mol CH}_4$

(b) for H_2: $\dfrac{-285.83 \text{ kJ}}{1 \text{ mol H}_2} \times \dfrac{1 \text{ mol H}_2}{2.0159 \text{ g H}_2} = -141.79 \text{ kJ/g H}_2$

for CH_4: $\dfrac{-890.4 \text{ kJ}}{1 \text{ mol CH}_4} \times \dfrac{1 \text{ mol CH}_4}{16.043 \text{ g CH}_4} = -55.50 \text{ kJ/g CH}_4$

(c) Find the number of moles of gas that occupy 1 m³ at STP:

$n = \dfrac{1 \text{ atm} \times 1 \text{ m}^3}{273 \text{ K}} \times \dfrac{1 \text{ mol-K}}{0.08206 \text{ L-atm}} \times \left[\dfrac{100 \text{ cm}}{1 \text{ m}}\right]^3 \times \dfrac{1 \text{ L}}{10^3 \text{ cm}^3} = 44.64 \text{ mol}$

for H_2: $\dfrac{-285.83 \text{ kJ}}{1 \text{ mol H}_2} \times \dfrac{44.64 \text{ mol H}_2}{1 \text{ m}^3 \text{ H}_2} = -1.276 \times 10^4 \text{ kJ/m}^3 \text{ H}_2$

for CH_4: $\dfrac{-890.4 \text{ kJ}}{1 \text{ mol CH}_4} \times \dfrac{44.64 \text{ mol CH}_4}{1 \text{ m}^3 \text{ CH}_4} = -3.975 \times 10^4 \text{ kJ/m}^3 \text{ CH}_4$

22.95   (a) $SO_2(g) + 2H_2S(aq) \rightarrow 3S(s) + 2H_2O(g)$ or, if we assume S_8 is the product,

$8SO_2(g) + 16H_2S(aq) \rightarrow 3S_8(s) + 16H_2O(g)$.

(b)   Assume that all S in the coal becomes SO_2 upon combustion, so that

1 mol S (coal) = 1 mol SO_2; 1 ton = 2000 lb; 760 torr = 1.00 atm

$$4000 \text{ lb coal} \times \frac{0.035 \text{ lb S}}{1 \text{ lb coal}} \times \frac{453.6 \text{ g S}}{1 \text{ lb S}} \times \frac{1 \text{ mol S (coal)}}{32.07 \text{ g S}} \times \frac{1 \text{ mol } SO_2}{1 \text{ mol S (coal)}} \times \frac{2 \text{ mol } H_2S}{1 \text{ mol } SO_2}$$

$$= 3960 = 4.0 \times 10^3 \text{ mol } H_2S$$

$$V = \frac{3960 \text{ mol} \times (0.08206 \text{ L - atm/mol - K}) \times 300 \text{ K}}{1.00 \text{ atm}} = 97,496 = 9.7 \times 10^4 \text{ L}$$

(c) $3960 \text{ mol } H_2S \times \dfrac{3 \text{ mol S}}{2 \text{ mol } H_2S} \times \dfrac{32.07 \text{ g S}}{1 \text{ mol S}} = 1.9 \times 10^5 \text{ g S}$

This is about 210 lb S per ton of coal combusted. (However, two-thirds of this comes from the H_2S, which presumably at some point was also obtained from coal.)

22.97   The reactions can be written as follows:

$$H_2(g) + X(\text{std state}) \rightarrow H_2X(g) \qquad \Delta H_f^\circ$$
$$2H(g) \rightarrow H_2(g) \qquad \Delta H_f^\circ(H-H)$$
$$X(g) \rightarrow X(\text{std state}) \qquad \Delta H_3$$

$$\overline{\text{Add}: 2H(g) + X(g) \rightarrow H_2X(g) \qquad \Delta H = \Delta H_f^\circ + \Delta H_f^\circ (H-H) + \Delta H_3}$$

These are all the necessary ΔH values. Thus,

Compound	$\underline{\Delta H}$	$\underline{D\ H-X}$
H_2O	$\Delta H = -241.8 \text{ kJ} - 436 \text{ kJ} - 248 \text{ kJ} = -926 \text{ kJ}$	463 kJ
H_2S	$\Delta H = -20.17 \text{ kJ} - 436 \text{ kJ} - 277 \text{ kJ} = -733 \text{ kJ}$	367 kJ
H_2Se	$\Delta H = +29.7 \text{ kJ} - 436 \text{ kJ} - 227 \text{ kJ} = -633 \text{ kJ}$	317 kJ
H_2Te	$\Delta H = +99.6 \text{ kJ} - 436 \text{ kJ} - 197 \text{ kJ} = -533 \text{ kJ}$	267 kJ

The average $H-X$ bond energy in each case is just half of ΔH. The $H-X$ bond energy decreases steadily in the series. The origin of this effect is probably the increasing size of the orbital from X with which the hydrogen 1s orbital must overlap.

22.101   First write the balanced equation to give the number of moles of gaseous products per mole of hydrazine.

(A) $(CH_3)_2NNH_2 + 2N_2O_4 \rightarrow 3N_2(g) + 4H_2O(g) + 2CO_2(g)$

(B) $(CH_3)HNNH_2 + 5/4 \, N_2O_4 \rightarrow 9/4 \, N_2(g) + 3H_2O(g) + CO_2(g)$

In case (A) there are nine moles gas per one mole $(CH_3)_2NNH_2$ plus two moles N_2O_4. The total mass of reactants is 60 + 2(92) = 244 g. Thus, there are

$$\frac{9 \text{ mol gas}}{244 \text{ g reactants}} = \frac{0.0369 \text{ mol gas}}{1 \text{ g reactants}}$$

In case (B) there are 6.25 moles of gaseous product per one mole $(CH_3)HNNH_2$ plus 1.25 moles N_2O_4. The total mass of this amount of reactants is 46.0 + 1.25(92.0) = 161 g.

$$\frac{6.25 \text{ mol gas}}{161 \text{ g reactants}} = \frac{0.0388 \text{ mol gas}}{1 \text{ g reactants}}$$

Thus the methylhydrazine (B) has marginally greater thrust.

22.103   (a) $3B_2H_6(g) + 6NH_3(g) \rightarrow 2(BH)_3(NH)_3(l) + 12H_2(g)$

           $3LiBH_4(s) + 3NH_4Cl(s) \rightarrow 2(BH)_3(NH)_3(l) + 9H_2(g) + 3LiCl(s)$

   (b)   30 valence e^-, 15 e^- pairs

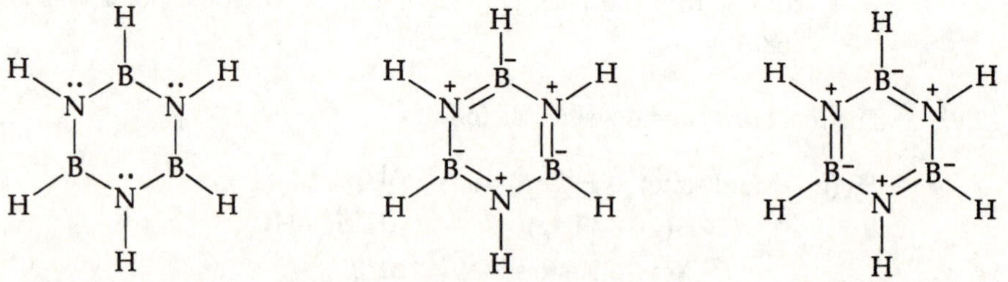

The structure with nonbonded pairs minimizes formal charge, but these electrons are almost certainly delocalized about the six-membered ring, mimicking the bonding in benzene.

   (c) $n = \dfrac{PV}{RT} = \dfrac{1.00 \text{ atm} \times 2.00 \text{ L}}{273 \text{ K}} \times \dfrac{\text{mol-K}}{0.08206 \text{ L-atm}} = 0.08929 = 8.93 \times 10^{-2} \text{ mol NH}_3$

$0.08929 \text{ mol NH}_3 \times \dfrac{2 \text{ mol (BH)}_3(NH)_3}{6 \text{ mol NH}_3} \times \dfrac{80.50 \text{ g (BH)}_3(NH)_3}{1 \text{ mol (BH)}_3(NH)_3}$

$$= 2.3956 = 2.40 \text{ g (BH)}_3(NH)_3$$

23 Transition Metals and Coordination Chemistry

Visualizing Concepts

23.2 *Analyze.* Given the formula of a coordination compound, determine the coordination geometry, coordination number, and oxidation state of the metal.

Plan. From the formula, determine the identity of the ligands and the number of coordination sites they occupy. From the total coordination number, decide on a likely geometry. Use ligand and overall complex charges to calculate the oxidation number of the metal.

Solve.

(a) The ligands are $2Cl^-$, one coordination site each, and en, ethylenediamine, two coordination sites, for a coordination number of 4. This coordination number has two possible geometries, tetrahedral and square planar. Pt is one of the metals known to adopt square planar geometry when CN = 4.

(b) CN = 4, coordination geometry = square planar

(c) $Pt(en)Cl_2$ is a neutral compound, the en ligand is neutral, and the $2Cl^-$ ligands are each –1, so the oxidation state of Pt must be +2, Pt(II).

23.4 *Analyze.* Given a ball-and-stick structure, name the complex ion, which has a 1– charge.

Plan. Write the chemical formula of the complex ion, determine the oxidation state of the metal, and name the complex.

Solve. $[Pt(NH_3)Cl_3]^-$. Oxidation numbers: $[Pt + 0 + 3(-1) = -1, Pt = +2, Pt(II)$

Arrange the ligands alphabetically, followed y the metal. Since the complex is an anion, add the suffix -ate, then the oxidation state of the metal: aminotrichloroplatinate(II)

23.6 *Analyze.* Given four structures, decide which are chiral.

Plan. Chiral molecules have nonsuperimposable mirror images. Draw the mirror image of each molecule and visualize whether it can be rotated into the original molecule. If so, the complex is not chiral. If the original orientation cannot be regenerated by rotation, the complex is chiral. *Solve.*

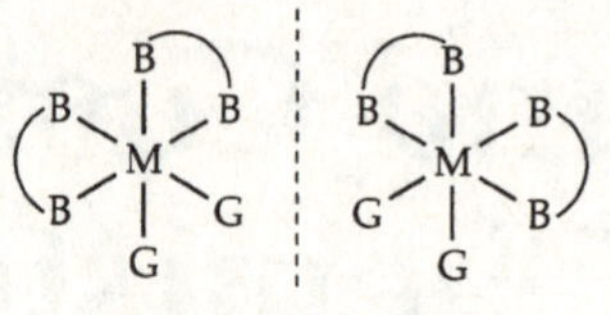

The two orientations are not superimposable and molecule (1) is chiral.

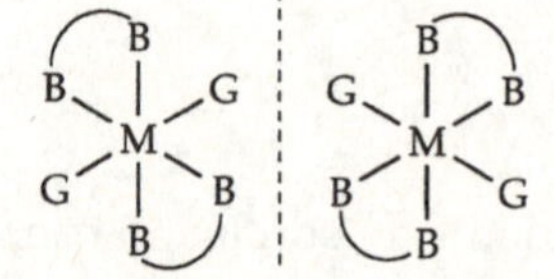

The two orientations are superimposible. Rotate the right-most structure 90° counterclockwise about the B-M-B axis to align the G's; the bidentate ligands then also overlap. Molecule (2) is not chiral.

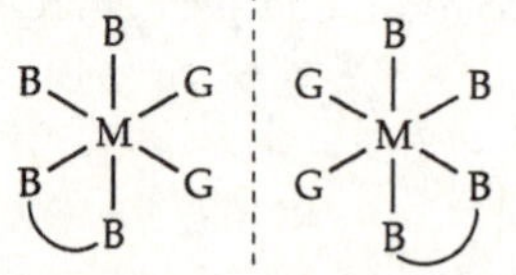

The two orientations are not superimposable and molecule (3) is chiral.

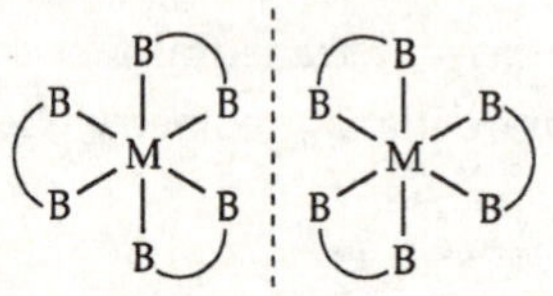

The two orientations are not superimposable and molecule (4) is chiral.

23.8   *Analyze.* Fit the crystal field splitting diagram to the complex description in each part.

Plan. Determine the number of d-electrons in each transition metal. On the splitting diagrams match the d-orbital splitting patterns to complex geometry and electron pairing to the definition of high-spin and low-spin.

Solve. Octahedral complexes have the 3 lower, 2 higher splitting pattern, while tetrahedral complexes have the opposite 2 lower, 3 higher pattern. Low spin complexes favor electron pairing because of large d-orbital splitting. High-spin complexes have maximum occupancy because of small orbital splitting.

(a) Fe^{3+}, 5 d-electrons; weak field: spins unpaired; octahedral: 3 lower, 2 higher d-splitting ∴ diagram (4)

(b) Fe^{3+}, 5 d-electrons; strong field: spins paired; octahedral: 3 lower, 2 higher d-splitting ∴ diagram (1)

(c) Fe^{3+}, 5 d-electrons; tetrahedral: 2 lower, 3 higher d-splitting ∴ diagram (3)

(d) Ni^{2+}, 8 d-electrons; tetrahedral: 2 lower, 3 higher d-splitting ∴ diagram (2)

Check. Diagram (2) was the remaining choice for (d) and it fits the description.

23 Transition Metals and Coordination Chemistry

The Transition Metals (section 23.1)

23.11   *Analyze/Plan.* Define lanthanide contraction. Based on the definition, describe how it affects properties of the transition elements. *Solve.*

The *lanthanide contraction* is the name given to the decrease in atomic size due to the build-up in effective nuclear charge as we move through the lanthanides (elements 58–71) and beyond them. This effect offsets the expected increase in atomic size, decrease in ionization energy and increase in electron affinity going from period 5 to period 6 transition elements; it causes density to increase more dramatically than expected. Because of its effect on size, ionization energy and electron affinity, it causes the chemical properties of period 5 and period 6 elements in the same family to have chemical properties even more similar than we would expect.

23.13   All transition metal atoms have two s-electrons in their valence shell. Because these s-electrons are, on average, farther from the nucleus, they are the first electrons lost to ionization. Loss of these two s-electrons leads to the +2 oxidation state common for most of the transition metals.

23.15   (a) Ti^{3+}, $[Ar]3d^1$ (b) Ru^{2+}, $[Kr]4d^6$ (c) Au^{3+}, $[Xe]4f^{14}5d^8$ (d) Mn^{4+}, $[Ar]3d^3$

23.17   *Analyze/Plan.* Consider the definitions of paramagnetic and diamagnetic. *Solve.*

The unpaired electrons in a *paramagnetic* material cause it to be weakly attracted into a magnetic field. A *diamagnetic* material, where all electrons are paired, is very weakly repelled by a magnetic field.

23.19   *Analyze/Plan.* Consider the orientation of spins in various types of magnetic materials as shown in Figure 23.5.

The diagram shows a material with misaligned spins that become aligned in the direction of an applied magnetic field. This is a paramagnetic material.

Transition-Metal Complexes (section 23.2)

23.21   (a)   In Werner's theory, *primary valence* is the charge of the metal cation at the center of the complex. *Secondary valence* is the number of atoms bound or coordinated to the central metal ion. The modern terms for these concepts are oxidation state and coordination number, respectively. (Note that "'oxidation state" is a broader term than ionic charge, but Werner's complexes contain metal ions where cation charge and oxidation state are equal.)

(b)   Ligands are the Lewis base in metal-ligand interactions. As such, they must possess at least one unshared electron pair. NH_3 has an unshared electron pair but BH_3, with less than 8 electrons about B, has no unshared electron pair and cannot act as a ligand. In fact, BH_3 acts as a Lewis acid, an electron pair acceptor, because it is electron-deficient.

23.23   *Analyze/Plan.* Follow the logic in Sample Exercises 23.1 and 23.2.   *Solve.*

(a)   This compound is electrically neutral, and the NH_3 ligands carry no charge, so the charge on Ni must balance the –2 charge of the 2 Br^- ions. The charge and oxidation state of Ni is +2.

(b)   Since there are 6 NH_3 molecules in the complex, the likely coordination number is 6. In some cases Br^- acts as a ligand, so the coordination number could be other than 6.

(c)   Assuming that the 6 NH_3 molecules are the ligands, 2 Br^- ions are not coordinated to the Ni^{2+}, so 2 mol AgBr(s) will precipitate. (If one or both of the Br^- act as a ligand, the mol AgBr(s) would be different.)

23.25   *Analyze/Plan.* Count the number of donor atoms in each complex, taking the identity of polydentate ligands into account. Follow the logic in Sample Exercise 23.2 to obtain oxidation numbers of the metals.   Use Tables 23.4 and 23.5 to determine the number and kinds of donor atoms in the ligands of the complexes.

(a)   Coordination number = 4, oxidation number = +2; 4 Cl^-

(b)   5, +4; 4 Cl^-, 1 O^{2-}

(c)   6, +3; 4 N, 2 Cl^-

(d)   5, +2; 5 C. In CN^-, both C and N have an unshared electron pair. C is less electronegative and more likely to donate its unshared pair.

(e)   6, +3; 6 O. $C_2O_4^{2-}$ is a bidentate ligand; each ion is bound through 2 O atoms for a total of 6 O donor atoms.

(f)   4, +2; 4 N. en is a bidentate ligand bound through 2 N atoms.

Common Ligands in Coordination Chemistry (section 23.3)

23.27   (a)   A monodendate ligand binds to a metal in through one atom; a bidendate ligand binds through two atoms.

(b)   If a bidentate ligand occupies two coordination sites, three bidentate ligands fill the coordination sphere of a six-coordinate complex.

(c)   A tridentate ligand has at least three atoms with unshared electron pairs in the correct orientation to simultanously bind one or more metal ions.

23.29   *Analyze/Plan.* Given the formula of a coordination compound, determine the number of coordination sites occupied by the polydentate ligand. The coordination number of the complexes is either 4 or 6. Note the number of monodentate ligands and determine the number of coordination sites occupied by the polydentate ligands.   *Solve.*

(a)   *ortho*-phenanthroline, *o*-phen, is bidentate

(b)   oxalate, $C_2O_4^{2-}$, is bidentate

(c)   ethylenediaminetetraacetate, EDTA, is pentadentate

(d)   ethylenediamine, en, is bidentate

23 Transition Metals and Coordination Chemistry

Solutions to Red Exercises

23.31 (a) The term *chelate effect* means there is a special stability associated with formation of a metal complex containing a polydentate (chelate) ligand relative to a complex containing only monodentate ligands.

(b) When a single chelating ligand replaces two or more monodendate ligands, the number of free molecules in the system increases and the entropy of the system increases. Chemical reactions with $+\Delta S$ tend to be spontaneous, have negative ΔG, and large positive values of K.

(c) Polydentate ligands can be used to bind metal ions and prevent them from undergoing unwanted chemical reactions without removing them from solution. The polydentate ligand thus hides or *sequesters* the metal ion.

23.33 *Analyze/Plan.* Consider the structural or steric requirements for chelating ligands. Analyze the structure of the ligand in the figure to determine whether it meets these requirements. *Solve.*

The ligand shown in the figure is not typically a chelate. The entire molecule is planar; there is no "bend" in the central 6-membered ring that includes the two N atoms. The benzene rings on either side of the two N atoms inhibit their approach in the correct orientation for chelation.

Nomenclature and Isomerism in Coordination Chemistry (section 23.4)

23.35 *Analyze/Plan.* Given the name of a coordination compound, write the chemical formula. Refer to Tables 23.4 and 23.5 to find ligand formulas. Place the metal complex (metal ion + ligands) inside square brackets and the counter ion (if there is one) outside the brackets. *Solve.*

(a) $[Cr(NH_3)_6](NO_3)_3$ (b) $[Co(NH_3)_4CO_3]_2SO_4$ (c) $[Pt(en)_2Cl_2]Br_2$

(d) $K[V(H_2O)_2Br_4]$ (e) $[Zn(en)_2][HgI_4]$

23.37 *Analyze/Plan.* Follow the logic in Sample Exercise 23.4, paying attention to naming rules in Section 23.4. *Solve.*

(a) tetraamminedichlororhodium(III) chloride

(b) potassium hexachlorotitanate(IV)

(c) tetrachlorooxomolybdenum(VI)

(d) tetraaqua(oxalato)platinum(IV) bromide

23.39 *Analyze/Plan.* Consider the definitions of the various types of isomerism, and which of the complexes could exhibit isomerism of the specified type. *Solve.*

(a)

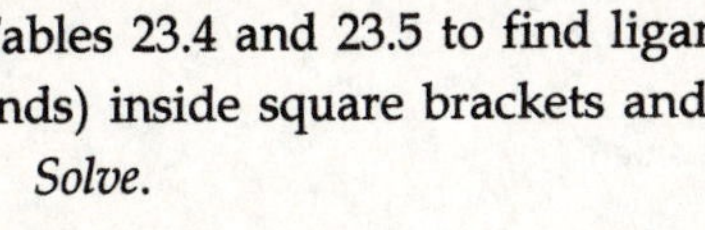

cis trans

(b) $[Pd(NH_3)_2(ONO)_2]$, $[Pd(NH_3)_2(NO_2)_2]$

(c)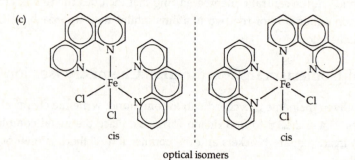

(d) $[Co(NH_3)_4Br_2]Cl$, $[Co(NH_3)_4BrCl]Br$

23.41 Yes. A tetrahedral complex of the form MA_2B_2 would have neither structural nor stereoisomers. For a tetrahedral complex, no differences in connectivity are possible for a single central atom, so the terms *cis* and *trans* do not apply. No optical isomers with tetrahedral geometry are possible because M is not bound to four different groups. The complex must be square planar with *cis* and *trans* geometric isomers.

23.43 *Analyze/Plan.* Follow the logic in Sample Exercise 23.5 and 23.6. *Solve.*

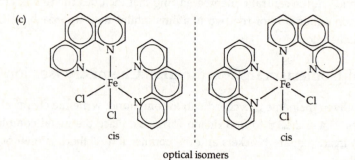

(The three isomeric complex ions in part (c) each have a 1+ charge.)

Color and Magnetism in Coordination Chemistry; Crystal-Field Theory
(sections 23.5 and 23.6)

23.45 (a) Visible light has wavelengths between 400 and 700 nm. We cannot see the light with 300 nm wavelength, but we can see the 500 nm light.

(b) *Complementary* colors are opposite each other on a color wheel such as Figure 23.25.

(c) A colored metal complex absorbs visible light of its complementary color. For example, a red complex absorbs green light.

(d) $E(J/photon) = hv = hc/\lambda$. Change $J/photon$ to kJ/mol.

$$E = \frac{6.626 \times 10^{-34} \text{ J-s}}{610 \text{ nm}} \times \frac{3.00 \times 10^8 \text{ m}}{\text{s}} \times \frac{1 \text{ nm}}{1 \times 10^{-9} \text{ m}} = 3.259 \times 10^{-19} = 3.26 \times 10^{-19} \text{ J}$$

$$\frac{3.259 \times 10^{-19} \text{ J}}{\text{photon}} \times \frac{1 \text{ kJ}}{1000 \text{ J}} \times \frac{6.022 \times 10^{23} \text{ photons}}{\text{mol}} = 196 \text{ kJ/mol}$$

23.47   No. The Fe^{2+} ion has 6 d-electrons. In an octahedral crystal field, the energies of the d orbitals are split three and two. That is, the d_{xy}, d_{xz} and d_{yz} orbitals are lower in energy than the free ion, and the $d_{x^2-y^2}$ and d_{z^2} orbitals are higher in energy. In a low-spin complex, the d-electrons are paired to the maximum possible extent. All 6 d-electrons in a low-spin octahedral complex will pair and occupy the low energy d_{xy}, d_{xz} and d_{yz} orbitals. With no unpaired electrons, the complex cannot be paramagnetic.

23.49   Most of the electrostatic interaction between a metal ion and a ligand is the attractive interaction between a positively charged metal cation and the full negative charge of an anionic ligand or the partial negative charge of a polar covalent ligand. Whether the interaction is ion-ion or ion-dipole, the ligand is strongly attracted to the metal center and can be modeled as a point negative charge.

23.51   (a)

(b)   The magnitude of Δ and the energy of the d-d transition for a d^1 complex are equal.

(c) $$\frac{6.626 \times 10^{-34} \text{ J-s}}{545 \text{ nm}} \times \frac{3.00 \times 10^8 \text{ m}}{\text{s}} \times \frac{1 \text{ nm}}{1 \times 10^{-9} \text{ m}} \times \frac{1 \text{ kJ}}{1000 \text{ J}} \times \frac{6.022 \times 10^{23} \text{ photons}}{\text{mol}}$$
$$= 220 \text{ kJ/mol}$$

23.53   *Analyze/Plan.* Consider the relationship between the color of a complex, the wavelength of absorbed light, and the position of a ligand in the spectrochemical series.   *Solve.*

A yellow color corresponds to absorption of a photon in the violet region of the visible spectrum, between 430 and 400 nm. The blue or green colors of aqua complexes correspond to absorptions in the region of 620 nm. The shorter wavelength corresponds to a higher-energy electron transition and larger Δ value. Cyanide is a stronger-field ligand, and its complexes are expected to have larger Δ values than aqua complexes. These are very general comparisons; other factors are involved, including whether the complex is high spin or low spin.

23.55   *Analyze/Plan.* Determine the charge on the metal ion, subtract it from the row number (3-12) of the transition metal, and the remainder is the number of d-electrons.   *Solve.*

(a) Ti^{3+}, d^1 (b) Co^{3+}, d^6 (c) Ru^{3+}, d^5

(d) Mo^{5+}, d^1 (e) Re^{3+}, d^4

23.57   Yes. A weak-field ligand leads to a small Δ value and a small d-orbital splitting energy. If the splitting energy of a complex is smaller than the energy required to pair electrons in an orbital, the complex is high-spin.

23.59   *Analyze/Plan.* Follow the logic in Sample Exercise 23.9.   *Solve.*

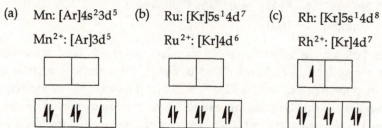

(a) Mn: $[Ar]4s^2 3d^5$ (b) Ru: $[Kr]5s^1 4d^7$ (c) Rh: $[Kr]5s^1 4d^8$

  Mn^{2+}: $[Ar]3d^5$ Ru^{2+}: $[Kr]4d^6$ Rh^{2+}: $[Kr]4d^7$

1 unpaired electron   0 unpaired electrons   1 unpaired electron

23.61   *Analyze/Plan.* All complexes in this exercise are six-coordinate octahedral. Use the definitions of high-spin and low-spin along with the orbital diagram from Sample Exercise 23.9 to place electrons for the various complexes.   *Solve.*

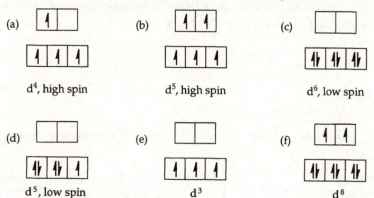

(a)   (b)   (c)

d^4, high spin d^5, high spin d^6, low spin

(d)   (e)   (f)

d^5, low spin d^3 d^8

23.63   *Analyze/Plan.* Follow the ideas but reverse the logic in Sample Exercise 24.9.   *Solve.*

high spin

Additional Exercises

23.67 $[Pt(NH_3)_6]Cl_4$; $[Pt(NH_3)_4Cl_2]Cl_2$; $[Pt(NH_3)_3Cl_3]Cl$; $[Pt(NH_3)_2Cl_4]$; $K[Pt(NH_3)Cl_5]$

23.71   (a)   Valence electrons: $2P + 6C + 16H = 10 + 24 + 16 = 50\ e^-$, 25 e$^-$ pr

$$
\begin{array}{c}
\quad\; H \qquad\quad H\;\; H \qquad\quad H \\
\quad\; | \qquad\qquad | \;\; | \qquad\qquad | \\
H-C-\ddot{P}-C-C-\ddot{P}-C-H \\
\quad\; | \qquad\quad | \;\; | \qquad\quad | \\
\quad\; H \qquad\quad H\; H \qquad\quad H \\
\quad\; H-C-H \;\; H-C-H \\
\qquad | \qquad\qquad\; | \\
\qquad H \qquad\qquad\; H
\end{array}
$$

Both dmpe and en are bidentate ligands. The dmpe ligand binds through P, while en binds through N. Phosphorus is less electronegative than N, so dmpe is a stronger electron pair donor and Lewis base than en. Dmpe creates a stronger ligand field and is higher on the spectrochemical series.

Structurally, P has a larger covalent radius than N, so M–P bonds are longer than M–N bonds. This is convenient because the two –CH₃ groups on each P atom in dmpe create more steric hindrance (bumping with adjacent atoms) than the H atoms on N in en.

(b)   CO and dmpe are neutral, $2CN^- = 2^-$, $2Na^+ = 2+$. The ion charges balance, so the oxidation state of Mo is zero.

(c)   The symbol $\overset{\frown}{P\ \ P}$ represents the bidentate dmpe ligand.

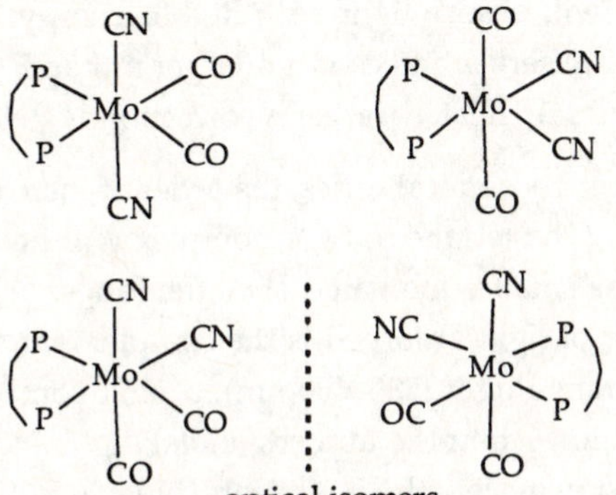

optical isomers

23.74   (a)   Hemoglobin is the iron-containing protein that transports O_2 in human blood.

(b)   Chlorophylls are magnesium-containing porphyrins in plants. They are the key components in the conversion of solar energy into chemical energy that can be used by living organisms.

(c)   Siderophores are iron-binding compounds or ligands produced by a microorganism. They compete on a molecular level for iron in the medium outside the organism and carry needed iron into the cells of the organism.

23.76   (a)   pentacarbonyliron(0)

(b)   Since CO is a neutral molecule, the oxidation state of iron must be zero.

(c) $[Fe(CO)_4CN]^-$ has two geometric isomers. In a trigonal bipyramid, the axial and equatorial positions are not equivalent and not superimposable. One isomer has CN in an axial position and the other has it in an equatorial position.

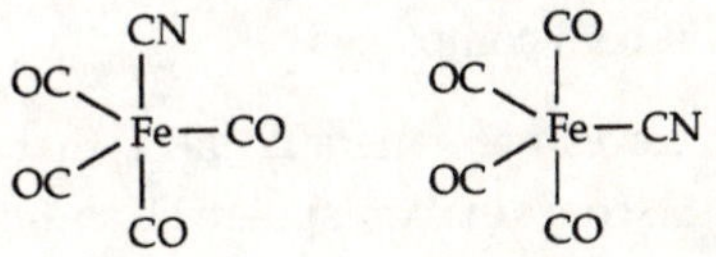

23.78 (a)

d^2

(b) These complexes are colored because the crystal-field splitting energy, Δ, is in the visible portion of the electromagnetic spectrum. Visible light with $\lambda = hc/\Delta$ is absorbed, promoting one of the d-electrons into a higher energy d-orbital. The remaining wavelengths of visible light are reflected or transmitted; the combination of these wavelengths is the color we see.

(c) $[V(H_2O)_6]^{3+}$ will absorb light with higher energy. H_2O is in the middle of the spectrochemical series, and causes a larger Δ than F^-, a weak-field ligand. Since Δ and λ are inversely related, larger Δ corresponds to higher energy and shorter λ.

23.80 According to the spectrochemical series, the order of increasing Δ for the ligands is $Cl^- < H_2O < NH_3$. (The tetrahedral Cl^- complex will have an even smaller Δ than an octahedral one.) The smaller the value of Δ, the longer the wavelength of visible light absorbed. The color of light absorbed is the complement of the observed color. A blue complex absorbs orange light (580–650 nm), a pink complex absorbs green light (490–560 nm) and a yellow complex absorbs violet light (400–430 nm). Since $[CoCl_4]^{2-}$ absorbs the longest wavelength, it appears blue. $[Co(H_2O)_6]^{2+}$ absorbs green and appears pink, and $[Co(NH_3)_6]^{3+}$ absorbs violet and appears yellow.

23.85 (a)

$$\left[\begin{array}{c} CO \\ NC \underset{\underset{CO}{|}}{\overset{|}{\underset{NC}{Fe}}} CN \\ CN \end{array}\right]^{2-}$$

(b) sodium dicarbonyltetracyanoferrate(II)

(c) +2, 6 d-electrons

(d) We expect the complex to be low spin. Cyanide (and carbonyl) are high on the spectrochemical series, which means the complex will have a large Δ splitting characteristic of low spin complexes.

Integrative Exercises

23.91 (a) Both compounds have the same general formulation, so Co is in the same (+3) oxidation state in both complexes.

(b) Cobalt(III) complexes are generally inert; that is, they do not rapidly exchange ligands inside the coordination sphere. Therefore, the ions that form precipitates in these two cases are probably outside the coordination sphere. The dark violet compound A forms a precipitate with $BaCl_2(aq)$ but not $AgNO_3(aq)$, so it has SO_4^{2-} outside the coordination sphere and coordinated Br^-, $[Co(NH_3)_5Br]SO_4$. The red-violet compound B forms a precipitate with $AgNO_3(aq)$ but not $BaCl_2(aq)$ so it has Br^- outside the coordination sphere and coordinated SO_4^{2-}, $[Co(NH_3)_5SO_4]Br$.

Compound A, dark violet   Compound B, red-violet

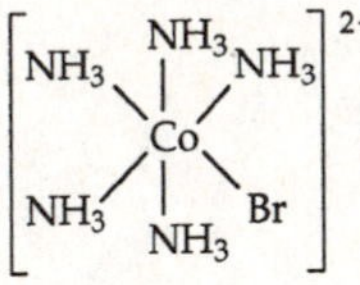

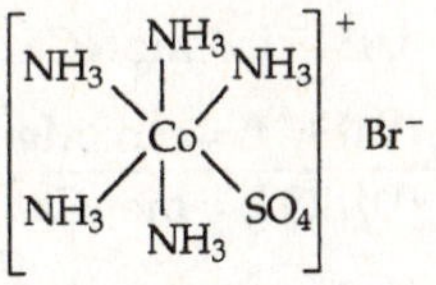

$$\left[\begin{array}{c} NH_3 \\ NH_3 \underset{Co}{\overset{NH_3}{\diagup}} NH_3 \\ NH_3 \quad Br \end{array}\right]^{2+} SO_4^{2-} \qquad \left[\begin{array}{c} NH_3 \\ NH_3 \underset{Co}{\overset{NH_3}{\diagup}} NH_3 \\ NH_3 \quad SO_4 \end{array}\right]^{+} Br^-$$

(c)   Compounds A and B have the same formula but different properties (color, chemical reactivity), so they are isomers. They vary by which ion is inside the coordination sphere, so they are *coordination sphere isomers*.

(d)   Compound A is an ionic sulfate and compound B is an ionic bromide, so both are strong electrolytes. According to the solubility rules in Table 4.1, both should be water-soluble.

23.94   First determine the empirical formula, assuming that the remaining mass of complex is Pd.

$$37.6 \text{ g Br} \times \frac{1 \text{ mol Br}}{79.904 \text{ g Br}} = 0.4706 \text{ mol Br}; \; 0.4706 / 0.2361 = 2$$

$$28.3 \text{ g C} \times \frac{1 \text{ mol C}}{12.01 \text{ g C}} = 2.356 \text{ mol C}; \; 2.356 / 0.2361 = 10$$

$$6.60 \text{ g N} \times \frac{1 \text{ mol N}}{14.01 \text{ g N}} = 0.4711 \text{ mol N}; \; 0.4711 / 0.2361 = 2$$

$$2.37 \text{ g H} \times \frac{1 \text{ mol H}}{1.008 \text{ g H}} = 2.351 \text{ mol H}; \; 2.351 / 0.2361 = 10$$

$$25.13 \text{ g Pd} \times \frac{1 \text{ mol Pd}}{106.42 \text{ g Pd}} = 0.2361 \text{ mol Pd}; \; 0.2361 / 0.2361 = 1$$

The chemical formula is $[Pd(NC_5H_5)_2Br_2]$. This should be a neutral square-planar complex of Pd(II), a nonelectrolyte. Because the dipole moment is zero, we can infer that it must be the trans isomer.

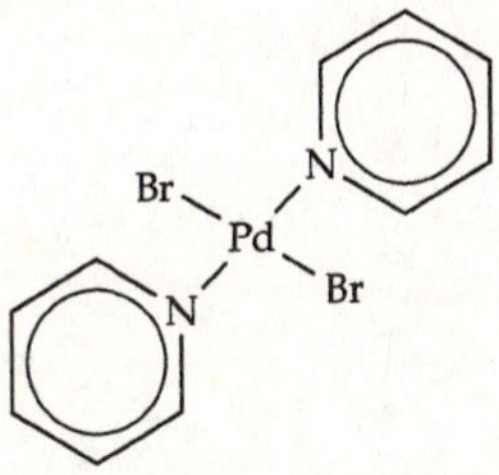

23.96   Calculate the concentration of Mg^{2+} alone, and then the concentration of Ca^{2+} by difference. $M \times L = mol$

$$\frac{0.0104 \text{ mol EDTA}}{1 \text{ L}} \times 0.0187 \text{ L} \times \frac{1 \text{ mol Mg}^{2+}}{1 \text{ mol EDTA}} \times \frac{24.31 \text{ g Mg}^{2+}}{1 \text{ mol Mg}^{2+}} \times \frac{1000 \text{ mg}}{\text{g}}$$

$$\times \frac{1}{0.100 \text{ L H}_2\text{O}} = 47.28 = 47.3 \text{ mg Mg}^{2+}/\text{L}$$

$0.0104 \, M \text{ EDTA} \times 0.0315 \text{ L} = \text{mol} (Ca^{2+} + Mg^{2+})$

$0.0104 \, M \text{ EDTA} \times 0.0187 \text{ L} = \text{mol Mg}^{2+}$

$0.0104 \, M \text{ EDTA} \times 0.0128 \text{ L} = \text{mol Ca}^{2+}$

$$0.0104 \text{ M EDTA} \times 0.0128 \text{ L} \times \frac{1 \text{ mol Ca}^{2+}}{1 \text{mol EDTA}} \times \frac{40.08 \text{ g Ca}^{2+}}{1 \text{ mol Ca}^{2+}} \times \frac{1000 \text{ mg}}{\text{g}} \times \frac{1}{0.100 \text{ L H}_2\text{O}}$$

$$= 53.35 = 53.4 \text{ mg Ca}^{2+}/\text{L}$$

23.99 $$\frac{182 \times 10^3 \text{ J}}{1 \text{ mol}} \times \frac{1 \text{ mol}}{6.022 \times 10^{23} \text{ molecules}} = 3.022 \times 10^{-19} = 3.02 \times 10^{-19} \text{ J/photon}$$

$\Delta E = h\nu = 3.02 \times 10^{-19} \text{ J}; \; \nu = \Delta E/h$

$\nu = 3.022 \times 10^{-19} \text{ J}/6.626 \times 10^{-34} \text{ J-s} = 4.561 \times 10^{14} = 4.56 \times 10^{14} \text{ s}^{-1}$

$$\lambda = \frac{2.998 \times 10^8 \text{ m/s}}{4.561 \times 10^{14} \text{ s}^{-1}} = 6.57 \times 10^{-7} \text{ m} = 657 \text{ nm}$$

We expect that this complex will absorb in the visible, at around 660 nm. It will thus exhibit a blue-green color (Figure 23.25).

24 The Chemistry of Life: Organic and Biological Chemistry

Visualizing Concepts

24.1 *Analyze/Plan.* Follow the logic in Sample Exercise 24.1 to name each compound. Decide which structures are the same compound. *Solve.*

(a) 2,2,4-trimethylpentane

(b) 3-ethyl-2-methylpentane

(c) 2,3,4-trimethylpentane

(d) 2,3,4-trimethylpentane

Structures (c) and (d) are the same molecule.

24.4 *Analyze/plan.* Given structural formulas, predict which molecule will have the highest boiling point. Boiling point is determined by strength of intermolecular forces; for neutral molecules with similar molar masses, the strongest intermolecular force is hydrogen bonding. The molecule that experiences hydrogen bonding will have the highest boiling point. *Solve.*

Only $F-H$, $O-H$, and $N-H$ bonds fit the strict definition of hydrogen bonding. Molecule (b), an alcohol, forms hydrogen bonds with like molecules; it has the highest boiling point.

[Molecules (a) and (d) have dipole-dipole and dispersion forces. The greater molar mass of (d) probably means it has stronger dispersion forces and a slightly higher boiling point than molecule (a). Molecule (c) has only dispersion forces and the lowest boiling point. The probable order of strength of forces and boiling points is: (b) > (d) > (a) > (c).]

Introduction to Organic Compounds; Hydrocarbons
(sections 24.1 and 24.2)

24.7 (a) sp^3 (b) sp^2 (c) sp^2 (d) sp

24.9 *Analyze/Plan.* Given a condensed structural formula, determine the bond angles and hybridization about each carbon atom in the molecule. Visualize the number of electron domains about each carbon. State the bond angle and hybridization based on electron domain geometry. *Solve.*

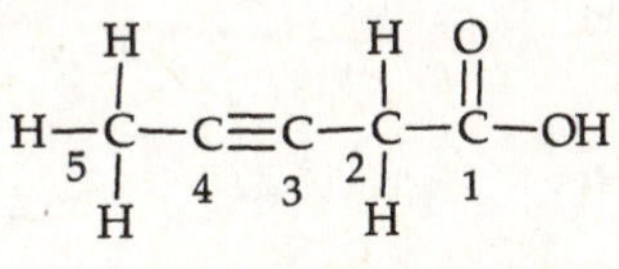

$$\underset{5}{\overset{\overset{\displaystyle H}{|}}{\underset{\underset{\displaystyle H}{|}}{C}}} - \underset{4}{C} \equiv \underset{3}{C} - \underset{2}{\overset{\overset{\displaystyle H}{|}}{\underset{\underset{\displaystyle H}{|}}{C}}} - \underset{1}{\overset{\overset{\displaystyle O}{||}}{C}} - OH$$

C1 has triagonal planar electron domain geometry, 120° bond angles and sp^2 hybridization. C3 and C4 have linear electron domain geometry, 180° bond angles and sp hybridization. C2 and C5 both have tetra-hedral electron domain geometry, 109° bond angles and sp^3 hybridization.

24.11 Ammonia, NH_3, contains no carbon, so it is not, strictly speaking, considered an organic molecule. Carbon monoxide contains a carbon atom that does not form four bonds, so it certainly is not a typical organic molecule.

24.13 (a) A *straight-chain alkane* has all carbon atoms connected in a continuous chain; no carbon atom is bound to more than two other carbon atoms. Carbon forms only single (sigma) bonds to other carbon or hydrogen atoms. A *branched-chain alkane* has a branch; at least one carbon atom is bound to three or more carbon atoms.

(b) An *alkane* is a complete molecule composed of carbon and hydrogen in which all bonds are single (sigma) bonds. An *alkyl group* is a substituent formed by removing a hydrogen atom from an alkane.

24.15 *Analyze/Plan.* Follow the rules for naming alkanes given in Section 24.2 and illustrated in Sample Exercise 24.1. *Solve.*

(a) 2-methylhexane

(b) 4-ethyl-2,4-dimethyldecane

(c)

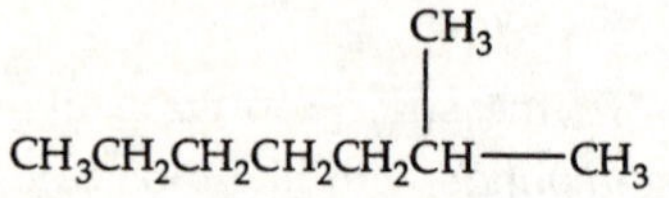

(d)

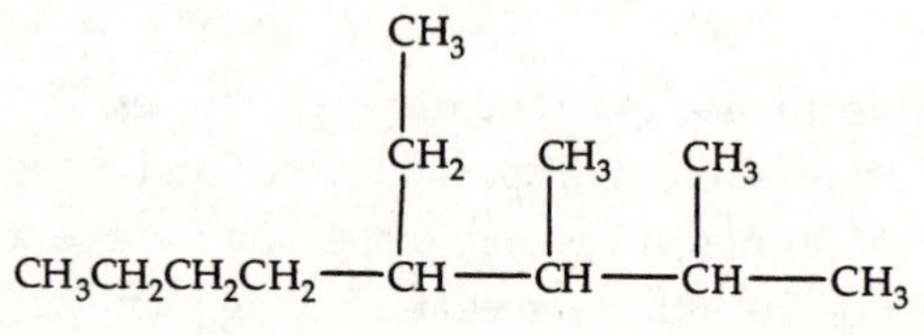

(e)

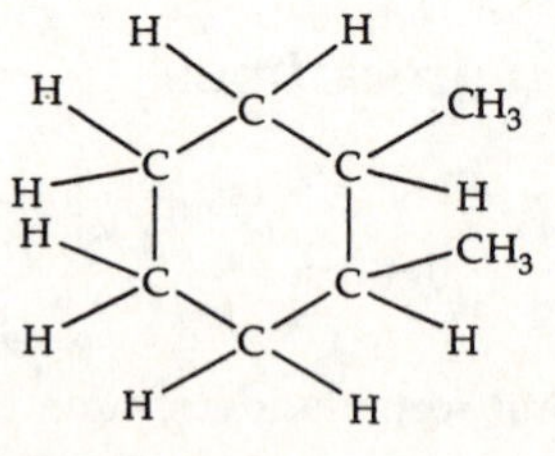

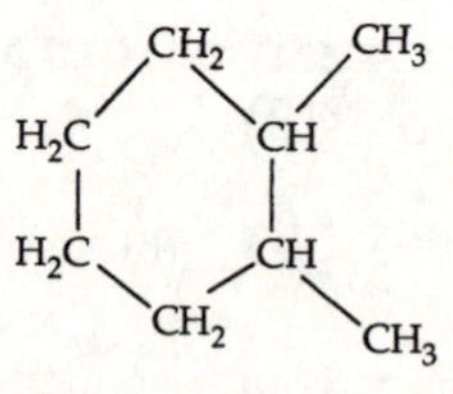

24.17 *Analyze/Plan.* Follow the rules for naming alkanes given in Section 24.2 and illustrated in Sample Exercise 24.1. *Solve.*

(a) 2,3-dimethylheptane

(b)

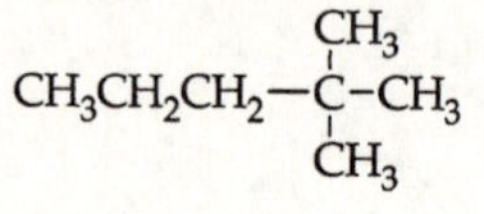

(c)

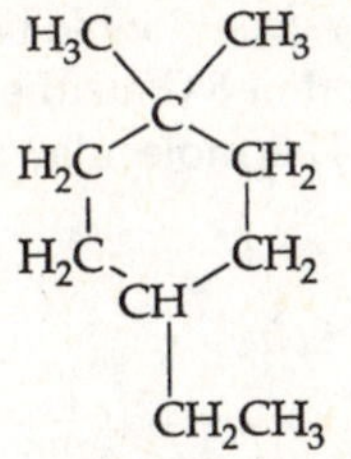

(d)   2,2,5-trimethylhexane

(e)   3-ethylheptane

24.19   Assuming that each component retains its effective octane number in the mixture (and this isn't always the case), we obtain: octane number = 0.35(0) + 0.65(100) = 65.

Alkenes, Alkynes, and Aromatic Hydrocarbons (section 24.3)

24.21   (a)   Alkanes are said to be *saturated* because they contain only single bonds. Multiple bonds that enable addition of H_2 or other substances are absent. The bonding capacity of each carbon atom is fulfilled with single bonds to C or H.

(b)   No, C_4H_6 is not saturated. Alkanes are saturated hydrocarbons. The maximum number of hydrogen atoms for 4 C atoms in an alkane is [(2×4)+2] = 10. C_4H_6 does not contain the maximum possible hydrogen atoms and is unsaturated.

24.23   *Analyze/Plan.* Consider the definition of the stated classification and apply it to a compound containing five C atoms.   *Solve.*

(a) $CH_3CH_2CH_2CH_2CH_3$, C_5H_{12}

(b) 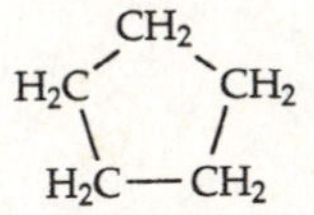
, C_5H_{12}

(c) $CH_2=CHCH_2CH_2CH_3$, C_5H_{10}

(d) $HC\equiv CCH_2CH_2CH_3$, C_5H_8 saturated: (a), (b); unsaturated: (c), (d)

24.25   *Analyze/Plan.* We are given the class of compounds "enediyne". Based on organic nomenclature, determine the structural features of an enediyne. Construct a molecule with 6 C atoms that has these features. *Solve.*

The term "enediyne" contains the suffixes –ene and –yne. The suffix –ene is used to name alkenes, molecules with one double bond. The suffix –yne is used to name alkynes, molecules with one triple bond. An enediyne then features one double and two triple bonds. Possible arrangements of these bonds involving 6 C atoms are:

$CH_2=CH-C\equiv C-C\equiv CH$ $CH\equiv C-CH=CH-C\equiv CH$

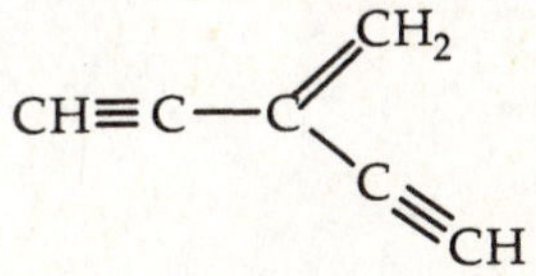

Check. The formula of a saturated alkane is C_nH_{2n+2}. For each double bond subtract 2 H atoms, for each triple bond subtract 4 H atoms. A saturated 6 C alkane has 14 H atoms. For the enediyne, subtract (2 + 4 + 4 =) 10 H atoms. The molecular formula is C_6H_4. That is the molecular formula of each structure above.

24.27 *Analyze/Plan.* Follow the logic in Sample Exercise 24.3. *Solve.*

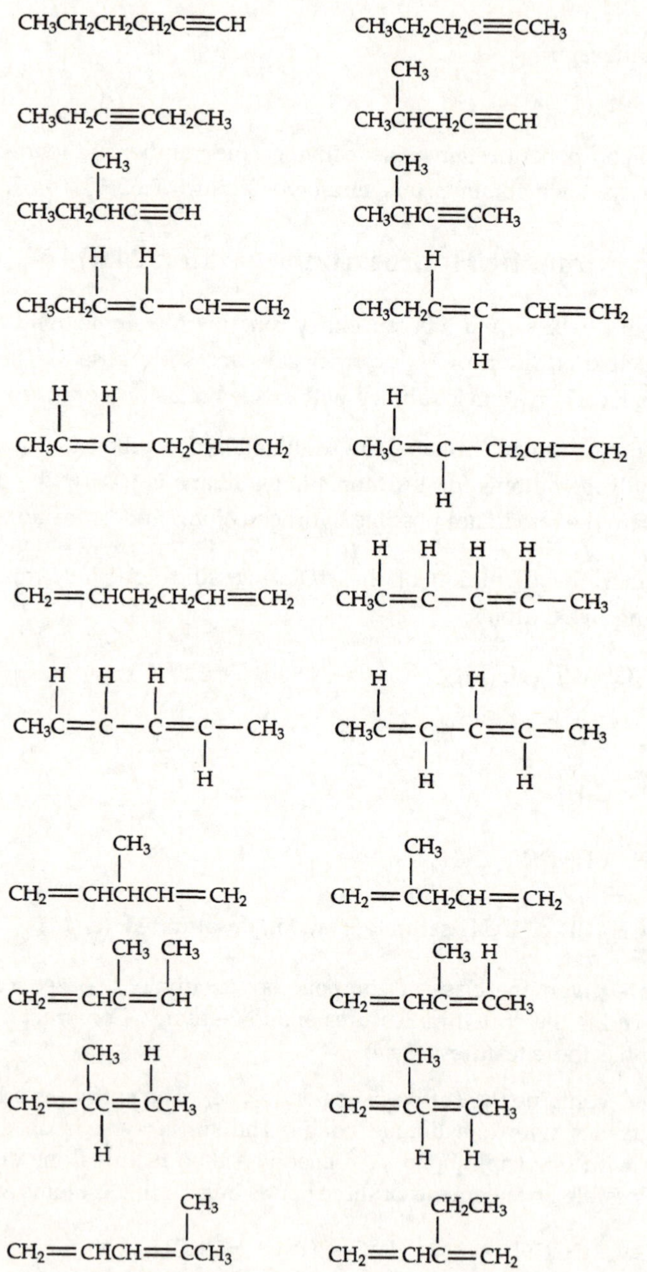

$CH_3CH_2CH_2CH_2C\equiv CH$ $CH_3CH_2CH_2C\equiv CCH_3$

$CH_3CH_2C\equiv CCH_2CH_3$

$$CH_3\overset{\displaystyle CH_3}{\underset{\displaystyle |}{CH}}CH_2C\equiv CH$$

$$CH_3CH_2\overset{\displaystyle CH_3}{\underset{\displaystyle |}{CH}}C\equiv CH$$

$$CH_3\overset{\displaystyle CH_3}{\underset{\displaystyle |}{CH}}C\equiv CCH_3$$

$$CH_3CH_2\overset{\displaystyle \overset{H}{|}}{C}=\overset{\overset{H}{|}}{C}-CH=CH_2$$

$$CH_3CH_2\overset{\displaystyle \overset{H}{|}}{C}=\underset{\underset{H}{|}}{C}-CH=CH_2$$

$$CH_3\overset{\displaystyle \overset{H}{|}}{C}=\overset{\overset{H}{|}}{C}-CH_2CH=CH_2$$

$$CH_3\overset{\displaystyle \overset{H}{|}}{C}=\underset{\underset{H}{|}}{C}-CH_2CH=CH_2$$

$CH_2=CHCH_2CH_2CH=CH_2$

$$CH_3\overset{\displaystyle \overset{H}{|}}{C}=\overset{\overset{H}{|}}{C}-\overset{\overset{H}{|}}{C}=\overset{\overset{H}{|}}{C}-CH_3$$

$$CH_3\overset{\displaystyle \overset{H}{|}}{C}=\overset{\overset{H}{|}}{C}-\underset{\underset{H}{|}}{C}=\overset{\overset{H}{|}}{C}-CH_3$$

$$CH_3\overset{\displaystyle \overset{H}{|}}{C}=\underset{\underset{H}{|}}{C}-\overset{\overset{H}{|}}{C}=\underset{\underset{H}{|}}{C}-CH_3$$

$$CH_2=CHCH\overset{\displaystyle CH_3}{\underset{\displaystyle |}{}}CH=CH_2$$

$$CH_2=C\overset{\displaystyle CH_3}{\underset{\displaystyle |}{}}CH_2CH=CH_2$$

$$CH_2=CHC\overset{\displaystyle CH_3}{\underset{\displaystyle |}{}}\overset{\displaystyle CH_3}{\underset{\displaystyle |}{}}\equiv CH$$

$$CH_2=CHC\overset{\displaystyle CH_3}{\underset{\displaystyle |}{}}=\overset{\displaystyle H}{\underset{\displaystyle}{}}CCH_3$$

$$CH_2=C\overset{\displaystyle CH_3}{\underset{\displaystyle |}{}}C\underset{\underset{H}{|}}{}=CHCH_3$$

$$CH_2=CC=CCH_3$$ with CH₃ and H groups

$$CH_2=CHCH=C\overset{\displaystyle CH_3}{\underset{\displaystyle |}{}}CH_3$$

$$CH_2=CHC\overset{\displaystyle CH_2CH_3}{\underset{\displaystyle |}{}}=CH_2$$

$$CH_2=\overset{\overset{\displaystyle CH_3}{|}}{C}-\overset{\overset{\displaystyle CH_3}{|}}{C}=CH_2$$

$$\underset{H}{\overset{CH_3}{\diagup}}C=C\underset{CH_3}{\overset{\diagup CH_3}{\diagdown}}$$

$$\underset{H}{\overset{CH_3}{\diagup}}C=C\underset{H}{\overset{\diagup CH_2CH_3}{\diagdown}}$$

$$\underset{H}{\overset{H}{\diagup}}C=C\underset{H}{\overset{\diagup CH_2CH_2CH_3}{\diagdown}}$$

$$\underset{H}{\overset{H}{\diagup}}C=C\underset{CH_3}{\overset{\diagup CH_2CH_3}{\diagdown}}$$

$$\begin{array}{c} HC \!=\! CH_2 \\ | \qquad | \\ H_2C \quad CH_2 \\ \diagdown \; CH_2 \diagup \end{array}$$

$$\begin{array}{c} CH\!=\!C\!-\!CH_3 \\ | \qquad | \\ H_2C \quad CH_2 \\ \diagdown \; CH_2 \diagup \end{array}$$

$$\begin{array}{c} CH\!=\!CH \\ | \qquad | \\ H_2C \quad C\!-\!CH_3 \\ \diagdown \; CH_2 \diagup H \end{array}$$

$$\begin{array}{c} CH\!=\!CH \\ | \qquad | \\ H_2C \quad CH_2 \\ \diagdown C \diagup \\ H \; CH_3 \end{array}$$

$$\begin{array}{c} H_2C\!-\!C\!-\!CH_3 \\ | \qquad \| \\ H_2C \quad C \\ \diagdown \; CH_3 \end{array}$$

$$\begin{array}{c} H \\ C\!-\!CH \\ H\,/H \quad \| \\ C\!-\!CH_3 \\ CH_3 \end{array}$$

$$\begin{array}{c} CH_3 \\ C\!-\!CH \\ | \qquad \| \\ H \\ H_2C\!-\!C\!-\!CH_3 \end{array}$$

$$\begin{array}{c} CH_3 \\ C\!-\!CH \\ | \qquad \| \\ CH_3 \\ H_2C\!-\!CH \end{array}$$

$$\begin{array}{c} CH_3 \\ CH_2\,C\!-\!CH \\ | \qquad \| \\ H \\ C\!-\!CH \\ H \end{array}$$

$$\begin{array}{c} CH_3 \\ C\!-\!CH \\ H\,/H \quad \| \\ C\!-\!CH \\ CH_3 \end{array}$$

$$\begin{array}{c} H_2C\!-\!C\!-\!CH_2CH_3 \\ | \qquad \| \\ H_2C\!-\!CH \end{array}$$

$$\begin{array}{c} CH_2CH_3 \\ C\!-\!CH \\ | \qquad \| \\ H \\ H_2C\!-\!CH \end{array}$$

$$\begin{array}{c} CH \\ \diagup \diagdown \\ H_2C\!-\!CCH_2CH_2CH_3 \end{array}$$

$$\begin{array}{c} CH_3 \\ C \\ \diagup \diagdown \\ H_2C\!-\!CCH_2CH_3 \end{array}$$

$$\begin{array}{c} CH \\ \diagup \diagdown \\ CH_3\!-\!CH\!-\!CCH_2CH_3 \end{array}$$

$$\begin{array}{c} CH \\ \diagup \diagdown \\ CH_3CH_2CH_2CH\!-\!CH \end{array}$$

$$\begin{array}{c} CH \\ \diagup \diagdown \\ CH_3CH_2CH\!-\!C\!-\!CH_3 \end{array}$$

$$\begin{array}{c} CH_3 \\ C \\ \diagup \diagdown \\ CH_3CH\!-\!CCH_3 \end{array}$$

$$\begin{array}{c} CH \\ CH_3 \diagup \diagdown \\ C\!-\!CCH_3 \\ CH_3 \end{array}$$

24.29　*Analyze/Plan.* Follow the logic in Sample Exercises 24.1 and 24.4.　*Solve.*

(a)

$$H\text{—} \underset{CH_3CH_2}{\overset{}{C}} = \underset{H}{\overset{CH_3}{C}}$$

(b)

$$CH_3CH_2CH_2\underset{}{\overset{CH_3}{C}} = CHCH_2CH_2 \text{—} \underset{\overset{|}{CH_3}}{\overset{CH_3}{CH}}$$

(c)　*cis*-6-methyl-3-octene

(d)　*para*-dibromobenzene

(e)　4,4-dimethyl-1-hexyne

24.31　Each doubly bound carbon atom in an alkene has two unique sites for substitution. These sites cannot be interconverted because rotation about the double bond is restricted; *geometric isomerism* results. In an alkane, carbon forms only single bonds, so the three remaining sites are interchangeable by rotation about the single bond. Although there is also restricted rotation around the triple bond of an alkyne, there is only one additional bonding site on a triply bound carbon, so no isomerism results.

24.33　*Analyze/Plan.* In order for geometrical isomerism to be possible, the molecule must be an alkene with two different groups bound to each of the alkene C atoms.　*Solve.*

(a)

$$Cl \text{—} \underset{}{\overset{Cl}{C}} = \underset{}{\overset{H}{C}} \text{—} CH_2 \text{—} CH_3, \text{ no}$$

(b)

$$\underset{H_3C}{\overset{Cl}{C}} = \underset{CH_2Cl}{\overset{H}{C}} \qquad \underset{Cl}{\overset{H_3C}{C}} = \underset{CH_2Cl}{\overset{H}{C}}$$

(c)　no, not an alkene

(d)　no, not an alkene

24.35　(a)　An *addition reaction* is the addition of some reagent to the two atoms that form a multiple bond. In a *substitution reaction*, one atom or group of atoms replaces (substitutes for) another atom or group of atoms. In an addition reaction, two atoms and a multiple bond on the target molecule are altered; in a substitution reaction, the environment of one atom in the target molecule changes. Alkenes typically undergo addition, while aromatic hydrocarbons usually undergo substitution.

(b)　*Plan.* Consider the general form of addition across a double bond. The π bond is broken and one new substituent (in this case two Br atoms) adds to each of the C atoms involved in the π bond.　*Solve.*

$$CH_3CH_2CH=CH\text{–}CH_3 + Br_2 \rightarrow CH_3CH_2CH(Br)CH(Br)CH_3$$

2-pentene　　　　　　　　　　　　　　　2,3-dibromopentane

(c)   *Plan.* Consider the general form of a *substitution* reaction. Two Cl atoms will replace two of the H atoms on the benzene ring. The term *para* means the Cl atoms will be opposite each other across the benzene ring in the product.

Solve.

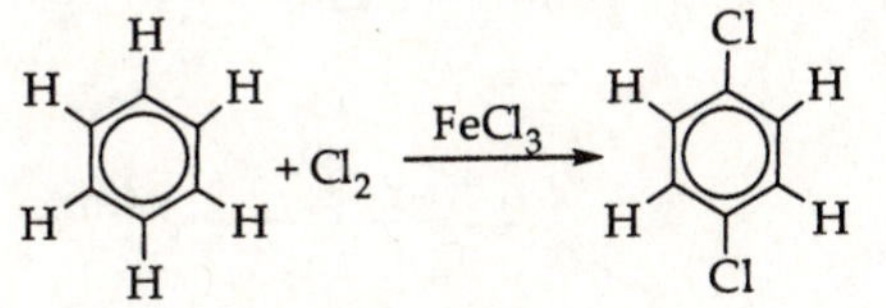

$$C_6H_6 + Cl_2 \xrightarrow{FeCl_3} C_6H_4Cl_2$$

24.37   (a)   *Plan.* Consider the structures of cyclopropane, cyclopentane, and cyclohexane.

Solve.

The small 60° C—C—C angles in the cyclopropane ring cause strain that provides a driving force for reactions that result in ring opening. There is no comparable strain in the five- or six-membered rings.

(b)   *Plan.* First form an alkyl halide: $C_2H_4(g) + HBr(g) \rightarrow CH_3CH_2Br(l)$; then carry out a Friedel-Crafts reaction.

Solve.

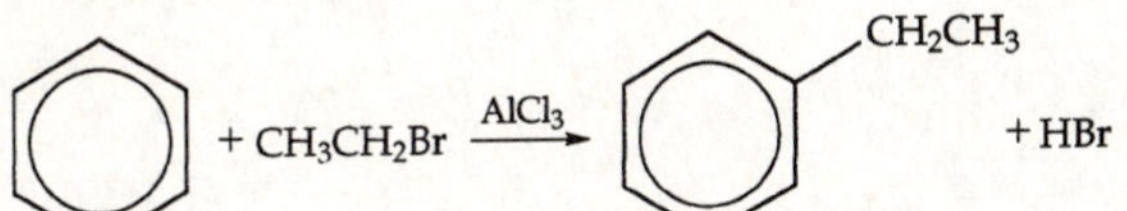

24.39   Not necessarily. That the rate laws are both first order in both reactants and second order overall indicates that the activated complex in the rate-determining step in each mechanism is bimolecular and contains one molecule of each reactant. This is usually an indication that the mechanisms are the same, but it does not rule out the possibility of different fast steps, or a different order of elementary steps.

24.41   *Analyze/Plan.* Both combustion reactions produce CO_2 and H_2O:

$$C_3H_6(g) + 9/2\,O_2(g) \rightarrow 3CO_2(g) + 3H_2O(l)$$

$$C_5H_{10}(g) + 15/2\,O_2(g) \rightarrow 5CO_2(g) + 5H_2O(l)$$

Thus, we can calculate the ΔH_{comb} / CH_2 group for each compound.   *Solve.*

$$\frac{\Delta H_{comb}}{CH_2\ group} = \frac{2089\ kJ/mol\ C_3H_6}{3\ CH_2\ groups} = \frac{696.3\ kJ}{mol\ CH_2}; \frac{3317\ kJ/mol\ C_5H_{10}}{5\ CH_2\ groups} = 663.4\ kJ/mol\ CH_2$$

$\Delta H_{comb}/CH_2$ group for cyclopropane is greater because C_3H_6 contains a strained ring. When combustion occurs, the strain is relieved and the stored energy is released during the reaction.

Functional Groups and Chirality (sections 24.4 and 24.5)

24.43   *Analyze/Plan.* Match the structural features of various functional groups shown in Table 24.6 to the molecular structures in this exercise. *Solve.*

(a)   −OH, alcohol

(b)   −NH−, amine; −C=C−, alkene

(c)   −O−, ether

(d) 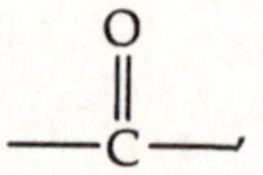 $\underset{\text{C}}{\overset{\text{O}}{\|}}$, ketone; −C=C−, alkene

(e) 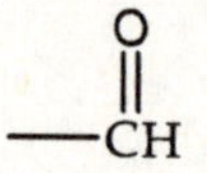 $-\overset{\text{O}}{\overset{\|}{\text{CH}}}$ (−CHO), aldehyde

(f)   −C≡C−(−CC−), alkyne; −COOH, carboxylic acid

24.45   *Analyze/Plan.* Given the name of a molecule, write the structural formula of an isomer molecular formula of the given molecule, draw the structural formula of a molecule with the same formula that contains the specified functional group.   *Solve.*

(a)   The formula of acetone is C_3H_6O. An aldehyde contains the group $-C\!\!\underset{H}{\overset{O}{\diagup}}$

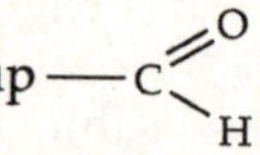

An aldehyde that is an isomer of acetone is propionaldehyde (or propanal):

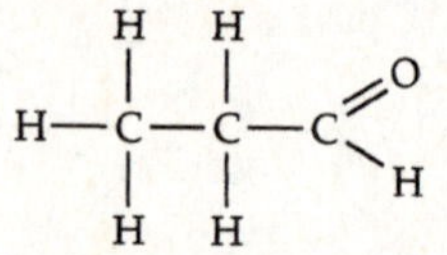

(b)   The formula of 1-propanol is C_3H_8O. An ether contains the group −O−. An ether that is an isomer of 1-propanol is ethylmethyl ether:

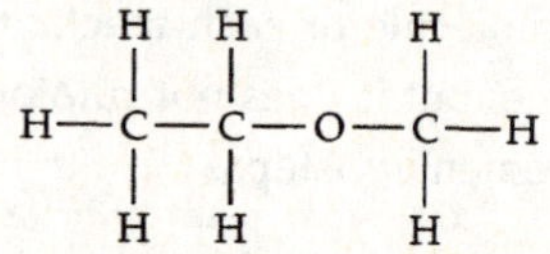

24.47   Analyze/Plan. From the hydrocarbon name, deduce the number of C atoms in the acid; one carbon atom is in the carboxyl group. *Solve.*

(a)   meth = 1 C atom

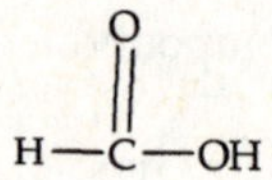

(b)   pent = 5 C atoms

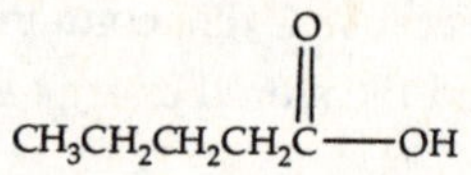

$CH_3CH_2CH_2CH_2\overset{O}{\overset{\|}{C}}\!-\!OH$ or

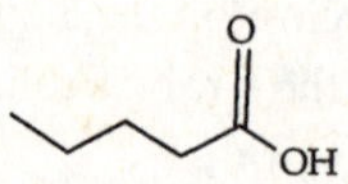

(c)   dec = 10 C atoms in backbone

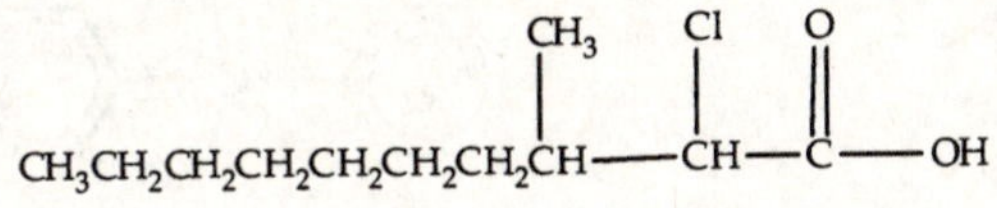

$$CH_3CH_2CH_2CH_2CH_2CH_2CH \overset{\overset{\displaystyle CH_3}{|}}{} - \overset{\overset{\displaystyle Cl}{|}}{CH} - \overset{\overset{\displaystyle O}{\|}}{C} - OH$$

or

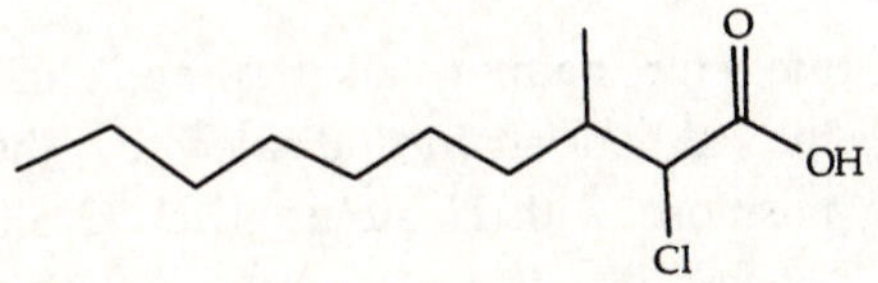

24.49   *Analyze/Plan.* In a condensation reaction between an alcohol and a carboxylic acid, the alcohol loses its —OH hydrogen atom and the acid loses its —OH group. The alkyl group from the acid is attached to the carbonyl group and the alkyl group from alcohol is attached to the ether oxygen of the ester. The name of the ester is the alkyl group from the alcohol plus the alkyl group from the acid plus the suffix *-oate*. *Solve.*

(a)

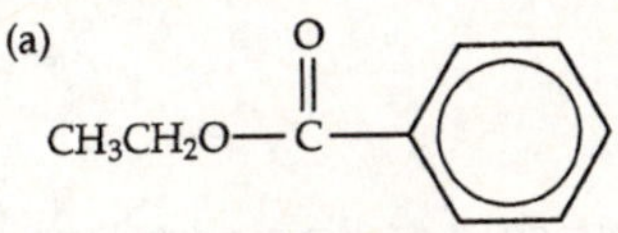

$$CH_3CH_2O - \overset{\overset{\displaystyle O}{\|}}{C} - \bigcirc$$

ethylbenzoate

(b)

$$CH_3N \overset{\overset{\displaystyle H}{|}}{} - \overset{\overset{\displaystyle O}{\|}}{C}CH_3$$

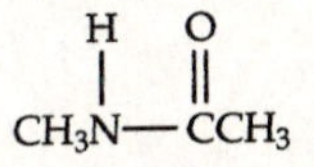

N-methylethanamide
or N-methylacetamide

(c)

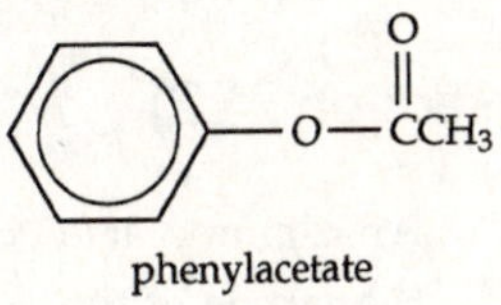

$$\bigcirc - O - \overset{\overset{\displaystyle O}{\|}}{C}CH_3$$

phenylacetate

24.51   *Analyze/Plan.* Follow the logic in Sample Exercise 24.6.   *Solve.*

(a)

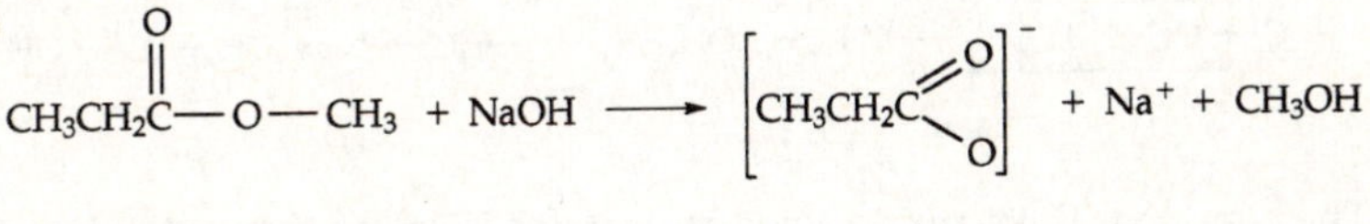

$$CH_3CH_2\overset{\overset{\displaystyle O}{\|}}{C} - O - CH_3 + NaOH \longrightarrow \left[CH_3CH_2C\overset{\diagup O}{\diagdown_O}\right]^- + Na^+ + CH_3OH$$

(b)

$$CH_3\overset{\overset{\displaystyle O}{\|}}{C} - O - \bigcirc + 2NaOH \longrightarrow \left[CH_3C\overset{\diagup O}{\diagdown_O}\right]^- + 2Na^+ + \left[\overset{\displaystyle O^-}{\bigcirc}\right]^-$$

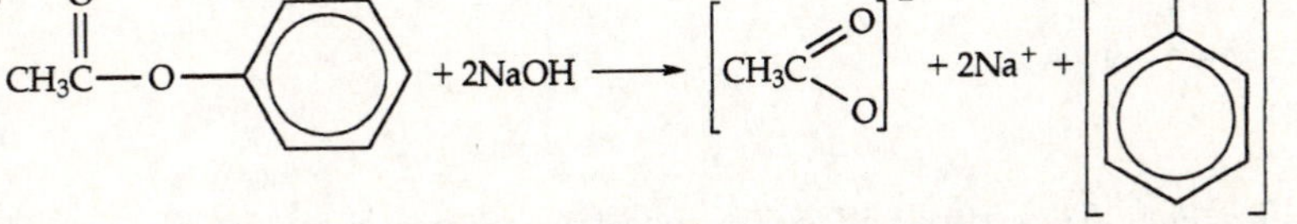

24.53   Yes, we expect acetic acid to be a strongly hydrogen-bonded substance. The carboxyl group has —OH, which acts as a donor, and —C=O, which acts as an acceptor in hydrogen bonding. The boiling point of acetic acid is higher than that of water (118°C vs. 100°C), indicating that hydrogen-bonding in acetic acid is even stronger than that of water (Figure 11.11). The melting points, 16.7°C for acetic acid and 0°C for water, show a similar trend.

24.55   *Analyze/Plan.* Follow the logic in Sample Exercise 24.2, incorporating functional group information from Table 24.6.   *Solve.*

(a) $CH_3CH_2CH_2CH(OH)CH_3$ (b) $CH_3CH(OH)CH_2OH$

(c)
$$CH_3COCH_2CH_3$$

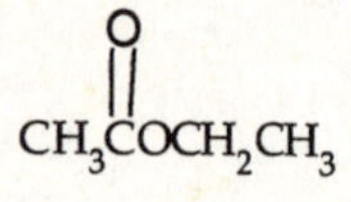

(d)

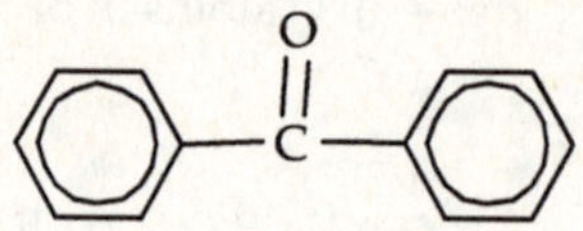

(e) $CH_3OCH_2CH_3$

24.57 *Analyze/Plan.* Review the rules for naming alkanes and haloalkanes; draw the structures. That is, draw the carbon chain indicated by the root name, place substituents, fill remaining positions with H atoms. Each C atom attached to four different groups is chiral. *Solve.*

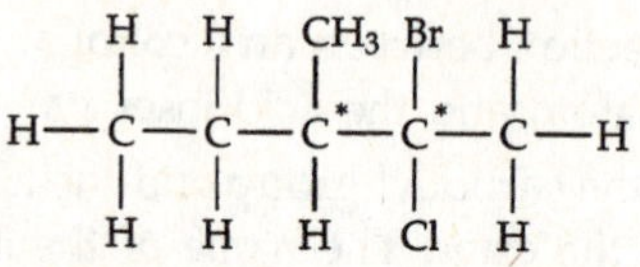

* chiral C atoms

C2 is obviously attached to four different groups. C3 is chiral because the substituents on C2 render the C1-C2 group different than the C4-C5 group.

Proteins (section 24.7)

24.59 (a) An α-amino acid contains an NH_2 group attached to the carbon that is bound to the carbon of the carboxylic acid function.

(b) In forming a protein, amino acids undergo a condensation reaction between the amino group and carboxylic acid:

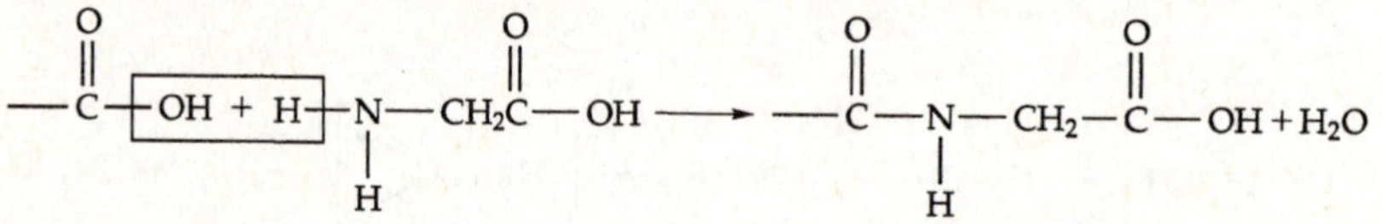

(c) The bond that links amino acids in proteins is called the *peptide* bond.

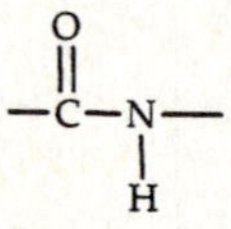

24.61 *Analyze/Plan.* Two dipeptides are possible. Either peptide can have the terminal carboxyl group or the terminal amino group. *Solve.*

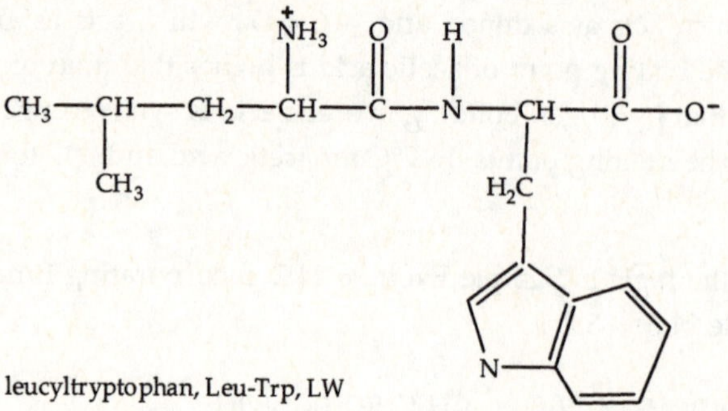

leucyltryptophan, Leu-Trp, LW

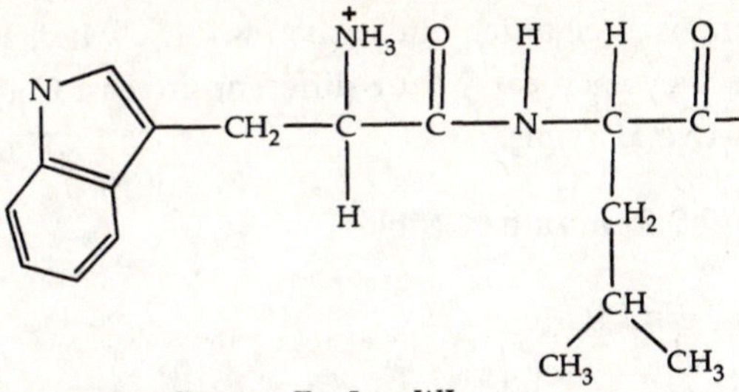

tryptophanylleucine, Trp-Leu, WL

24.63   *Analyze/Plan.* Follow the logic in Sample Exercise 24.7.   *Solve.*

(a)

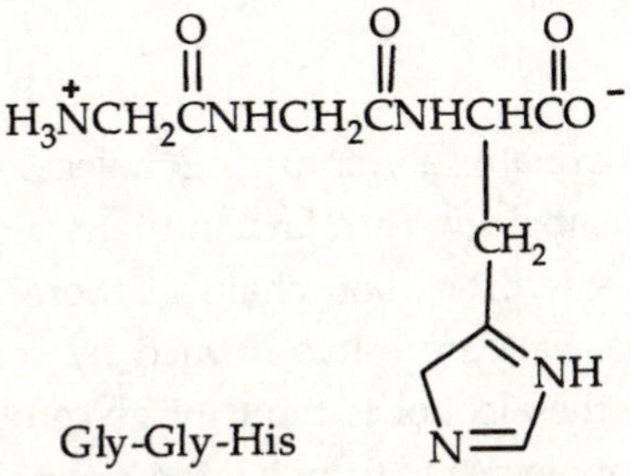

Gly-Gly-His

(b)   Three tripeptides are possible: Gly-Gly-His, GGH; Gly-His-Gly, GHG; His-Gly-Gly, HGG

24.65   (a)   The *primary structure* of a protein refers to the sequence of amino acids in the chain. Along any particular section of the protein chain the configuration may be helical, may be an open chain, or arranged in some other way. This is called the *secondary structure*. The overall shape of the protein molecule is determined by the way the segments of the protein chain fold together, or pack. The interactions which determine the overall shape are referred to as the *tertiary structure*.

(b)   X-ray crystallography is the primary and preferred technique for determining protein structure.

Carbohydrates and Lipids (sections 24.8 and 24.9)

24.67   (a)   *Carbohydrates*, or sugars, are composed of carbon, hydrogen, and oxygen. From a chemical viewpoint, they are polyhydroxyaldehydes or ketones. Carbohydrates are primarily derived from plants and are a major food source for animals.

(b)   A *monosaccharide* is a simple sugar molecule that cannot be decomposed into smaller sugar molecules by (acid) hydrolysis.

(c)   A *disaccharide* is a carbohydrate composed of two simple sugar units. Hydrolysis breaks the disaccharides into two monosaccharides.

(d)   A *polysaccharide* is a polymer composed of many simple sugar units.

24.69   The empirical formula of cellulose is $C_6H_{10}O_5$. As in glycogen, the six-membered ring form of glucose forms the monomer unit that is the basis of the polymer cellulose. In cellulose, glucose monomer units are joined by β linkages.

24.71 (a) In the linear form of mannose, the aldehydic carbon is C1. Carbon atoms 2, 3, 4, and 5 are chiral because they each carry four different groups. Carbon 6 is not chiral because it contains two H atoms.

(b) Both the α (left) and β (right) forms are possible.

24.73 The term *lipid* refers to the broad class of naturally-occurring molecules that are soluble in nonpolar solvents. Two important subgroups are fats and fatty acids. Structurally, fatty acids are carboxylic acids with a hydrocarbon chain of more than four carbon atoms (typically 16-20 carbon atoms). Fats are esters formed by condensation of an alcohol and a fatty acid. In animals, the alcohol is the triol glycerol; three fatty acid molecules condense with one glycerol molecule to form a large, nonpolar molecule. For both fatty acids and fats it is the long hydrocarbon chains that define the polarity, solubility and other physical properties of the molecules.

Phospholipids are glycerol esters formed from one phosphoric acid ($RPO(OH)_2$ and two fatty acid (RCOOH) molecules. At body pH, the phosphate group is depronated and has a negative charge. It is the juxtaposition of two long, nonpolar hydrocarbon chains with the charged phosphate "head" that causes phospholipids to form bilayers in water. The nonpolar chains do not readily mix with polar water. They do interact with the nonpolar chains of other phospholipids molecules on the inside of the bilayer. The charged phosphate heads interact with polar water molecules on the outsides of the bilayer.

Nucleic Acids (section 24.10)

24.75 Dispersion forces increase as molecular size (and molar mass) increases. The larger purines (2 rings vs. 1 ring for pyrimidines) have larger dispersion forces.

24.77 *Analyze/Plan.* Consider the structures of the organic bases in Section 24.10. The first base in the sequence is attached to the sugar with the free phosphate group in the 5′ position. The last base is attached to the sugar with a free −OH group in the 3′ position. *Solve.*

The DNA sequence is 5′−TACG−3′.

24.79 *Analyze/Plan.* Recall that there is complimentary base pairing in nucleic acids because of hydrogen bond geometry. The DNA pairs are A−T and G−C. (The RNA pairs are A−U and G−C.)

Solve. From the single strand sequence, formulate the complimentary strand. Note that 3′ of the complimentary strand aligns with 5′ of the parent strand.

5′–GCATTGGC–3′

3′–CGTAACCG–5′

Additional Exercises

24.81

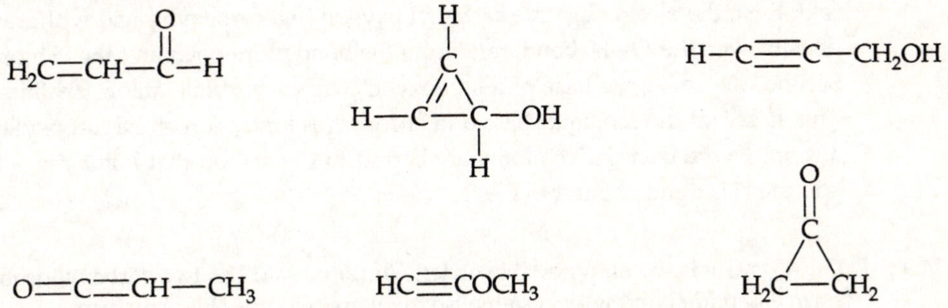

$H_2C=CH-\overset{\overset{\displaystyle O}{\|}}{C}-H$

H–C≡C–CH$_2$OH

O=C=CH–CH$_3$

HC≡COCH$_3$

Structures with the —OH group attached to an alkene carbon atom are not included. These molecules are called "vinyl alcohols" and are not the major form at equilibrium.

24.83

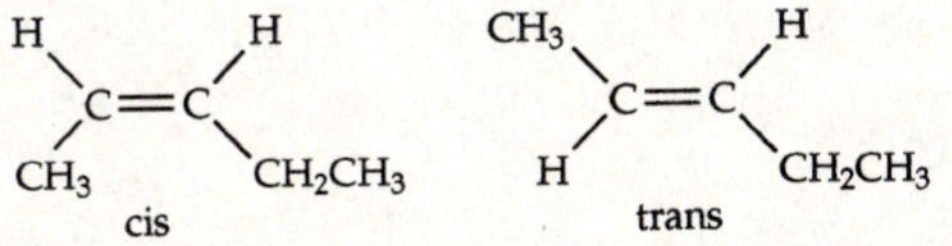

cis   trans

Cyclopentene does not show cis-trans isomerism because the existence of the ring demands that the C—C bonds be cis to one another.

24.86 (a)

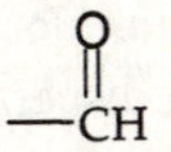

$-\overset{\overset{\displaystyle O}{\|}}{C}H$, aldehyde;    , *trans*-alkene; 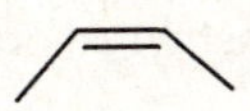   , *cis*-alkene

(b)   —O— , ether;   —OH , alcohol; C=C , alkene;

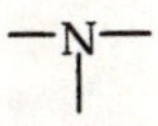

—N— , amine (two of these, one aliphatic and one aromatic);

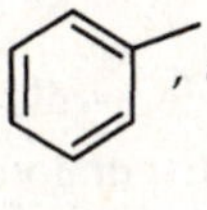

, aromatic (phenyl) ring

(c)

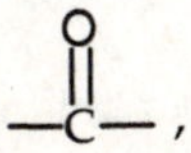

$-\overset{\overset{\displaystyle O}{\|}}{C}-$, ketone (2 of these);   —N— , amine (2 of these)

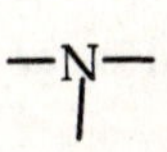

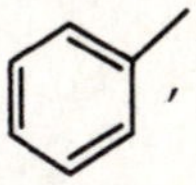

, aromatic (phenyl) ring (2 of these)

(d)

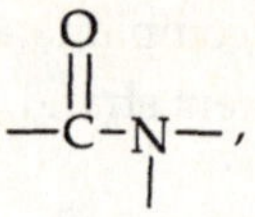

$-\overset{\overset{\displaystyle O}{\parallel}}{C}-\overset{|}{N}-$, amide;

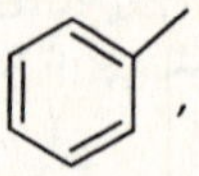

, aromatic (phenyl) ring;

$-OH$, alcohol (aromatic alcohol, phenol)

24.88 The difference between an alcoholic hydrogen and a carboxylic acid hydrogen is two-fold. First, the electronegative carbonyl oxygen in a carboxylic acid withdraws electron density from the O—H bond, rendering the bond more polar and the H more ionizable. Second, the conjugate base of a carboxylic acid, carboxylate anion, exhibits resonance. This stabilizes the conjugate base and encourages ionization of the carboxylic acid. In an alcohol no electronegative atoms are bound to the carbon that holds the —OH group, and the H is tightly bound to the O.

24.92 Glu-Cys-Gly is the only possible order. Glutamic acid has two carboxyl groups that can form a peptide bond with cysteine, so there are two possible structures.

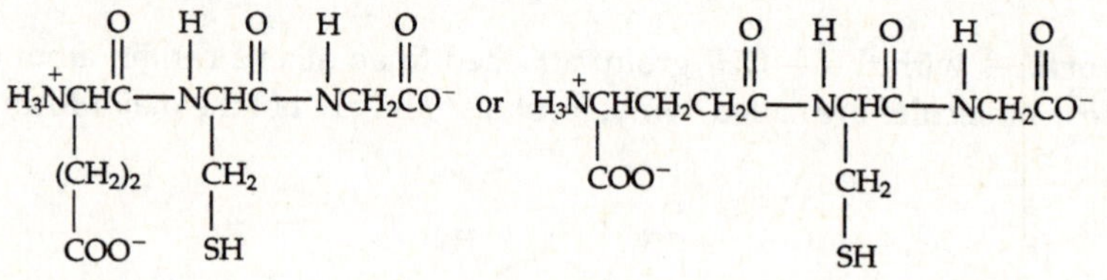

Integrative Exercises

24.95 CH_3CH_2OH CH_3-O-CH_3
     ethanol   dimethyl ether

Ethanol contains —O—H bonds which form strong intermolecular hydrogen bonds, while dimethyl ether experiences only weak dipole-dipole and dispersion forces.

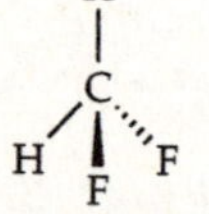

difluoromethane 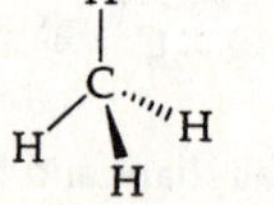   methane

CH_2F_2 has much greater molar mass and is a polar molecule, while CH_4 is lighter and nonpolar. CH_2F_2 experiences dipole-dipole and stronger dispersion forces, while CH_4 experiences only weaker dispersion forces.

In both cases, stronger intermolecular forces lead to the higher boiling point.

24.97 Determine the empirical formula, molar mass, and thus molecular formula of the compound. Confirm with physical data.

$$66.7 \text{ g C} \times \frac{1 \text{ mol C}}{12.01 \text{ g C}} = 5.554 \text{ mol C}; \ 5.554/1.381 = 4.021 = 4$$

$$11.2 \text{ g H} \times \frac{1 \text{ mol H}}{1.008 \text{ g H}} = 11.11 \text{ mol H}; \ 11.11/1.381 = 8.043 = 8$$

$$22.1 \text{ g O} \times \frac{1 \text{ mol O}}{16.00 \text{ g O}} = 1.381 \text{ mol O}; \ 1.381/1.381 = 1$$

The empirical formula is C_4H_8O. Using Equation 10.11 (MM = molar mass):

$$\text{MM} = \frac{(2.28 \text{ g}/\text{L})(0.08206 \text{ L-atm}/\text{mol-K})(373\text{K})}{0.970 \text{-atm}} = 71.9 \text{ g}/\text{mol}$$

The formula weight of C_4H_8O is 72, so the molecular formula is also C_4H_8O. Since the compound has a carbonyl group and cannot be oxidized to an acid, the only possibility is 2-butanone.

$$\overset{\displaystyle O}{\underset{\displaystyle CH_3CCH_2CH_3}{\|}}$$

The boiling point of 2-butanone is 79.6°C, confirming the identification.

24.99 The reaction is: $2NH_2CH_2COOH(aq) \rightarrow NH_2CH_2CONHCH_2COOH(aq) + H_2O(l)$

$\Delta G° = (-488) + (-237.13) - 2(-369) = 12.87 = 13 \text{ kJ}$